T0329303

CHEMOSENSORY TRANSDUCTION

CHEMOSENSORY TRANSDUCTION

THE DETECTION OF ODORS, TASTES, AND OTHER CHEMOSTIMULI

Edited by

FRANK ZUFALL

University of Saarland School of Medicine, Homburg, Germany

STEVEN D. MUNGER

University of Florida, Gainesville, FL, USA

AMSTERDAM • BOSTON • HEIDELBERG • LONDON
NEW YORK • OXFORD • PARIS • SAN DIEGO
SAN FRANCISCO • SINGAPORE • SYDNEY • TOKYO
Academic Press is an imprint of Elsevier

Academic Press is an imprint of Elsevier
125 London Wall, London EC2Y 5AS, UK
525 B Street, Suite 1800, San Diego, CA 92101-4495, USA
50 Hampshire Street, 5th Floor, Cambridge, MA 02139, USA
The Boulevard, Langford Lane, Kidlington, Oxford OX5 1GB, UK

British Library Cataloguing-in-Publication Data
A catalogue record for this book is available from the British Library

Library of Congress Cataloging-in-Publication Data
A catalog record for this book is available from the Library of Congress

ISBN: 978-0-12-801694-7

For information on all Academic Press publications
visit our website at https://www.elsevier.com/

Working together
to grow libraries in
developing countries

www.elsevier.com • www.bookaid.org

Typeset by TNQ Books and Journals
www.tnq.co.in

Cover Image: Original artwork by Dr. Stephan Vigues showing a cartoon rendering of the canonical mammalian olfactory transduction cascade.

Contents

6. Insect Olfactory Receptors: An Interface between Chemistry and Biology

GREGORY M. PASK AND ANANDASANKAR RAY

7. Cyclic AMP Signaling in the Main Olfactory Epithelium

CHRISTOPHER H. FERGUSON AND HAIQING ZHAO

8. Cyclic GMP Signaling in Olfactory Sensory Neurons

TRESE LEINDERS-ZUFALL AND PABLO CHAMERO

9. Ciliary Trafficking of Transduction Molecules

JEREMY C. McINTYRE AND JEFFREY R. MARTENS

10. Vomeronasal Receptors: V1Rs, V2Rs, and FPRs

IVAN RODRIGUEZ

Contributors

Barry W. Ache Whitney Laboratory for Marine Bioscience, St. Augustine, FL, USA; Departments of Biology and Neuroscience, Gainesville, FL, USA; Center for Smell and Taste, McKnight Brain Institute, University of Florida, Gainesville, FL, USA

Hubert Amrein Department of Molecular and Cellular Medicine, College of Medicine, Texas A&M Health Science Center, Bryan, TX, USA

Mari Aoki Department of Pharmacology and Toxicology, University of Saarland School of Medicine, Homburg, Germany

Maik Behrens Department of Molecular Genetics, German Institute of Human Nutrition Potsdam-Rehbruecke, Nuthetal, Germany

Ulrich Boehm Department of Pharmacology and Toxicology, University of Saarland School of Medicine, Homburg, Germany

Pablo Chamero Department of Physiology, Center for Integrative Physiology and Molecular Medicine, University of Saarland School of Medicine, Homburg, Germany

Elizabeth A. Corey Whitney Laboratory for Marine Bioscience, St. Augustine, FL, USA; Center for Smell and Taste, McKnight Brain Institute, University of Florida, Gainesville, FL, USA

Sami Damak Nestlé Research Center, Lausanne, Switzerland

Christopher H. Ferguson Department of Biology, The Johns Hopkins University, Baltimore, MD, USA

Bill Hansson Department Evolutionary Neuroethology, Max Planck Institute for Chemical Ecology, Jena, Germany

Sayoko Ihara Department of Applied Biological Chemistry, Graduate School of Agricultural and Life Sciences, The University of Tokyo, Tokyo, Japan; ERATO Touhara Chemosensory Signal Project, JST, The University of Tokyo, Tokyo, Japan

Sue C. Kinnamon Department of Otolaryngology, University of Colorado Medical School, Aurora, CO, USA

Sigrun Korsching Institute of Genetics, Biocenter, University at Cologne, Cologne, Germany

Tsung-Han Kuo Department of Molecular and Cellular Neuroscience, The Scripps Research Institute, La Jolla, CA, USA

Trese Leinders-Zufall Department of Physiology, Center for Integrative Physiology and Molecular Medicine, University of Saarland School of Medicine, Homburg, Germany

Qian Li Department of Cell Biology, Harvard Medical School, Boston, MA, USA

Stephen D. Liberles Department of Cell Biology, Harvard Medical School, Boston, MA, USA

Jeffrey R. Martens Department of Pharmacology and Therapeutics, Center for Smell and Taste, University of Florida, College of Medicine, Gainesville, FL, USA

Jeremy C. McIntyre Department of Pharmacology and Therapeutics, Center for Smell and Taste, University of Florida, College of Medicine, Gainesville, FL, USA

Wolfgang Meyerhof Department of Molecular Genetics, German Institute of Human Nutrition Potsdam-Rehbruecke, Nuthetal, Germany

Steven D. Munger Center for Smell and Taste, University of Florida, Gainesville, FL, USA; Department of Pharmacology and Therapeutics, University of Florida, Gainesville, FL, USA; Department of Medicine, Division of Endocrinology, Diabetes and Metabolism, University of Florida, Gainesville, FL, USA

Yoshihito Niimura Department of Applied Biological Chemistry, Graduate School of Agricultural and Life Sciences, The University of Tokyo, Tokyo, Japan; ERATO Touhara Chemosensory Signal Project, JST, The University of Tokyo, Tokyo, Japan

Yuzo Ninomiya Section of Oral Neuroscience, Graduate School of Dental Sciences, Kyushu University, Fukuoka, Japan; Division of Sensory Physiology, Research and Development Center for Taste and Odor Sensing, Kyushu University, Fukuoka, Japan

Gregory M. Pask Department of Entomology, University of California Riverside, Riverside, CA, USA

Nanduri R. Prabhakar Institute for Integrative Physiology and Center for Systems Biology of O_2 Sensing, Biological Sciences Division, University of Chicago, Chicago, IL, USA

Anandasankar Ray Department of Entomology, University of California Riverside, Riverside, CA, USA; Institute for Integrative Genome Biology, University of California Riverside, Riverside, CA, USA

Ivan Rodriguez Department of Genetics and Evolution, University of Geneva, Geneva, Switzerland; Geneva Neuroscience Center, University of Geneva, Geneva, Switzerland

Noriatsu Shigemura Section of Oral Neuroscience, Graduate School of Dental Sciences, Kyushu University, Fukuoka, Japan

Jay P. Slack Department of Science and Technology, Givaudan, Cincinnati, OH, USA

Marc Spehr Department of Chemosensation, Institute for Biology II, RWTH Aachen University, Aachen, Germany

Lisa Stowers Department of Molecular and Cellular Neuroscience, The Scripps Research Institute, La Jolla, CA, USA

Shingo Takai Section of Oral Neuroscience, Graduate School of Dental Sciences, Kyushu University, Fukuoka, Japan

Kazushige Touhara Department of Applied Biological Chemistry, Graduate School of Agricultural and Life Sciences, The University of Tokyo, Tokyo, Japan; ERATO Touhara Chemosensory Signal Project, JST, The University of Tokyo, Tokyo, Japan

Shuping Wen Department of Pharmacology and Toxicology, University of Saarland School of Medicine, Homburg, Germany

Dieter Wicher Department Evolutionary Neuroethology, Max Planck Institute for Chemical Ecology, Jena, Germany

Ryusuke Yoshida Section of Oral Neuroscience, Graduate School of Dental Sciences, Kyushu University, Fukuoka, Japan

Haiqing Zhao Department of Biology, The Johns Hopkins University, Baltimore, MD, USA

Foreword

Virtually everything that we know about the world comes to us through the little holes in our head—eyes, ears, nose, mouth. Our brains sit protected inside a hard skull bathed in warm salty solution—and completely in the dark, as they say. It/we depend on our sensory organs to tell us about the world. Inside each of those holes in our heads is a specialized piece of tissue that has evolved to be especially sensitive to a particular physical stimulus—electromagnetic waves for vision, mechanical air pressure changes for sound, and chemicals of an immense variety for tastes and smells.

But most important, none of these physical stimuli themselves—radiation, pressure, or environmental chemicals—ever reach the brain itself. Only a signal that they are present is sent to the brain. This process of changing a physical stimulus into a neural signal is the job of the specialized cells in the tissues of your sense organs. It is a process we call *transduction*, and it is just this side of magic. All the wonderful sensations connected with smells and tastes—from food to sex to the aesthetics of incense and the remarkably compelling memories of emotional moments in our lives—are first transduced by these hard-working little cellular machines. Evolution has hit on numerous ways to get this to happen. Most of them involve various sorts of proteins that work together in a carefully coordinated set of steps that transform a captured molecule into a small change in voltage that can be read by the brain.

All of this science will be treated in great detail and clarity in the chapters of this remarkable collection. What makes the chemical sense so intriguing as a subject for study is that almost all the mechanisms we know of for performing this transduction function are embodied in one or the other of the sensory systems devoted to sensing environmental chemicals. Classically, we think of these as just smell and taste, but really there is much more diversity. What we call the senses of smell and taste are in fact a collection of senses that work in many different ways. Those different ways are the subjects of the chapters in this book.

A little bit of history. Studies of olfactory and taste transduction were in many ways responsible for ushering in the modern era of chemical senses research. Studies using the most sophisticated techniques from molecular biology and electrophysiology throughout the 1980s uncovered the first complex mechanisms of chemical transduction. These studies were critical because they showed that the chemical senses worked like many other signaling systems in the brain and that they were not some idiosyncratic and strange island of neuroscience. What we learned from other brain and sensory systems could be applied to olfaction and taste and vice versa. We were in the mainstream of neuroscience. All this led up to the landmark discovery of the mammalian odor receptors in 1991, the insect receptors in 1999, and the basic taste receptors in the early 2000s. In a shockingly short period for science, the chemical senses went from a cul-de-sac of neuroscience to one of the most exciting frontiers of neuroscience. And there it has remained, as this volume will demonstrate to the reader.

In many ways, though, what is most remarkable to this old hand is how different a book on olfactory and taste transduction would have been a mere decade ago. By my count, at least 10 of the chapters in this volume would not have even appeared a decade ago—we just did not know about many of these things. And, of course, all the chapters have content not even imagined a decade past.

Will progress continue at this rate? Will this book be out of date soon? One hopes so. I imagine a young reader, graduate student, or postdoctoral fellow being excited by this material, seeing in it endless opportunities, and setting out to make the next decade of unimagined discoveries. I look forward to the next volume.

Stuart Firestein
New York, 2015

Preface

Chemical stimuli can inform us about the palatability, safety, and nutritional value of food; alert us to the presence of potential predators or other dangers; and guide our social interactions. Animals of all types employ a variety of detection mechanisms that recognize these chemical cues present in the external or internal environment and convert that stimulus detection into neural or endocrine responses on which the organism can act. Over the past 30 years, the details of this process, known as *chemosensory transduction*, have come into sharper focus; exploring these details and how they link us to our chemical world is the purpose of this book.

Each of us has a longstanding interest in understanding how the diversity of chemosensory transduction mechanisms enables both vertebrates and invertebrates to sense the complexity of the chemical world. Our individual research programs (and our 15-plus years of collaboration) focus on the sensory transduction mechanisms that are critical for mammalian chemosensation. However, we have each spent time studying olfaction in arthropods and retain a keen interest in differences and commonalities of smell and taste function across the animal kingdom. When Melanie Tucker (Senior Acquisitions Editor at Academic Press/Elsevier) first suggested to one of us (Zufall) that a book focused on olfaction might be a timely addition to the literature, we quickly recognized that what was lacking was an inclusive view of how odors, tastes, and other chemostimuli are recognized and encoded by the diverse array of chemosensory systems in a variety of animals. Thus, we embarked on this 2-year effort, with the help of many of the leaders in chemosensory science, to produce the book you see here. We hope that you, the reader, will find that it informs you about the complexity of chemosensation, guides you to the intricate and revealing scientific studies that we can only touch on here, and inspires you to investigate any of the many unanswered questions about how we sense our chemical world.

As with so many major undertakings, this book could not have been started (let alone published) without the invaluable contributions of many. We would like to thank Melanie Tucker, Kristi Anderson, and the production and marketing staff at AP/Elsevier for helping us to refine our ideas for this volume, recruit the many chapter authors, and navigate all the unfamiliar processes that are part of bringing a book to print. Both were helpful and patient but not afraid to prod a bit when it was needed. We thank Stephan Vigues, an accomplished chemosensory researcher in his own right with more than a little artistic talent, who created the cover art for the book. This picture not only represents many of the molecules that can be found in a typical transduction cascade, but also conveys the activity of this living cellular machinery. We are indebted to Stuart Firestein for providing an insightful and entertaining Foreword. Stuart's own research has been instrumental in illuminating the molecular and physiological mechanisms by which the mammalian nose detects and encodes odors. He has mentored, either formally or informally, many of the authors of this book as well as both of us. More recently, he has been quite successfully and effectively engaging in the public communication of science. We encourage you to read his fascinating insights on the importance of both *Ignorance* and *Failure* in his books by those names. And of course, we heartily thank the many authors that contributed to this book. Book chapters are often thankless work and may not seem to have the payoff of other activities, such as writing a review paper in a top journal. But these friends and colleagues joined us in this endeavor anyway, and the book is all the better for it. We hope that the participation, and the final product, has been rewarding for them.

In addition, we each have our own thanks to express.

Zufall: First and foremost, I would like to thank my academic mentors who guided my scientific journey from Germany to the United States and back: Randolf Menzel (Berlin); Hanns Hatt and Josef Dudel (Munich); Gordon Shepherd, Stuart Firestein, and Charles Greer (New Haven); and Michael Shipley (Baltimore). I'd like to express my gratitude to all present and former members of our laboratories who contributed to building a lasting research record in this field as well as the numerous individuals who teamed up with us on this exciting endeavor. I am grateful to the Deutsche Forschungsgemeinschaft for funding much of my own research and for supporting the national research program "Integrative Analysis of Olfaction" for the past 6 years. Finally, I'd like to thank Trese Leinders-Zufall, my wife and long-term collaborator, and our daughter Nicola who have been a constant source of inspiration.

Munger: My coeditor Frank Zufall deserves special thanks. As mentioned, Frank and I have been collaborators (and friends and colleagues) for more than 15 years, a relationship I hope will continue for many more. I'd like to express my deep appreciation to the many other colleagues and mentors I have encountered over the nearly 30 years I have spent studying the chemical senses (many of which are chapter authors), including those who have worked with me in my laboratory. Scientists in other disciplines may not appreciate the uniqueness of this community, including its strong support for its most junior members and enthusiastic appreciation of the power of fundamental science. There are many reasons why I have continued to work in this field since I first discovered it while I did undergraduate research in Mike Mellon's laboratory, but my colleagues are far from the least of them. I'd like to thank the National Institute on Deafness and Other Communication Disorders, which funds much of the research in the chemical senses (including in my own laboratory). Without their diehard support of this field, we would know very little about these senses and how they impact our daily lives. Finally, I would like to thank my very supportive family, especially my wife Caroline, son Garrett, and daughter Gwynn. They are more tolerant than I deserve and more inspiring than they know.

Frank Zufall, PhD
Homburg, Saarland, Germany

Steven D. Munger, PhD
Gainesville, Florida, USA

September 21, 2015

Introduction and Overview

Frank Zufall, Steven D. Munger

Hay smells different to lovers and horses. **Stanislaw Jerzy Lec**

Animals rely on their chemical senses to make their way in the world. The environment contains a complex mixture of chemicals that convey important, and sometimes critical, information that can influence what we eat, affect our interactions with others, help protect us from dangers, and impact our feelings and behaviors. As is suggested in the quotation, the meaning of a chemical stimulus can vary based on our experience: although the smell of hay may trigger hunger in a horse, that same smell evokes a very different feeling for two young people seeking a place for a private rendezvous. The aversive bitterness of beer or coffee may be off-putting with the first taste, but you can learn to appreciate it once it is paired with the pharmacological effects of alcohol or caffeine or the pleasure you feel when drinking these beverages in the company of friends. In other words, context matters when it comes to the meaning of odors or tastes.

The neural circuitry that conveys and processes chemosensory information can also dictate its meaning. Many animals have specialized chemosensory subsystems that mediate narrowly circumscribed behaviors that are essential to health, reproduction, survival, or even social relationships. For example, although normal mice may cower when they smell TMT, a component of predator urine, mice that lack the most dorsal aspects of the olfactory system (that near the top of the head) can still smell the TMT, but no longer show signs of fear. The ability of certain odor blends to act as attractive signals for conspecific insects (but not for other insect species) is a dramatic example of how a compound can carry a very specific meaning based on the receiver's ability to perceive it.

But no matter what information a chemical stimulus may convey, it is useless if it cannot be detected by a chemosensory organ and communicated to the nervous or endocrine systems to evoke a perception, behavior, or physiological change. This process by which odors, tastes, and other chemical stimuli are detected and converted into a cellular signal is known as *chemosensory transduction*, and is the subject of this book. The study of chemosensory transduction has seen an explosion of knowledge in recent years about the molecular biological, genetic, and physiological mechanisms that convert chemical information to cellular, neural, and endocrine responses, and has thus propelled this small field into the mainstream of membrane signaling and placed it at the forefront of elucidating the cellular and molecular logic of the nervous system. Therefore, this book will not only explore the machinery of chemosensory transduction in both vertebrates and invertebrates but will highlight the organizational principles underlying the recognition of chemical stimuli.

WHAT IS CHEMOSENSORY TRANSDUCTION?

Chemosensory transduction may be defined as the process by which chemical stimuli—including odors, tastes, nutrients, irritants, and even gases—are detected and converted into internal signals that elicit changes in cellular membrane properties or the release of transmitters or hormones. These processes usually take place within specialized cells, such as sensory neurons, that often contain dedicated subcellular compartments (such as cilia or microvilli) that are optimized for the transduction process. In most cases, chemosensory transduction is a multistep mechanism in which biochemical membrane signals will be converted into electrical signals—such as graded receptor potentials, action potential sequences, or both—a process that is known as chemoelectrical transduction. We distinguish between primary signal transduction (e.g., the initial detection and transduction steps taking place within the ciliary structures of an olfactory sensory neuron) and subsequent processes within the same cell that further shape and modulate the output signal of a given sensory neuron. As in sensory receptor cells from other modalities, a set of common operations can be defined in chemosensory cells that include the detection and discrimination of stimuli, amplification and sensory channel gating, adaptation, termination, and signal transmission to the brain. In combination, these distinct mechanisms will enable a chemosensory cell to transduce an external molecular cue into an internal signal that can be encoded, propagated, and processed by the nervous system.

LEVELS OF ANALYSIS IN CHEMOSENSORY SYSTEMS

The study of chemosensory transduction brings together people with diverse expertise: chemists, perfumers, and applied food scientists; geneticists and molecular biologists studying how the genome links to the unique chemosensory functions of an organism; neurobiologists and biophysicists interested in the function of the nervous system; psychophysicists that seek to understand how sensory stimuli influence behavior; behavioral endocrinologists and immunologists; and even clinicians interested in understanding the mechanistic basis of sensory disorders in humans and how to develop effective strategies for diagnosis and treatment. Accordingly, modern studies of chemosensory transduction include, but extend far beyond, mechanistic analyses of stimulus detection in sensory cells.

The individual chapters of this book, which are written by chemosensory scientists at the forefront of their field, will provide evidence that a rich diversity of chemosensory systems have evolved in both vertebrates and invertebrates to sense chemical (i.e., molecular) information. This diversity can even be found within a single chemosensory organ such as the mammalian olfactory epithelium (which resides in the nose) or gustatory epithelium (on the tongue and palate). An important development in the field has been the finding that the noncanonical expression of specific receptors outside the olfactory or gustatory systems is critical for sensing many internal chemostimuli, such as ingested nutrients and blood gases. Therefore, the principles obtained from an analysis of the olfactory and taste systems can be applied equally well to understanding the mechanisms of internal chemosensing and homeostatic regulation within the body.

SECTION I: SOCIAL ODORS AND CHEMICAL ECOLOGY

One important branch of modern chemosensory research aims at answering systems-level questions that are focused at understanding how the sensing of specific chemostimuli alters the behavioral response of a given organism. This endeavor includes a detailed analysis of the function of the neural circuits that connect a primary chemosensory activation to a specific behavioral output. To provide insight into these strategies and the current status of the field, we begin this book with two chapters that both offer a fascinating systems-level perspective of the chemical senses.

In **Chapter 1**, Lisa Stowers and Tsung-Han Kuo focus on specialized chemostimuli that impact social and survival behaviors. These compounds, collectively referred to as semiochemicals and including pheromones and kairomones, activate sensory circuits that are specialized to elicit a preset behavior without associative learning. As a whole, these studies will provide important insight into the question how the mammalian olfactory systems harnesses the brain to guide an individual's behavioral decision.

Chapter 2, from Bill Hansson and Dieter Wicher, illustrates how insects detect and process chemosignals throughout their life cycle. They include an overview of how insects use chemosignaling to interact with each other as well as with their environment. The use and meaning of the term "chemosignal" are put into an evolutionary perspective in the context of insect–plant interactions.

SECTION II: OLFACTORY TRANSDUCTION

The modern era of olfactory transduction began with work in the mid-1980s and early 1990s, revealing the basic principles of vertebrate odor transduction and uncovering a set of unifying principles. An early key step was the finding that odorants can stimulate a cAMP enzymatic cascade, including a GTP-binding protein and an adenylyl cyclase. These results highlighted a strong analogy of signal transduction events in the olfactory system with other known signaling processes related to the detection of hormones, neurotransmitters, and light, and thus provided a clear conceptual framework for moving forward to dissect the sensory transduction mechanisms in the olfactory system. The identification of a cAMP-gated cation channel in the ciliary membrane of vertebrate olfactory sensory neurons resolved the issue of how an odor-stimulated cAMP rise could produce a tonic membrane depolarization. The discovery and cloning of a large family of odorant receptor genes, leading to the 2004 Nobel Prize in Physiology and Medicine for Linda Buck and Richard Axel, not only solved the puzzle of how an almost unlimited number of odorant molecules can be detected by the nose, but was also the starting point for using modern gene targeting methods to unravel the molecular logic of smell and to map specific olfactory pathways from the periphery to the brain. During the subsequent years, it became clear that the olfactory system is actually composed of a number of subsystems that are anatomically segregated within the nasal cavity, make neural connections that project to distinct subregions of the olfactory forebrain, and use specialized detection and signaling mechanisms. Similar principles were applied in parallel to discover the odorant detection mechanisms in invertebrate species, specifically in insects and nematodes. Chapters 3–12 of this book highlight this fascinating diversity of chemosensory transduction mechanisms in the olfactory system of mammals, lower aquatic vertebrates, and insects.

In **Chapter 3**, Kazushige Touhara, Yoshihito Niimura, and Sayoko Ihara describe the discovery of the canonical odorant receptor (OR) gene family in 1991, the evolution of that family, and the structure and function of ORs in primates and other vertebrate species.

Chapter 4, by Stephen Liberles and Qian Li, summarizes the identification of a second family of G protein–coupled olfactory receptors in the mammalian main olfactory epithelium, the trace amine-associated receptors (TAARs). These receptors provide an excellent model for mechanistic studies of innately aversive behavioral responses and of odor valence encoding.

Sigrun Korsching explores the growing understanding of the olfactory receptors of aquatic vertebrates in **Chapter 5**. This chapter takes an evolutionary perspective to examine the expression and function of several classes of olfactory receptors in fish, amphibians, and aquatic mammals.

In **Chapter 6**, Anandasankar Ray and Gregory Pask discuss the identification and functional roles of three known families of insect olfactory receptors: odorant receptors (Ors), ionotropic receptors (Irs), and CO_2-sensitive gustatory receptors (Grs). The evolution, structure/function relationships, and role in insect behavior are reviewed for each of these chemoreceptor families.

Chapter 7, from Christopher H. Ferguson and Haiqing Zhao, focuses on the cAMP-mediated signaling cascade of canonical vertebrate olfactory sensory neurons (OSNs) and describes the core components of this "original" olfactory transduction pathway. Many of the mechanisms that regulate the transduction process are explored, including those that may regulate the size and duration of the cAMP transient. Together, these mechanisms govern changes in the sensitivity and response kinetics of the olfactory system, thereby allowing the system to accommodate highly variable environmental cues.

Chapter 8, by Trese-Leinders Zufall and Pablo Chamero, summarizes the evidence that cAMP is not the only significant cyclic nucleotide in odor transduction and describes how cGMP signaling also plays

important roles. Subsets of olfactory neurons in the mammalian nose, such as those that express the receptor guanylyl cyclase GC-D and other cells that are located in the Grueneberg ganglion, use cGMP signaling for chemosensory transduction. A strong case for cGMP signaling mechanisms has also been built in the nematode olfactory system.

Jeffrey Martens and Jeremy McIntyre provide insights into the mechanisms and mutations that affect olfactory cilia structure or function and that can have a profound impact on the sense of smell in **Chapter 9**. These mutations underlie a class of disorders termed ciliopathies that are often associated with anosmia, a loss of the sense of smell. Work on protein trafficking in olfactory cilia provides a basis for developing therapies that may be able to restore the sense of smell in ciliopathy patients.

Chapter 10, by Ivan Rodriguez, is aimed at describing the types of chemoreceptors found in the sensory part of the mammalian vomeronasal organ (VNO), also known as Jacobson's organ. This chapter summarizes the discovery, evolution, and function of three known types of receptor families in the VNO: type 1 and type 2 vomeronasal receptors (V1Rs and V2Rs, respectively) and formyl peptide receptors (FPRs).

In **Chapter 11**, Marc Spehr turns the focus to vomeronasal signaling mechanisms downstream of the receptors. Because the VNO is essential for many types of chemical communication in rodents, it has received specific interest over the past 20 years. This chapter summarizes our current knowledge of signaling and transduction mechanisms in vomeronasal sensory neurons (VSNs) with a particular emphasis on rodent models.

Finally in this section, Elizabeth A. Corey and Barry W. Ache present a comparative analysis of olfactory transduction mechanisms in a variety of animal models in **Chapter 12**. This approach helps to identify the important functional characteristics that define olfaction and suggest new avenues to be investigated.

SECTION III: GUSTATORY TRANSDUCTION

The sense of taste in vertebrates comprises the five basic taste qualities—salty, sour, sweet, umami, and bitter—and is required for detecting and evaluating the chemical composition of food. Following in the footsteps of the discovery of the odorant receptor genes, the cloning of several families of taste receptor genes enabled the generation of genetically modified mouse lines that, in turn, provided powerful tools to identify the cellular and molecular logic of taste coding at the periphery. As in olfaction, these genetic strategies are now providing the basis for a systems level approach that is aimed at understanding the neural representation of taste quality in the central nervous system to ultimately elucidate how behavioral decisions of an organism are driven by the detection of gustatory stimuli in specific taste receptor cells (TRCs).

Chapter 13, by Maik Behrens and Wolfgang Meyerhof, describes the structures and functions of G protein–coupled taste receptors (GPCRs) that respond to sweet-, umami-, and bitter-tasting stimuli. The potential existence of additional taste qualities represents a current major topic. Therefore, this chapter also includes a section on GPCRs responsive to free fatty acids.

Hubert Amrein provides a comprehensive discussion of insect taste detection mechanisms in **Chapter 14**. Although the chapter focuses on the adult fruit fly, it also explores the receptor basis of gustatory function in *Drosophila* larvae.

In **Chapter 15**, Sue C. Kinnamon details the canonical signal transduction events downstream of the mammalian taste GPCRs, including receptor activation of the heterotrimeric G protein, Gα-gustducin, and its βγ partners, Gβ1γ13, Gβγ activation of phospholipase C β2, production of the second messenger inositol trisphosphate (IP$_3$), the release of Ca^{2+} from intracellular stores, the activation of the transduction channel TRPM5, and the nonvesicular release of ATP as a transmitter to activate purinergic receptors on afferent nerve fibers.

Chapter 16, by Steven D. Munger, summarizes our current knowledge of the mechanisms underlying salt and acid (sour) taste that, unlike other taste qualities, relies on ion channels as receptors for their proximate stimuli. Special attention is payed to the differences in transduction strategies between sodium-specific and generalist salt responses and between weak and strong acids.

Finally, in **Chapter 17,** Yuzo Ninomiya, Shingo Takai, Ryusuke Yoshida, and Noriatsu Shigemura discuss the various functions of peptide signaling in the mammalian peripheral taste system. These peptides affect peripheral taste responsiveness of animals and play important roles in the regulation of feeding behavior and the maintenance of homeostasis.

SECTION IV: STIMULUS TRANSDUCTION IN OTHER CHEMODETECTION SYSTEMS

The unifying principles obtained from a systematic analysis of signaling mechanisms in the olfactory and taste systems are now also being applied to chemoreceptors that sense internal molecules important for the regulation of body homeostasis. Although this field is still in its infancy, the application of state-of-the-art, genetically based tools should help to target these cells within multiple organs of the body and enable better functional recording and manipulation of these detectors. Chapters 18–20 provide three examples of such internal chemodetection systems. A closely related case whereby a gustatory receptor functions as a brain nutrient sensor in insects is discussed in Chapter 14.

In **Chapter 18**, Nanduri R. Prabhakar focuses on the sensory organs for monitoring arterial blood O_2 and CO_2 levels in mammals, the carotid and aortic bodies. Emerging evidence suggests a complex interplay among three gases—oxygen, carbon monoxide, and hydrogen sulfide—and their interaction with K^+ channels and/or mitochondrial electron transport chain in carotid body sensory nerve excitation by hypoxia.

Chapter 19, by Shuping Wen, Mari Aoki, and Ulrich Boehm, highlights recent developments in chemosensing by cells of the ventricular system. These cells, including tanycytes, play important roles in different forms of internal chemosensation, from glucose sensing to the detection of hormones and many others factors that convey information on internal chemical status of the central nervous system.

Sami Damak summarizes our current understanding of the mechanisms underlying nutrient sensing in the gut in **Chapter 20**. In many cases, the same receptors that mediate the taste responses in the mouth are also present in the gastrointestinal tract and act as intestinal chemosensors.

The somatosensory system is another important sensory system by which chemical stimuli can be detected in the body. Chemodetection by somatosensory neurons in the skin or in the oral or nasal cavities mediates the process that we call "chemesthesis:" the detection of chemical irritants or toxins by cutaneous neurons. **Chapter 21**, by Jay P. Slack, describes our current understanding of chemesthesis, the somatosensory receptors that convey chemesthetic sensations and their relationships to pain and temperature sensing and the perception of flavor.

Although this book is inclusive, it is by no means exhaustive of the subject of chemosensory transduction. However, we hope this book serves as useful introduction to the subject and guide to the wealth of details and complexity that could not be adequately conveyed. Of particular note, there are other emerging topics in the chemical senses that could not be covered by this edition. For example, how do chemosensory cells detect specific pathogens and how is this information used to guide behavioral decisions (interestingly, these detection mechanisms seem to resemble the pathogen detection mechanisms known from the immune system)? We can only hope that this edition will attract a new generation of scientists who are fascinated by the chemosenses and will propel this field to the next level.

SOCIAL ODORS AND CHEMICAL ECOLOGY

Specialized Chemosignaling that Generates Social and Survival Behavior in Mammals

Lisa Stowers, Tsung-Han Kuo

Department of Molecular and Cellular Neuroscience, The Scripps Research Institute, La Jolla, CA, USA

WHAT ARE SPECIALIZED ODORS?

Upon detection, chemosensory ligands can generate two different categories of behavioral responses: associative or specialized. In associative olfaction, one experiences an odor and interacts with the environment to determine its meaning such as whether it is pleasant or aversive. This association is not fixed; its significance can change with each new encounter and therefore varies within and between individuals.[1] For example, the smell of cinnamon associatively coupled with a delicious cookie when one is hungry is likely to cause a positive appetitive behavior and enjoyment response. However, the same smell associatively coupled with the demands of shopping, planning, and hosting during winter holidays may lead to a negative avoidance response and the production of physiology-changing stress hormones. These flexible responses to the same odor cue enable one's behavior to be plastic and adapt to the immediate situation. Furthermore, a group of individuals sampling the same smell is likely to respond with a variety of different behaviors based on their past associative experiences. Because of this response variability, the underlying neural ensembles throughout the brain following repeated associative odor detection are expected to differ between and within individuals. Over time, and depending on the intensity of one's personal experiences with the environment and the odor, these responses may continue to be flexible or can become fixed as either positive or negative. However, even if they become fixed, an associative experience was required to generate the ultimate behavioral meaning.

In contrast, upon first detection of a specialized odor, most individuals will initiate a predictable behavior response or change in their neuroendocrine physiology.[2] Subsequent interactions with the odor are equally likely to generate the same outcome. Sensory neurons that detect this type of chemosignal have a high probability of generating a fixed, predetermined behavior. To generate this fixed outcome, the responding neural ensembles throughout the brain are unlikely to undergo the same plasticity observed in associative olfaction. Instead, it is thought that these chemosignals activate subsets of sensory receptors that are genetically determined, "hardwired," to elicit a preset behavior. Because these cues elicit predictable behavior throughout the social group, it is possible that the underlying neural ensembles that link olfaction to behavior are similar across many individuals.[3] Specialized odors, known as semiochemicals, are proposed to include pheromones and kairomones.[3,4] Pheromones are chemosensory ligands emitted by individuals that elicit behavior or neuroendocrine changes when detected by other members of the same species. Kairomones are ligands emitted from one species that generate behavior in another species (such as aversion upon detection by a prey species). These specialized chemosensory cues are proposed to be instrumental in generating social and survival behavior to enhance the fitness of an individual.

SEARCH FOR THE SENSORY NEURONS UNDERLYING SPECIALIZED OLFACTORY BEHAVIOR

Upon classifying olfactory behavior into two major types, associative and specialized, it follows that there may exist at least two different types of olfactory chemosensory detectors. One type that is able to generate flexible associative sensations by integrating past experience and using regions of the brain responsible for learning and memory. To activate another kind of response that is already preset with meaning the specialized detection system may be composed of sensory detectors tuned to different classes of ligands, that generate different physiological properties (perhaps in adaptation or temporal response), and/or they may project axons to targets in the brain that are preset and more resistant to plasticity.[2]

Corresponding to the duality of function, the nasal cavity of almost all terrestrial vertebrates contains two anatomically separate chemosensory organs: the vomeronasal organ

(VNO) and the main olfactory epithelium (MOE).[5] The functional significance of having two separate olfactory systems is not known. What is the MOE incapable of that the VNO evolved to provide? Not only are they physically independent, but, as detailed later, their sensory neurons largely express completely different repertoires of olfactory receptors and signal transduction components, they respond to different classes of ligands, and they each project axons to different regions of the brain. Noting the differences in anatomical projections Cajal first speculated that the accessory olfactory bulb (the AOB, the first relay of the VNO) "is a special center, perhaps differentiated for receiving impressions of some particular kind of olfactory excitation."[6]

The segregation of olfactory sensory components into two distinct sensory organs, as noted by the emergence of the AOB, occurred with the evolution of tetrapods.[7] The size and anatomy of the VNO and AOB vary dramatically across species,[7] and the variation largely correlates with the magnitudes of the vomeronasal receptor (*Vmn1r/Vmn2r*) gene repertoire[8,9] (see Chapter 10). Why do some animals have a more elaborate VNO, whereas in others it is much reduced? Does it correlate with the detection of specialized chemosignals? Based on known ethological characteristics, gain or loss of VNO receptors has not been found to correlate with body size, nocturnality, diet, sociality, or mating system.[9] This lack of correlation between sociality and VNO function is evidenced in dogs, which appear to readily sample and use olfactory information, yet their genome only contains nine VNO receptors.[9] It was believed that perhaps domestication altered the need for VNO-mediated behaviors and imposed selective pressure to favor VNO receptor inactivating mutations on this new species. It is now known that this is not the case because the VNO receptor repertoire of wild wolves is similar to the dog, indicating the loss of VNO receptors occurred before domestication.[9] Further, humans are extremely social, yet the VNO appears vestigial, the human genome only contains five VNO receptors that are expressed in the MOE, and the primary signal transduction channel of the VNO, transient receptor potential channel type C2 (TRPC2), underwent inactivating mutations in Old World primates.[10,11] This has led to the hypothesis that the evolution of trichromatic vision functioned to eliminate selective pressures on VNO signaling elements.[10] However, analysis of additional primate genomes indicates that VNO receptor loss began in the common ancestor of both New and Old World primates, occurring independently of the gain of color vision.[9] Analysis of species with a limited (or nonfunctional) VNO, such as humans and dogs, demonstrate that the MOE alone is sufficiently sensitive and complex to generate olfactory perceptions and guide behavior. Whether species without a significant VNO system such as humans (see Breakout Box) and dogs engage in specialized olfaction remains to be determined.

BREAKOUT BOX

HUMAN PHEROMONES

Is human behavior influenced by pheromones? Are specialized odorants clandestinely guiding one's choice of partner, provoking rage, or enabling infant bonding? Whether humans emit and detect pheromones is one of the great mysteries still facing research in social communication.[179] Scientists have been trying to identify and study human pheromones by focusing on a variety of social situations that depend upon olfaction. Some of these have identified olfactory-mediated behaviors conserved across the animal kingdom and investigated whether they extend to humans. An example is menstrual synchrony, which is well established in livestock. In humans, the menstrual cycles of women who live together tend to become synchronized over time because of stimulation of an unknown female-emitted pheromone.[180] However, the effect has not been rigorously replicated, specialized ligands have not been purified, and the mechanism behind this phenomenon is still unclear. Innate infant suckling is another behavior that may depend on specialized chemical communication in humans since all mammals must innately suckle milk to survive. Secretions from areola skin glands of human mothers have been shown to elicit suckling and nipple-search behavior in newborns.[181,182] A rabbit suckling-promoting pheromone was rigorously purified from rabbit milk and it is possible that specialized chemosensory cues equally govern this essential behavior in all mammals[183,184]; however, mice may use a biologically restricted associative olfaction to properly perform first suckling.[41] Other social behaviors that are specific to humans such as hand shaking or crying have prompted the investigation of human tears and axillary sweat for specialized chemosignals that alter human emotion, behavior, or brain activity.[185][−188] These studies have not been replicated, nor has it been determined if the cues are truly specialized or are associative.[179] Surprisingly, even without strong scientific evidence, androstenone, an axillary steroid, has been successfully sold as a potential sex-attractant indicating interest and desire of the general public to find human pheromones.

Without a validated human pheromone ligand, research on human pheromone perception is an even greater challenge. The human VNO is thought to be nonexistent or nonfunctional.[189] However, lack of this sensory organ does not preclude the existence of human pheromones because other mammals, including rabbits and pigs, use the MOE to detect pheromones.[190,191] Currently, the most promising candidate pheromone receptor in human is OR7D4, which has been shown to respond to androstenone, a component of axillary sweat, in vitro.[192] The human MOE also expresses VRs and TAARs,[193,194] whose functions in pheromone detection have been identified in mice,[5] and may serve as orthologs to identify potentially specialized chemoligands. In fact, use of a heterologous expression system to deorphanize chemosensory receptors has identified hedione, a commonly used compound in perfume, as a ligand for the VN1R1[195]; however, whether humans secrete hedione remains an open question. Future studies focusing on pheromone detection in the MOE in other animals may provide a testable model for pheromone sensing in humans. For now, whether human behavior is influenced by specialized chemosensation remains an unproven possibility.

Recent deep sequencing analysis has revealed dramatic molecular differences between sensory neurons located in the VNO and MOE.[12,13] The VNO of mice detects chemical ligands through the expression of 585 G protein—coupled receptors (GPCRs) of the Vmn1r (type 1 vomeronasal receptor, or V1R) and Vmn2R (type 2 vomeronasal receptor, or V2R) families.[12] These receptors are largely absent from the MOE, which instead expresses >1200 GPCRs of the odorant receptor family, ORs (see Chapter 3).[12,13] Although an MOE or VNO sensory neuron is thought to achieve tuning to a particular chemical moiety through the expression of either a single odorant receptor (OR) or V1R (respectively),[14,15] some VNO sensory neurons (VSNs) express as many as three different receptors: two from the V2R family as well as a nonclassical major histocompatibility receptor of the H2-Mv family (Figure 1).[16-19] The functional significance of expressing two VNO receptors remains unknown; however, chromosome engineering has succeeded to delete the entire family of H2-Mvs and found that these membrane proteins are necessary for the VNO to be ultrasensitive to minute quantities of ligands in the environment.[17] Because the MOE can be similarly sensitive to a very low concentration of ligands, it is unclear whether the expression of H2-Mvs constitutes a functional difference between the VNO and MOE. A striking difference between the two olfactory systems comes from sequence analysis of vomeronasal receptors (VRs). Unlike the MOE, where there are orthologous receptors across divergent species, the receptors of the VNO appear to rapidly evolve between species, creating VR repertoires that are largely specific to one species.[20,21] It is likely that this customized receptor expression enables each species to recognize chemical cues that are of particular relevance to their survival and fitness.

In addition to differences in expression of the ligand-detecting GPCRs, the VNO and MOE largely rely on different signaling mechanisms to depolarize the neuron (Figure 1). The MOE uses cyclic adenosine monophosphate—mediated signal transduction to activate a cyclic nucleotide—gated channel (see Chapter 7), whereas the VNO employs a phospholipase C-mediated pathway that leads to opening of the TRPC2 cation channel (see Chapter 11).[5] These differences in molecular expression profiles support the notion that the MOE and VNO are not simply an anatomically separated single chemosensory organ, but instead have evolved to serve different needs of the organism.

Molecular genetics has been used to evaluate the functional consequences of VNO and MOE sensory detection based on the differences in gene expression. Ablation of the primary signal transduction channel of the VNO, *Trpc2*, prevents VSNs from responding to many ligands naturally emitted from other mice (as expected of pheromones) and potential predator animals (as expected of kairomones). Male *Trpc2$^{-/-}$* mutants have no defects in courting and mating females, which is thought to be facilitated by pheromonal cues, yet are unable to initiate territorial aggression, establish dominance hierarchies, perform territorial scent marking, display appropriate parental behavior, identify diseased conspecifics, or respond to some predator odors, which are behaviors dependent on semiochemicals.[22-31] Mutant females show various defects in mating and maternal behavior.[28] Genetically ablating the genes encoding VNO signal transduction elements that are commonly expressed by many different VSN populations (such as *Gnao*, which encodes the G protein subunit Gα_o, and *Gnai*, which encodes Gα_{i2}) alters the ability to perform several pheromone-induced behaviors including parental behavior and male territory defense aggression.[32-35]

Complementary experiments ablating genes for the primary signal transduction elements of the MOE (the obligatory cyclic nucleotide channel subunit *Cnga2*; *Gnal*, which encodes

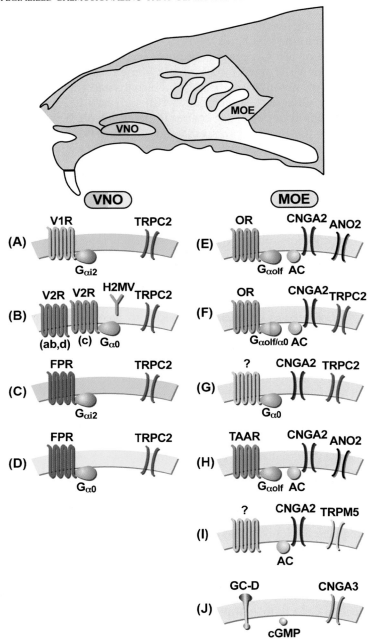

FIGURE 1   **Molecular identity of sensory neurons of the main and accessory olfactory systems implicated in specialized chemosignaling.** Top: anatomical separation of the two major olfactory sensory organs in the mouse; vomeronasal organ (VNO; yellow/orange) and main olfactory epithelium (MOE; blue). Left column: the mouse VNO is composed of at least four sensory neuron subtypes: (A) and (C), two express either a Vmn1r (V1R) or FPR as well as $G\alpha_{i2}$ and TRPC2 and are anatomically segregated to one region of the VNO (orange); and (B) and (D), two that express either two Vmn2r (V2R) or an FPR as well as $G\alpha_o$ and TRPC2 (yellow).[5] The VR-expressing neurons have been implicated in specialized olfaction. Right column: (E), canonical MOE neuron that has been implicated in associative olfaction expresses OR, $G\alpha_{olf}$, and CNGA2; (F–J), neurons that have been implicated in specialized olfaction are distinguished by either lacking OR or expressing different primary signal transduction elements.

$G\alpha_{olf}$; and the type 3 adenylyl cyclase, *Adcy3*) reveal that in addition to associative olfaction, the MOE is required to promote several pheromone-mediated innate behaviors.[36–40] When assayed for pheromone-mediated behavior, *Cnga2* null mutant mice are unable to initiate aggression, both males and females fail to mate, and newborn pups die of dehydration because they are unable to detect maternal olfactory cues necessary to initiate suckling.[37,40,41] It is possible that animals devoid of olfactory input may primarily lack motivation to investigate the environment, which is a necessary first step to engage in appropriate instinctive behavior. The sensation of odorant molecules by the MOE accelerates sniffing, a biophysical process that is postulated to be essential to pump behavior-inducing ligands into the blind-ended, fluid-filled VNO lumen.[37,42] Genetic mutants lacking MOE function have been shown to exhibit decreased sniffing rate[37]; therefore, it is possible that the decrease in social behavior in animals lacking CNGA2 might be a secondary effect because of ineffective ligand access to the VNO lumen. However, genetic mutants in which neurons expressing the *O-MACS* gene are ablated by the expression of diphtheria toxin (ΔD^{MUT})[36] more precisely indicate that the MOE directly functions to detect specialized odors. In these animals, the majority of the MOE is intact and functional; however, they specifically lack development of sensory receptor neurons that project to the dorsal region of the main olfactory bulb (MOB).[36] With most of their sensory receptors intact, these animals do not show any gross defects in general olfactory-mediated behavior such as feeding or investigation, and do not have general malaise. Yet, similar to *Cnga2* mutants, they fail to display a subset of innate avoidance behavior.[36] One interpretation of these experiments is that the MOB is preorganized with a functional topography where avoidance-generating neurons reside in its dorsal portion. Such a hardwired mechanism may provide a framework for the innate responses of specialized odors.

Together, many lines of evidence indicate that the VNO is very different from the MOE in a number of measures. However, both olfactory systems are equally capable of mediating an innate response to ethologically relevant odor cues that underlie social behavior. This suggests that neurons that mediate specialized olfaction are likely to be found in both the MOE and the VNO.

ARE CERTAIN TYPES OF OLFACTORY LIGANDS SPECIALIZED?

The MOE is located in the upper recesses of the nasal cavity bathed in a protective layer of mucus. To be sensed by the OR, an odor molecule must possess enough volatility to be sampled during normal breathing and then solubilize in the aqueous mucus. Consistent with this requirement, most known ligands that activate ORs have been found to be organic small molecules (<900D).[15] The only nonvolatile ligands shown to be detected by the MOE are major histocompatibility complex (MHC) peptide ligands which are detected by an undefined subset of MOE neurons, and guanylin peptides that activate a specialized population of sensory neurons that use a cyclic guanosine monophosphate—mediated signaling cascade (see Chapter 8).[43–45] Aside from these exceptions, if the MOE is detecting specialized odors, then they are likely to be small organic molecules.

The anatomy of the VNO is quite different from that of the MOE, which enables it to detect a different variety of chemosignals. The VNO is a blind-ended, mucus-filled tube at the base of the nasal cavity just above the palate of the oral cavity. Ligands from the oral cavity can access the nasal cavity through the nasal pharyngeal duct.[5] Nonvolatile ligands can access the VNO by direct contact, such as licking the emitter during social investigation, or through inhalation of aerosolized native secretions associated with the emitter or deposited in the environment. It has been proposed that larger molecules that are not a common by-product of general metabolism, although not so complex that synthesis becomes costly, are likely to serve as species-specific cues suitable to be pheromones.[46]

Six categories of chemosensory ligands that stimulate the VNO have now been identified: exocrine gland-secreting peptides (ESPs), MHC peptide ligands, major urinary proteins (MUPs), volatile small molecules, sulfated steroids, and formyl-peptides (Table 1). These known ligands fall into two major groups. ESPs, MHCs, formylated peptides, and MUPs are all peptides or small proteins. All except formylated peptides have undergone species-specific gene duplications in rodents.[47–49] The majority of these protein ligands have been shown to activate V2R-expressing VSNs (Figure 1).[25,47,50] The second group is made up of VNO-stimulating volatile small molecules and sulfated steroids (also small molecules), the majority of these small molecules are detected by V1R-expressing neurons,[55,56] many of which have been demonstrated to also activate MOE neurons (Figure 1).[51–54]

Examples from most of the VNO-stimulating ligand families have been shown to mediate social behavior. Sex-specific expression is one characteristic thought necessary to elicit social courtship and reproductive behaviors. The expression of ESP1 meets this criterion because it was found to be excreted only in male tears.[29] When detected by females, ESP1 specifically activates a single V2R (known as V2Rp5), whose action is necessary to stimulate female mating behavior.[57] Another characteristic of social behavior is the ability to recognize and appropriately respond to individuals. MHC class I peptides are thought to be excreted in urine throughout the life of both males and females, providing a stable representation of individuality. In addition to their ability to function in self-recognition in the immune system, they have also been shown to function as olfactory sensory stimuli.[50] They are detected by V2R-expressing neurons in the VNO as well as currently undefined neurons in the MOE.[44,50] Once detected by the VNO, they are required for a female to imprint the memory of her mating partner, which is essential to maintain a productive pregnancy.[50] The identification of an unrelated family of GPCRs, formyl-peptide receptors (FPRs), as exclusively expressed in VNO sensory neurons enabled the discovery of their cognate ligands, which include formylated peptides and inflammation-related ligands (Figure 1).[58–60] Formylated peptides are produced both by bacteria and mitochondria. Currently, little is known about the significance of detecting this ligand group, but they may signal the health status of an individual.[30,61]

MUPs have been proposed to function as pheromone carrier proteins serving to transmit the small volatile ligands found in urine into the mucus-filled VNO, and as stabilizers of these volatile ligands to extend their signaling potency by delaying their release after excretion into the environment.[62–64] Of the 21 MUP genes in the mouse genome, each individual only expresses four to 12 urine MUPs.[48,65,66] This personally tailored protein excretion has been proposed to differentially stabilize small molecules to facilitate individual recognition.[67–70] In addition, MUPs alone, without the associated small molecule ligands, have been shown to stimulate V2R2-expressing neurons.[24,25,27] Their detection promotes gender-specific behavior, including territorial countermarking and aggression when detected by other males

TABLE 1 Candidate Specialized Olfactory Ligands

Category	Ligand	Sources	Sensory organ	Behavior	Identified receptor	References
Urinary volatiles	2-Heptanone trans-5-Hepten-2-one trans-4-Hepten-2-one n-Pentyl acetate cis-2-Penten-1-yl-acetate 2,5-Dimethylpyrazine	Female urine	VNO	Puberty delay		199
Urinary volatiles	SBT	Male urine	VNO	Puberty acceleration		74
				Estrus synchronization		81
				Female attraction		74
				Intermale aggression		53
	Isobutylamine			Puberty acceleration		73
	Isoamylamine			Puberty acceleration		73
	DHB			Puberty acceleration		74
				Estrus synchronization		81, 53
				Intermale aggression		
				Female attraction		74
Urinary volatiles	α and β Farnesenes	Male urine	VNO	Puberty acceleration		74
				Intermale aggression		52,
				Female attraction		197
				Estrus synchronization		198

(Continued)

TABLE 1   Candidate Specialized Olfactory Ligands—cont'd

Category	Ligand	Sources	Sensory organ	Behavior	Identified receptor	References
Urinary volatiles	6-Hydroxy-6-methyl-3-heptanone	Male urine	VNO	Puberty acceleration	Olfr1509	74
Urinary volatiles	(Methylthio) methanethiol	Male urine	MOE	Female attraction	Olfr1509	120,196
Urinary volatiles	Trimethylamine	Male urine	MOE	Attraction	TAAR5	98
Major urinary proteins	MUPs	Male urine	VNO	Male countermarking	V2Rs	27
Major urinary proteins	Mup3	Male urine	VNO	Internale aggression	V2Rs	27
Major urinary proteins	Mup20	Male urine	VNO	Internale aggression Female attraction	V2Rs	27, 71
MHC peptides	MHC class I peptides	Urine	VNO and MOE	Individual recognition	V2R1b	44,50
Exocrine gland-secreting peptides	ESP1	Male tear	VNO	Female attraction	V2Rp5	57
Exocrine gland-secreting peptides	ESP22	Juvenile tear	VNO	Inhibiting mating		26
Sulfated steroids	Sulfated estrogens	Female urine	VNO	Male attraction	V1rj clade	51,86,92
Kairomones	MUPs	Cat saliva and rat urine	VNO	Avoidance	V2Rs	24
Kairomones	2-Phenylethylamine	Carnivore urine	MOE	Avoidance	TAAR4	39
Kairomones	2,5-dihydro-2,4,5-trimethylthiazoline	Fox odor	MOE	Avoidance		36

DHB, 3,4-dehydro-exo-brevicomin; MHC, major histocompatibility complex; MOE, main olfactory epithelium; MUPs, major urinary proteins; SBT, 2-sec-butyl-4,5-dihydrothiazole; VNO, vomeronasal organ.

and attraction when detected by females,[24,27,71] as well as gender-neutral behavior, acting to facilitate conditioned place preference.[72] Additionally, MUP orthologs emitted from heterospecifics are sufficient to signal VNO-mediated innate fear.[24]

To identify gender-specific bioactive ligands, subtractive gas chromatography-mass spectrometry methods were used to screen for compounds enriched in urine from intact males compared with androgen-negative castrated male urine. This approach identified several volatile VNO-activating ligands including isoamylamine, isobutylamine, 3,4-dehydro-exo-brevicomin (DHB), 2-sec-butyl-4,5-dihydrothiazole (SBT), 6-hydroxy-6-methyl-3-heptanone (HMH), and α/β-farnesene.[73–75] All of these small molecules bind MUPs, which likely act as carriers to transport those molecules into the VNO lumen.[76] Unbound, each agonist has been shown to directly activate V1R-expressing neurons, some at as little as 10^{-11} M concentration.[55,77] SBT, α/β-farnesene, and HMH are excreted in male urine and have been shown to promote female puberty acceleration[74,75,78] (although see[79]), whereas DHB and 2-heptanone are emitted by females to delay estrus or puberty.[74,80,81] DHB, α/β-farnesene, and 2-heptanone have been shown to direct social behavior in several evolutionarily diverse species, including insects.[82–85]

Sulfated steroids, including the androstan, androsten, cholestan, estratrien, pregnan, and pregnen families of sulfated steroids, collectively have the potential to transmit many qualities about an individual's physiological state that are expected of specialized ligands.[51,86,87] Within an individual, steroids modulate basic physiology as well as neural responses that together define state-specific social or survival behavior including developmental stage, social dominance, reproductive capability, and stress states. Sulfation promotes clearance of steroids to enable a change or maintenance of a particular steroid state.[88] Therefore, sulfated steroids may provide a reliable chemosignature of the internal state of the signaling individual to inform the receivers' behavior. Sulfated steroids strongly activate subsets of V1Rs in the VNO and may account for 80% of the VNO-mediated chemoligands in female urine.[51] Although most ligands are detected by 0.5–1% of the VSNs, individual sulfated steroids activate a greater number of VSNs (about 10%) and those sensory neurons respond with a larger depolarization than is seen with any other vomeronasal ligand.[86,89,90,91] Together, these characteristics indicate that they are of great importance to the VNO system; however, their meaning to the receiving animal is largely unknown.

In the mouse, a sulfated estrogen emitted by ovulating females has now been found to activate Vmnr1j clade-expressing VSNs and promote male courtship behavior, but only when detected in combination with a second, currently uncharacterized, female urine ligand that activates Vmn1re clade-expressing neurons.[92] Each of these ligands alone is not sufficient to induce the behavior, but may combine to form a multicomponent pheromone, or they may be transmitting different kinds of information about the emitter. Many social behaviors require the receiver to simultaneously confirm several independent characteristics of the emitter (such as their species, gender, dominance, age, and health) some of which are stable across an animal's lifetime and others that rapidly change with environment and experience. Sulfated steroids may provide a mechanism to transmit dynamic information about the emitter's internal state, which then may work in concert with other, more general, ligands that transmit stable information in order to appropriately guide the receiver's response.

The mouse VNO may also play an important role in detecting kairomones, which are ligands emitted by one species that elicit a behavior to advantage the receiver of another species.[24,86,93] Indeed, of 88 VRs found to be stimulated by ethologically relevant odors, only 17 are activated by cues emitted by other mice, whereas 71 have been found to be activated by scents emitted by other species such as birds and snakes.[86] Upon detection, kairomones function to elicit defensive behaviors such as avoidance and an increase in adrenocorticotropic hormone (ACTH) stress hormones.[24] Except for the cat- and rat-emitted MUP orthologs,[24] neither the structural identification, purification of the activating ligands, nor the associated behavior promoted by these cues have been determined, so it is not known if these ligands emitted from other species also promote a specialized function.[86] However, it is reasonable to speculate that many of the ligands emitted from potential predators that are detected by the VNO are preset to generate fear similar to the cat- and rat-purified cues.[24]

The mouse VNO expresses hundreds of different sensory receptors (V1Rs, V2Rs, and FPRs; Figure 1),[12] yet only a handful of salient ligands have been identified. Most of the known stimuli, except for MUPs and MHC peptides, have been shown to be detected by a single, highly tuned receptor.[27,51,55,57,94] If each receptor functions to detect a single salient cue, then the majority of VNO ligands, and potentially other ligand classes, are still largely unknown.

THE VNO DETECTS SPECIALIZED ODORS

To determine if chemosensory ligands and their receptor neurons are truly specialized (preprogrammed with meaning), one needs to test the innate response of animals upon their very first encounter with the ligand to ensure that there has been no prior associative learning. This has been impossible because the precise cues that generate behavioral responses are largely unknown. Without knowing the identity of the bioactive ligands, one cannot determine if an individual is truly naive at the time of behavioral analysis. Identification and purification of bioactive ligands that direct aggression, mating, and fear in the mouse have now made such experiments possible. The outcome of testing behavioral responses upon first exposure to purified, bioactive odors supports the existence of specialized semiochemicals.

For example, one of the MUPs excreted in male urine, MUP20 also known as darcin, activates V2R neurons and is specialized to promote rage when detected by another adult male.[25,27,71] The behavioral response to synthetic MUP20 is as robust as native sources indicating that it underlies the bioactivity; however, for many years, it was unknown whether the behavior was associative or specialized. Because all males emit MUP20, no individual male could be expected to be naive to the cue. MUP20 expression is testosterone- and growth hormone—dependent and therefore its production begins around 4 weeks of age,[27] whereas territorial aggression behavior is not displayed prior to sexual maturity (at least 8 weeks old). Therefore, each male experiences many weeks of exposure to MUP20 emitted by themselves and their brother cage-mates before they become sexually mature and display territorial aggression. Throughout this exposure period, associative learning of MUP20 as a rage-provoking cue may have occurred during developmental aggressive play. This confound was resolved by identifying BALB/c mice as a laboratory strain that fortuitously fails to express MUP20.[48,49] The repertoire of MUP proteins that BALB/c males do express and secrete lack aggression-promoting bioactivity when detected by other MUP20-naive BALB/c males or even males from the C57BL/6J strain, which robustly expresses MUP20.[27] BALB/c males therefore have no experience with MUP20 as an aggression-promoting chemoligand throughout their development. Yet, upon first exposure to MUP20 at maturity, they initiate the complete repertoire of aggressive behavior.[27] This indicates that MUP20 is an example of a specialized olfactory ligand; able to activate sensory detectors that are hardwired to trigger aggressive behavior initiating circuits without associative learning.

Specialized ligands have also been confirmed that are preset to direct appropriate mating. Males secrete the small peptide ESP1, which is detected by V2R-expressing neurons, in their tears.[29,47] Receptive females investigate the face of males and upon detection of ESP1 initiate lordosis (a female mating posture).[29] Analysis of females from a strain that does not produce ESP1 (C57BL/6J), and therefore are experientially naive to its detection and ensemble of neural activation, showed an increase in lordosis upon first exposure to males from an ESP1-producing strain (BALB/c) compared with C57BL/6J males.[57] This response confirms that experience with the ligand is not necessary for its specialized function.

Another small peptide (ESP22), which is secreted by juveniles and detected by V2R-expressing neurons, has also been shown to be a specialized mating cue.[26] Upon detection, the ligand inhibits mating advances from adult males, which display promiscuous mating behavior with juveniles if chemosignaling is rendered inoperative.[26] Analysis of males from a strain that lacks ESP22 expression (C3H) reveals that without this single sensory cue adult males fail to recognize the juvenile as an inappropriate mating target and instead target these immature mice with robust mating behavior. However, first detection of ESP22 by these adult C3H males rapidly and appropriately inhibits sexual behavior toward juveniles.[26]

Kairomones have also been shown to activate sensory receptors that are evolved to promote fixed, preprogrammed, responses upon first detection by mice without associative learning.[3] Some inbred strains of laboratory mice (such as C57BL/6J) have been housed for hundreds of generations in isolation from other predator species. Strikingly, first exposure to a cotton swab infused with just the odor of cats, rats, or snakes, and thereby not coupled to the threatening environment of the predator itself, is sufficient to promote a repertoire of fear-like responses, including avoidance and cautious stretch-attend behavior as well as a rapid and robust increase in levels of the stress hormone ACTH.[24] Identification of the bioactive ligands produced by cats and rats shows that they are MUP orthologs and has enabled their artificial synthesis. In turn, these were used to demonstrate that their activation of chemosensory neurons is sufficient to generate fear-like behaviors upon an animal's first encounter.[24]

When considered together, these experiments reveal that mice do indeed detect specialized odors. First experience with these odors generates a defined, consistent meaning among many individuals. This indicates that subsets of chemosensory ligands emitted by mice are specialized to activate neurons that are preset to initiate social and survival behavior.

THE MOE ALSO DETECTS SPECIALIZED LIGANDS

With the immense diversity of odor space and an intractable number of orphan ORs in the MOE, discovering the minor subset that underlies a specialized response has proven remarkably difficult.[12] The first candidates have now been identified through the assumption that the transcriptome of specialized sensory neurons may differ from those that perform associative olfaction (Figure 1). A subset of MOE neurons has been found to lack the expression of canonical ORs and instead express one of nine trace amine-associated receptors (TAARs)[95,96] (see Chapter 4). These olfactory-specific receptors detect ultra-low concentrations of volatile amines as well as unknown ligands in predator urine.[96–98] Upon first exposure to this class of ligands, individuals largely respond with innate aversion,[39,97] with the exception of trimethylamine, which promotes innate attraction.[98] In addition to TAAR-expressing neurons, canonical OR-expressing neurons may also function as detectors of amines in the MOE. ORs do not recognize entire ligands; instead, they detect molecular features (parts of the molecule).[15] This means that many ORs recognize each ligand and each ligand is detected by several ORs; this is termed "combinatorial coding."[15] Following this principle, there likely exist ORs capable of detecting molecular features of TAAR-activating ligands and the coincidental activity of ORs and TAARs may be expected to cooperatively generate the aversive behavior. Remarkably, genetic deletion of a single TAAR, TAAR4, entirely eliminates the aversive behavior evoked by a subset of amines and even converts the response toward predator odor from innately aversive to attractive.[39] This indicates that the TAAR4 expressing neurons (see Chapter 4) are powerfully specialized and confirms that individual MOE neuron types can be preset with meaning to instruct an individual's behavior.

In addition to identification of the TAARs, molecular profiling has discovered several other noncanonical MOE neuron types (Figure 1). One type expresses the primary sensory transduction channel found in VNO neurons: TRPC2.[22,23,99,100] TRPC2 neurons located in the MOE appear to be hybrids of canonical VNO and MOE neurons. Some TRPC2 neurons express the MOE-specific ORs and G protein ($G\alpha_{olf}$), whereas in others of this type the sensory receptors have not been found but they have been shown to express VNO-type G proteins ($G\alpha_o$).[100] Although these cells are excellent candidates to participate in specialized olfaction, there is currently no report on their function. Most studies with the TRPC2 mutant generally used multiple approaches to ensure the underlying function relies on the VNO.[24–29] However, two early studies attributed multiple phenotypes of specialized function to the VNO based solely on analysis of TRPC2 mutants.[22,23] Of these reported phenotypes, male aggression and countermarking, have since been validated to be VNO-mediated,[25,27] but other reported phenotypes[22] may have been due, partially or completely, to the loss of this new class of MOE sensory detection.

Subsets of MOE neurons, defined by the expression of another channel TRPM5, are also implicated to generate specialized function (Figure 1).[101–103] TRPM5 has been shown to be necessary for detection of 2-heptanone and 2-DMP at ultrasensitive concentrations.[101,104] These ligands are secreted in the urine of adult males and have been reported to promote social behavior, which is consistent with pheromone activity. However, 2-heptanone and 2-DMP also robustly activate VNO neurons.[55] Therefore, whether TRPM5 expressing neurons located in the MOE contribute to innate behavioral responses to these odors remains to be determined.

Another molecularly distinct cell type in the MOE is characterized by the expression of a guanylyl cyclase D (GC-D) membrane receptor and a lack of expression of other canonical olfactory sensory transduction elements (Figure 1).[43,105,106,107] These neurons have been implicated in the detection of peptide hormones (uroguanylin and guanylin), which may enable the MOE to play a role in salt uptake or thirst regulation[43,105]; the detection of carbon disulfide (CS_2),[108] a component of breath; and in CO_2 detection.[109] Functionally, CS_2 detected from the breath of another mouse in conjunction with a nonspecialized odor enables a naive social partner to acquire the preference for novel food. Molecular analysis has now confirmed that the expression of GC-D in these neurons is essential to generate this social transmission of food preference.[108,110] Though this is a learned behavior, its display likely depends upon specialized mechanisms.

NEURAL CIRCUITS THAT PROCESS SPECIALIZED LIGANDS

Comparison of the circuits activated by specialized olfactory ligands shows that MOE sensory neurons directly excite cortical centers,[111–113] which mediate associative olfaction,[1] whereas those from the VNO activate limbic centers, which are necessary to display proper social behavior.[111,114,115] The lack of connections to a known cortical region may indicate that VNO-mediated olfaction provides a "direct" route into the limbic system to faithfully and rapidly initiate specialized behavior. These separate processing streams reinforce the notion that each system is designed to produce a different function. However, this interpretation must be approached cautiously until the entire circuit has been functionally studied.

In the MOE, neurons expressing the same OR converge on one to three glomeruli in MOB.[116] MOB output neurons, the mitral/tufted cells, project to all olfactory centers in the brain including the anterior olfactory nucleus (AON), the piriform cortex (PC), the olfactory tubercle, the cortical amygdala (CoA), and the lateral entorhinal cortex.[112,117] Functionally, the PC is thought to house neurons that enable associative learning and a plastic response to chemostimuli, both within and across individuals[1,113] and projection of the mitral cell axons to the PC does not appear to follow a preset, topographic map as may be expected to generate specialized function.[112,117,118] In contrast, anatomical tracing shows that the AON is organized topographically and the CoA primarily receives inputs from the dorsal MOB.[112,117] The CoA is composed of subsets of neurons that direct either positive or negative valence associated with odor-mediated behavior.[119] Hardwired organization and inclusion of regions thought to generate emotional valence are circuit characteristics thought to underlie specialized chemosensory behavior. Detailed functional analysis of circuit activity beyond the MOB[120] following exposure to specialized MOE remains to be investigated.

Like neurons of the MOE, VSNs coalesce in 6–10 glomeruli across the AOB. Further, not all are composed of axons from cells expressing a single type of VR; some glomeruli are composed of sensory axons from different, but apparently closely related, VSNs.[121,122] This organization into mixed glomeruli likely serves to integrate incoming sensory information.[123,124] Glomeruli formed from V1R- and V2R-expressing neurons are spatially segregated in the AOB. V1R neurons project to the rostral region, whereas V2R neurons target the caudal bulb (Figure 2).[125] The logic behind this separation of information remains unclear. From the AOB, subsets of second order mitral and tufted cells directly project to multiple amygdala nuclei including the medial amygdala (MeA), CoA, bed nucleus of the stria terminalis (BST), nucleus of the lateral olfactory tract, bed nucleus of the accessory olfactory tract (BAOT), and the anterior amygdaloid area (AA),[126] several of which have

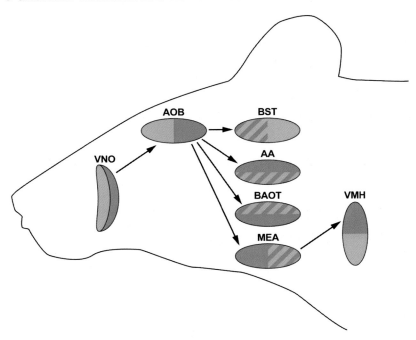

FIGURE 2   **Anatomical and functional division of vomeronasal organ (VNO)-activated circuits.** It appears that the VNO may direct two different processing streams: (pink) one that is primarily initiated by V1R-expressing neurons and (blue) another that is instigated by type 2 vomeronasal receptor—expressing neurons. Initial nodes in the circuit, from the VNO through the accessory olfactory bulb (AOB) and to the bed nucleus of the stria terminalis (BST), anterior amygdaloid area (AA), bed nucleus of the accessory olfactory tract (BAOT), and medial amygdala (MeA), have been defined anatomically. The segregation has also been supported functionally by comparing activity generated by conspecific mouse odors (pink) to that of potential kairomones from cat odor (blue).

been functionally implicated in mediating VNO signaling.[127–130] In general, the amygdala is thought to serve as a vigilance system, to determine whether incoming sensory information contains information that is positive or negative to an individual's fitness or survival, and then signaling this assessment to the rest of the brain.[131–134] Notably, although the mitral/tufted neurons stimulated by either V1R- or V2R-expressing VSNs both equally project to most of these amygdalar nuclei,[135,136] anatomical and functional evidence suggest that the V1R/V2R segregation established in the AOB is maintained in the anterior MeA and BST (Figure 2).[137,138] Specifically, cells from the anterior AOB project to the medial-posterior-medial BST, with no cells from the posterior part projecting to this subnucleus, whereas cells from the posterior AOB project to the dorsal AA and deep layers of the BAOT and anteroventral MeA, with no cells from the anterior AOB projecting to these regions.[137,139] One clue that may help solve the puzzle of this anatomical distinction may come from functional analyses revealing that subnuclei of the MeA are activated by distinct stimuli. Odors from the same species (such as pheromones) activate both the anterior and posterior MeA, whereas odors from different species (such as kairomones) have been found to activate only the anterior MeA.[130] Consistent with mediating biologically relevant stimuli, anterior MeA inactivation results in deficits in mating, social memory, and social chemo investigative behavior.[140–144]

The hypothalamus is thought to be the next known processing module for VNO stimuli. The hypothalamus is composed of multiple nuclei that function to maintain homeostasis (such as hunger, thirst, temperature, or blood pressure) by releasing hormones to direct the actions of the pituitary.[145] It also functions to release the motor patterns that underlie behavioral responses including fear, aggression, mating, and sleep.[111,131,139,146–151] Neural activity from the MeA can directly engage the hypothalamus.[111,131,139,146–151] Alternatively, the signal can be further processed by either the BST, a region thought to integrate homeostatic and sensory information[152,153] or by the hippocampus, which may serve to prioritize behavioral response to multiple, behaviorally conflicting, VNO stimuli.[151] Notably, the ventromedial hypothalamus has been proposed to be anatomically segregated into two functional zones. A dorsal medial region that is activated by kairomones and generates defensive fear behavior, and a ventrolateral region that is activated by other mouse odors and generates reproductive and aggressive social behavior.[149,154–157]

Neurons in both the amygdala and the hypothalamus express receptors that detect and respond to peptides and other neural modulators. These include steroid hormones (estrogen, testosterone),[158–160] peptide hormones (NPY, substance P, oxytocin, GNRH, ACTH),[140,161–163] and small molecule neurotransmitters (dopamine, serotonin).[164–166] It is known that these receptors enable VNO-activated circuits to modulate hardwired aspects of their response to account for an individual's state (dominance, gender, hunger, and stress levels). However, precisely how neuromodulators alter circuit activity is currently unknown.

It is well established that the behavior elicited by VNO signaling varies depending on the gender of the receiving animal.[167,168] For example, male-emitted cues elicit aggression when detected by other males but not females. Consistent with these observations, several nuclei within the amygdala and hypothalamus are sexually dimorphic in volume, usually larger in males (based on differences in absolute number, density, and size of individual neurons).[169–172] VNO-activated circuits are also functionally dimorphic.[28,129,158,173–175] Current studies have begun to determine the molecular mechanisms that underlie gender-dimorphic differences in VNO-activated circuits.[28,176–178]

TRANSFORMING SPECIALIZED LIGANDS INTO SOCIAL BEHAVIOR

Although recent advances in molecular biology and biochemistry have enabled the identification of ligands that promote specialized behavior, determined characteristics of the cognate sensory neurons, and have begun to identify a framework of the responding circuits, how the brain transforms a specialized chemosensory ligand into innate social behavior still remains a mystery. What is unique about the activated circuits that ultimately command such dynamic behaviors such as rage, fear, and courtship? Why does the organism rely on two distinct olfactory organs to mediate social behavior? How does the internal state and previous experience of an individual alter the probability of specialized ligands to generate social behavior? Further study is necessary to determine exactly how the olfactory system harnesses the brain to guide an individual's behavioral decisions.

References

1. Choi GB, Stettler DD, Kallman BR, Bhaskar ST, Fleischmann A, Axel R. Driving opposing behaviors with ensembles of piriform neurons. *Cell*. 2011;146(6):1004–1015.
2. Stowers L, Logan DW. Olfactory mechanisms of stereotyped behavior: on the scent of specialized circuits. *Curr Opin Neurobiol*. 2010;20(3):274–280.
3. Wyatt TD. Pheromones and signature mixtures: defining species-wide signals and variable cues for identity in both invertebrates and vertebrates. *J Comp Physiol A Neuroethol Sens Neural Behav Physiol*. 2010;196(10):685–700.
4. Karlson P, Luscher M. Pheromones: a new term for a class of biologically active substances. *Nature*. 1959;183(4653):55–56.
5. Liberles SD. Mammalian pheromones. *Annu Rev Physiol*. 2014;76:151–175.
6. Cajal SR. Textura del sistema nervioso del hombre y de los vertebrados. *Revista Española de Patología*. 1904;Vol. 2. Madrid.
7. Meisami E, Bhatnagar KP. Structure and diversity in mammalian accessory olfactory bulb. *Microsc Res Tech*. 1998;43(6):476–499.
8. Grus WE, Shi P, Zhang YP, Zhang J. Dramatic variation of the vomeronasal pheromone receptor gene repertoire among five orders of placental and marsupial mammals. *Proc Natl Acad Sci USA*. 2005;102(16):5767–5772.
9. Young JM, Massa HF, Hsu L, Trask BJ. Extreme variability among mammalian V1R gene families. *Genome Res*. 2010;20(1):10–18.
10. Liman ER, Innan H. Relaxed selective pressure on an essential component of pheromone transduction in primate evolution. *Proc Natl Acad Sci USA*. 2003;100(6):3328–3332.
11. Zhang J, Webb DM. Evolutionary deterioration of the vomeronasal pheromone transduction pathway in catarrhine primates. *Proc Natl Acad Sci USA*. 2003;100(14):8337–8341.
12. Ibarra-Soria X, Levitin MO, Saraiva LR, Logan DW. The olfactory transcriptomes of mice. *PLoS Genet*. 2014;10(9):e1004593.
13. Pascarella G, Lazarevic D, Plessy C, et al. NanoCAGE analysis of the mouse olfactory epithelium identifies the expression of vomeronasal receptors and of proximal LINE elements. *Front Cell Neurosci*. 2014;8:41.
14. Khan M, Vaes E, Mombaerts P. Regulation of the probability of mouse odorant receptor gene choice. *Cell*. 2011;147(4):907–921.
15. Malnic B, Hirono J, Sato T, Buck LB. Combinatorial receptor codes for odors. *Cell*. 1999;96(5):713–723.
16. Ishii T, Mombaerts P. Coordinated coexpression of two vomeronasal receptor V2R genes per neuron in the mouse. *Mol Cell Neurosci*. 2011;46(2):397–408.
17. Leinders-Zufall T, Ishii T, Chamero P, et al. A family of nonclassical class I MHC genes contributes to ultrasensitive chemodetection by mouse vomeronasal sensory neurons. *J Neurosci*. 2014;34(15):5121–5133.
18. Silvotti L, Cavalca E, Gatti R, Percudani R, Tirindelli R. A recent class of chemosensory neurons developed in mouse and rat. *PloS One*. 2011;6(9):e24462.
19. Loconto J, Papes F, Chang E, et al. Functional expression of murine V2R pheromone receptors involves selective association with the M10 and M1 families of MHC class Ib molecules. *Cell*. 2003;112(5):607–618.
20. Grus WE, Zhang J. Rapid turnover and species-specificity of vomeronasal pheromone receptor genes in mice and rats. *Gene*. 2004;340(2):303–312.
21. Grus WE, Zhang J. Distinct evolutionary patterns between chemoreceptors of 2 vertebrate olfactory systems and the differential tuning hypothesis. *Mol Biol Evol*. 2008;25(8):1593–1601.
22. Leypold BG, Yu CR, Leinders-Zufall T, Kim MM, Zufall F, Axel R. Altered sexual and social behaviors in trp2 mutant mice. *Proc Natl Acad Sci USA*. 2002;99(9):6376–6381.
23. Stowers L, Holy TE, Meister M, Dulac C, Koentges G. Loss of sex discrimination and male-male aggression in mice deficient for TRP2. *Science*. 2002;295:1493–1500.
24. Papes F, Logan DW, Stowers L. The vomeronasal organ mediates interspecies defensive behaviors through detection of protein pheromone homologs. *Cell*. 2010;141(4):692–703.
25. Chamero P, Marton TF, Logan DW, et al. Identification of protein pheromones that promote aggressive behaviour. *Nature*. 2007;450(7171):899–902.
26. Ferrero DM, Moeller LM, Osakada T, et al. A juvenile mouse pheromone inhibits sexual behaviour through the vomeronasal system. *Nature*. 2013;502(7471):368–371.
27. Kaur AW, Ackels T, Kuo TH, et al. Murine pheromone proteins constitute a context-dependent combinatorial code governing multiple social behaviors. *Cell*. 2014;157(3):676–688.

28. Kimchi T, Xu J, Dulac C. A functional circuit underlying male sexual behaviour in the female mouse brain. *Nature.* 2007;448(7157):1009–1014.

29. Kimoto H, Haga S, Sato K, Touhara K. Sex-specific peptides from exocrine glands stimulate mouse vomeronasal sensory neurons. *Nature.* 2005;437(7060):898–901.

30. Boillat M, Challet L, Rossier D, Kan C, Carleton A, Rodriguez I. The vomeronasal system mediates sick conspecific avoidance. *Curr Biol.* 2015;25(2):251–255.

31. Wu Z, Autry AE, Bergan JF, Watabe-Uchida M, Dulac CG. Galanin neurons in the medial preoptic area govern parental behaviour. *Nature.* 2014;509(7500):325–330.

32. Chamero P, Katsoulidou V, Hendrix P, et al. G protein G(alpha)o is essential for vomeronasal function and aggressive behavior in mice. *Proc Natl Acad Sci USA.* 2011;108(31):12898–12903.

33. Montani G, Tonelli S, Sanghez V, et al. Aggressive behaviour and physiological responses to pheromones are strongly impaired in mice deficient for the olfactory G-protein subunit G8. *J Physiol.* 2013;591 (Pt 16):3949–3962.

34. Norlin EM, Gussing F, Berghard A. Vomeronasal phenotype and behavioral alterations in G alpha i2 mutant mice. *Curr Biol.* 2003;13(14):1214–1219.

35. Oboti L, Perez-Gomez A, Keller M, et al. A wide range of pheromone-stimulated sexual and reproductive behaviors in female mice depend on G protein Galphao. *BMC Biol.* 2014;12:31.

36. Kobayakawa K, Kobayakawa R, Matsumoto H, et al. Innate versus learned odour processing in the mouse olfactory bulb. *Nature.* 2007;450(7169):503–508.

37. Mandiyan VS, Coats JK, Shah NM. Deficits in sexual and aggressive behaviors in Cnga2 mutant mice. *Nat Neurosci.* 2005;8(12):1660–1662.

38. Wang Z, Balet Sindreu C, Li V, Nudelman A, Chan GC, Storm DR. Pheromone detection in male mice depends on signaling through the type 3 adenylyl cyclase in the main olfactory epithelium. *J Neurosci.* 2006;26(28):7375–7379.

39. Dewan A, Pacifico R, Zhan R, Rinberg D, Bozza T. Non-redundant coding of aversive odours in the main olfactory pathway. *Nature.* 2013;497(7450):486–489.

40. Brunet LJ, Gold GH, Ngai J. General anosmia caused by a targeted disruption of the mouse olfactory cyclic nucleotide-gated cation channel. *Neuron.* 1996;17(4):681–693.

41. Logan DW, Brunet LJ, Webb WR, Cutforth T, Ngai J, Stowers L. Learned recognition of maternal signature odors mediates the first suckling episode in mice. *Curr Biol.* 2012;22(21):1998–2007.

42. Meredith M, O'Connell RJ. Efferent control of stimulus access to the hamster vomeronasal organ. *J Physiol.* 1979;286:301–316.

43. Leinders-Zufall T, Cockerham RE, Michalakis S, et al. Contribution of the receptor guanylyl cyclase GC-D to chemosensory function in the olfactory epithelium. *Proc Natl Acad Sci USA.* 2007;104(36):14507–14512.

44. Spehr M, Kelliher KR, Li XH, Boehm T, Leinders-Zufall T, Zufall F. Essential role of the main olfactory system in social recognition of major histocompatibility complex peptide ligands. *J Neurosci.* 2006;26(7):1961–1970.

45. Spehr M, Spehr J, Ukhanov K, Kelliher KR, Leinders-Zufall T, Zufall F. Parallel processing of social signals by the mammalian main and accessory olfactory systems. *Cell Mol Life Sci.* 2006;63(13):1476–1484.

46. Wilson EO. Pheromones. *Sci Am.* 1963;208:100–114.

47. Kimoto H, Sato K, Nodari F, Haga S, Holy TE, Touhara K. Sex- and strain-specific expression and vomeronasal activity of mouse ESP family peptides. *Curr Biol.* 2007;17(21):1879–1884.

48. Logan DW, Marton TF, Stowers L. Species specificity in major urinary proteins by parallel evolution. *PloS One.* 2008;3(9):e3280.

49. Mudge JM, Armstrong SD, McLaren K, et al. Dynamic instability of the major urinary protein gene family revealed by genomic and phenotypic comparisons between C57 and 129 strain mice. *Genome Biol.* 2008;9(5):R91.

50. Leinders-Zufall T, Brennan P, Widmayer P, et al. MHC class I peptides as chemosensory signals in the vomeronasal organ. *Science.* 2004;306(5698):1033–1037.

51. Nodari F, Hsu FF, Fu X, et al. Sulfated steroids as natural ligands of mouse pheromone-sensing neurons. *J Neurosci.* 2008;28(25):6407–6418.

52. Novotny M, Harvey S, Jemiolo B. Chemistry of male dominance in the house mouse, Mus domesticus. *Experientia.* 1990;46(1):109–113.

53. Novotny M, Harvey S, Jemiolo B, Alberts J. Synthetic pheromones that promote inter-male aggression in mice. *Proc Natl Acad Sci USA.* 1985;82(7):2059–2061.

54. Novotny M, Schwende FJ, Wiesler D, Jorgenson JW, Carmack M. Identification of a testosterone-dependent unique volatile constituent of male mouse urine: 7-exo-ethyl-5-methyl-6,8-dioxabicyclo[3.2.1]-3-octene. *Experientia*. 1984;40(2):217–219.

55. Leinders-Zufall T, Lane AP, Puche AC, et al. Ultrasensitive pheromone detection by mammalian vomeronasal neurons. *Nature*. 2000;405(6788):792–796.

56. Meeks JP, Arnson HA, Holy TE. Representation and transformation of sensory information in the mouse accessory olfactory system. *Nat Neurosci*. 2010;13(6):723–730.

57. Haga S, Hattori T, Sato T, et al. The male mouse pheromone ESP1 enhances female sexual receptive behaviour through a specific vomeronasal receptor. *Nature*. 2010;466(7302):118–122.

58. Liberles SD, Horowitz LF, Kuang D, et al. Formyl peptide receptors are candidate chemosensory receptors in the vomeronasal organ. *Proc Natl Acad Sci USA*. 2009;106(24):9842–9847.

59. Riviere S, Challet L, Fluegge D, Spehr M, Rodriguez I. Formyl peptide receptor-like proteins are a novel family of vomeronasal chemosensors. *Nature*. 2009;459(7246):574–577.

60. Ackels T, von der Weid B, Rodriguez I, Spehr M. Physiological characterization of formyl peptide receptor expressing cells in the mouse vomeronasal organ. *Front Neuroanat*. 2014;8:134.

61. Arakawa H, Cruz S, Deak T. From models to mechanisms: odorant communication as a key determinant of social behavior in rodents during illness-associated states. *Neurosci Biobehav Rev*. 2011;35(9):1916–1928.

62. Beynon RJ, Hurst JL. Multiple roles of major urinary proteins in the house mouse, *Mus domesticus*. *Biochem Soc Trans*. 2003;31(Pt 1):142–146.

63. Beynon RJ, Hurst JL. Urinary proteins and the modulation of chemical scents in mice and rats. *Peptides*. 2004;25(9):1553–1563.

64. Hurst JL, Robertson DHL, Tolladay U, Beynon RJ. Proteins in urine scent marks of male house mice extend the longevity of olfactory signals. *Anim Behav*. 1998;55(5):1289–1297.

65. Robertson DH, Cox KA, Gaskell SJ, Evershed RP, Beynon RJ. Molecular heterogeneity in the major urinary proteins of the house mouse *Mus musculus*. *Biochem J*. 1996;316(Pt 1):265–272.

66. Robertson DH, Hurst JL, Bolgar MS, Gaskell SJ, Beynon RJ. Molecular heterogeneity of urinary proteins in wild house mouse populations. *Rapid Commun Mass Spectrom*. 1997;11(7):786–790.

67. Armstrong CM, DeVito LM, Cleland TA. One-trial associative odor learning in neonatal mice. *Chem Senses*. 2006;31(4):343–349.

68. Cheetham SA, Smith AL, Armstrong SD, Beynon RJ, Hurst JL. Limited variation in the major urinary proteins of laboratory mice. *Physiol Behav*. 2009;96(2):253–261.

69. Hurst JL, Beynon RJ. Scent wars: the chemobiology of competitive signalling in mice. *BioEssays*. 2004;26(12):1288–1298.

70. Hurst JL, Payne CE, Nevison CM, et al. Individual recognition in mice mediated by major urinary proteins. *Nature*. 2001;414(6864):631–634.

71. Roberts SA, Simpson DM, Armstrong SD, et al. Darcin: a male pheromone that stimulates female memory and sexual attraction to an individual male's odour. *BMC Biol*. 2010;8:75.

72. Roberts SA, Davidson AJ, McLean L, Beynon RJ, Hurst JL. Pheromonal induction of spatial learning in mice. *Science*. 2012;338(6113):1462–1465.

73. Nishimura K, Utsumi K, Yuhara M, Fujitani Y, Iritani A. Identification of puberty-accelerating pheromones in male mouse urine. *J Exp Zool*. 1989;251(3):300–305.

74. Novotny MV, Ma W, Wiesler D, Zidek L. Positive identification of the puberty-accelerating pheromone of the house mouse: the volatile ligands associating with the major urinary protein. *Proc R Soc Lond B Biol Sci*. 1999;266(1432):2017–2022.

75. Schwende FJ, Wiesler D, Jorgenson JW, Carmack M, Novotny M. Urinary volatile constituents of the house mouse, *Mus musculus*, and their endocrine dependency. *J Chem Ecol*. 1986;12(1):277–296.

76. Novotny MV. Pheromones, binding proteins and receptor responses in rodents. *Biochem Soc Trans*. 2003;31 (Pt 1):117–122.

77. Boschat C, Pelofi C, Randin O, et al. Pheromone detection mediated by a V1r vomeronasal receptor. *Nat Neurosci*. 2002;5(12):1261–1262.

78. Novotny MV, Soini HA, Koyama S, Wiesler D, Bruce KE, Penn DJ. Chemical identification of MHC-influenced volatile compounds in mouse urine. I: quantitative proportions of major chemosignals. *J Chem Ecol*. 2007;33(2):417–434.

79. Flanagan KA, Webb W, Stowers L. Analysis of male pheromones that accelerate female reproductive organ development. *PloS One.* 2011;6(2):e16660.
80. Jemiolo B, Andreolini F, Xie TM, Wiesler D, Novotny M. Puberty-affecting synthetic analogs of urinary chemosignals in the house mouse, *Mus domesticus. Physiol Behav.* 1989;46(2):293–298.
81. Jemiolo B, Harvey S, Novotny M. Promotion of the Whitten effect in female mice by synthetic analogs of male urinary constituents. *Proc Natl Acad Sci USA.* 1986;83(12):4576–4579.
82. Goodwin TE, Eggert MS, House SJ, Weddell ME, Schulte BA, Rasmussen LE. Insect pheromones and precursors in female African elephant urine. *J Chem Ecol.* 2006;32(8):1849–1853.
83. Papachristoforou A, Kagiava A, Papaefthimiou C, et al. The bite of the honeybee: 2-heptanone secreted from honeybee mandibles during a bite acts as a local anaesthetic in insects and mammals. *PloS One.* 2012;7(10):e47432.
84. Schulz S, Kruckert K, Weldon PJ. New terpene hydrocarbons from the alligatoridae (crocodylia, reptilia). *J Nat Prod.* 2003;66(1):34–38.
85. Silk PJ, Lemay MA, LeClair G, Sweeney J, MaGee D. Behavioral and electrophysiological responses of *Tetropium fuscum* (Coleoptera: Cerambycidae) to pheromone and spruce volatiles. *Environ Entomol.* 2010;39(6):1997–2005.
86. Isogai Y, Si S, Pont-Lezica L, et al. Molecular organization of vomeronasal chemoreception. *Nature.* 2011;478(7368):241–245.
87. Turaga D, Holy TE. Organization of vomeronasal sensory coding revealed by fast volumetric calcium imaging. *J Neurosci.* 2012;32(5):1612–1621.
88. Hanson SR, Best MD, Wong CH. Sulfatases: structure, mechanism, biological activity, inhibition, and synthetic utility. *Angew Chem Int Ed Engl.* 2004;43(43):5736–5763.
89. Arnson HA, Holy TE. Robust encoding of stimulus identity and concentration in the accessory olfactory system. *J Neurosci.* 2013;33(33):13388–13397.
90. Celsi F, D'Errico A, Menini A. Responses to sulfated steroids of female mouse vomeronasal sensory neurons. *Chem Senses.* 2012;37(9):849–858.
91. Hammen GF, Turaga D, Holy TE, Meeks JP. Functional organization of glomerular maps in the mouse accessory olfactory bulb. *Nat Neurosci.* 2014;17(7):953–961.
92. Haga-Yamanaka S, Ma L, He J, et al. Integrated action of pheromone signals in promoting courtship behavior in male mice. *Elife.* 2014;3:e03025.
93. Ben-Shaul Y, Katz LC, Mooney R, Dulac C. In vivo vomeronasal stimulation reveals sensory encoding of conspecific and allospecific cues by the mouse accessory olfactory bulb. *Proc Natl Acad Sci USA.* 2010;107(11):5172–5177.
94. Leinders-Zufall T, Ishii T, Mombaerts P, Zufall F, Boehm T. Structural requirements for the activation of vomeronasal sensory neurons by MHC peptides. *Nat Neurosci.* 2009;12(12):1551–1558.
95. Liberles SD, Buck LB. A second class of chemosensory receptors in the olfactory epithelium. *Nature.* 2006;442(7103):645–650.
96. Pacifico R, Dewan A, Cawley D, Guo C, Bozza T. An olfactory subsystem that mediates high-sensitivity detection of volatile amines. *Cell Rep.* 2012;2(1):76–88.
97. Ferrero DM, Lemon JK, Fluegge D, et al. Detection and avoidance of a carnivore odor by prey. *Proc Natl Acad Sci USA.* 2011;108(27):11235–11240.
98. Li Q, Korzan WJ, Ferrero DM, et al. Synchronous evolution of an odor biosynthesis pathway and behavioral response. *Curr Biol.* 2013;23(1):11–20.
99. Liman ER, Corey DP, Dulac C. TRP2: a candidate transduction channel for mammalian pheromone sensory signaling. *Proc Natl Acad Sci USA.* 1999;96(10):5791–5796.
100. Omura M, Mombaerts P. Trpc2-expressing sensory neurons in the main olfactory epithelium of the mouse. *Cell Rep.* 2014;8(2):583–595.
101. Lin W, Margolskee R, Donnert G, Hell SW, Restrepo D. Olfactory neurons expressing transient receptor potential channel M5 (TRPM5) are involved in sensing semiochemicals. *Proc Natl Acad Sci USA.* 2007;104(7):2471–2476.
102. Thompson JA, Salcedo E, Restrepo D, Finger TE. Second-order input to the medial amygdala from olfactory sensory neurons expressing the transduction channel TRPM5. *J Comp Neurol.* 2012;520(8):1819–1830.
103. Lopez F, Delgado R, Lopez R, Bacigalupo J, Restrepo D. Transduction for pheromones in the main olfactory epithelium is mediated by the Ca^{2+}-activated channel TRPM5. *J Neurosci.* 2014;34(9):3268–3278.

104. Rolen SH, Salcedo E, Restrepo D, Finger TE. Differential localization of NT-3 and TrpM5 in glomeruli of the olfactory bulb of mice. *J Comp Neurol*. 2014;522(8):1929–1940.

105. Cockerham RE, Leinders-Zufall T, Munger SD, Zufall F. Functional analysis of the guanylyl cyclase type D signaling system in the olfactory epithelium. *Ann N Y Acad Sci*. 2009;1170:173–176.

106. Fulle HJ, Vassar R, Foster DC, Yang RB, Axel R, Garbers DL. A receptor guanylyl cyclase expressed specifically in olfactory sensory neurons. *Proc Natl Acad Sci USA*. 1995;92(8):3571–3575.

107. Juilfs DM, Fulle HJ, Zhao AZ, Houslay MD, Garbers DL, Beavo JA. A subset of olfactory neurons that selectively express cGMP-stimulated phosphodiesterase (PDE2) and guanylyl cyclase-D define a unique olfactory signal transduction pathway. *Proc Natl Acad Sci USA*. 1997;94(7):3388–3395.

108. Munger SD, Leinders-Zufall T, McDougall LM, et al. An olfactory subsystem that detects carbon disulfide and mediates food-related social learning. *Curr Biol*. 2010;20(16):1438–1444.

109. Hu J, Zhong C, Ding C, et al. Detection of near-atmospheric concentrations of CO_2 by an olfactory subsystem in the mouse. *Science*. 2007;317(5840):953–957.

110. Arakawa H, Kelliher KR, Zufall F, Munger SD. The receptor guanylyl cyclase type D (GC-D) ligand uroguanylin promotes the acquisition of food preferences in mice. *Chem Senses*. 2013;38(5):391–397.

111. Lo L, Anderson DJ. A cre-dependent, anterograde transsynaptic viral tracer for mapping output pathways of genetically marked neurons. *Neuron*. 2011;72(6):938–950.

112. Miyamichi K, Amat F, Moussavi F, et al. Cortical representations of olfactory input by trans-synaptic tracing. *Nature*. 2011;472(7342):191–196.

113. Stettler DD, Axel R. Representations of odor in the piriform cortex. *Neuron*. 2009;63(6):854–864.

114. Meredith M. Sensory processing in the main and accessory olfactory systems: comparisons and contrasts. *J Steroid Biochem Mol Biol*. 1991;39(4B):601–614.

115. Scalia F, Winans SS. The differential projections of the olfactory bulb and accessory olfactory bulb in mammals. *J Comp Neurol*. 1975;161(1):31–55.

116. Mombaerts P, Wang F, Dulac C, et al. Visualizing an olfactory sensory map. *Cell*. 1996;87(4):675–686.

117. Sosulski DL, Bloom ML, Cutforth T, Axel R, Datta SR. Distinct representations of olfactory information in different cortical centres. *Nature*. 2011;472(7342):213–216.

118. Ghosh S, Larson SD, Hefzi H, et al. Sensory maps in the olfactory cortex defined by long-range viral tracing of single neurons. *Nature*. 2011;472(7342):217–220.

119. Root CM, Denny CA, Hen R, Axel R. The participation of cortical amygdala in innate, odour-driven behaviour. *Nature*. 2014;515(7526):269–273.

120. Lin DY, Zhang SZ, Block E, Katz LC. Encoding social signals in the mouse main olfactory bulb. *Nature*. 2005;434(7032):470–477.

121. Del Punta K, Puche A, Adams NC, Rodriguez I, Mombaerts P. A divergent pattern of sensory axonal projections is rendered convergent by second-order neurons in the accessory olfactory bulb. *Neuron*. 2002;35(6):1057–1066.

122. Wagner S, Gresser AL, Torello AT, Dulac C. A multireceptor genetic approach uncovers an ordered integration of VNO sensory inputs in the accessory olfactory bulb. *Neuron*. 2006;50(5):697–709.

123. Belluscio L, Koentges G, Axel R, Dulac C. A map of pheromone receptor activation in the mammalian brain. *Cell*. 1999;97(2):209–220.

124. Rodriguez I, Feinstein P, Mombaerts P. Variable patterns of axonal projections of sensory neurons in the mouse vomeronasal system. *Cell*. 1999;97(2):199–208.

125. Dulac C, Torello AT. Molecular detection of pheromone signals in mammals: from genes to behaviour. *Nat Rev Neurosci*. 2003;4(7):551–562.

126. Gutierrez-Castellanos N, Martinez-Marcos A, Martinez-Garcia F, Lanuza E. Chemosensory function of the amygdala. *Vitam Horm*. 2010;83:165–196.

127. Bressler SC, Baum MJ. Sex comparison of neuronal Fos immunoreactivity in the rat vomeronasal projection circuit after chemosensory stimulation. *Neuroscience*. 1996;71(4):1063–1072.

128. Butler RK, Sharko AC, Oliver EM, et al. Activation of phenotypically-distinct neuronal subpopulations of the rat amygdala following exposure to predator odor. *Neuroscience*. 2011;175:133–144.

129. Kang N, Janes A, Baum MJ, Cherry JA. Sex difference in Fos induced by male urine in medial amygdala-projecting accessory olfactory bulb mitral cells of mice. *Neurosci Lett*. 2006;398(1–2):59–62.

130. Meredith M, Westberry JM. Distinctive responses in the medial amygdala to same-species and different-species pheromones. *J Neurosci*. 2004;24(25):5719–5725.

131. LeDoux J. Rethinking the emotional brain. *Neuron.* 2012;73(4):653—676.
132. LeDoux JE. Emotion circuits in the brain. *Annu Rev Neurosci.* 2000;23:155—184.
133. Pessoa L, Adolphs R. Emotion processing and the amygdala: from a 'low road' to 'many roads' of evaluating biological significance. *Nat Rev Neurosci.* 2010;11(11):773—783.
134. Salzman CD, Fusi S. Emotion, cognition, and mental state representation in amygdala and prefrontal cortex. *Annu Rev Neurosci.* 2010;33:173—202.
135. Martinez-Marcos A, Halpern M. Differential centrifugal afferents to the anterior and posterior accessory olfactory bulb. *Neuroreport.* 1999;10(10):2011—2015.
136. von Campenhausen H, Mori K. Convergence of segregated pheromonal pathways from the accessory olfactory bulb to the cortex in the mouse. *Eur J Neurosci.* 2000;12(1):33—46.
137. Mohedano-Moriano A, Pro-Sistiaga P, Ubeda-Banon I, Crespo C, Insausti R, Martinez-Marcos A. Segregated pathways to the vomeronasal amygdala: differential projections from the anterior and posterior divisions of the accessory olfactory bulb. *Eur J Neurosci.* 2007;25(7):2065—2080.
138. Mohedano-Moriano A, Pro-Sistiaga P, Ubeda-Banon I, de la Rosa-Prieto C, Saiz-Sanchez D, Martinez-Marcos A. V1R and V2R segregated vomeronasal pathways to the hypothalamus. *Neuroreport.* 2008;19(16):1623—1626.
139. Pardo-Bellver C, Cadiz-Moretti B, Novejarque A, Martinez-Garcia F, Lanuza E. Differential efferent projections of the anterior, posteroventral, and posterodorsal subdivisions of the medial amygdala in mice. *Front Neuroanat.* 2012;6:33.
140. Ferguson JN, Aldag JM, Insel TR, Young LJ. Oxytocin in the medial amygdala is essential for social recognition in the mouse. *J Neurosci.* 2001;21(20):8278—8285.
141. Kondo Y. Lesions of the medial amygdala produce severe impairment of copulatory behavior in sexually inexperienced male rats. *Physiol Behav.* 1992;51(5):939—943.
142. Kondo Y, Sachs BD. Disparate effects of small medial amygdala lesions on noncontact erection, copulation, and partner preference. *Physiol Behav.* 2002;76(4—5):443—447.
143. Lehman MN, Winans SS. Vomeronasal and olfactory pathways to the amygdala controlling male hamster sexual behavior: autoradiographic and behavioral analyses. *Brain Res.* 1982;240(1):27—41.
144. Liu YC, Salamone JD, Sachs BD. Lesions in medial preoptic area and bed nucleus of stria terminalis: differential effects on copulatory behavior and noncontact erection in male rats. *J Neurosci.* 1997;17(13):5245—5253.
145. Besser GM, Mortimer CH. Hypothalamic regulatory hormones: a review. *J Clin Pathol.* 1974;27(3):173—184.
146. Adams DB. Brain mechanisms of aggressive behavior: an updated review. *Neurosci Biobehav Rev.* 2006;30(3):304—318.
147. Canteras NS, Chiavegatto S, Ribeiro do Valle LE, Swanson LW. Severe reduction of rat defensive behavior to a predator by discrete hypothalamic chemical lesions. *Brain Res Bull.* 1997;44(3):297—305.
148. Choi GB, Dong HW, Murphy AJ, et al. Lhx6 delineates a pathway mediating innate reproductive behaviors from the amygdala to the hypothalamus. *Neuron.* 2005;46(4):647—660.
149. Lin D, Boyle MP, Dollar P, et al. Functional identification of an aggression locus in the mouse hypothalamus. *Nature.* 2011;470(7333):221—226.
150. Motta SC, Goto M, Gouveia FV, Baldo MV, Canteras NS, Swanson LW. Dissecting the brain's fear system reveals the hypothalamus is critical for responding in subordinate conspecific intruders. *Proc Natl Acad Sci USA.* 2009;106(12):4870—4875.
151. Petrovich GD, Canteras NS, Swanson LW. Combinatorial amygdalar inputs to hippocampal domains and hypothalamic behavior systems. *Brain Res Brain Res Rev.* 2001;38(1—2):247—289.
152. Dong HW, Petrovich GD, Swanson LW. Topography of projections from amygdala to bed nuclei of the stria terminalis. *Brain Res Brain Res Rev.* 2001;38(1—2):192—246.
153. Dumont EC. What is the bed nucleus of the stria terminalis? *Prog Neuropsychopharmacol Biol Psychiatry.* 2009;33(8):1289—1290.
154. Falkner AL, Dollar P, Perona P, Anderson DJ, Lin D. Decoding ventromedial hypothalamic neural activity during male mouse aggression. *J Neurosci.* 2014;34(17):5971—5984.
155. Kunwar PS, Zelikowsky M, Remedios R, et al. Ventromedial hypothalamic neurons control a defensive emotion state. *Elife.* 2015;4.
156. Lee H, Kim DW, Remedios R, et al. Scalable control of mounting and attack by Esr1$^+$ neurons in the ventromedial hypothalamus. *Nature.* 2014;509(7502):627—632.

157. Wang L, Chen IZ, Lin D. Collateral pathways from the ventromedial hypothalamus mediate defensive behaviors. *Neuron*. 2015;85(6):1344−1358.
158. Blake CB, Meredith M. Change in number and activation of androgen receptor-immunoreactive cells in the medial amygdala in response to chemosensory input. *Neuroscience*. 2011;190:228−238.
159. Merchenthaler I, Lane MV, Numan S, Dellovade TL. Distribution of estrogen receptor alpha and beta in the mouse central nervous system: in vivo autoradiographic and immunocytochemical analyses. *J Comp Neurol*. 2004;473(2):270−291.
160. Mitra SW, Hoskin E, Yudkovitz J, et al. Immunolocalization of estrogen receptor beta in the mouse brain: comparison with estrogen receptor alpha. *Endocrinology*. 2003;144(5):2055−2067.
161. Jennes L, Conn PM. Gonadotropin-releasing hormone and its receptors in rat brain. *Front Neuroendocrinol*. 1994;15(1):51−77.
162. Van Pett K, Viau V, Bittencourt JC, et al. Distribution of mRNAs encoding CRF receptors in brain and pituitary of rat and mouse. *J Comp Neurol*. 2000;428(2):191−212.
163. Wood RI, Swann JM. The bed nucleus of the stria terminalis in the Syrian hamster: subnuclei and connections of the posterior division. *Neuroscience*. 2005;135(1):155−179.
164. Fremeau Jr RT, Duncan GE, Fornaretto MG, et al. Localization of D1 dopamine receptor mRNA in brain supports a role in cognitive, affective, and neuroendocrine aspects of dopaminergic neurotransmission. *Proc Natl Acad Sci USA*. 1991;88(9):3772−3776.
165. Lemberger T, Parlato R, Dassesse D, et al. Expression of Cre recombinase in dopaminoceptive neurons. *BMC Neurosci*. 2007;8:4.
166. Schiller L, Donix M, Jahkel M, Oehler J. Serotonin 1A and 2A receptor densities, neurochemical and behavioural characteristics in two closely related mice strains after long-term isolation. *Prog Neuropsychopharmacol Biol Psychiatry*. 2006;30(3):492−503.
167. Baum MJ. Sexual differentiation of pheromone processing: links to male-typical mating behavior and partner preference. *Horm Behav*. 2009;55(5):579−588.
168. Stowers L, Logan DW. Sexual dimorphism in olfactory signaling. *Curr Opin Neurobiol*. 2010;20(6):770−775.
169. Dulac C, Kimchi T. Neural mechanisms underlying sex-specific behaviors in vertebrates. *Curr Opin Neurobiol*. 2007;17(6):675−683.
170. Guillamon A, Segovia S. Sex differences in the vomeronasal system. *Brain Res Bull*. 1997;44(4):377−382.
171. Kawata M. Roles of steroid hormones and their receptors in structural organization in the nervous system. *Neurosci Res*. 1995;24(1):1−46.
172. Yang CF, Shah NM. Representing sex in the brain, one module at a time. *Neuron*. 2014;82(2):261−278.
173. Halem HA, Baum MJ, Cherry JA. Sex difference and steroid modulation of pheromone-induced immediate early genes in the two zones of the mouse accessory olfactory system. *J Neurosci*. 2001;21(7):2474−2480.
174. Wu MV, Manoli DS, Fraser EJ, et al. Estrogen masculinizes neural pathways and sex-specific behaviors. *Cell*. 2009;139(1):61−72.
175. Wu MV, Shah NM. Control of masculinization of the brain and behavior. *Curr Opin Neurobiol*. 2011;21(1):116−123.
176. Boehm U, Zou Z, Buck LB. Feedback loops link odor and pheromone signaling with reproduction. *Cell*. 2005;123(4):683−695.
177. Gregg C, Zhang J, Weissbourd B, et al. High-resolution analysis of parent-of-origin allelic expression in the mouse brain. *Science*. 2010;329(5992):643−648.
178. Xu X, Coats JK, Yang CF, et al. Modular genetic control of sexually dimorphic behaviors. *Cell*. 2012;148(3):596−607.
179. Wyatt TD. The search for human pheromones: the lost decades and the necessity of returning to first principles. *Proc Biol Sci*. 2015;282(1804).
180. McClintock MK. Menstrual synchorony and suppression. *Nature*. 1971;229(5282):244−245.
181. Doucet S, Soussignan R, Sagot P, Schaal B. The secretion of areolar (Montgomery's) glands from lactating women elicits selective, unconditional responses in neonates. *PloS One*. 2009;4(10):e7579.
182. Doucet S, Soussignan R, Sagot P, Schaal B. An overlooked aspect of the human breast: areolar glands in relation with breastfeeding pattern, neonatal weight gain, and the dynamics of lactation. *Early Hum Dev*. 2012;88(2):119−128.

183. Charra R, Datiche F, Casthano A, Gigot V, Schaal B, Coureaud G. Brain processing of the mammary pheromone in newborn rabbits. *Behav Brain Res*. 2012;226(1):179–188.

184. Schaal B, Coureaud G, Langlois D, Ginies C, Semon E, Perrier G. Chemical and behavioural characterization of the rabbit mammary pheromone. *Nature*. 2003;424(6944):68–72.

185. Frumin I, Perl O, Endevelt-Shapira Y, et al. A social chemosignaling function for human handshaking. *Elife*. 2015;4.

186. Gelstein S, Yeshurun Y, Rozenkrantz L, et al. Human tears contain a chemosignal. *Science*. 2011;331(6014):226–230.

187. Preti G, Wysocki CJ, Barnhart KT, Sondheimer SJ, Leyden JJ. Male axillary extracts contain pheromones that affect pulsatile secretion of luteinizing hormone and mood in women recipients. *Biol Reprod*. 2003;68(6):2107–2113.

188. Savic I, Berglund H, Gulyas B, Roland P. Smelling of odorous sex hormone-like compounds causes sex-differentiated hypothalamic activations in humans. *Neuron*. 2001;31(4):661–668.

189. Smith TD, Laitman JT, Bhatnagar KP. The shrinking anthropoid nose, the human vomeronasal organ, and the language of anatomical reduction. *Anat Rec Hob*. 2014;297(11):2196–2204.

190. Dorries KM, Adkins-Regan E, Halpern BP. Sensitivity and behavioral responses to the pheromone androstenone are not mediated by the vomeronasal organ in domestic pigs. *Brain Behav Evol*. 1997;49:53–62.

191. Hudson R, Distel H. Pheromonal release of suckling in rabbits does not depend on the vomeronasal organ. *Physiol Behav*. 1986;37(1):123–128.

192. Keller A, Zhuang H, Chi Q, Vosshall LB, Matsunami H. Genetic variation in a human odorant receptor alters odour perception. *Nature*. 2007;449(7161):468–472.

193. Carnicelli V, Santoro A, Sellari-Franceschini S, Berrettini S, Zucchi R. Expression of trace amine-associated receptors in human nasal mucosa. *Chem Percept*. 2010;3:99–107.

194. Rodriguez I, Greer CA, Mok MY, Mombaerts P. A putative pheromone receptor gene expressed in human olfactory mucosa. *Nat Genet*. 2000;26(1):18–19.

195. Wallrabenstein I, Gerber J, Rasche S, et al. The smelling of Hedione results in sex-differentiated human brain activity. *Neuroimage*. 2015;113.

196. Duan X, Block E, Li Z, et al. Crucial role of copper in detection of metal-coordinating odorants. *Proc Natl Acad Sci USA*. 2012;109(9):3492–3497.

197. Jemiolo B, Xie TM, Novotny M. Socio-sexual olfactory preference in female mice: attractiveness of synthetic chemosignals. *Physiol Behav*. 1991;50(6):1119–1122.

198. Ma W, Miao Z, Novotny MV. Induction of estrus in grouped female mice (*Mus domesticus*) by synthetic analogues of preputial gland constituents. *Chem Senses*. 1999;24(3):289–293.

199. Novotny M, Jemiolo B, Harvey S, Wiesler D, Marchlewska-Koj A. Adrenal-mediated endogenous metabolites inhibit puberty in female mice. *Science*. 1986;231(4739):722–725.

2

Chemical Ecology in Insects

Bill Hansson, Dieter Wicher

Department Evolutionary Neuroethology, Max Planck Institute for Chemical Ecology,
Jena, Germany

OUTLINE

INTRODUCTION

Chemosensation (i.e., the perception of environmental chemical compounds and the subsequent processing of this information) is probably the oldest way for a living being to get information from the outside world.[1] Based on whether or not molecules have a meaning for the emitter and the receiver, we can distinguish various levels of interactions. The most common case is probably that an emitted molecule does not carry any useful information for either the emitter or the receiver (Figure 1). For an olfactory cue, the recipient obtains useful information about the emitter (e.g., a food source). In this case, the emission is usually not intended to serve as information for the receiver. Representing a third level, semiochemicals are molecules released as part of a voluntary or involuntary communication between sender and receiver.[2]

Chemosensory Transduction
http://dx.doi.org/10.1016/B978-0-12-801694-7.00002-0

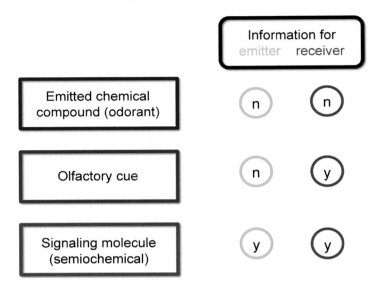

FIGURE 1   Chemicals emitted from a source may (y) or may not (n) convey information to the emitter and/or the receiver. An olfactory cue contains valuable information just for the receiver, whereas a semiochemical is used for the purpose of communication between sender and receiver.

Among the chemosignals, hormones transport information within organisms, whereas semiochemicals transfer information between organisms (Figure 2). Semiochemicals mediating communication within one species are pheromones, whereas those responsible for signaling between different species are called allelochemicals. The allelochemicals are grouped into four categories according to the benefit an emitter and/or receiver can get from the chemosignal: allomones, kairomones, synomones, and apneumones.[3,4] For allomones, only the emitter benefits from the released chemosignal. Sent by a plant, it may repel a herbivorous insect or attract an insect into a trap. By contrast, for kairomones, the information is positive just for the receiver, which may be, for example, a predator that is attracted by the smell of the prey. Synomones have a beneficial effect for both parties, which is the case when a plant pollinator is rewarded with nectar. Apneumones are released by nonliving matter and may contain beneficial information for the receiver.[4]

Pheromones are signals released for the purpose of communication between individuals within a species (Figure 2). The term was created to indicate that an external chemosignal is carrying (Greek "pherein") an exciting (Greek "hormone") message.[5,6] The first pheromone to be discovered was the sexual attractant bombykol produced by the female silk moth *Bombyx mori*.[7] The pheromones are usually beneficial for both emitter and receiver. Depending on the time domain in which they act, pheromones are distinguished as releasers or as primers.[8] Releasers activate a short and immediate behavioral response in the receiver, whereas primers lead to long-lasting physiological changes or prepare the receiver for later

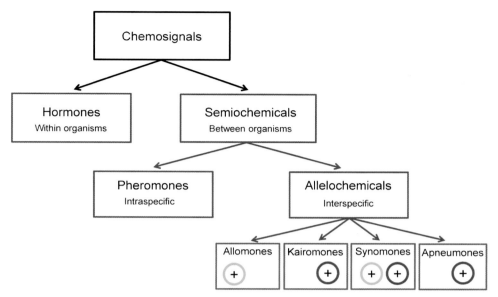

FIGURE 2   Hierarchical scheme of chemosignals according to the relationship between emitter and receiver. +, benefit for emitter (yellow) or receiver (red).

signaling. Examples for releasers are sex pheromones, alarm pheromones, or recruitment pheromones, whereas primers regulate developmental processes such as the growth of workers in a beehive.

In this chapter, we will present examples of chemosignaling related to the life cycle of insects. It involves finding suitable oviposition sites and the search for suitable food to avoiding intoxication or becoming a victim of predator or parasitoid attack. Closing the circle, the process of finding a mate to reproduce is also mediated by chemosignaling/ olfaction. We will give an overview of the different levels of chemosignaling-mediated interactions between insects as well as between insects and their environment. In these contexts, we will present illustrative examples, often from the model insect *Drosophila melanogaster* (that is, the fruit fly or vinegar fly), and discuss evolutionary aspects of the use and meaning of chemosignals.

CHEMOSTIMULI AND RECEPTORS

Chemical signals (that is, chemostimuli) are molecules emitted from various sources such as soil, plants, or animals. The molecules emitted can be volatile or in solution. Volatiles are typically detected by olfactory receptors (see Chapter 6), whereas molecules in solution are recognized by gustatory receptors (GRs) (see Chapter 14). Here we will focus on volatile chemostimuli. In insects, these signals are detected by olfactory sensory neurons (OSNs) that express olfactory receptors in their dendrites and are localized in fore-head appendages, the antennae, and the maxillary palps. There are three classes of receptor proteins that form olfactory receptors. First, the odorant receptors (ORs) that detect food odors and pheromones are heterodimeric constructs composed of an odor-specific receptor protein and a ubiquitous coreceptor protein (Orco) (see Chapter 6). Although the OR proteins are heptahelical proteins, such as G protein–coupled receptors (GPCRs), they differ from GPCRs in respect to the orientation within the plasma membrane: for insect OR proteins, the N-terminus is cytoplasmic, whereas the C-terminus is extracellular.[9] The ORs operate as ligand-gated ion channels,[10,11] the sensitivity of which can be tuned by intracellular signaling (for a review, see Ref. 12). Second, receptors homologous to ionotropic glutamate receptors called "ionotropic receptors" (IRs)[13] respond to compounds such as acids and amines (see Chapter 6). The odor spectrum detected by the IRs is complementary to that of the ORs.[14] IRs form heterotetramers of odor-specific and coreceptor proteins. Third, special GR proteins form heterodimers to sense carbon dioxide[15,16] (see Chapter 14).

A comprehensive odor panel covering different chemical classes (Table 1), including food odors of different quality ranging from fresh fruit odors to odors from rotten goods, was tested for *Drosophila*.[17] The receptive range of the olfactory receptors filled a continuum from narrowly to broadly tuned receptors. Some receptors were as broadly tuned as responding to about 20% of the odors tested. These receptors were most sensitive to structurally similar odorants. However, because of the immense number of possible stimuli, a caveat in olfactory science is always that the key ligand might have been overlooked. The processing of the odor information reflects the valence of the odor for the animal. There is a clear segregation in the representation of attractive and aversive odors among the olfactory glomeruli present in the primary olfactory center of the insect brain, the antennal lobe.[18,19]

TABLE 1   Odor Classes Tested to Characterize the Detection Profile of *Drosophila* Olfactory Receptors[17]

Acids

Alcohols

Aldehydes

Amines

Aromatics

Esters

Ketones

Lactones

Sulfur compounds

Terpenes

INSECT HOMEOSTASIS

A permanent and necessary task to maintain homeostasis during insect life is the search for food. Most important for this search is the sense of smell. Food odors are mainly detected by OSNs endowed with ORs. These neurons are localized in basiconic and (rarely) coeloconic sensilla (see Chapter 6). Whereas the IRs and the GRs emerged early in evolution and show a broad occurrence in the animal kingdom, the heterodimeric ORs are restricted to insects.[20,21] In fact, the heterodimeric ORs evolved together with emerging insect flight.[22] One might speculate that ORs provided flying insects with a benefit for fast and sensitive detection of odors during flight. In this situation, insects are faced with two challenges. First, they have to track odors during their own motion, and, second, odors dissipate because of air motion that might be turbulent and form shifting, filamentous plumes.[23] The perception of such information clearly benefits from fast and sensitive receptors.[24] But what makes ORs so special in contrast to IRs? The answer might be that the sensitivity of ORs can be tuned, whereas this does not happen with IRs.[25] When flies are stimulated with a subthreshold odor concentration that fails to initiate the ionotropic OR response, the electrical activity of the OSN does not change. It was previously observed that heterologously expressed Orco proteins form ion channels[11,26] that can be activated by cyclic nucleotides.[11] In case the sub-threshold odor stimulus was sufficient to enhance cyclic adenosine monophosphate produc-tion to activate Orco channels, this sensitized ORs via a so-far unknown mechanism.[25] A second presentation of the weak stimulus within a time window between a couple of seconds and a few minutes produced an ionotropic response and enhanced the activity of the OSN. This autoregulatory tuning of OR sensitivity may thus adjust the dynamic range of an OR according to odor presentations in a short time range.

After finding a food source, it is necessary to assess its nutritional value. In case of a contamination with harmful, toxin-producing microbes, the animals should be alerted and repelled, just like you would be after opening the fridge and smelling the dinner you forgot 2 weeks earlier. This is especially important for insects such as the vinegar fly, *D. melanogaster*, which prefers fermenting fruits. For these flies, it is vital to detect when nutri-tious yeast has been covered with dangerous microbes. One key volatile compound produced by some of these toxic fungi and bacteria is geosmin. In *D. melanogaster*, which is highly attracted by the smell of vinegar,[27] geosmin was seen to dramatically reduce the attraction toward vinegar.[28] The basis for the processing of this behaviorally relevant olfactory signal was investigated in a recent study,[29] which revealed an extraordinarily efficient chemosignal processing at different levels (Figure 3). In the scheme given in Figure 3, the signal geosmin emitted by toxin-producing microbes is a cue that informs the receiver *Drosophila* about poten-tial danger. Because of the very high relevance of this information for the survival of the receiver, an appropriate signal-processing mechanism has evolved. At the level of OSNs, only a single class of cells (ab4B) is activated by geosmin. The ab4B neurons express the OR protein Or56a. Intriguingly, the OR formed by Or56a and Orco is exclusively tuned to geo-smin. The ab4B neurons send their axons to the DA2 glomerulus in the antennal lobe, the first olfactory information-processing unit in the central nervous system. In the glomeruli, OSNs form synapses with projection neurons, which in turn transfer the sensory input to higher brain centers. The information transfer is usually modified by local interneurons that connect

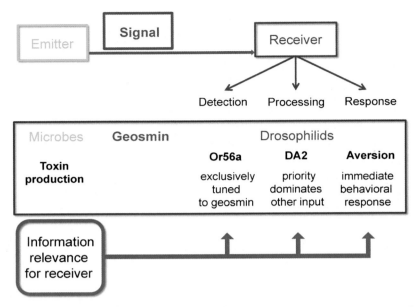

FIGURE 3   Processing by drosophilid flies of the geosmin signal emitted by toxin-producing microbes. The odor is detected by the receptor Or56a; this information is then sent to glomerulus DA2 in the antennal lobe. Activation of neurons innervating DA2 is sufficient to induce aversive behavior. The pathway is organized as an ecologically labeled line, keeping the geosmin information strictly separated from other inputs.

various glomeruli, thereby contributing to odor coding, signal amplification, and weighting. The projection neurons getting input from the DA2 glomerulus, however, respond exclusively to geosmin just like the OSNs providing the input. The strict specificity observed for the OSN is thus conserved for the projection neurons transferring the geosmin information (i.e., there is a straightforward, segregated pathway for processing the alerting chemosignal). Activation of neurons innervating the DA2 glomerulus induces an aversive behavior. DA2 signaling is highly prioritized because it dominates input from other olfactory pathways. The aversive response blocks positive chemotaxis, feeding, and oviposition. Among drosophilid flies, the specific geosmin detection system is highly conserved. It is somewhat difficult to categorize the smell of geosmin. The fly clearly benefits from detecting it, but the role for the microbe remains unknown. If the microbe benefits by not being consumed, the geosmin odor should be termed a reciprocally beneficial synomone. If the microbe indeed wants to be consumed, the aversive odor is a kairomone, benefitting the receiver.

Another kind of food quality control is related to the search for especially beneficial contents. High metabolic activity is accompanied by production of reactive oxygen species (ROSs) that may damage tissue and accelerate aging.[30] Antioxidants can protect against ROSs. The detection of such compounds in food helps to prevent oxidative stress that may also be induced by infection, cold exposure, or toxin ingestion. Vinegar flies are able to detect the presence of such an antioxidant, hydroxycinnamic acid (HCA), in fruits via olfactory cues.[31] The flies do not detect HCA directly but perceive ethylphenols that yeast produce as a metabolite derived from HCA. These compounds activate one class of OSNs, the Pb1B neurons, which are localized on the maxillary palps and express Or71a. Stimulation of Pb1B neuron activity by ethylphenols induces positive chemotaxis, feeding, and oviposition. Fly larvae are not equipped with Or71a, but they are able to detect ethylphenols via the larval-specific receptor Or94b. Thus, flies at adult and larval stages are able to detect food enriched with antioxidants to protect themselves from oxidative stress, but do so via different receptors. In this case, the odors should be considered kairomones because the fly consumes the yeast after smelling its bouquet. However, the yeast might benefit by being spread by spores on the fly body.[32] If this relationship outweighs the negative aspects of being grazed upon by the flies, the odor becomes a synomone.

Blood-feeding mosquitos orient toward humans by integration of various sensory cues with olfaction as the most important input.[33] The human body odors that guide mosquitos to their hosts are typical examples of kairomones (Figure 2).[34] The attractiveness of these body odors is enhanced by carbon dioxide detected by specific GRs.[35,33] For the mosquito *Aedes aegypti*, the preference for humans represents an evolutionary shift because of altered sensitivity for chemosignals. In addition to the human blood-sucking "domestic" form of *A. aegypti*, there is an ancestral "forest" form that lives in Kenya and prefers nonhuman animals.[36] This difference in preference is related to the expression and odor sensitivity of the receptor AaegOr4. This receptor is activated by sulcatone (6-methyl-5-hepten-2-one), a component of human body odor, and the expression of AaegOr4 is upregulated in the domestic form of the mosquito. Although nonhuman animals also release sulcatone, the amount is much greater in humans. Furthermore, differences in the Or4 sequence between the domestic and the forest forms of *A. aegypti* lead to differences in sulcatone sensitivity. Taken together, a change in the level of OR expression combined with altered OR sensitivity accompanied a change in mosquito host preference.

An interesting illustration to how a species becomes a food choice specialist is found in the island-living *Drosophila sechellia*. These flies are endemic to the Seychelles and are specialists to *Morinda citrifolia*, which is their only host. The flies feed from the fruits of the Morinda tree (Figure 4) and use it for egg-laying. All other drosophilid species are killed by contact with the toxic, acid-producing fruits. *D. sechellia* flies carry a mutation in the *Catsup* gene.[37] This mutation causes a reduced egg production in female *D. sechellia* flies because of a low level of L-DOPA, a precursor of dopamine. The Morinda fruit is rich in L-DOPA, and by dietary L-DOPA uptake, the flies can compensate for the impaired fertility. Moreover, the *Catsup* mutation provides the *D. sechellia* flies with an enhanced resistance to the toxic acids present in Morinda fruits, allowing survival in this hostile environment. A scheme of the chemo-signaling that attracts *D. sechellia* and repels all other drosophilids is given in Figure 5.

The attractive odor emitted by the Morinda fruit is methyl hexanoate. This odor stimulates ab3A neurons that express the Or22a receptor and innervate the DM2 glomerulus in the antennal lobe. Activation of these OSNs induces positive chemotaxis.[38] For *D. melanogaster* and other drosophilids, ethyl hexanoate is the best ligand for Or22a. The Or22a in *D. sechellia*, however, differs in nine amino acids from that in *D. melanogaster*. This change

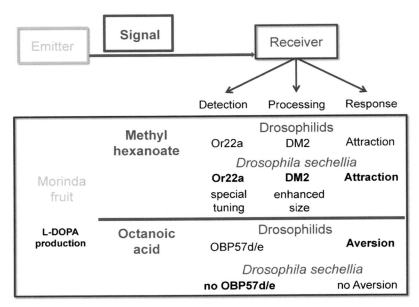

FIGURE 4   Odor signals emanating from the Morinda fruit are differently processed by *Drosophila sechellia* compared with other drosophilid flies. Methyl hexanoate activates the odorant receptor Or22a. In *D. sechellia*, Or22a is highly expressed and specifically tuned to this odor. The odor information converges in the DM2 glomerulus, which is enlarged in *D. sechellia*. Activation of neurons innervating DM2 mediates attraction, which is especially strong for *D. sechellia*. Octanoic acid is transferred to the receptor by the odor-binding proteins OBP57d/e. The perception of octanoic acid leads to strong aversion. *Drosophila sechellia* lacks OBP57d/e, which may explain why *D. sechellia* is not displaying aversive behavior toward the acids.

has caused a shift of the preferred ligand to methyl hexanoate.[38] In addition, the *D. sechellia* antennae are equipped with more basiconic sensillae housing ab3A neurons. Presumably, because of this enhanced input, the DM2 glomeruli in *D. sechellia* are nearly three times larger than those found in *D. melanogaster*. The aversive behavior of non-sechellia drosophilids toward Morinda fruits is related to the emission of octanoic acid. The receptor activated by this volatile is as yet unknown, but the perception requires specific odorant-binding proteins (OBP57d/e) that are not found in *D. sechellia*. This fact might be one factor explaining why the acid aversion is restricted to other drosophilids. Methyl hexanoate is thus highly beneficial to the *D. sechellia* flies. The fitness effect for *M. citrifolia* is, however, unclear. The seeds have already been produced and the fruit is more or less decaying when it becomes extremely attractive to the flies. It thus seems to be of little harm to the plant to have the flies feeding on its fruit. Still, the fruit odors should be considered kairomones.

    About one-third of the insect orders are herbivorous. The evolution of such a lifestyle involves transitions in physiology and behavior. A recent study in drosophilid flies of the genus *Scaptomyza* has shown that the transition from yeast-feeding to plant-feeding was accompanied by marked adaptations.[39] Orthologs of ORs used in *D. melanogaster* to detect yeast odors were either deleted (Or22a, Or85d) or appeared as pseudogenes (Or9a, Or42b).

FIGURE 5   (A) Noni fruit (*Morinda citrifolia*); (B) titan arum (*Amorphophallus titanum*); (C) bee orchid (*Ophrys apifera*). *Photo credits: M. Knaden (A); D. Wicher (B, C).*

In parallel, there was a gene duplication in paralogs of the *D. melanogaster* Or67b receptor, which detects green-leaf volatiles. These changes on the receptor level are mirrored by behavioral changes, where the *Scaptomyza* flies lost their attraction to yeast volatiles. In parallel, they became attracted to green leaf volatiles, which are of negative valence in other drosophilids. Synomones keeping *D. melanogaster* and its relatives away from green plants have thus become kairomones in *Scaptomyza* flies.

INSECT REPRODUCTION

Sex pheromones produced by female moths are mixtures of long-chained alcohols, their esters, and/or aldehydes. In the silk moth *B. mori*, the mixture is simple and contains bombykol (*E,Z*-10,12-hexadecadien-1-ol) as the main component and bombykal (*E,Z*-10,12-hexadecadien-1-al) as a minor component in a ratio of 10:1.[7,40] Typically, pheromones are more complicated blends of several components. The species specificity is obtained by variation of double bond position and isomeric form of the components along with the mixture ratio of those components (see, for example, the review by Zhang et al.[41]).

Even within a species, differences in pheromone signaling can produce a reproductive barrier, as is seen with the European corn borer moth (*Ostrinia nubilalis*). There are two races in this species, called E and Z, which produce pheromone blends characterized by opposite mixture ratios of the components E-11-14-acetate and Z-11-14-acetate.[42] For the E-strain, the mixture ratio E:Z is 98:2, whereas for the Z-strain it is 3:97. The reason for the different production of pheromone precursors in these two races is an allelic variation in a fatty-acyl reductase.[43] The male response on the other hand is under very complicated genetic control. The sensillum setup and the ORs expressed provide input to glomeruli of the male-specific macroglomerular complex. The innervation pattern is mirror imaged in the two strains,[44] as is the OSN setup in the sensilla.[45,46] Recent investigations have also revealed a very complicated pattern of receptor expression on the male antenna.[47]

In some moth species, including *B. mori*, pheromones produced by females are perceived exclusively by males. Activation of the bombykol receptor BmOR1, which is expressed in male antennae, initiates innate motor programs leading to copulation.[48] This gender-restricted production and reception is, however, not a general rule within the insects. For example, in *D. melanogaster* the male-specific pheromone 11-cis-vaccenyl acetate activates the receptor DmOr67d, which is expressed both by females and males.[49,50] Receptor activation regulates the mating behavior of both sexes. In males, it inhibits male-to-male courtship. For recent reviews on insect pheromone signaling, see Refs 51–54,41.

Apart from selecting mating partners, reproductive behaviors require a series of important decisions.[55] One of these is the selection of suitable oviposition sites that reflect the lifestyle of a given insect species. For herbivores, this choice is largely guided by plant volatiles.[56] These volatiles are used to distinguish host plants from nonhost plants and to evaluate state and suitability of possible host plants. Quite important is to assess whether a plant is already attacked by other herbivores (see the following section). *Drosophila melanogaster* prefers fermenting fruits, while avoiding those infected by geosmin-producing bacteria or slime molds.[29] Among the fruits, there is a clear preference for *Citrus* fruits.[57] This preference is an ancestral trait and may reflect an adaptation to fruits present in the native African habitat. Terpenes such as limonene and valencene released from *Citrus* fruits are detected by OSNs housed in an antennal intermediate sensillum type containing ai2A neurons that express the receptor DmOr19a. This receptor is necessary and sufficient to mediate a *Citrus* preference. Intriguingly, the *Citrus* odor valencene is a strong repellent for the endoparasitic wasp *Leptopilina boulardi*, which is a very important larval parasite of *D. melanogaster*. Parasitation with this wasp is a major cause of mortality in *Drosophila* flies. Oviposition on *Citrus* fruits thus reduces the risk for larvae being attacked by *L. boulardi*. The semiochemical interaction is thus very complicated. The *Citrus* odor is a kairomone benefiting the fly receiver and not the fruit. In the case of the parasitoid, it is less obvious because the behavioral significance of valencene is unclear.

INSECT–PLANT INTERACTIONS: EXPLOITING THE SYSTEM

Plant metabolism produces volatile cues that are used to detect a food source for herbivores[58] (Figure 6). In response to such challenges, plants have evolved chemical defense systems for protection against herbivory that are comprehensively reviewed elsewhere.[59] Here, we focus on mechanisms involving chemical signaling, which is either direct or

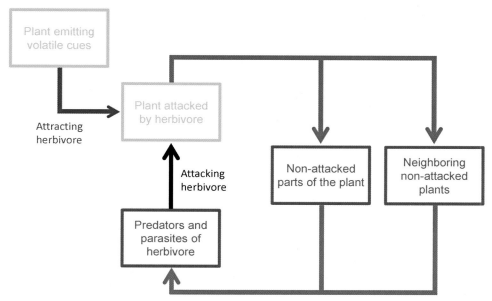

FIGURE 6   Reaction chain of plant—insect interaction. Volatile cues emitted by a plant (kairomones) attract herbivorous insects. The attacked plant releases alarm hormones/pheromones to inform nonattacked parts of the plant and neighboring plants. Odors are also released that attract predators and parasitoids of the herbivores. The predators and parasitoids attack the herbivorous insects thereby counteracting the herbivore attack.

indirect. A direct defense includes the release of toxins or substances that act within the insect nervous system such as nicotine, caffeine, cocaine or other neuroactive substances. Biogenic amines, for example, act as competitive Orco antagonists.[60] In case of an indirect defense, the emitted compounds attract parasitoids or predators of the feeding insects and can considerably reduce the herbivore number.[61] In addition to a constitutive type of defense, herbivory attack leads to induced defense responses. This induction on demand helps to save energy required to synthesize compounds. Figure 6 illustrates the insect—plant interaction following an herbivore attack. Volatiles emitted from plant parts on which an insect has fed can alert other parts to prepare a response to a potential attack. Such signals can evolve to a communication with neighboring plants and may be designed to attract predators or parasitoids of the attacking herbivore. Such functional shifts of chemical signals released by plants have been recently reviewed.[62]

Chemosignals can change their functional role. For example, an agent that originally served a defensive role, such as a resin secreted by a wounded plant, may turn into an attractant that rewards a pollinator.[63] To attract insects as pollinators, plants often release odors with a flowery note. Although attraction is predominantly mediated via olfaction, other sensory modalities are also involved. For example, colorful flowers can act as visual landmarks indicating nectar reward. In addition to reward-connected attraction, some plant groups have evolved to attract pollinators by deception.[64] They do so, by promising a reward without providing it. To attract drosophilid flies an arum species, *Arum palestinum*, releases fruity-yeasty odors,[27] thereby activating ORs specifically tuned to attractive food odors. Some arum species such as the dead horse arum (*Helicodiceros muscivorus*) or the titan arum (*Amorphophallus titanum*) (Figure 5(B)) emit the smell of rotting meat, thereby mimicking an oviposition site for flesh feeding flies.[65,66] Other odors attract insects such as flies and dung beetles by producing fecal or urine odors. It is also possible that these odors evolved to repel herbivores and that the attraction of pollinators reflects a functional shift of the chemosignal.[67] Taken together, the deceptive plants emit odors as allomones (Figure 2) to induce a behavior in the attracted insects that is beneficial for the plant. To be convincing, however, it is sometimes necessary to complement chemosignaling with other sensory cues.[64] This is especially the case for sexually deceptive orchids that attract male pollinators by emitting female-specific sex pheromones. The visual signal from flowers that mimic insects (Figure 5(C)) enhances the reproductive success.[68–70]

Normally, floral scents released to attract pollinators act as synomones (Figure 2), where the pollinator is rewarded (e.g., by nectar). An example for such an interaction is the attraction of the hawkmoth *Manduca sexta* by the odor plume from flowers of the jimsonweed *Datura wrightii*. The challenge for the moths is to discriminate the *D. wrightii* odor plume from the mélange of other odors.[71] *Datura wrightii* often coexists with other plants that release volatiles such as benzaldehyde, which also occurs in the *D. wrightii* bouquet. In wind tunnel experiments, it was shown that the ability of moths to correctly navigate to the source of the *D. wrightii* mixture was decreased in the presence of a benzaldehyde background.[71] Nevertheless, the moth could cope with this situation by changing the processing of the olfactory information within the antennal lobe. Disturbing this flexibility in neuronal representation by inhibiting local interneuron signaling in the antennal lobe disrupted successful odor navigation.[71] Whereas *M. sexta* has an innate preference for certain night-blooming plants including *D. wrightii*,[72] the moths can learn to use other flowers for nectar feeding. These learning processes, including octopaminergic signaling, modify the neuronal representation of the previously unattractive scent bouquets but they do not override the representation of the innately preferred odor plumes.[72]

In addition to nectar feeding, *D. wrightii* serves *M. sexta* also for oviposition.[73] Interestingly, there is a difference in the floral scents inducing both behaviors. The *D. wrightii* odor bouquet contains linalool, which occurs as the (+) and/or (−) enantiomer. Although feeding is independent of the enantiomeric form of linalool, oviposition is preferred for plants emitting (+) linalool.[73]

EVOLUTIONARY ASPECTS

Evolution of chemical signaling may start from the release of metabolic by-products that contain useful information for a receiver.[62] As we have seen in the examples presented previously, the use and meaning of chemosignals are not erratic or subject to rapid changes. A metabolic by-product can become an attractant, whereas a defense signal can become an alarm signal or become attractive.[62] Plants produce toxins to repel herbivores (see Breakout Box), and herbivorous insects develop detoxification systems.[74]

BREAKOUT BOX

PLANT PRODUCTS AFFECTING ION CHANNELS IN ANIMALS

As part of their defense system against herbivory, plants produce neuroactive substances that disturb the neuronal signaling in insects feeding on them. Preferential targets of these plant products are ion channels and receptors. Because of the conservation of these proteins during evolution, these plant products are sometimes also toxic for mammals. Voltage-gated sodium channels are necessary for electrical signaling in neurons. Accordingly, they are targets for a broad range of neurotoxins.[79] Among the six neurotoxin receptor sites, two bind plant-derived toxins, the alkaloids veratridine (from lilies) and aconitine (from helmet flower, *Aconitum napellus*) as well as grayanotoxins (from rhododendrons and *Ericaceae* plants). These lipid-soluble toxins are allosteric modulators of channel function and affect voltage-dependent gating. They switch the activation threshold toward negative voltages, thereby leading to persistent channel activation and flooding of neurons with Na^+, which ultimately kills the cells. Pyrethrins are toxins from flowers of *Chrysanthemum* plants.[80] They are relatively specific for insect sodium channels and gave rise to the production of synthetic insecticides (pyrethroids, DDT). Specific mutations in the *para* sodium channels produce a resistance against these insecticides, known as knockdown resistance (kdr).

Ionotropic receptors (ligand-gated ion channels) are targets of many defensive plant products.[81] Nicotine is an alkaloid from tobacco plants that acts as agonist for the nicotinic acetylcholine receptor (nAChR). nAChRs are excitatory cation channels and are of special importance at the neuromuscular junction. A well-known competitive antagonist is D-tubocurarine (from *Chondrodendron tomentosum*), which has been used as a paralyzing arrow poison. Examples for noncompetitive antagonists for nAChRs are cocaine (from *Erythroxylaceae* plants) and strychnine (from *Strychnos nux-vomica*). Type A γ-aminobutyric acid receptors (GABA$_A$Rs) are ligand-gated chloride channels and confer inhibitory activity in the brain. Competitive antagonists of GABA$_A$Rs such as bicuculline (from *Dicentra cucullaria*) lead to convulsions. Although GABA$_A$Rs are the most important inhibitory receptors in the central nervous system, the glycine receptors (also chloride channels) play this role in the periphery. Here, the plant compound strychnine acts as antagonist and produces convulsion and asphyxia.

Optimizing the success of a released signal can include simple changes in the quantity released. Considering a plant that attracts pollinators: such a strategy raises the question whether the energetic costs for the production of the attractant are balanced by a higher number of visitors, which might be restricted in a given habitat. On the other hand, the quantity of an emitted chemosignal may directly inform about the state of the sender. For example, the moth *Utetheisa ornatrix* perceives pyrrolizidine alkaloids from its host plants that are toxic for its predators. Female moths choose mating partners according to the quantity of released male courtship pheromones, which directly reflects the quantity of the protecting alkaloids acquired and thus the expected benefit from mating.[62]

For the investigation of various aspects of chemical ecology, the availability of systems that allow genetic modification has a great value. Among the insects the vinegar fly *D. melanogaster*, for which there is a myriad of various mutants (FlyBase, flybase.org), offers such a system. The use of genetics in conjunction with molecular, cellular, electrophysiological, optical, and behavioral approaches allows us to understand how the flies interact with their environment.[75] This approach is complemented by the availability of insect host plants in which metabolic pathways can be genetically manipulated to allow their role in insect–plant interaction to be studied.[76,77] In the next generation of experiments, we see genome editing approaches allowing for genetic manipulations in new species.[33]

Insect chemical ecology presents a fascinating area of research. The multitude of interactions between insects, between insects and their hosts or prey, between insects and their enemies, or between insects and their general environment present a nearly inexhaustible source of scientific topics. This multitude, in combination with modern technology in chemistry, molecular biology, physiology, and behavioral investigations, has allowed major advances in recent years. Here, we focused on chemical signaling with relevance to the life cycle of insects. However, we largely ignored the impact of other sensory inputs that are known to interact strongly with the perception of chemical stimuli. Circadian rhythms, for example, control the sensitivity of insects for sex pheromones.[51] Also, abiotic factors can influence interactions within ecosystems.[78] These important aspects must always be considered when dissecting the chemical life of insects.

Acknowledgments

The writing of this study was financed by the Max Planck Society.

References

1. Wyatt TD. Introduction to chemical signaling in vertebrates and invertebrates. In: Mucignat-Caretta C, ed. *Neurobiology of Chemical Communication*. Boca Raton, FL: CRC Press; 2014 Chapter 1.
2. Law JH, Regnier FE. Pheromones. *Annu Rev Biochem*. 1971;40:533–548.
3. Brown WLJ, Eisner T, Whittacker RH. Allomones and kairomones: transpecific chemical messengers. *Bioscience*. 1970;20:21–22.
4. Nordlund DA, Lewis WJ. Terminology of chemical releasing stimuli in intraspecific and interspecific interactions. *J Chem Ecol*. 1976;2:211–220.
5. Karlson P, Butenandt A. Pheromones (ectohormones) in insects. *Annu Rev Entomol*. 1959;4:39–58.
6. Karlson P, Lüscher M. 'Pheromones': a new term for a class of biologically active substances. *Nature*. 1959;183(4653):55–56.
7. Butenandt A, Beckmann R, Stamm D, Hecker E. Über den Sexuallockstoff des Seidenspinners *Bombyx mori*. Reindarstellung und Konstitution. *Z Naturforsch*. 1959;14b:283–284.
8. Regnier FE, Law JH. Insect pheromones. *J Lipid Res*. 1968;9(5):541–551.
9. Benton R, Sachse S, Michnick SW, Vosshall LB. Atypical membrane topology and heteromeric function of *Drosophila* odorant receptors in vivo. *PLoS Biol*. 2006;4(2):e20.
10. Sato K, Pellegrino M, Nakagawa T, Nakagawa T, Vosshall LB, Touhara K. Insect olfactory receptors are heteromeric ligand-gated ion channels. *Nature*. 2008;452:1002–1006.
11. Wicher D, Schäfer R, Bauernfeind R, et al. *Drosophila* odorant receptors are both ligand-gated and cyclic-nucleotide-activated cation channels. *Nature*. 2008;452:1007–1011.
12. Wicher D. Olfactory signaling in insects. *Prog Mol Biol Transl Sci*. 2015;130:37–54.

13. Benton R, Vannice KS, Gomez-Diaz C, Vosshall LB. Variant ionotropic glutamate receptors as chemosensory receptors in *Drosophila*. *Cell*. 2009;136(1):149—162.
14. Rytz R, Croset V, Benton R. Ionotropic receptors (IRs): chemosensory ionotropic glutamate receptors in *Drosophila* and beyond. *Insect Biochem Mol Biol*. 2013;43(9):888—897.
15. Jones WD, Cayirlioglu P, Kadow IG, Vosshall LB. Two chemosensory receptors together mediate carbon dioxide detection in *Drosophila*. *Nature*. 2007;445(7123):86—90.
16. Kwon JY, Dahanukar A, Weiss LA, Carlson JR. The molecular basis of CO_2 reception in *Drosophila*. *Proc Natl Acad Sci USA*. 2007;104(9):3574—3578.
17. Hallem EA, Carlson JR. Coding of odors by a receptor repertoire. *Cell*. 2006;125(1):143—160.
18. Semmelhack JL, Wang JW. Select *Drosophila* glomeruli mediate innate olfactory attraction and aversion. *Nature*. 2009;459(7244):218—223.
19. Knaden M, Strutz A, Ahsan J, Sachse S, Hansson BS. Spatial representation of odorant valence in an insect brain. *Cell Rep*. 2012;1(4):392—399.
20. Croset V, Rytz R, Cummins SF, et al. Ancient protostome origin of chemosensory ionotropic glutamate receptors and the evolution of insect taste and olfaction. *PLoS Genet*. 2010;6(8):e1001064.
21. Penalva-Arana DC, Lynch M, Robertson HM. The chemoreceptor genes of the waterflea *Daphnia pulex*: many Grs but no Ors. *BMC Evol Biol*. 2009;9(79):79.
22. Missbach C, Dweck HK, Vogel H, et al. Evolution of insect olfactory receptors. *Elife*. 2014;3(3):e02115.
23. Carde RT, Willis MA. Navigational strategies used by insects to find distant, wind-borne sources of odor. *J Chem Ecol*. 2008;34(7):854—866.
24. Szyszka P, Gerkin RC, Galizia CG, Smith BH. High-speed odor transduction and pulse tracking by insect olfactory receptor neurons. *Proc Natl Acad Sci USA*. 2014;111(47):16925—16930.
25. Getahun MN, Olsson SB, Lavista-Llanos S, Hansson BS, Wicher D. Insect odorant response sensitivity is tuned by metabotropically autoregulated olfactory receptors. *PLoS One*. 2013;8(3):e58889.
26. Jones PL, Pask GM, Rinker DC, Zwiebel LJ. Functional agonism of insect odorant receptor ion channels. *Proc Natl Acad Sci USA*. 2011;108(21):8821—8825.
27. Stökl J, Strutz A, Dafni A, et al. A deceptive pollination system targeting drosophilids through olfactory mimicry of yeast. *Curr Biol*. 2010;20(20):1846—1852.
28. Becher PG, Bengtsson M, Hansson BS, Witzgall P. Flying the fly: long-range flight behavior of *Drosophila melanogaster* to attractive odors. *J Chem Ecol*. 2010;36(6):599—607.
29. Stensmyr MC, Dweck HK, Farhan A, et al. A conserved dedicated olfactory circuit for detecting harmful microbes in *Drosophila*. *Cell*. 2012;151(6):1345—1357.
30. Le Bourg E. Oxidative stress, aging and longevity in *Drosophila melanogaster*. *FEBS Lett*. 2001;498(2—3): 183—186.
31. Dweck HK, Ebrahim SA, Farhan A, Hansson BS, Stensmyr MC. Olfactory proxy detection of dietary antioxidants in *Drosophila*. *Curr Biol*. 2015;21(14):01555—01556.
32. Stamps JA, Yang LH, Morales VM, Boundy-Mills KL. *Drosophila* regulate yeast density and increase yeast community similarity in a natural substrate. *PLoS One*. 2012;7(7):e42238.
33. McMeniman CJ, Corfas RA, Matthews BJ, Ritchie SA, Vosshall LB. Multimodal integration of carbon dioxide and other sensory cues drives mosquito attraction to humans. *Cell*. 2014;156(5):1060—1071.
34. Zwiebel LJ, Takken W. Olfactory regulation of mosquito—host interactions. *Insect Biochem Mol Biol*. 2004;34(7):645—652.
35. Robertson HM, Kent LB. Evolution of the gene lineage encoding the carbon dioxide receptor in insects. *J Insect Sci*. 2009;9(19):19.
36. McBride CS, Baier F, Omondi AB, et al. Evolution of mosquito preference for humans linked to an odorant receptor. *Nature*. 2014;515(7526):222—227.
37. Lavista-Llanos S, Svatos A, Kai M, et al. Dopamine drives *Drosophila sechellia* adaptation to its toxic host. *Elife*. 2014;3(3):03785.
38. Dekker T, Ibba I, Siju KP, Stensmyr MC, Hansson BS. Olfactory shifts parallel superspecialism for toxic fruit in *Drosophila melanogaster* sibling, *D. sechellia*. *Curr Biol*. 2006;16(1):101—109.
39. Goldman-Huertas B, Mitchell RF, Lapoint RT, Faucher CP, Hildebrand JG, Whiteman NK. Evolution of herbivory in Drosophilidae linked to loss of behaviors, antennal responses, odorant receptors, and ancestral diet. *Proc Natl Acad Sci USA*. 2015;26:201424656.

40. Kaissling KE, Kasang G, Bestmann HJ, Stransky W, Vostrowsky O. A new pheromone of the silkworm moth *Bombyx mori*. *Naturwissenschaften*. 1978;65:382–384.

41. Zhang J, Walker WB, Wang G. Pheromone reception in moths: from molecules to behaviors. *Prog Mol Biol Transl Sci*. 2015;130:109–128.

42. Klun JA, Chapman OL, Mattes KC, Wojtkowski PW, Beroza M, Sonnet PE. Insect sex pheromones: minor amount of opposite geometrical isomer critical to attraction. *Science*. 1973;181(4100):661–663.

43. Lassance JM, Groot AT, Lienard MA, et al. Allelic variation in a fatty-acyl reductase gene causes divergence in moth sex pheromones. *Nature*. 2010;466(7305):486–489.

44. Karpati Z, Dekker T, Hansson BS. Reversed functional topology in the antennal lobe of the male European corn borer. *J Exp Biol*. 2008;211(Pt 17):2841–2848.

45. Hansson BS, Löfstedt C, Roelofs W. Inheritance of olfactory response to sex pheromone components in *Ostrinia nubilalis*. *Naturwissenschaften*. 1987;74:497–499.

46. Olsson SB, Kesevan S, Groot AT, Dekker T, Heckel DG, Hansson BS. *Ostrinia* revisited: evidence for sex linkage in European corn borer *Ostrinia nubilalis* (Hübner) pheromone reception. *BMC Evol Biol*. 2010;10(285):285.

47. Koutroumpa FA, Kárpáti Z, Monsempes C, et al. Shifts in sensory neuron identity parallel differences in pheromone preference in the European corn borer. *Front Ecol Evol*. 2014;2:69.

48. Sakurai T, Mitsuno H, Haupt SS, et al. A single sex pheromone receptor determines chemical response specificity of sexual behavior in the silkmoth *Bombyx mori*. *PLoS Genet*. 2011;7(6):e1002115.

49. Ha TS, Smith DP. A pheromone receptor mediates 11-cis-vaccenyl acetate-induced responses in *Drosophila*. *J Neurosci*. 2006;26(34):8727–8733.

50. Kurtovic A, Widmer A, Dickson BJ. A single class of olfactory neurons mediates behavioural responses to a *Drosophila* sex pheromone. *Nature*. 2007;446(7135):542–546.

51. Stengl M. Pheromone transduction in moths. *Front Cell Neurosci*. 2010;4:133.

52. Gomez-Diaz C, Benton R. The joy of sex pheromones. *EMBO Rep*. 2013;14(10):874–883.

53. Kaissling KE. Pheromone reception in insects: the example of silk moths. In: Mucignat-Caretta C, ed. *Neurobiology of Chemical Communication*. Boca Raton, FL: CRC Press; 2014 Chapter 4.

54. Sakurai T, Namiki S, Kanzaki R. Molecular and neural mechanisms of sex pheromone reception and processing in the silk moth *Bombyx mori*. *Front Physiol*. 2014;5(125):125.

55. Laturney M, Billeter JC. Neurogenetics of female reproductive behaviors in *Drosophila melanogaster*. *Adv Genet*. 2014;85:1–108.

56. Anderson P, Anton S. Experience-based modulation of behavioural responses to plant volatiles and other sensory cues in insect herbivores. *Plant Cell Environ*. 2014;37(8):1826–1835.

57. Dweck HK, Ebrahim SA, Kromann S, et al. Olfactory preference for egg laying on citrus substrates in *Drosophila*. *Curr Biol*. 2013;23(24):2472–2480.

58. Baldwin IT. Plant volatiles. *Curr Biol*. 2010;20(9):R392–R397.

59. Mithöfer A, Boland W. Plant defense against herbivores: chemical aspects. *Annu Rev Plant Biol*. 2012;63:431–450.

60. Chen S, Luetje CW. Trace amines inhibit insect odorant receptor function through antagonism of the co-receptor subunit. *F1000Res*. 2014;3(84):84.

61. Kessler A, Baldwin IT. Defensive function of herbivore-induced plant volatile emissions in nature. *Science*. 2001;291(5511):2141–2144.

62. Steiger S, Schmitt T, Schaefer HM. The origin and dynamic evolution of chemical information transfer. *Proc Biol Sci*. 2011;278(1708):970–979.

63. Armbruster WS, Lee J, Baldwin BG. Macroevolutionary patterns of defense and pollination in *Dalechampia* vines: adaptation, exaptation, and evolutionary novelty. *Proc Natl Acad Sci USA*. 2009;106(43):18085–18090.

64. Urru I, Stensmyr MC, Hansson BS. Pollination by brood-site deception. *Phytochemistry*. 2011;72(13):1655–1666.

65. Stensmyr MC, Urru I, Collu I, Celander M, Hansson BS, Angioy AM. Pollination: rotting smell of dead-horse arum florets. *Nature*. 2002;420(6916):625–626.

66. Shirasu M, Fujioka K, Kakishima S, et al. Chemical identity of a rotting animal-like odor emitted from the inflorescence of the titan arum (*Amorphophallus titanum*). *Biosci Biotechnol Biochem*. 2010;74(12):2550–2554.

67. Lev-Yadun S, Ne'eman G, Shanas U. A sheep in wolf's clothing: do carrion and dung odours of flowers not only attract pollinators but also deter herbivores? *Bioessays*. 2009;31(1):84–88.

68. Gaskett AC. Orchid pollination by sexual deception: pollinator perspectives. *Biol Rev Camb Philos Soc*. 2011;86(1):33–75.

69. Ayasse M, Stökl J, Francke W. Chemical ecology and pollinator-driven speciation in sexually deceptive orchids. *Phytochemistry*. 2011;72(13):1667−1677.
70. Rakosy D, Streinzer M, Paulus HF, Spaethe J. Floral visual signal increases reproductive success in a sexually deceptive orchid. *Arthropod Plant Interact*. 2012;6(4):671−681.
71. Riffell JA, Shlizerman E, Sanders E, et al. Sensory biology. Flower discrimination by pollinators in a dynamic chemical environment. *Science*. 2014;344(6191):1515−1518.
72. Riffell JA, Lei H, Abrell L, Hildebrand JG. Neural basis of a pollinator's buffet: olfactory specialization and learning in *Manduca sexta*. *Science*. 2013;339(6116):200−204.
73. Reisenman CE, Riffell JA, Bernays EA, Hildebrand JG. Antagonistic effects of floral scent in an insect-plant interaction. *Proc Biol Sci*. 2010;277(1692):2371−2379.
74. Pauchet Y, Wilkinson P, Vogel H, et al. Pyrosequencing the *Manduca sexta* larval midgut transcriptome: messages for digestion, detoxification and defence. *Insect Mol Biol*. 2010;19(1):61−75.
75. Mansourian S, Stensmyr MC. The chemical ecology of the fly. *Curr Opin Neurobiol*. 2015;34C:95−102.
76. Kessler D, Baldwin IT. Back to the past for pollination biology. *Curr Opin Plant Biol*. 2011;14(4):429−434.
77. Gase K, Baldwin IT. Transformational tools for next-generation plant ecology: manipulation of gene expression for the functional analysis of genes. *Plant Ecol Divers*. 2012;5:485−490.
78. Erb M, Lu J. Soil abiotic factors influence interactions between belowground herbivores and plant roots. *J Exp Bot*. 2013;64(5):1295−1303.
79. Cestele S, Catterall WA. Molecular mechanisms of neurotoxin action on voltage-gated sodium channels. *Biochimie*. 2000;82(9−10):883−892.
80. Davies TG, Field LM, Usherwood PN, Williamson MS. DDT, pyrethrins, pyrethroids and insect sodium channels. *IUBMB Life*. 2007;59(3):151−162.
81. Nasiripourdori A, Taly V, Grutter T, Taly A. From toxins targeting ligand gated ion channels to therapeutic molecules. *Toxins (Basel)*. 2011;3(3):260−293.

OLFACTORY TRANSDUCTION

Vertebrate Odorant Receptors

Kazushige Touhara[1,2], Yoshihito Niimura[1,2], Sayoko Ihara[1,2]

[1]Department of Applied Biological Chemistry, Graduate School of Agricultural and Life Sciences, The University of Tokyo, Tokyo, Japan; [2]ERATO Touhara Chemosensory Signal Project, JST, The University of Tokyo, Tokyo, Japan

OUTLINE

TOWARD THE DISCOVERY OF THE OR GENES

Odor Recognition Theory

It has been estimated that 0.4–0.5 million odorants exist in our external environment. This rough estimate is based on the observation that 20–25% of the 2 million possible low-molecular-weight (<300 MW) organic compounds are volatile and odorous. Even if this number is an overestimate, it remains that an enormous number of odorants can be detected, and thus each animal must have a sophisticated olfactory system that recognizes and discriminates a large set of odorant molecules.

The emotional associations humans have with smell have likely contributed to our historical fascination with olfactory physiology. The first description positing a mechanism for our sense of smell was provided by Titus Lucretius Carus, a Roman poet and philosopher. He proposed in his book *On the Nature of Things* (50 BC) that we can detect a variety of odors in our lives because each odorant possesses a different structure that determines the odor's quality.[1] In the mid-twentieth century, this concept was extended by Amoore as the stereospecific receptor theory.[2] The receptor theory speculated that there are receptor sites for odorants, and that odor perception occurs only when the structure of an odorant molecule and its binding site fit each other. As long as many receptors exist, this theory could explain molecular mechanisms underlying a remarkable olfactory sensing system. Based on the studies on specific anosmia, Amoore proposed structures of receptor sites for seven primary odors,[2] and he estimated the number of human primary odors is at least 32.[3]

Other theories, including vibrational theory, puncturing theory, radiational theory, and absorption theory, were also proposed to explain the mechanisms of olfaction. The vibrational theory, originally proposed by Dyson in 1928[4] and expanded by Wright in 1954,[5] was refined by Luca Turin in the mid-1990s.[6] Turin proposed that after odorants bind their receptor sites based on shape, the receptor measures the molecular vibrational energy of the molecule to convey a particular olfactory signal for that odorant. This theory was proposed to explain aspects of the phenomena that could not be explained by the stereospecific receptor theory. Although it is theoretically supported by physicists, experiments so far have failed to support the prediction, and thus this theory remains controversial.

Discovery of Putative OR Genes

Evidence accumulated in the mid-1980s suggesting the involvement of second messenger signaling and channels in olfactory signal transduction.[7–10] Based on these observations, Buck and Axel hypothesized that the receptors must be expressed in the olfactory epithelium and must belong to a family of G-protein—coupled receptors (GPCRs). By applying a homology cloning strategy with the new polymerase chain reaction (PCR) technology, they successfully identified a multigene family that appeared to encode candidate receptors for odorants in rats.[11] The encoded proteins were named olfactory receptors or odorant receptors (ORs): olfactory receptor because the proteins are expressed in olfactory sensory neurons (OSNs), and odorant receptor because the proteins were thought to recognize odorant molecules. Neither turned out to be strictly the case, because ORs are now known to be expressed in tissues other than the olfactory epithelium (see "Ectopic Expression of OR Genes" later in the chapter) and some ORs have been found to detect chemicals other than odorants.

The identified ORs belong to the class A rhodopsin-like GPCR family that includes rhodopsin and the β2-adrenergic receptor. The ORs possess structural features in common with GPCRs: seven transmembrane (TM) domains, a conserved glycosylation site in the N-terminal region, and a disulfide bond between the conserved cysteines in the extracellular loops. ORs can be distinguished from other GPCRs by several conserved OR family-specific motifs including an LHTPMY motif within the first intracellular loop, an MAYDRYVAIC

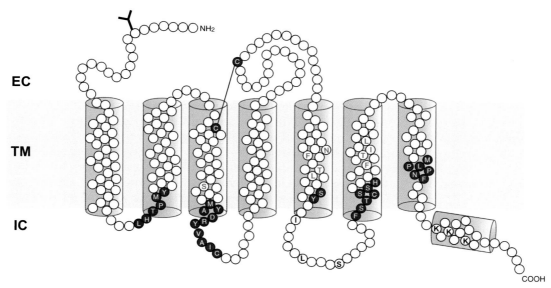

FIGURE 1 **Schematic representation of the seven transmembrane structure of an OR.** OR-specific conserved amino acids described in the text are shaded black. Amino acids involved in ligand binding and G protein-coupling in a mouse OR, mOR-EG, are shown in red and black, respectively.[51] The glycosylation site is indicated by "Y". EC, extracellular; TM, transmembrane; IC, intracellular.

motif at the end of TM3, an SY motif at the end of TM5, FSTCSSH at the beginning of TM6, and PMLNPF in TM7[12,13] (Figure 1). Hypervariable regions reside in the extracellular side of several TM domains, suggesting a role of these regions in ligand recognition (see "Structure and Function" later in the chapter).

Proven to Be an OR

After the discovery of the putative OR genes in 1991, a long time passed before functional evidence was obtained that their products were indeed odorant-binding receptors. The major challenge was in getting the ORs to be expressed in heterologous mammalian cell lines.[14] ORs expressed in these systems were not being properly trafficked to the cell surface, but rather were retained in the endoplasmic reticulum and degraded by proteasomes. Another reason for the delay was because there was no certainty that the ligand was in fact in the odorant repertoire being used for screening.

A strategy for functional OR expression eventually succeeded. In a strategy developed by Zhao et al.,[15] the OSNs themselves were targeted for exogenous OR expression since OSNs should be equipped with the cellular machinery required for proper OR membrane trafficking. This homologous expression system was possible by using an adenovirus-mediated gene transfer technique. Viral infection of mouse OSNs resulted in expression of an OR delivered by the engineered viral vector in addition to the native OR being expressed endogenously by the neuron. Since many neurons expressing different endogenous ORs were infected with the same exogenous OR gene, the investigators could identify the cognate ligand's signal for the introduced OR against the background of diverse endogenous ORs. Indeed, when adenoviruses encoding a rat OR gene named I7 infected the experimental olfactory epithelium, robust responses to the cognate odor ligand were observed in electro-olfactograms. This was the first OR to be paired with its cognate ligand.

At the same time, Touhara et al. successfully identified a mouse OR they named MOR23.[16] In their study, OR activation was measured as increases in intracellular Ca^{2+} in single adenovirus-infected neurons labeled by GFP. The choice of ORs was different in these two approaches. Zhao's approach selected an OR randomly and applied various odorants during screening, whereas Touhara's approach identified the OR from a single OSN responding to a certain odorant. The latter strategy of Touhara was based on the fact that each olfactory neuron expressed a single OR according to in situ hybridization studies. Thus, the response assay for the GFP-labeled adenovirus-infected neurons was carried out with the odorant used to isolate the OR from the single responsive neuron. Touhara et al. successfully recapitulated the odorant response of a cell from which the receptor gene was functionally cloned. This study proved that the OR functions in odorant recognition, but also supported the single cell—single OR rule.

This adenovirus approach has been abandoned since these early studies because it was technically laborious. Instead, an improved heterologous expression system has been used to identify ligands for other ORs. The first successful attempt fused 20 amino acids from the N-terminal end of rhodopsin (Rho-tag) to the N-terminal end of the OR, resulting in successful trafficking of the OR to the cell surface in nonolfactory cell lines.[17] These intensive trials in the late 1990s[15–18] confirmed that the putative OR genes discovered by Buck and Axel indeed encode odorant receptor proteins. Subsequent studies of functional expression of ORs will be discussed in "Functional Assay Methods" later in the chapter.

THE OR GENE FAMILY

Bioinformatic Analysis of Vertebrate OR Genes

GPCRs have a common structural feature of seven alpha-helical transmembrane regions. GPCR genes are classified into five or six groups based on sequence similarities. Vertebrate OR genes belong to the rhodopsin-like GPCR superfamily. This superfamily includes opsin genes for detecting light, and genes for receptors for neurotransmitters, peptide hormones, chemokines, lipids, nucleotides, and others.[19] All vertebrate OR genes form a monophyletic clade in a phylogenetic tree of the rhodopsin-like GPCR superfamily; therefore their sequence similarities make them clearly distinguishable from other non-OR GPCR genes.[20]

In general, vertebrate OR genes do not contain any introns in their coding regions. However, the number of exons in the 5′ untranslated region often varies among OR genes, and these noncoding exons can be alternatively spliced to generate multiple mRNA isoforms that nonetheless result in the same protein.[21] The biological significance of these multiple isoforms is still unknown.

Thanks to the advent of next-generation sequencers, we now have access to whole genome sequences from diverse organisms. Due to sequence homologies among vertebrate OR genes and their intron-less gene structure, identifying all putative OR genes by repeated homology searches is straightforward.[22,23] By comparing OR gene repertoires of various species, we can gain some insight into the evolution of olfaction in view of an adaptation to each species' living environment.

OR Genes in Humans

Approximately 820 OR genes have been found in the human genome (Figure 2).[24,25] Among them, ~400 are intact, and more than half (~52%) are pseudogenes. Intact genes have been presumed to be functional. Note, however, that the distinction between intact genes and pseudogenes is based only on their sequences; in most cases there has been no experimental validation of their expression. Even if a gene retains an intact open-coding sequence, it may not be actually expressed if a yet-undiscovered mutation in its regulatory region disrupts transcription. It is also possible that a sequence variation leads to a posttranslational conformation that is inactive. Verbeurgt et al.[26] investigated the expression of 356 human intact OR genes in whole olfactory mucosa tissues from 26 individuals and confirmed the expression of 273 OR genes on average, 77% of the intact genes tested.

OR genes reside in genomic clusters in human chromosomes. Every chromosome, except chromosome 20 and the Y chromosome, encodes OR genes, and chromosome 11 contains >40% of all human OR genes.[24] Mammalian OR genes are readily classified into two groups by sequence similarity. All Class I genes are found in a single cluster on chromosome 11, whereas Class II genes are dispersed in clusters throughout the genome. The largest human OR gene cluster contains ~100 Class II genes (intact genes or pseudogenes) and occupies a ~2 Mb genomic region also on chromosome 11.

OR gene loci exhibit remarkably high between-individual diversity, and are among the most diverse regions of the human genome. Olender et al.[27] used data from the 1000 Genomes Project to investigate the diversity of OR gene repertoires among individuals. They identified 244 segregating OR pseudogenes, for which both intact and pseudogene forms are present in a population. They also found 63 OR loci exhibiting deletion copy number variation (CNV); such loci are present in some individuals, but not in others. In all, 66% of the ~400 human intact OR loci are affected by nonfunctional single-nucleotide polymorphisms (SNPs), insertion-deletions (indels), and/or CNVs. Therefore, each individual has a unique set of functional OR genes.

Olfactory perception differs considerably among individuals. Individuals who lack the ability to perceive a particular odor, though they generally have a good sense of smell, are said to have specific anosmia.[3] For example, hydrogen cyanide is an extremely poisonous gas having a faint almond-like sweet odor for most people, but one in 10 individuals cannot perceive this odor. One in 1000 does not smell butyl mercaptan, a main component of skunk spray that is artificially added to natural gas because it can be detected at very low concentrations by most people.

So far, associations between differences in odor perception and SNPs in OR gene loci have been shown for several OR-odorant combinations. Androstenone, a pig pheromone,

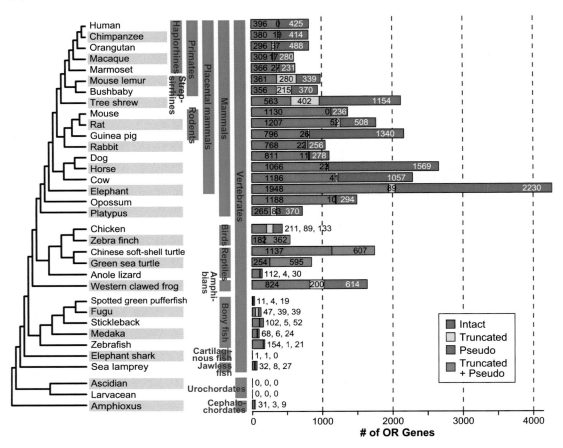

FIGURE 2   **Phylogeny and numbers of the OR genes.** OR genes are classified into three categories: intact gene, truncated gene, and pseudogene. An intact gene is defined as a sequence starting from an initiation codon and ending with a stop codon that does not contain any disrupting mutations. A pseudogene is defined as a sequence with a nonsense mutation, frameshift, deletion within conserved regions, or some combination thereof. A truncated gene is defined as a partial, intact sequence located at a contig end. As the quality of genome sequence data improves, truncated genes become reassigned as an intact gene or a pseudogene. For a low-coverage genome sequence (e.g., mouse lemur, bushbaby, or tree shrew), the fraction of truncated genes tends to become higher because of shorter contig lengths. Truncated genes and pseudogenes were not distinguished for the zebra finch and two turtle species. The numbers of OR genes were taken from Refs 20,25,35,91.

is subject to specific anosmia. People report one of three different types of perception of this molecule: offensive (sweaty or urinous), pleasant (sweet or floral), or odorless. Keller et al.[28] revealed that perception of androstenone is associated with the SNPs in an OR gene named *OR7D4*. Variants of this locus include two nonsynonymous SNPs linked to each other, R88W and T133M, and subjects having an RT/WM or WM/WM genotype were less sensitive and had a less unpleasant perception of androstenone than did RT/RT subjects.

Several other studies have shown associations between odor perception and SNPs in OR gene loci: *OR11H7P* is associated with isovaleric acid (sweaty) perception,[29] *OR2J3* with *cis*-3-hexen-1-ol (grassy) perception,[30] *OR5A1* with β-ionone (floral) perception,[31] and *OR10G4* with guaiacol (smoky) perception.[32]

OR Genes in Primates

OR gene repertoires are highly diverse among individuals within the human population. The differences in OR gene repertoires become much larger when different species are compared. Genome analyses of our closest relative, the chimpanzee, demonstrated that the numbers of both intact OR genes and OR pseudogenes in chimpanzees are nearly the same as those in humans, but only ~75% of the OR genes show orthologous relationships between the two species.[33] In other words, approximately one-fourth of the OR gene repertoires in humans and chimpanzees are specific to each species. Therefore, chimpanzees' olfactory perception might be quite different from humans.

Higher primates generally have much smaller numbers (300–400) of intact OR genes than do most other mammals (~1000) (Figure 2). Matsui et al.[25] showed that the most recent common ancestor of hominoids (human, chimpanzee, orangutan; Figure 2), Old World monkeys (OWMs; macaque), and New World monkeys (NWMs; marmoset) had ~550 functional OR genes, and each species has lost >200 OR genes up to the present. This observation is thought to reflect that higher primates rely more heavily on vision than olfaction; thus primate olfaction may have degenerated due to relaxed selective pressure.

Within the order Primates, two suborders based on the morphology of nostrils are recognized: (1) strepsirrhines, which means "twisted nose" and includes lemurs and lorises; and (2) haplorhines, which means "simple nose" and includes tarsiers, NWMs, OWMs, and hominoids (Figure 2). Strepsirrhines and haplorhines are characterized by the presence and absence of the rhinarium, respectively. The rhinarium is a moist and hairless surface at the tip of the nose, and is used to detect the directional origin of odorants. Many mammalian species, including cats and dogs, have a rhinarium. In general, the relative size of the olfactory bulb is smaller in haplorhines than in strepsirrhines.[34] Furthermore, most strepsirrhines are nocturnal, while most haplorhines are diurnal, and color vision is well developed only in haplorhines. Therefore reliance on olfaction has apparently decreased more among haplorhines than strepsirrhines. To determine which factors led to this shrinkage of OR gene repertoires during primate evolution, a wide variety of primate species living in a wide range of ecological niches should be examined.

OR Genes in Mammals

Mammals are extremely diverse in size, shape, and habitat use. Mammals occupy all habitats on earth: terrestrial, fossorial, arboreal, volant, and aquatic. Their feeding habitats are also highly diversified. Insectivores, herbivores, carnivores, and omnivores are found among mammals; some feed on fish, others leaves, yet other on grains or seeds; some are even ant specialists. Predictably, OR gene repertoires are highly variable among mammals and reflect their ecological diversity.

The number of OR genes varies widely among mammal species (Figure 2). Notably, African elephants have a surprisingly large number of intact OR genes, ~2000.[35] This number is by far the largest ever characterized; it is more than twice that found in dogs, and five times more than in humans. The African elephant genome contains an even larger number of OR pseudogenes (>2200). The fraction of OR pseudogenes in African elephants is 52%, which is similar to that in humans. This is notable because in general the fraction of OR pseudogenes within a genome does not correlate with the number of intact OR genes.[35]

Although the African elephant OR sequences have not been assigned to chromosome loci, it is possible to reconstruct the organization of OR gene clusters by comparing them with other species. The number of OR gene clusters in the African elephant genome containing 5 or more OR genes was estimated to be 34, which is exactly the same as that in mice and humans.[36] Because African elephants have a much larger number of OR genes than mice or humans but they are organized into a similar number of clusters, each OR gene cluster seems to have become expanded in the African elephant genome. The largest African elephant OR gene cluster has been estimated to contain 353 Class II OR genes, spanning a 5.42-Mb region.

Do elephants have a superior sense of smell? It is known that elephants heavily rely on olfaction in various contexts, including foraging, social communication, and reproduction.[37,38] Neuroanatomical studies also indicate that elephants have well-developed olfactory systems with relatively large olfactory bulbs and large olfactory areas in the brain.[39] Recently, Rizvanovic et al.[40] conducted behavioral tests using another elephant species, Asian elephants, to assess their olfactory ability. They found that Asian elephants can discriminate between 12 enantiomeric odor pairs and between 12 odor pairs of aliphatic alcohols, aldehydes, etc., each having only a one-carbon difference within a pair. This indicates that elephants perform at least as well as mice and clearly better than humans, pigtail macaques, or squirrel monkeys. Moreover, African elephants can reportedly distinguish between two Kenyan ethnic groups—the Maasai, whose young men demonstrate virility by spearing elephants, and the Kamba, an agricultural group who pose little threat to elephants—by using olfactory cues.[41]

OR genes with different sequences are assumed to bind different sets of odorants. Therefore, it is reasonable to assume that a species with a larger OR gene repertoire can discriminate among more subtle differences in structurally related odorants, and that the number of OR genes is positively correlated with resolution in their olfactory world. On the other hand, sensitivity to a given odorant should be higher for an animal expressing more copies of the receptor for that odorant. For example, dogs are famous for a keen sense of smell, but their OR gene repertoire is not particularly large (Figure 2). They are very sensitive to particular odors such as fatty acids, a major component of human body odor, but they may not need to distinguish among many different types of odors.

It has been estimated that the most recent common ancestor of placental mammals, which lived ~100 million years ago, had 781 functional OR genes.[35] Hundreds of OR gene gains and losses have occurred in a lineage-specific manner during mammalian evolution to shape each species' OR gene repertoire. The estimated gene birth and death rates are considerably higher than the average rate among all mammalian gene families.[35] These studies provide evidence that the OR multigene family is one of the most extreme examples of "birth-and-death evolution" in which new genes are created by gene duplication, some of which have been maintained in the genome for a long time and some of which have been deleted or became nonfunctional due to deleterious mutations.[42]

OR Genes in Nonmammalian Vertebrates

Fish, like mammals, use olfactory cues to find food, avoid danger, and identify conspecific individuals. Olfactory information is also used to recognize places within an organism's environment. Salmon have a remarkable homing ability: in adulthood they return to the river where they were spawned, and this behavior depends on olfaction. Salmon imprint to place-specific odors during a sensitive developmental period, and adults use the odorant memory to return to their natal streams. Fish can detect four main groups of water-soluble molecules as odorants: amino acids, gonadal steroids, bile acids, and prostaglandins. These odorants are nonvolatile, and humans cannot smell them.

Fish have much smaller numbers of OR genes than mammals (Figure 2) (see Chapter 5). However, OR gene repertoires in fish are more diverse than those in mammals. Phylogenetic analyses have revealed that the most recent common ancestor of tetrapods (including mammals, birds, reptiles, and amphibians) and bony fish had at least seven ancestral OR genes.[20,43] OR genes in tetrapods and bony fish can be classified into seven groups, designated $\alpha-\eta$, each of which corresponds to one of the ancestral OR genes. (For comparison to mammalian OR gene classes, group γ corresponds to Class II, and groups α and β together correspond to Class I.) The distribution of OR genes in various vertebrate groups reveals an intriguing pattern[20,43]: Groups α and γ are well-represented in tetrapods, but they are absent from all bony fish (with one exception in zebrafish). Conversely, groups δ, ε, ζ, and η are found in bony fish and amphibians, but amniotes (reptiles, birds, and mammals) completely lack these groups. Therefore, group α and γ genes are considered to be terrestrial-type genes, while groups δ, ε, ζ, and η are regarded as aquatic-type genes. Interestingly, only amphibians have both types. These observations suggest that terrestrial-type genes function in detecting volatile odorants while aquatic-type genes detect water-soluble ones.

Vertebrates belong to the phylum Chordata. Chordates include two invertebrate subphyla, cephalochordates (amphioxus) and urochordates (tunicates). Amphioxi have a fish-like appearance but lack any distinctive sensory apparatus corresponding to the eyes, ears, or nose. Thus, amphioxus is considered "acraniate," or headless. Nonetheless, more than 30 vertebrate-like OR genes were identified from the amphioxus genome sequence (Figure 2).[20] Amphioxus OR genes are highly divergent from vertebrate OR genes in amino acid sequences, but they are more closely related to vertebrate OR genes than any other non-OR GPCR genes. The olfactory system of amphioxus is not well understood, and the function of these genes remains unclear. Because amphioxi are the most basal chordates, the origin of vertebrate OR gene families can be traced back to the most recent common ancestor of all chordate species, which is thought to have lived ~700 million years ago.

FUNCTIONAL ASPECT OF ORs

Functional Assay Methods

Coexpression of chaperones or other proteins that behave as cofactors with ORs is associated with efficient membrane expression of ORs in heterologous cells. RTP1 (receptor transporter protein 1), a one-transmembrane-domain protein expressed in mouse OSNs, was identified as a chaperone protein that facilitates membrane targeting of ORs.[44] Indeed, many ORs have been deorphanized in cultured cells that coexpress RTP1. Ric8B, a putative GEF (guanine nucleotide exchange factor) for $G\alpha_{olf}$ and $G\alpha_s$, was also effective in amplifying OR signaling in cultured cells.[45] Subsequently, myristoylated Ric8A, a Ric8B homologue that functions as a GEF for a $G\alpha_q$-type G protein, was reported to enhance $G\alpha_{15}$-mediated signaling greatly.[46] Recently, a cleavable 17-amino acid leucine-rich N-terminal signal peptide called Lucy tag was shown to promote cell surface expression of an OR in HEK293T cells.[47] The effect of Lucy tag was enhanced synergistically in the presence of Rho-tag and the chaperones described above (see "Proven to be an OR" earlier in the chapter). Collectively, introduction of these cofactors in combination with the addition of the N-terminal tag paved the way for overcoming the difficulties in the functional expression of ORs in heterologous systems.

Activation of a functionally expressed OR upon odorant stimulation can be detected by measuring the intracellular concentration of a second messenger, cAMP or Ca^{2+} (Figure 3).[14,48] In heterologous cells, ectopically expressed OR activates endogenous $G\alpha_s$, resulting in an increase in the level of cAMP in response to odorant. In cultured cells such as HEK293, COS7, or CHO-K1 cells, cAMP is often measured directly by an enzyme-linked immunoassay or indirectly by luciferase assay under the control of CRE (cAMP-responsive element). When a *Xenopus laevis* oocyte expression system is used, an increase in cAMP can be detected electrophysiologically by coexpressing a cAMP-dependent Cl^- channel, the cystic fibrosis transmembrane conductance regulator (CFTR). On the contrary, a Ca^{2+} increase cannot be detected unless the OR of interest is coupled to $G\alpha_q$. Therefore, the more promiscuous G protein α, $G\alpha_{15/16}$, is often used to force activation of the inositol phosphate-mediated pathway, leading to Ca^{2+} influx from the endoplasmic reticulum. In cultured cells, changes in intracellular calcium can be monitored by fluorescence of a calcium indicator such as Fura-2 preloaded into cells. In *X. laevis* oocytes, Ca^{2+} increases can be detected electrophysiologically by changes in membrane currents due to activation of endogenous Ca^{2+}-dependent Cl^- channels.

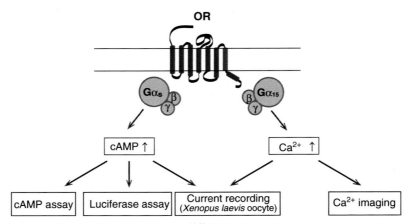

FIGURE 3   **Assays used to monitor OR activation.**[14,48] ORs activate $G\alpha_s$ in heterologous cells, resulting in cAMP increases that could be directly measured or monitored by the luciferase reporter gene assay. ORs also activate $G\alpha_{15}$ in HEK293 cells, resulting in intracellular Ca^{2+} increases. Odorant-induced cAMP or Ca^{2+} increases can be monitored in *X. laevis* oocytes electrophysiologically by changes in membrane currents.

These functional assays using a heterologous expression system are quite useful for studying various aspects of ORs, such as deorphanization of receptors and detailed analysis of structure—activity relationships. However, several cautionary cases of assay-dependent bias should be considered when evaluating results obtained from heterologous expression systems. One such case arose when G protein was forced to couple with an OR. Previous studies reported a difference in ligand responsiveness when measured by $G\alpha_{15}$-mediated Ca^{2+} elevation or $G\alpha_{s/olf}$-mediated cAMP increase, invalidating the assumption that the responsiveness measured by either method should reflect the one under physiological conditions.[49,50] This is another reminder that odorant responsiveness obtained in a heterologous expression system may not necessarily reflect the in vivo response of an OR in an olfactory epithelium context in which mucus may affect reception.[49]

Structure and Function

The binding pocket of ORs has been studied using computational homology modeling with mutational analysis in an in vitro functional assay. This approach revealed that the ligand-binding sites in a murine OR, mOR-EG (also known as MOR174-9 or Olfr73), is formed by amino acids residing in TM3, TM5, and TM6[51] (Figure 1). Most of the amino acids were hydrophobic, suggesting that the binding of an odorant to its receptor is mediated mainly by relatively weak hydrophobic interactions.[51] This is in contrast with nonolfactory GPCRs that bind to their ligand primarily via hydrogen or ionic interactions. EC_{50} values for OR-ligand pairs are larger than those for nonolfactory GPCR-ligand pairs. The lower affinity of ORs to their ligands is likely explained by this ligand-binding property. A similar homology modeling study of MOR42-3 validated by functional assay gave consistent results that the essential amino acids required for ligand binding resided in TM3, TM5, and TM6.[52]

In addition to the ligand-binding domains, the regions required for $G\alpha_{olf}$ coupling were also studied by site-directed mutagenesis.[53] Several amino acids in the third intracellular loop and C-terminal domain were shown to be involved in $G\alpha_{olf}$ coupling[53] (Figure 1).

More structural details about OR receptor function await determination of crystal structures of the ORs. Although no OR crystal structure has been reported thus far, there has been recent progress in the development of methodologies for crystallizing a growing number of GPCRs,[54] which may facilitate solving the structure of an OR in the near future.

Physiological (Natural) Ligand Pair

Improvements in the use of heterologous expression systems for functional assays of ORs have facilitated the pairing of ORs to their ligands. To date, $\sim 10\%$ of murine ORs have been deorphanized.[32] However, most of these ligands were screened from a collection of synthetic odorants that are not necessarily present in the natural mouse habitat. Since the olfactory system was originally equipped with sensors to perceive essential cues from the external world such as foods, predators, poisons, potential mates, and competitors, the physiological ligands for ORs may be quite different from the odorant collection used in a laboratory setting.

Recently, OR-physiological ligand pairs have been reported. MOR244-3 (alternatively MOR83 or Olfr1509) was identified as a receptor for (methylthio)methanethiol (MTMT),[55] an organosulfur odorant found in the urine of fertile male mice that attracts females.[56] The activation of this receptor expressed in HEK293T cells was greatly enhanced in the presence of copper ion at a physiological concentration, 30 μM, suggesting a requirement of copper ion for the physiological detection of MTMT. Consistently, discrimination of MTMT was impaired in mice injected with a copper ion chelator.

Another chemical substance in mouse urine, (Z)-5-tetradecen-1-ol (Z5-14:OH), was identified as a physiological ligand for Olfr288 by activity-guided purification from the preputial gland.[57] Z5-14:OH is excreted in adult male mouse urine in a testosterone-dependent manner and contributes to the attractiveness of the male urine to female mice. Whereas Olfr288 recognizes a broad range of synthetic lactone-derivatives, it appears to be specifically tuned to Z5-14:OH in the natural environment. Z5-14:OH has not been included in any previous collection of synthetic odorants used in ligand screening, suggesting a need to seek natural sources for physiological ligands in future work.

Physiological ligands have been detected not only in urine but also in feces. A small group of ORs found exclusively in mammals, called the OR37 family, was shown to recognize long-chain aliphatic aldehydes.[58] Recently, hexadecanal, a ligand for an OR in the OR37B family, was reported to be produced in anal glands and deposited with feces into the environment.[59]

These results suggest the possibility that there may be many more physiological ligands for each OR found in context-dependent secretions from various exocrine glands. Identifying them will aid our understanding of chemical communication between animals via ORs.

Odor Discrimination Mechanisms

A combinatorial coding strategy has been considered to be the general principle operating in the olfactory system. A single odorant can be recognized by several ORs, and a single OR can recognize multiple odorants.[60–62] This strategy enables discrimination of odorants by a combination of ORs that are activated, accounting for the vast number of odor qualities that can be distinguished beyond the number of ORs. While studies pairing ORs with synthetic odorants support this notion,[63] the specific responsiveness of Olfr288 to Z5-14:OH among natural ligands raises the question of whether the combinatorial coding strategy alone can explain the physiological process of olfaction. Since whether an OR is broadly tuned or narrowly tuned depends largely on the size and variety of odorants in the repertoire utilized for screening, as is the case for Olfr288, the words "narrow" and "broad" should be carefully used especially with respect to the physiological role of the OR.

Odorants encoded by a limited number of specific ORs have been reported. Large-scale in vitro OR screening suggested that androstenone (5α-androst-16-en-3-one), an odorous steroid derived from testosterone, is mainly detected by OR7D4 in humans.[28] Genetic variations due to SNPs in this gene were shown to affect OR activity in vitro, which correlated with the sensitivity of the odor perception in human subjects.[28] This suggests that the contribution of this OR's activity to perception is very large. Another example of a specific odor–receptor pair is MOR215-1, a mouse OR for muscone, a fascinating odor secreted from stink glands of animals.[64] MOR215-1 has been considered the sole or one of the few receptors for muscone.[64] Some odorants activate only a small subset of ORs, and the ORs responsible for perception of that odor are very few. In such cases, an SNP in a single OR gene can affect perception of that odorant significantly.

Desensitization

The olfactory system has a mechanism for desensitization following prolonged exposure to an odorant, a process called adaptation. At the level of the OSN, OR-mediated signaling declines after odorant exposure.[65] Ca^{2+} elevated by OR activation through cyclic nucleotide-gated (CNG) channels has been shown to mediate the inactivation of CNG channels and type III adenylyl cyclase while activating phosphodiesterase, resulting in desensitization of OR signaling (see Chapter 7).[66]

Regarding nonolfactory GPCRs, it is generally accepted that the desensitization mechanism operates through phosphorylation of the receptor. Second messenger-dependent protein kinases such as cAMP-dependent protein kinase A (PKA) and calcium-dependent protein kinase C (PKC), and G protein-coupled receptor kinases (GRKs), are known to phosphorylate serine/threonine residues within the intracellular loops and C-terminal regions of GPCRs. Phosphorylation of GPCRs by second messenger-dependent protein kinases decreases the efficiency of coupling to the G protein. GPCRs phosphorylated by GRKs recruit β-arrestin to uncouple the G proteins from their GPCRs, which induces internalization of the GPCRs. Based on these observations, the role of these kinases in OR desensitization has been examined. In particular, much attention has been paid to the role of GRK3, a GRK enriched in OSNs. Since the early 1990s, several studies have suggested that GRK3 is involved in olfactory desensitization.[67–69] However, these studies were based on in vitro experiments using cilia preparations, and the in vivo role of this kinase in desensitization of a specific receptor has not been examined. A recent study examined the role of GRK3 on desensitization of a specific murine OR, mOR-EG.[70] Although attenuation of OR signaling upon ligand stimulation on a 15–30 min scale was observed in GRK3-expressing HEK293 cells, neither the rapid desensitization nor internalization of the OR was observed. Furthermore, the kinetics and amplitude of ligand responsiveness of the mOR-EG-expressing OSNs were not significantly affected in GRK3-deficient mice, suggesting that GRK3 may not have an essential role in a short-term desensitization of ORs.

As for involvement of second messenger-dependent protein kinases, it has been suggested that two human ORs, OR2AG1 and OR17-4, may be phosphorylated by PKA and that the phosphorylation may be required for internalization.[71,72] However, precise roles of the PKA-mediated phosphorylation of these receptors in OSNs remain unclear. In any case, the desensitization mechanisms for ORs seem to be more varied and not as straightforward as those of nonolfactory GPCRs.

Ectopic Expression of OR Genes

OR genes are not expressed exclusively in olfactory sensory neurons. Ectopic OR expression has been detected in a variety of tissues, including the male germ line,[73] the spleen and insulin-secreting cells,[74] lingual epithelial cells,[75,76] ganglia of the autonomic nervous system,[77] pyramidal neurons in the cerebral cortex,[78] colon,[79] myocardial cells in developing heart,[80,81] prostate gland,[79] and erythroid cells.[82–85] A comprehensive study carried out recently revealed that more than 100 ORs are ectopically expressed, and ORs expressed in nonolfactory tissue were more highly conserved in animals than ORs expressed only in the olfactory epithelium.[86]

These OR transcripts were detected primarily by RT-PCR; therefore the expression level might be too low to be considered of physiological significance. However, there are several reports that suggest functions for ORs in nonolfactory tissues. The human OR, OR1D2, is expressed in testis, and is proposed to be involved in sperm chemotaxis.[87] Similar findings have been observed for the mouse MOR23 (Olfr16) expressed in testis and found to regulate sperm motility.[88] ORs expressed in mouse myocytes during unique stages of muscle regeneration appear to be involved in regulating cellular migration.[89] The human OR51E2, first named as a prostate-specific GPCR (PSGR), regulates proliferation of human prostate cancer cells.[90] The endogenous ligands for these ectopic ORs, however, have not been identified. Nonetheless, these ORs appear to function in sensing not only environmental volatile cues but also endocrine cues.

CONCLUSIONS

Genome projects have revealed that OR genes comprise the largest gene family in the vertebrate lineage and that the number and variety of ORs in the olfactory repertoire have changed dramatically during the course of evolution. The change appears to be due to selective pressures derived from which sensory modalities are dominant in communication and what environment they inhabit. Accumulated evidence suggests that ORs recognize the shape, size, and functional groups of odorant molecules at a binding site formed by transmembrane domains, and that many odorants are recognized by ORs using a combinatorial coding strategy. Some odorants, however, are detected by a small subset of ORs, and in such a case, SNPs in these ORs can affect odor perception at the level of both sensitivity and quality. Although more and more ORs have recently been deorphanized, we do not have much information about physiological odor—receptor pairs. In addition, endogenous ligands for ORs expressed in various nonolfactory tissues are yet to be identified. In this regard, a combination of chemical and biological approaches to identify the natural ligands for these ORs may reveal how animals acquire information for survival and reproduction in the complex world of external chemical signaling.

References

1. Lucretius. *On the Nature of Things: De rerum natura.* Johns Hopkins University Press; 1995.
2. Amoore JE. Stereochemical theory of olfaction. *Nature.* 1963;199:912—913.
3. Amoore JE. Specific anosimia and the concept of primary odors. *Chem Senses.* 1977;2(3):267—281.
4. Dyson GM. Some aspects of the vibration theory of odor. *Perfum Essent Oil Rec.* 1928;19:456—459.
5. Wright RH. Odour and chemical constitution. *Nature.* 1954;173:831.
6. Turin L. A spectroscopic mechanism for primary olfactory reception. *Chem Senses.* 1996;21(6):773—791.
7. Kurihara K, Koyama N. High activity of adenyl cyclase in olfactory and gustatory organs. *Biochem Biophys Res Commun.* 1972;48(1):30—34.
8. Pace U, Hanski E, Salomon Y, Lancet D. Odorant-sensitive adenylate cyclase may mediate olfactory reception. *Nature.* 1985;316(6025):255—258.
9. Sklar PB, Anholt RR, Snyder SH. The odorant-sensitive adenylate cyclase of olfactory receptor cells. Differential stimulation by distinct classes of odorants. *J Biol Chem.* 1986;261(33):15538—15543.
10. Nakamura T, Gold GH. A cyclic nucleotide-gated conductance in olfactory receptor cilia. *Nature.* 1987;325(6103):442—444.

11. Buck L, Axel R. A novel multigene family may encode odorant receptors: a molecular basis for odor recognition. *Cell*. 1991;65(1):175−187.
12. Touhara. K, Structure. expression, and function of olfactory receptors. In: Firestein S, Beaucham GK, eds. *The Senses: A Comprehensive Reference*. New York: Academic Press; 2008:527−544.
13. Fleischer J, Breer H, Strotmann J. Mammalian olfactory receptors. *Front Cell Neurosci*. 2009;3:9.
14. Touhara K. Deorphanizing vertebrate olfactory receptors: recent advances in odorant-response assays. *Neurochem Int*. 2007;51(2−4):132−139.
15. Zhao H, Ivic L, Otaki JM, Hashimoto M, Mikoshiba K, Firestein S. Functional expression of a mammalian odorant receptor. *Science*. 1998;279(5348):237−242.
16. Touhara K, Sengoku S, Inaki K, et al. Functional identification and reconstitution of an odorant receptor in single olfactory neurons. *Proc Natl Acad Sci USA*. 1999;96(7):4040−4045.
17. Krautwurst D, Yau KW, Reed RR. Identification of ligands for olfactory receptors by functional expression of a receptor library. *Cell*. 1998;95(7):917−926.
18. Wetzel CH, Oles M, Wellerdieck C, Kuczkowiak M, Gisselmann G, Hatt H. Specificity and sensitivity of a human olfactory receptor functionally expressed in human embryonic kidney 293 cells and *Xenopus laevis* oocytes. *J Neurosci*. 1999;19(17):7426−7433.
19. Fredriksson R, Lagerstrom MC, Lundin LG, Schioth HB. The G-protein-coupled receptors in the human genome form five main families. Phylogenetic analysis, paralogon groups, and fingerprints. *Mol Pharmacol*. 2003;63(6):1256−1272.
20. Niimura Y. On the origin and evolution of vertebrate olfactory receptor genes: comparative genome analysis among 23 chordate species. *Genome Biol Evol*. 2009;1(1):34−44.
21. Young JM, Shykind BM, Lane RP, et al. Odorant receptor expressed sequence tags demonstrate olfactory expression of over 400 genes, extensive alternate splicing and unequal expression levels. *Genome Biol*. 2003;4(11):R71.
22. Niimura Y. Identification of olfactory receptor genes from mammalian genome sequences. *Methods Mol Biol*. 2013;1003:39−49.
23. Niimura Y. Identification of chemosensory receptor genes from vertebrate genomes. *Methods Mol Biol*. 2013;1068:95−105.
24. Niimura Y, Nei M. Evolution of olfactory receptor genes in the human genome. *Proc Natl Acad Sci USA*. 2003;100(21):12235−12240.
25. Matsui A, Go Y, Niimura Y. Degeneration of olfactory receptor gene repertories in primates: no direct link to full trichromatic vision. *Mol Biol Evol*. 2010;27(5):1192−1200.
26. Verbeurgt C, Wilkin F, Tarabichi M, Gregoire F, Dumont JE, Chatelain P. Profiling of olfactory receptor gene expression in whole human olfactory mucosa. *PLoS One*. 2014;9(5):e96333.
27. Olender T, Waszak SM, Viavant M, et al. Personal receptor repertoires: olfaction as a model. *BMC Genomics*. 2012;13:414.
28. Keller A, Zhuang H, Chi Q, Vosshall LB, Matsunami H. Genetic variation in a human odorant receptor alters odour perception. *Nature*. 2007;449(7161):468−472.
29. Menashe I, Abaffy T, Hasin Y, et al. Genetic elucidation of human hyperosmia to isovaleric acid. *PLoS Biol*. 2007;5(11):e284.
30. McRae JF, Mainland JD, Jaeger SR, Adipietro KA, Matsunami H, Newcomb RD. Genetic variation in the odorant receptor OR2J3 is associated with the ability to detect the "grassy" smelling odor, cis-3-hexen-1-ol. *Chem Senses*. 2012;37(7):585−593.
31. Jaeger SR, McRae JF, Bava CM, et al. A mendelian trait for olfactory sensitivity affects odor experience and food selection. *Curr Biol*. 2013;23(16):1601−1605.
32. Mainland JD, Keller A, Li YR, et al. The missense of smell: functional variability in the human odorant receptor repertoire. *Nat Neurosci*. 2014;17(1):114−120.
33. Go Y, Niimura Y. Similar numbers but different repertoires of olfactory receptor genes in humans and chimpanzees. *Mol Biol Evol*. 2008;25(9):1897−1907.
34. Barton RA. Olfactory evolution and behavioral ecology in primates. *Am J Primatol*. 2006;68(6):545−558.
35. Niimura Y, Matsui A, Touhara K. Extreme expansion of the olfactory receptor gene repertoire in African elephants and evolutionary dynamics of orthologous gene groups in 13 placental mammals. *Genome Res*. 2014;24(9):1485−1496.

II. OLFACTORY TRANSDUCTION

36. Niimura Y, Nei M. Comparative evolutionary analysis of olfactory receptor gene clusters between humans and mice. *Gene*. 2005;346:13−21.
37. Rasmussen LEL, Krishnamurthy V. How chemical signals integrate Asian elephant society: the known and the unknown. *Zoo Biol*. 2000;19(5):405−423.
38. Langbauer WR. Elephant communication. *Zoo Biol*. 2000;19(5):425−445.
39. Shoshani J, Kupsky WJ, Marchant GH. Elephant brain. Part I: gross morphology, functions, comparative anatomy, and evolution. *Brain Res Bull*. 2006;70(2):124−157.
40. Rizvanovic A, Amundin M, Laska M. Olfactory discrimination ability of Asian elephants (*Elephas maximus*) for structurally related odorants. *Chem Senses*. 2013;38(2):107−118.
41. Bates LA, Sayialel KN, Njiraini NW, Moss CJ, Poole JH, Byrne RW. Elephants classify human ethnic groups by odor and garment color. *Curr Biol*. 2007;17(22):1938−1942.
42. Nei M, Rooney AP. Concerted and birth-and-death evolution of multigene families. *Annu Rev Genet*. 2005;39:121−152.
43. Niimura Y, Nei M. Evolutionary dynamics of olfactory receptor genes in fishes and tetrapods. *Proc Natl Acad Sci USA*. 2005;102(17):6039−6044.
44. Saito H, Kubota M, Roberts RW, Chi Q, Matsunami H. RTP family members induce functional expression of mammalian odorant receptors. *Cell*. 2004;119(5):679−691.
45. Von Dannecker LE, Mercadante AF, Malnic B. Ric-8B promotes functional expression of odorant receptors. *Proc Natl Acad Sci USA*. 2006;103(24):9310−9314.
46. Yoshikawa K, Touhara K. Myr-Ric-8A enhances G(alpha15)-mediated Ca^{2+} response of vertebrate olfactory receptors. *Chem Senses*. 2009;34(1):15−23.
47. Shepard BD, Natarajan N, Protzko RJ, Acres OW, Pluznick JL. A cleavable N-terminal signal peptide promotes widespread olfactory receptor surface expression in HEK293T cells. *PLoS One*. 2013;8(7):e68758.
48. Katada S, Nakagawa T, Kataoka H, Touhara K. Odorant response assays for a heterologously expressed olfactory receptor. *Biochem Biophys Res Commun*. 2003;305(4):964−969.
49. Oka Y, Katada S, Omura M, Suwa M, Yoshihara Y, Touhara K. Odorant receptor map in the mouse olfactory bulb: in vivo sensitivity and specificity of receptor-defined glomeruli. *Neuron*. 2006;52(5):857−869.
50. Shirokova E, Schmiedeberg K, Bedner P, et al. Identification of specific ligands for orphan olfactory receptors. G protein-dependent agonism and antagonism of odorants. *J Biol Chem*. 2005;280(12):11807−11815.
51. Katada S, Hirokawa T, Oka Y, Suwa M, Touhara K. Structural basis for a broad but selective ligand spectrum of a mouse olfactory receptor: mapping the odorant-binding site. *J Neurosci*. 2005;25(7):1806−1815.
52. Abaffy T, Malhotra A, Luetje CW. The molecular basis for ligand specificity in a mouse olfactory receptor: a network of functionally important residues. *J Biol Chem*. 2007;282(2):1216−1224.
53. Kato A, Katada S, Touhara K. Amino acids involved in conformational dynamics and G protein coupling of an odorant receptor: targeting gain-of-function mutation. *J Neurochem*. 2008;107(5):1261−1270.
54. Venkatakrishnan AJ, Deupi X, Lebon G, Tate CG, Schertler GF, Babu MM. Molecular signatures of G-protein-coupled receptors. *Nature*. 2013;494(7436):185−194.
55. Duan X, Block E, Li Z, et al. Crucial role of copper in detection of metal-coordinating odorants. *Proc Natl Acad Sci USA*. 2012;109(9):3492−3497.
56. Lin DY, Zhang SZ, Block E, Katz LC. Encoding social signals in the mouse main olfactory bulb. *Nature*. 2005;434(7032):470−477.
57. Yoshikawa K, Nakagawa H, Mori N, Watanabe H, Touhara K. An unsaturated aliphatic alcohol as a natural ligand for a mouse odorant receptor. *Nat Chem Biol*. 2013;9(3):160−162.
58. Bautze V, Bar R, Fissler B, et al. Mammalian-specific OR37 receptors are differentially activated by distinct odorous fatty aldehydes. *Chem Senses*. 2012;37(5):479−493.
59. Bautze V, Schwack W, Breer H, Strotmann J. Identification of a natural source for the OR37B ligand. *Chem Senses*. 2014;39(1):27−38.
60. Kajiya K, Inaki K, Tanaka M, Haga T, Kataoka H, Touhara K. Molecular bases of odor discrimination: reconstitution of olfactory receptors that recognize overlapping sets of odorants. *J Neurosci*. 2001;21(16):6018−6025.
61. Malnic B, Hirono J, Sato T, Buck LB. Combinatorial receptor codes for odors. *Cell*. 1999;96(5):713−723.
62. Touhara K. Odor discrimination by G protein-coupled olfactory receptors. *Microsc Res Tech*. 2002;58(3):135−141.
63. Saito H, Chi Q, Zhuang H, Matsunami H, Mainland JD. Odor coding by a mammalian receptor repertoire. *Sci Signal*. 2009;2(60):ra9.

64. Shirasu M, Yoshikawa K, Takai Y, et al. Olfactory receptor and neural pathway responsible for highly selective sensing of musk odors. *Neuron*. 2014;81(1):165—178.

65. Kato A, Touhara K. Mammalian olfactory receptors: pharmacology, G protein coupling and desensitization. *Cell Mol Life Sci*. 2009;66(23):3743—3753.

66. Zufall F, Leinders-Zufall T. The cellular and molecular basis of odor adaptation. *Chem Senses*. 2000;25(4):473—481.

67. Dawson TM, Arriza JL, Jaworsky DE, et al. Beta-adrenergic receptor kinase-2 and beta-arrestin-2 as mediators of odorant-induced desensitization. *Science*. 1993;259(5096):825—829.

68. Boekhoff I, Inglese J, Schleicher S, Koch WJ, Lefkowitz RJ, Breer H. Olfactory desensitization requires membrane targeting of receptor kinase mediated by beta gamma-subunits of heterotrimeric G proteins. *J Biol Chem*. 1994;269(1):37—40.

69. Peppel K, Boekhoff I, McDonald P, Breer H, Caron MG, Lefkowitz RJ. G protein-coupled receptor kinase 3 (GRK3) gene disruption leads to loss of odorant receptor desensitization. *J Biol Chem*. 1997;272(41):25425—25428.

70. Kato A, Reisert J, Ihara S, Yoshikawa K, Touhara K. Evaluation of the role of G protein-coupled receptor kinase 3 in desensitization of mouse odorant receptors in a mammalian cell line and in olfactory sensory neurons. *Chem Senses*. 2014;39(9):771—780.

71. Mashukova A, Spehr M, Hatt H, Neuhaus EM. β-arrestin2-mediated internalization of mammalian odorant receptors. *J Neurosci*. 2006;26(39):9902—9912.

72. Neuhaus EM, Mashukova A, Barbour J, Wolters D, Hatt H. Novel function of β-arrestin2 in the nucleus of mature spermatozoa. *J Cell Sci*. 2006;119(Pt 15):3047—3056.

73. Parmentier M, Libert F, Schurmans S, et al. Expression of members of the putative olfactory receptor gene family in mammalian germ cells. *Nature*. 1992;355(6359):453—455.

74. Blache P, Gros L, Salazar G, Bataille D. Cloning and tissue distribution of a new rat olfactory receptor-like (OL2). *Biochem Biophys Res Commun*. 1998;242(3):669—672.

75. Durzynski L, Gaudin JC, Myga M, Szydlowski J, Gozdzicka-Jozefiak A, Haertle T. Olfactory-like receptor cDNAs are present in human lingual cDNA libraries. *Biochem Biophys Res Commun*. 2005;333(1):264—272.

76. Gaudin JC, Breuils L, Haertle T. Mouse orthologs of human olfactory-like receptors expressed in the tongue. *Gene*. 2006;381:42—48.

77. Weber M, Pehl U, Breer H, Strotmann J. Olfactory receptor expressed in ganglia of the autonomic nervous system. *J Neurosci Res*. 2002;68(2):176—184.

78. Otaki JM, Yamamoto H, Firestein S. Odorant receptor expression in the mouse cerebral cortex. *J Neurobiol*. 2004;58(3):315—327.

79. Yuan TT, Toy P, McClary JA, Lin RJ, Miyamoto NG, Kretschmer PJ. Cloning and genetic characterization of an evolutionarily conserved human olfactory receptor that is differentially expressed across species. *Gene*. 2001;278(1—2):41—51.

80. Drutel G, Arrang JM, Diaz J, Wisnewsky C, Schwartz K, Schwartz JC. Cloning of OL1, a putative olfactory receptor and its expression in the developing rat heart. *Recept Channels*. 1995;3(1):33—40.

81. Ferrand N, Pessah M, Frayon S, Marais J, Garel JM. Olfactory receptors, Golf alpha and adenylyl cyclase mRNA expressions in the rat heart during ontogenic development. *J Mol Cell Cardiol*. 1999;31(5):1137—1142.

82. Feingold EA, Penny LA, Nienhuis AW, Forget BG. An olfactory receptor gene is located in the extended human beta-globin gene cluster and is expressed in erythroid cells. *Genomics*. 1999;61(1):15—23.

83. Feldmesser E, Olender T, Khen M, Yanai I, Ophir R, Lancet D. Widespread ectopic expression of olfactory receptor genes. *BMC Genomics*. 2006;7:121.

84. Zhang X, Rogers M, Tian H, et al. High-throughput microarray detection of olfactory receptor gene expression in the mouse. *Proc Natl Acad Sci USA*. 2004;101(39):14168—14173.

85. Pluznick JL, Zou DJ, Zhang X, et al. Functional expression of the olfactory signaling system in the kidney. *Proc Natl Acad Sci USA*. 2009;106(6):2059—2064.

86. De la Cruz O, Blekhman R, Zhang X, Nicolae D, Firestein S, Gilad Y. A signature of evolutionary constraint on a subset of ectopically expressed olfactory receptor genes. *Mol Biol Evol*. 2009;26(3):491—494.

87. Spehr M, Gisselmann G, Poplawski A, et al. Identification of a testicular odorant receptor mediating human sperm chemotaxis. *Science*. 2003;299(5615):2054—2058.

88. Fukuda N, Yomogida K, Okabe M, Touhara K. Functional characterization of a mouse testicular olfactory receptor and its role in chemosensing and in regulation of sperm motility. *J Cell Sci*. 2004;117(Pt 24):5835—5845.

II. OLFACTORY TRANSDUCTION

89. Griffin CA, Kafadar KA, Pavlath GK. MOR23 promotes muscle regeneration and regulates cell adhesion and migration. *Dev Cell*. 2009;17(5):649–661.
90. Neuhaus EM, Zhang W, Gelis L, Deng Y, Noldus J, Hatt H. Activation of an olfactory receptor inhibits proliferation of prostate cancer cells. *J Biol Chem*. 2009;284(24):16218–16225.
91. Wang Z, Pascual-Anaya J, Zadissa A, et al. The draft genomes of soft-shell turtle and green sea turtle yield insights into the development and evolution of the turtle-specific body plan. *Nat Genet*. 2013;45(6):701–706.

Odor Sensing by Trace Amine-Associated Receptors

Qian Li, Stephen D. Liberles

Department of Cell Biology, Harvard Medical School, Boston, MA, USA

INTRODUCTION

The olfactory system is a powerful sensory modality that detects a tremendous diversity of environmental odors and translates them into varied perceptions and behaviors. The main olfactory system contains a large repertoire of sensory receptors for recognition of diverse odors. Among these are the odorant receptors (ORs), a family of >350 intact GPCRs in humans, and >1000 in mice and rats[1-6] (see Chapter 3). OR (*Olfr*) genes encompass 1–5% of genes in mammalian genomes, and are found on almost every chromosome in clusters varying in size from one gene to hundreds of genes.[7] Furthermore, ORs are used in combination to encode odor identity, further increasing the diversity of chemicals that can be detected and discriminated.[8]

The main olfactory system also contains a second family of olfactory receptors called trace amine-associated receptors, or TAARs.[9,10] TAARs were originally named based on the proposition that they function in the brain to detect low-abundance neurotransmitters termed trace amines[11]; however, all TAARs except TAAR1 instead serve predominantly or exclusively as chemosensory receptors in the olfactory system.[10] Significant progress has been made to uncover the roles of TAARs in the olfactory system—from ligand recognition to behaviors.[9] Here, we review the cellular neurobiology of TAAR-containing olfactory sensory neurons (OSNs).

THE TAAR FAMILY

TAARs are a family of GPCRs that derived early in the vertebrate lineage soon before or soon after the emergence of jawed fish,[12–14] likely from a serotonin receptor subtype (5-HT$_4$).[15] TAARs share homology with biogenic amine receptors (many retain a conserved amine recognition motif; see below), and are phylogenetically distinct from ORs. The size of the *Taar* gene family varies across vertebrates, with six in human, 15 in mouse, 17 in rat, and six in macaque.[12,15–18] There are many more *Taar* genes in teleosts; for example, there are 112 in zebrafish, with most forming a teleost-specific clade (see Chapter 5).[12] RNA splicing is predicted to occur in the 5′ untranslated region of *Taar* mRNAs, but not within the coding sequence, which is contained on a single exon (an exception is the *Taar2* coding sequence, which contains an intron). In mice and humans, genes encoding TAARs are found in a single genomic cluster that lacks interspersing genes and spans 100–200 kilobases on chromosomes 10 and 6, respectively (Figure 1(A)). *Taar* genes are numbered based on their chromosomal order in mouse, from *Taar1* to *Taar9*, with mice having five full-length *Taar7* genes (*Taar7a, Taar7b, Taar7d, Taar7e,* and *Taar7f; Taar7c* is the only *Taar* pseudogene in the mouse cluster) and three full-length *Taar8* genes (*Taar8a, Taar8b,* and *Taar8c*). Rats and mice have orthologous *Taar* repertoires, except for lineage-specific expansions in the *Taar7* (seven in rat) and *Taar8* (three in rat) groups, which occurred after divergence from the last common ancestor. The six full-length *TAAR* genes in humans are *TAAR1, TAAR2, TAAR5, TAAR6, TAAR8,* and *TAAR9*.

TAAR EXPRESSION PATTERNS

All TAARs, with the exception of TAAR1, are expressed in the main olfactory epithelium (MOE), where they function as chemosensory receptors (Figure 1(B)). TAAR expression is observed in the olfactory systems of mice, zebrafish, and macaques,[10,12,17] consistent with their function as olfactory receptors being widely conserved across vertebrates. In mouse, six TAARs are also expressed in sensory neurons of the Grueneberg ganglion,[19] a small olfactory organ located at the anterior tip of the mouse nose (Figure 1(B)). TAAR expression in Grueneberg ganglion sensory neurons is developmentally regulated, with maximal expression in embryonic and neonatal mice but decreasing with age. In contrast to other TAARs, TAAR1 is expressed at low levels in the pancreas and brain, including in monoamine neurons of the ventral tegmental area and dorsal raphe nucleus.[20,21]

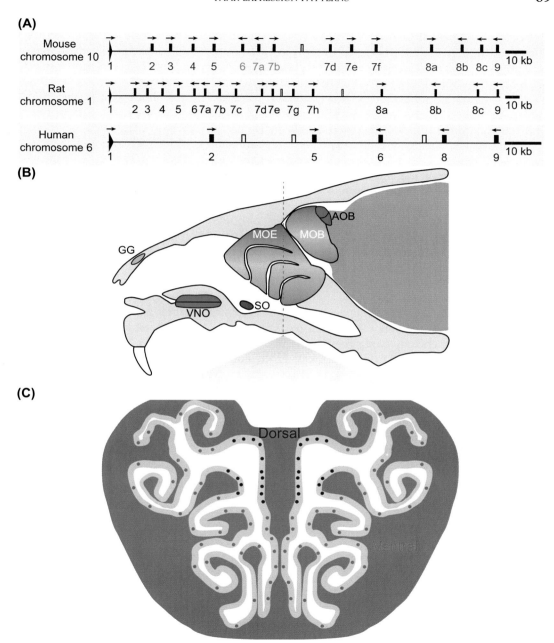

FIGURE 1 **TAAR-expressing sensory neurons in the olfactory system.** (A) Genomic organization of the *Taar* cluster in mouse, rat, and human indicating full-length genes (black-filled rectangles) and *Taar* pseudogenes (white-filled rectangles). Arrows indicate direction of sense strand, and colors of mouse genes indicate dorsal (blue) or ventral (red) expression, with data for mouse *Taar7d* ambiguous. *Taar1* (triangle) is not expressed in the olfactory system. (B) Anatomy of the mouse nose, including the main olfactory epithelium (MOE), vomeronasal organ (VNO), Grueneberg ganglion (GG), septal organ (SO), main olfactory bulb (MOB), and accessory olfactory bulb (AOB). *(Image adapted from Ref. 61.)* (C) Depiction of dorsal (blue) and ventral (red) *Taar* gene expression in a coronal section of the mouse MOE.

II. OLFACTORY TRANSDUCTION

Within the main olfactory system, TAAR-expressing OSNs are similar to OR-expressing OSNs in several regards, but also display some distinct and characteristic features. ORs are expressed in a strikingly dispersed pattern in the mouse MOE, as each OSN expresses one and only one of >2000 receptor alleles.[22,23] Furthermore, once a functional receptor is selected, receptor gene choice is stable for the lifetime of the neuron.[24,25] This expression control ensures that information about activation of a particular receptor is transmitted with precision to the central nervous system, and is stable over time. Transcription factors, locus control regions, and heterochromatin marks involved in *Olfr* gene choice have been identified.[24–30]

Olfactory TAARs are similarly expressed in small subsets of OSNs scattered in an epithelial zone[10] (Figure 1(C)). In mouse, all *Taar* genes are expressed in a dorsal zone of the MOE, with the exception of *Taar6* and some *Taar7s*, which are expressed more ventrally.[10,31,32] Each TAAR defines a unique OSN population that does not contain other TAARs or ORs. Thus, TAAR expression is governed by the canonical "one receptor-one neuron" rule.[10] Only one *Taar* allele is transcribed in a given OSN,[32,33] although epigenetic signatures present on silenced *Olfr* genes are lacking on *Taar* genes,[33] raising the possibility that *Taar* genes are selected through a different monoallelic choice mechanism. Locus control regions similar to those involved in *Olfr* gene choice have not been identified near the *Taar* gene cluster. Furthermore, expression of a *Taar* pseudogene leads to preferential expression of a second *Taar* gene rather than an *Olfr* gene,[32,33] suggesting that TAAR-expressing OSNs undergo a committed cell fate decision prior to receptor choice. Future studies are needed to understand the cellular basis of gene choice and cell maturation in TAAR-expressing OSNs.

Immunohistochemistry experiments revealed that three mouse TAARs (TAAR4, TAAR5, and TAAR6) are localized to olfactory cilia,[33] the site of odor detection. TAAR proteins are also found on OSN axons in the main olfactory bulb (MOB),[33] where they may participate in axon guidance (see below for organization of TAAR inputs in the brain). TAAR-expressing OSNs contain key olfactory signaling molecules, including the G-protein subunit $G\alpha_{olf}$, type III adenylyl cyclase (ACIII), and a cyclic nucleotide gated ion channel (CNGA2)[10,34] (see Chapter 7). Furthermore, pharmacological inhibition of adenylyl cyclase blocks odor responses in TAAR-expressing OSNs.[34] Together, these findings indicate that TAAR-expressing OSNs likely utilize the canonical cAMP-based signaling cascade that responds to GPCR activation and results in membrane depolarization and action potential generation.

TAAR LIGANDS AND BEHAVIORS

Before TAAR ligands were discovered, many TAARs were predicted to recognize amines based on sequence conservation with biogenic amine receptors. Serotonin, histamine, dopamine, acetylcholine and adrenaline receptors have a key aspartic acid on the third transmembrane α-helix that forms a direct ionic contact with the ligand amino group.[35] This aspartic acid is retained in all human TAARs (6/6), most mouse TAARs (13/15), and some zebrafish TAARs (27/112). It was thus proposed that many TAARs would detect amines. Interestingly, an entire clade of teleost TAARs lacks the conserved aspartic acid; while ligands for these fish TAARs are unknown, it is possible that they display distinct ligand recognition properties.

Among TAARs, ligands were first identified for TAAR1.[36,37] TAAR1 detects several biogenic amines that were proposed to function as low-abundance neurotransmitters (so-called trace amines for which this receptor family was named), including tyramine (EC_{50}: 200 nM), 2-phenylethylamine (EC_{50}: 300 nM), tryptamine (EC_{50}: >6 μM), and octopamine (EC_{50}: 4 μM).[36,37] TAAR1 also detects other structurally diverse amines, including amphetamines and thyronamines.[37−39] Structural modeling and mutagenesis studies predicted that like other biogenic amine receptors, TAAR1 binds ligands in the membrane plane with contacts distributed across several transmembrane α-helices, including the conserved salt bridge on transmembrane α-helix 3.[38] Synthetic and highly specific TAAR1 agonists and antagonists modulate the excitability, synaptic release, and neurotransmitter uptake properties of midbrain dopamine neurons and brainstem serotonin neurons.[40,41] Deletion of TAAR1 increases activity of these central monoamine neurons and also enhances amphetamine-induced locomotor activity in mice.[20] TAAR1-induced inhibition of the basal firing frequency of monoaminergic neurons is mediated by a pathway involving $G\alpha_s$ and inward rectifying potassium channels.[40] Based on these findings, TAAR1 may be a promising therapeutic target for neuropsychiatric disorders caused by dopamine and serotonin system dysfunction. Some other physiological effects of TAAR1 ligands, such as thyronamine-evoked hypothermia, persist in TAAR1 knockout mice and are thus likely mediated by TAAR1-independent mechanisms.[42]

The finding that other TAARs function as olfactory receptors enabled ligand search efforts involving small, volatile molecules rather than brain-derived biogenic amines. TAARs were transiently expressed in HEK-293 cells together with a cAMP-dependent reporter gene that enabled a high throughput chemical screening approach.[10,43] Using this or related heterologous cell systems, ligands were discovered for six mouse TAARs, seven rat TAARs, one human TAAR, one macaque TAAR, and one zebrafish TAAR, with each of these 16 receptors detecting different volatile amines (Figure 2).[10,17,43−45] Some of these amines are ecological odors that evoke behavioral responses (see below for a discussion of each TAAR).[31,44,46,47] TAARs generally prefer the same ligands when expressed in HEK-293 cells and OSNs,[34,47] although sensitivities vary, likely due to optimization of signaling components in the native context. GPCRs are typically thought to recognize the same ligands across cell types, with GPCR folding thought to be an intrinsic property determined by receptor sequence. Therefore, it is not surprising that TAARs prefer the same ligands in HEK-293 cells and OSNs.

Mapping the ligand binding preference of different TAARs onto the TAAR phylogenetic tree reveals branch-specific ligand preferences (Figure 2).[43] TAARs on one branch of the phylogenetic tree prefer primary amines, several of which are produced by amino acid decarboxylation. This branch includes TAAR1 (see above), TAAR2, TAAR3, TAAR4, and several teleost TAARs without mammalian orthologs, such as TAAR13c. Another branch includes TAAR5 through TAAR9, with several members preferring tertiary amines such as trimethylamine and N,N-dimethylamines. A third branch occurs in teleosts and has widespread mutation of the amine-contacting aspartic acid; it is unclear what these receptors detect.

TAAR3

Mouse TAAR3 detects several small primary amines, including isopentylamine (also known as isoamylamine, EC_{50}: 10 μM), cyclohexylamine (EC_{50}: 20−30 μM), and 2-methylbutylamine (EC_{50}: 100 μM).[10] TAAR3 is not similarly activated by the corresponding

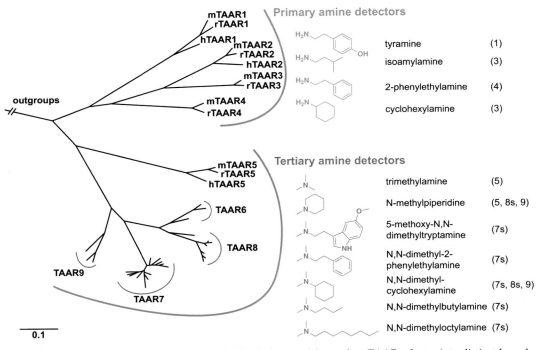

FIGURE 2   **Ligand preferences across the TAAR phylogeny.** Mammalian TAARs cluster into distinct branches that prefer primary amines (top) or tertiary amines (bottom). Zebrafish TAAR13c, which is not depicted, recognizes cadaverine and clusters phylogenetically with TAARs that detect primary amines. *Image adapted from Ref. 43.*

alcohols, isopentyl alcohol or cyclohexanol, and so the ligand amino group is essential for recognition.[10] Isopentylamine activates TAAR3 in HEK-293 cells, TAAR3-containing OSNs, and MOB glomeruli within the TAAR innervation domain (see below).[10,34,46] Rat TAAR3 also detects isopentylamine but does so with reduced sensitivity.[43] Isopentylamine and 2-methylbutylamine are decarboxylated amino acids derived from leucine and isoleucine, respectively, and may be components of natural odor sources such as rotting meat.[48] Isopentylamine was also proposed to function as a puberty-accelerating mouse pheromone,[49] although this effect was debated.[50] Isopentylamine evokes an innate aversion response in mice,[51] and this avoidance behavior is lost in knockout mice lacking the entire cluster of olfactory *Taars*.[46]

TAAR4

Mouse and rat TAAR4 detect 2-phenylethylamine (EC$_{50}$: 1–10 μM),[10,31,36] a phenylalanine derivative also produced by amino acid decarboxylation. Mouse TAAR4 is narrowly selective for 2-phenylethylamine and does not similarly respond to related chemicals such as benzylamine, phenylalanine, tyramine, dopamine, phenylethanol, or other phenylalanine metabolites.[31] In mouse, 2-phenylethylamine is detected by a sizable cohort of

OSNs ($\sim$5% at physiological levels),[31] with TAAR4-containing OSNs displaying unusually sensitive responses.[34] Furthermore, in the MOB, 2-phenylethylamine apparently evokes responses specific to the TAAR4 glomerulus (of glomeruli accessible for imaging),[46] perhaps due to lateral inhibition that suppresses responses in other glomeruli.

There are many natural odor sources where 2-phenylethylamine is found; it is particularly abundant in the urine of certain carnivores.[31] Mouse TAAR4 was found to be a direct and sensitive detector of carnivore odor, and chemical fractionation of bobcat urine as well as enzymatic depletion of lion urine revealed 2-phenylethylamine to be the natural TAAR4 agonist in each specimen.[31] Quantitative high-performance liquid chromatography (HPLC) studies across 38 zoo-derived specimens revealed a general enrichment of 2-phenylethylamine in many carnivore urines.[31] Some species (lions, servals, and tigers) produced large amounts of 2-phenylethylamine that were >100- to 1000-fold greater than in non-carnivore urines examined.[31] Innate avoidance responses are invoked by 2-phenylethylamine in mice and rats,[31] although high 2-phenylethylamine concentrations are required when this odor is presented individually. One possibility is that 2-phenylethylamine is a component of a predator odor blend. Consistent with this model, lion urine depleted of 2-phenylethylamine is a weaker repellant.[31] Mouse avoidance responses to 2-phenylethylamine are lost in *Taar4* knockout mice,[46] and these findings together indicate that TAAR4 functions as a kairomone receptor in mice (Figure 3(A)) that could facilitate avoidance of dangerous predators. Interestingly, TAAR4 is highly conserved in carnivore species,[52] where it may mediate a different behavioral response.

TAAR5

Trimethylamine is a high-affinity agonist of mouse (EC$_{50}$: 300 nM) and rat TAAR5 (EC$_{50}$: 1 μM) and a lower affinity agonist of macaque (EC$_{50}$: 100 μM) and human TAAR5 (EC$_{50}$: 100 μM).[10,17,45,47] Mouse TAAR5 also detects structurally similar tertiary amines (dimethylethylamine, *N*-methylpiperidine) with slightly reduced affinity, but has a striking preference for tertiary amines compared with related primary (methylamine), secondary (dimethylamine), or quaternary (tetramethylammonium chloride) amines.[10] Trimethylamine activates TAAR5 (from mouse, rat, human, and macaque) when expressed in HEK-293 cells as well as TAAR5-containing mouse OSNs.[10,47]

Trimethylamine is present in mouse urine-derived scent marks used for social behavior.[47] NMR spectroscopy indicates that trimethylamine is abundant (5—10 mM) and sexually dimorphic ($\sim$20-fold male enriched) in mouse urine.[47] Consistent with these findings, mouse TAAR5 is an exquisitely sensitive detector of male mouse urine (half maximal response occurs at $\sim$30,000-fold diluted urine),[10] and enzymatic depletion of trimethylamine from mouse urine eliminates TAAR5 responses.[47] Interestingly, abundant and sexually dimorphic trimethylamine production arose recently in the *Mus* lineage,[47] with some related rodent species, such as *Mus caroli*, failing to produce trimethylamine at detectable levels (<10 μM). Trimethylamine is initially generated by gut bacteria as a by-product of the metabolism of dietary nutrients such as choline.[53] Trimethylamine is then eliminated in many host species (including humans and rats) by a liver and kidney catabolism enzyme (flavin monooxygenase 3 or FMO3).[53] Mutations in FMO3 are the basis for the human disease trimethylaminuria, which causes a profoundly aversive body odor.[54] In *Mus musculus*, FMO3 expression is

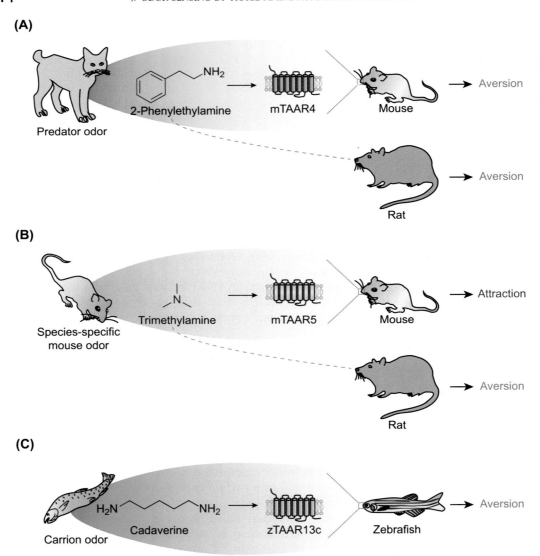

FIGURE 3 **Ecologically salient TAAR responses.** (A) TAAR4 detects the carnivore odor 2-phenylethylamine, which evokes aversion in mice and rats. (B) TAAR5 detects the male mouse odor trimethylamine, which evokes species-specific behavioral responses, including TAAR5-dependent attraction in mice. (C) TAAR13c detects the carrion odor cadaverine, which is aversive to zebrafish.

reduced in kidney, and strikingly sex-dependent in liver, with male-specific *Fmo3* gene repression decreasing expression levels over 1000-fold after puberty.[47] Thus, recent changes in *Fmo3* gene control in the *Mus* lineage provide a molecular basis for the evolution of a species-specific scent.

(A)

Dorsal   Ventral

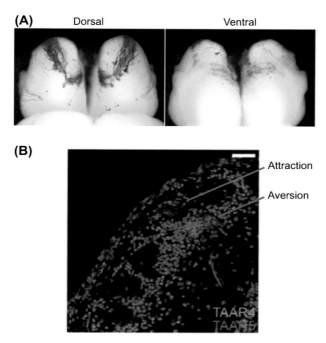

(B)

Attraction

Aversion

FIGURE 4   **Projections of TAAR-expressing OSNs in the main olfactory bulb.** (A) Dorsal and ventral views of MOBs in which all types of TAAR-expressing OSNs are genetically labeled with β-galactosidase. Briefly, the endogenous *Taar5* gene is replaced by *LacZ*, and choice of this allele, presumably like a *Taar* pseudogene, leads to secondary expression of other *Taar* genes. (B) TAAR4-expressing (green) and TAAR5-expressing (red) OSNs project to adjacent glomeruli in the MOB, as revealed by antibody staining. *Images are from Ref. 33.*

Trimethylamine evokes species-specific behaviors[47] (Figure 3(B)) and is profoundly aversive to humans and rats.[47,53] In contrast, mice are attracted to trimethylamine at levels found in scent depositions.[47] Furthermore, *Taar5* knockout mice lose behavioral attraction to trimethylamine.[47] It is possible that the ability of trimethylamine to repel other species provided the selective pressure for enriched production in urinary odor. It is unclear how trimethylamine evokes species-specific behavioral responses, which perhaps involves changes in the trimethylamine-activated receptor repertoire, changes in TAAR5-coupled neural circuitry, or learned override of a developmentally determined aversion circuit. It is also unclear how TAAR4 and TAAR5 mediate opposing behaviors, even though they are encoded by adjacent genes in the mouse genome and localize to adjacent glomeruli in the MOB (Figure 4(B)).[33]

TAAR7s

The TAAR7 subfamily was lost in humans, but has expanded in rodents (five in mice, seven in rats). The last common ancestor of mice and rats likely had a single *Taar7* gene,[43] so rodent TAAR7s are recently derived and highly related (>87% identical). Ligands were found for six TAAR7s (TAAR7b, TAAR7e, and TAAR7f in mouse; and TAAR7b, TAAR7d, and TAAR7h in rats), and each detects various N,N-dimethylalkylamines.[10,43] The mouse TAAR7e agonist is a reported urinary tryptophan metabolite, 5-methoxy-N,N-dimethyltryptamine,[43,55] but the ethological salience of this or other TAAR7 ligands is unclear. Mouse TAAR7f is directly activated by a urinary constituent common to several mammalian species,[31] with the chemical identity of the natural agonist unknown.

The TAAR7 subfamily contains highly related receptors with different ligand preferences, providing a unique opportunity to probe TAAR recognition properties.[43] For example, mouse TAAR7e and TAAR7f display strikingly distinct odor response profiles that are due to variations in two adjacent amino acid side chains that line the ligand-binding pocket on transmembrane α-helix 3.[43] Structural modeling of TAAR7e and TAAR7f, as well as mutagenesis studies, predicted that the bulkier side chain at one position in TAAR7f sterically clashes with larger molecules, and thus limits ligand size.[43] The TAAR7 subfamily provides a direct example of how the olfactory receptor repertoire can functionally expand through a pattern of gene duplication and subsequent mutation.

TAAR8c, TAAR9

Rat TAAR9 and rat TAAR8c are activated by the same agonists, N-methylpiperidine and N,N-dimethylcyclohexylamine, with TAAR8c showing higher affinity responses.[43] The ecological importance of these ligands is unclear, but both receptors also directly detect the urine of humans, mice, rats, and cougars.[31] The natural rat TAAR8c and rat TAAR9 ligand (or ligands) are unknown but likely not N-methylpiperidine or N,N-dimethylcyclohexylamine, as TAAR9 rather than TAAR8c shows enhanced sensitivity to mammalian urine. N-methylpiperidine is also a ligand for several other TAARs[43] and is aversive to mice, with responses lost in TAAR cluster knockouts.[46]

TAAR13c

Ligands are known for only one zebrafish TAAR, TAAR13c, which detects cadaverine and other diamines.[44] Structure-activity analysis indicates a preference for unbranched diamines of medium size and odd chain length.[44] Removing one amino group of the ligand, or replacing it with an alcohol, eliminates TAAR13c agonism, suggesting that both amine groups are specifically recognized in the ligand-binding pocket.[44] Zebrafish likely contain other olfactory diamine receptors that prefer even-chained diamines such as putrescine, based on results from odor cross-adaptation studies.[56] Cadaverine evokes immediate early gene expression in zebrafish OSNs, and at lower concentrations, the majority of responsive neurons contain TAAR13c.[44]

Cadaverine is a carrion odor produced by microbial metabolism of the amino acid lysine. Like other TAAR ligands, cadaverine reportedly evokes species-specific behavioral responses such as attraction in goldfish and aversion in zebrafish.[44,57] TAAR13c is an excellent candidate to mediate cadaverine aversion in zebrafish (Figure 3(C)), but loss-of-function studies are needed. Mice also avoid cadaverine, a behavior lost in *Taar* cluster knockouts.[46] However, TAAR13c is not conserved in mouse, and a cadaverine-detecting mouse TAAR has not been identified.

TAAR NEURON PROJECTIONS TO THE MAIN OLFACTORY BULB IN MOUSE

OSNs send long-range projections to the MOB, where they form synaptic connections with second-order neurons termed mitral and tufted cells. OSN synapses are localized to specialized spherical structures near the MOB surface termed glomeruli. There are ~2000 glomeruli in a mouse MOB, and each is a coalescence point for axons containing a single receptor type.[58,59] The position of a glomerulus for a particular OR is stereotyped in different individuals, and is determined, in part, by axon guidance cues whose expression is controlled by receptor-mediated cAMP signaling and OSN position.[60] Through this anatomical organization, the dispersed pattern of odorant evoked-neural activity in the MOE is converted into a highly organized, spatially stereotyped neural activity map in the MOB.

TAAR-expressing OSNs also project to the MOB with stereotypy.[32,33] Each class of TAAR-containing OSN innervates a small number of dedicated glomeruli,[32,33] with TAAR proteins localized to axon terminals.[33] Most TAAR projections are clustered in a hotspot in the dorsomedial MOB (Figure 4(A)),[32,33] including projections of TAAR neurons that mediate different behaviors (e.g., TAAR4 and TAAR5).[33] Glomeruli in the TAAR innervation zone of the MOB respond robustly to TAAR ligands identified in cell culture studies, with responses lost in *Taar* cluster knockouts.[46]

CONCLUSIONS AND FUTURE PERSPECTIVES

The TAAR olfactory subsystem provides a valuable model for understanding how a sensory system can drive divergent behaviors. Basic questions remain about how the TAAR olfactory subsystem forms and functions. What transcription factors, epigenetic determinants, and regulators of cell differentiation control the maturation of TAAR-expressing OSNs? What mechanisms underlie the formation of TAAR-coupled neural circuits in the MOB, and how are TAAR inputs organized in higher-order olfactory centers? How do TAAR4 and TAAR5 mediate different species-specific behaviors, what is different about TAAR4- and TAAR5-coupled neural circuits, and how are such differences established during development or learning? Future studies of olfactory TAARs may provide basic insights into how neural circuits process sensory inputs to ensure ethologically appropriate behavioral responses.

References

1. Buck L, Axel R. A novel multigene family may encode odorant receptors: a molecular basis for odor recognition. *Cell*. April 5, 1991;65(1):175–187.
2. Glusman G, Yanai I, Rubin I, Lancet D. The complete human olfactory subgenome. *Genome Res*. May 2001;11(5):685–702.
3. Godfrey PA, Malnic B, Buck LB. The mouse olfactory receptor gene family. *Proc Natl Acad Sci USA*. February 17, 2004;101(7):2156–2161.
4. Malnic B, Godfrey PA, Buck LB. The human olfactory receptor gene family. *Proc Natl Acad Sci USA*. February 24, 2004;101(8):2584–2589.
5. Young JM, Trask BJ. The sense of smell: genomics of vertebrate odorant receptors. *Hum Mol Genet*. May 15, 2002;11(10):1153–1160.

6. Zhang X, Firestein S. The olfactory receptor gene superfamily of the mouse. *Nat Neurosci*. February 2002;5(2):124–133.

7. Sullivan SL, Adamson MC, Ressler KJ, Kozak CA, Buck LB. The chromosomal distribution of mouse odorant receptor genes. *Proc Natl Acad Sci USA*. January 23, 1996;93(2):884–888.

8. Malnic B, Hirono J, Sato T, Buck LB. Combinatorial receptor codes for odors. *Cell*. March 5, 1999;96(5):713–723.

9. Liberles SD. Trace amine-associated receptors: ligands, neural circuits, and behaviors. *Curr Opin Neurobiol*. January 20, 2015;34C:1–7.

10. Liberles SD, Buck LB. A second class of chemosensory receptors in the olfactory epithelium. *Nature*. August 10, 2006;442(7103):645–650.

11. Lindemann L, Ebeling M, Kratochwil NA, Bunzow JR, Grandy DK, Hoener MC. Trace amine-associated receptors form structurally and functionally distinct subfamilies of novel G protein-coupled receptors. *Genomics*. March 2005;85(3):372–385.

12. Hussain A, Saraiva LR, Korsching SI. Positive Darwinian selection and the birth of an olfactory receptor clade in teleosts. *Proc Natl Acad Sci USA*. March 17, 2009;106(11):4313–4318.

13. Libants S, Carr K, Wu H, et al. The sea lamprey *Petromyzon marinus* genome reveals the early origin of several chemosensory receptor families in the vertebrate lineage. *BMC Evol Biol*. 2009;9:180.

14. Tessarolo JA, Tabesh MJ, Nesbitt M, Davidson WS. Genomic organization and evolution of the trace amine-associated receptor (TAAR) repertoire in Atlantic salmon (*Salmo salar*). *G3 (Bethesda)*. June 2014;4(6):1135–1141.

15. Lindemann L, Hoener MC. A renaissance in trace amines inspired by a novel GPCR family. *Trends Pharmacol Sci*. May 2005;26(5):274–281.

16. Hashiguchi Y, Nishida M. Evolution of trace amine associated receptor (TAAR) gene family in vertebrates: lineage-specific expansions and degradations of a second class of vertebrate chemosensory receptors expressed in the olfactory epithelium. *Mol Biol Evol*. September 2007;24(9):2099–2107.

17. Horowitz LF, Saraiva LR, Kuang D, Yoon KH, Buck LB. Olfactory receptor patterning in a higher primate. *J Neurosci*. September 10, 2014;34(37):12241–12252.

18. Nei M, Niimura Y, Nozawa M. The evolution of animal chemosensory receptor gene repertoires: roles of chance and necessity. *Nat Rev*. December 2008;9(12):951–963.

19. Fleischer J, Schwarzenbacher K, Breer H. Expression of trace amine-associated receptors in the Grueneberg ganglion. *Chem Senses*. July 2007;32(6):623–631.

20. Lindemann L, Meyer CA, Jeanneau K, et al. Trace amine-associated receptor 1 modulates dopaminergic activity. *J Pharmacol Exp Ther*. March 2008;324(3):948–956.

21. Regard JB, Sato IT, Coughlin SR. Anatomical profiling of G protein-coupled receptor expression. *Cell*. October 31, 2008;135(3):561–571.

22. Ressler KJ, Sullivan SL, Buck LB. A zonal organization of odorant receptor gene expression in the olfactory epithelium. *Cell*. May 7, 1993;73(3):597–609.

23. Vassar R, Ngai J, Axel R. Spatial segregation of odorant receptor expression in the mammalian olfactory epithelium. *Cell*. July 30, 1993;74(2):309–318.

24. Serizawa S, Miyamichi K, Nakatani H, et al. Negative feedback regulation ensures the one receptor-one olfactory neuron rule in mouse. *Science (New York, NY)*. December 19, 2003;302(5653):2088–2094.

25. Shykind BM, Rohani SC, O'Donnell S, et al. Gene switching and the stability of odorant receptor gene choice. *Cell*. June 11, 2004;117(6):801–815.

26. Hirota J, Mombaerts P. The LIM-homeodomain protein Lhx2 is required for complete development of mouse olfactory sensory neurons. *Proc Natl Acad Sci USA*. June 8, 2004;101(23):8751–8755.

27. Lyons DB, Allen WE, Goh T, Tsai L, Barnea G, Lomvardas S. An epigenetic trap stabilizes singular olfactory receptor expression. *Cell*. July 18, 2013;154(2):325–336.

28. Magklara A, Yen A, Colquitt BM, et al. An epigenetic signature for monoallelic olfactory receptor expression. *Cell*. May 13, 2011;145(4):555–570.

29. Nishizumi H, Kumasaka K, Inoue N, Nakashima A, Sakano H. Deletion of the core-H region in mice abolishes the expression of three proximal odorant receptor genes in cis. *Proc Natl Acad Sci USA*. December 11, 2007;104(50):20067–20072.

30. Wang SS, Lewcock JW, Feinstein P, Mombaerts P, Reed RR. Genetic disruptions of O/E2 and O/E3 genes reveal involvement in olfactory receptor neuron projection. *Development*. March 2004;131(6):1377–1388.

31. Ferrero DM, Lemon JK, Fluegge D, et al. Detection and avoidance of a carnivore odor by prey. *Proc Natl Acad Sci USA*. July 5, 2011;108(27):11235–11240.

32. Pacifico R, Dewan A, Cawley D, Guo C, Bozza T. An olfactory subsystem that mediates high-sensitivity detection of volatile amines. *Cell Rep*. July 26, 2012;2(1):76–88.

33. Johnson MA, Tsai L, Roy DS, et al. Neurons expressing trace amine-associated receptors project to discrete glomeruli and constitute an olfactory subsystem. *Proc Natl Acad Sci USA*. August 14, 2012;109(33):13410–13415.

34. Zhang J, Pacifico R, Cawley D, Feinstein P, Bozza T. Ultrasensitive detection of amines by a trace amine-associated receptor. *J Neurosci*. February 13, 2013;33(7):3228–3239.

35. Shi L, Javitch JA. The binding site of aminergic G protein-coupled receptors: the transmembrane segments and second extracellular loop. *Annu Rev Pharmacol Toxicol*. 2002;42:437–467.

36. Borowsky B, Adham N, Jones KA, et al. Trace amines: identification of a family of mammalian G protein-coupled receptors. *Proc Natl Acad Sci USA*. July 31, 2001;98(16):8966–8971.

37. Bunzow JR, Sonders MS, Arttamangkul S, et al. Amphetamine, 3,4-methylenedioxymethamphetamine, lysergic acid diethylamide, and metabolites of the catecholamine neurotransmitters are agonists of a rat trace amine receptor. *Mol Pharmacol*. December 2001;60(6):1181–1188.

38. Reese EA, Norimatsu Y, Grandy MS, Suchland KL, Bunzow JR, Grandy DK. Exploring the determinants of trace amine-associated receptor 1's functional selectivity for the stereoisomers of amphetamine and methamphetamine. *J Med Chem*. January 23, 2014;57(2):378–390.

39. Scanlan TS, Suchland KL, Hart ME, et al. 3-Iodothyronamine is an endogenous and rapid-acting derivative of thyroid hormone. *Nat Med*. June 2004;10(6):638–642.

40. Bradaia A, Trube G, Stalder H, et al. The selective antagonist EPPTB reveals TAAR1-mediated regulatory mechanisms in dopaminergic neurons of the mesolimbic system. *Proc Natl Acad Sci USA*. November 24, 2009;106(47):20081–20086.

41. Revel FG, Moreau JL, Gainetdinov RR, et al. TAAR1 activation modulates monoaminergic neurotransmission, preventing hyperdopaminergic and hypoglutamatergic activity. *Proc Natl Acad Sci USA*. May 17, 2011;108(20):8485–8490.

42. Panas HN, Lynch LJ, Vallender EJ, et al. Normal thermoregulatory responses to 3-iodothyronamine, trace amines and amphetamine-like psychostimulants in trace amine associated receptor 1 knockout mice. *J Neurosci Res*. July 2010;88(9):1962–1969.

43. Ferrero DM, Wacker D, Roque MA, Baldwin MW, Stevens RC, Liberles SD. Agonists for 13 trace amine-associated receptors provide insight into the molecular basis of odor selectivity. *ACS Chem Biol*. July 20, 2012;7(7):1184–1189.

44. Hussain A, Saraiva LR, Ferrero DM, et al. High-affinity olfactory receptor for the death-associated odor cadaverine. *Proc Natl Acad Sci USA*. November 26, 2013;110(48):19579–19584.

45. Wallrabenstein I, Kuklan J, Weber L, et al. Human trace amine-associated receptor TAAR5 can be activated by trimethylamine. *PLoS One*. 2013;8(2):e54950.

46. Dewan A, Pacifico R, Zhan R, Rinberg D, Bozza T. Non-redundant coding of aversive odours in the main olfactory pathway. *Nature*. May 23, 2013;497(7450):486–489.

47. Li Q, Korzan WJ, Ferrero DM, et al. Synchronous evolution of an odor biosynthesis pathway and behavioral response. *Curr Biol*. January 7, 2013;23(1):11–20.

48. Barger G, Walpole GS. Isolation of the pressor principles of putrid meat. *J Physiol*. March 22, 1909;38(4):343–352.

49. Nishimura K, Utsumi K, Yuhara M, Fujitani Y, Iritani A. Identification of puberty-accelerating pheromones in male mouse urine. *J Exp Zool*. September 1989;251(3):300–305.

50. Price MA, Vandenbergh JG. Analysis of puberty-accelerating pheromones. *J Exp Zool*. October 15, 1992;264(1):42–45.

51. Kobayakawa K, Kobayakawa R, Matsumoto H, et al. Innate versus learned odour processing in the mouse olfactory bulb. *Nature*. November 22, 2007;450(7169):503–508.

52. Staubert C, Boselt I, Bohnekamp J, Rompler H, Enard W, Schoneberg T. Structural and functional evolution of the trace amine-associated receptors TAAR3, TAAR4 and TAAR5 in primates. *PLoS One*. 2010;5(6):e11133.

53. Mitchell SC, Smith RL. Trimethylaminuria: the fish malodor syndrome. *Drug Metab Dispos*. April 2001;29(4 Pt 2):517–521.

54. Dolphin CT, Janmohamed A, Smith RL, Shephard EA, Phillips IR. Missense mutation in flavin-containing mono-oxygenase 3 gene, FMO3, underlies fish-odour syndrome. *Nat Genet*. December 1997;17(4):491–494.

55. Sitaram BR, Lockett L, Blackman GL, McLeod WR. Urinary excretion of 5-methoxy-N,N-dimethyltryptamine, N,N-dimethyltryptamine and their N-oxides in the rat. *Biochem Pharmacol*. July 1, 1987;36(13):2235–2237.

56. Michel WC, Sanderson MJ, Olson JK, Lipschitz DL. Evidence of a novel transduction pathway mediating detection of polyamines by the zebrafish olfactory system. *J Exp Biol*. May 2003;206(Pt 10):1697–1706.

57. Rolen SH, Sorensen PW, Mattson D, Caprio J. Polyamines as olfactory stimuli in the goldfish *Carassius auratus*. *J Exp Biol*. May 2003;206(Pt 10):1683—1696.

58. Ressler KJ, Sullivan SL, Buck LB. Information coding in the olfactory system: evidence for a stereotyped and highly organized epitope map in the olfactory bulb. *Cell*. December 30, 1994;79(7):1245—1255.

59. Vassar R, Chao SK, Sitcheran R, Nunez JM, Vosshall LB, Axel R. Topographic organization of sensory projections to the olfactory bulb. *Cell*. December 16, 1994;79(6):981—991.

60. Sakano H. Neural map formation in the mouse olfactory system. *Neuron*. August 26, 2010;67(4):530—542.

61. Li Q, Liberles SD. Aversion and attraction through olfaction. *Curr Biol*. February 2, 2015;25(3):R120—R129.

Aquatic Olfaction

Sigrun Korsching

Institute of Genetics, Biocenter, University at Cologne, Cologne, Germany

81

INTRODUCTION

The vertebrate olfactory system, like the independently developed olfactory system of arthropods, has its origins in aquatic species, and the first vertebrate olfactory receptor genes evolved to detect water borne substances (see Chapter 12 for a comparative discussion of olfactory transduction). Those receptors and receptor families, which were kept and expanded in terrestrial olfaction, initially must have had properties useful for aquatic olfaction. Currently, not enough is known about ligand specificities of aquatic olfactory receptors to understand this transition.

Yet, because of the recent availability of sequenced genomes for many aquatic species, there is a large body of knowledge about the evolution of olfactory receptor repertoires in aquatic vertebrates. Moreover, there exists a wealth of information about aquatic odors, odor-induced neuronal activity, and innate behavioral odor responses. Aquatic vertebrates ranging from jawless fish to teleost fish and amphibians can detect a wide array of chemically diverse odors with remarkable specificity and sensitivity and can exhibit complex odor-guided social behaviors including the timing of reproductive stages in female—male interaction.

To inform thinking about receptor/ligand pairing in aquatic olfaction, this chapter will discuss the evolution of all four main classes of olfactory receptors, beginning with a short overview of cephalochordate, jawless vertebrate, and cartilaginous vertebrate olfactory receptor repertoires. Next, the focus will be on coelacanths (*Latimeria*) as representative of lobe-finned fish, teleost fish as the major group of aquatic species, and clawed frog (*Xenopus*) as representative of amphibians, which are situated at the transition point between aquatic and terrestrial olfaction (Figure 1). Then the consequences of the inverse transition in secondarily aquatic mammals will be described.

An overview of aquatic odors known to activate olfactory sensory neurons (OSNs) and/or elicit olfactory-mediated behavior will be followed by a description of odor ligands for the handful of aquatic olfactory receptors deorphanized so far. Finally, aquatic olfactory signal transduction will be discussed and the different types of OSNs will be described. Two of the olfactory receptor families have as their closest relatives taste receptor families; the characteristic properties of these taste receptor families will be described in the breakout box.

THE EVOLUTIONARY ORIGIN OF VERTEBRATE OLFACTORY RECEPTOR GENES

The vertebrate olfactory system has its origins in aquatic species. The very first olfactory receptor genes had to detect water borne substances; consequently, the evolution of aquatic olfaction stretches over a much larger evolutionary period than that of airborne olfaction.

In fact, the origin of one of the four main olfactory receptor families, odorant receptors (ORs; see Chapter 3) predates the genesis of vertebrates. A family of about 50 OR-like genes was identified in the genome of amphioxus, a cephalochordate,[1,2] and expression in presumptive chemosensory cells of amphioxus was demonstrated.[3,4]

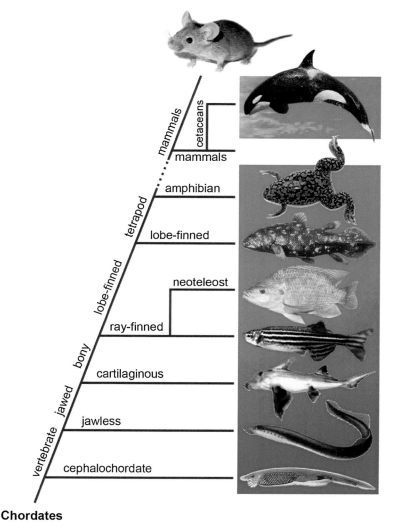

Chordates

FIGURE 1 **Aquatic species in chordate evolution.** A phylogenetic tree of chordate species. Pictures at the branches depict exemplary species for which the olfactory receptor repertoires have been studied in some detail: Cephalochordates, amphioxus; jawless vertebrates, lamprey; cartilaginous fish, elephant shark; earlier diverging ray-finned fish, zebrafish; neoteleosts, tilapia; lobe-finned fish, coelacanth; amphibian, *Xenopus*; cetaceans, *Orca* (representing several toothed and baleen whales); mammals, and mouse. Aquatic species are indicated by the blue background.

The amphioxus OR gene family lies basal to all vertebrate OR genes in the phylogenetic analysis,[2] and thus outside of the nomenclature suggested for vertebrate ORs.[5] Greek letters from α to η are used to designate the subfamilies, and the last subfamily (η) is classified as type 2 because of its larger phylogenetic distance from the remaining six subfamilies.

Originally, three further subfamilies were proposed but these were later found to be nonolfactory.[2]

Within the vertebrate lineage, OR genes have been found both in cartilaginous and jawless fish. Before whole genome sequences became available, an early study by Freitag et al.[6] had identified four lamprey OR genes. In total, about 50 lamprey OR genes are currently known. These group into three subfamilies, among them subfamily η with three genes, and another subfamily basal to all type 1 jawed vertebrate ORs.[7,8] Other putative OR genes in lamprey[9] turned out to be aminergic receptors, and as such most closely related to the second main family, trace amine-associated receptors (TAARs; see Chapter 4). Although they are sometimes considered TAARs,[7,10] these lamprey aminergic receptors are not monophyletic with mammalian, teleost, and shark TAARs, whose origin appears to be in the most recent common ancestor (MRCA) of cartilaginous and bony fish.[11]

The other two main families are the vomeronasal receptors type 1 (V1R) and type 2 (V2R; see Chapter 10). V2Rs belong to the G protein−coupled receptor (GPCR) class C, whereas ORs, TAARs, and V1Rs are all class A GPCRs. Four *Vmn1r* genes (encoding the V1Rs) have been identified in lamprey,[12,13] but none has been detected in amphioxus.[14] Thus, the V1R family appears to have its origin in the MRCA of jawed and jawless fish. The V2R family (encoded by the *Vmn2r* genes) is evolutionarily somewhat younger, as no representatives could be found in lamprey. However, a substantial group of *Vmn2r* genes was identified in elephant shark.[12]

Summarized according to evolutionary age: OR genes evolved in the MRCA of cephalochordates and vertebrates; *Vmn1r* genes originated in the MRCA of jawed and jawless vertebrates; and *Vmn2r* and *Taar* genes originated in the MRCA of cartilaginous and bony fish.

AQUATIC VERSUS TERRESTRIAL OLFACTION

The transition between aquatic and airborne olfaction has fascinated people for a long time. The difference between the respective odorants is extreme, with little overlap. The main odor classes in fish are charged molecules (amino acids, bile acids, nucleotides, steroids, and prostaglandins), whereas airborne odors of necessity have to be volatile and thus uncharged. Nevertheless, recognition of both types of odors is performed by the orthologous systems in all vertebrates, and indeed main principles of olfactory coding are preserved between zebrafish, the most commonly studied representative of teleost fish,[15,16] and mammals.[17]

However, there also exist quite a few differences beyond the disparity in odor classes, most notably the large size differences of orthologous receptor gene repertoires. Most families exhibit a much larger size, at least in some mammals, as compared with fish, but there is also a family for which the inverse holds true: the *Taar* genes.[11] An obvious morphological difference concerns the segregation of the mammalian olfactory system into several, molecularly and functionally different sensory subsystems,[18] whereas teleost fish possess a single sensory surface, in which at least four different types of OSNs intermingle.[19−21]

It appeared tempting to link these differences to the transition from aquatic to airborne olfaction. However, the evolutionary distance between zebrafish and mouse is considerable, with the MRCA living 0.43 billion years ago.[22] And, indeed, a comparison between aquatic tetrapods (the amphibian African clawed frog) and terrestrial tetrapods (mammals) shows somewhat larger similarities between olfactory characteristics of these two groups than between zebrafish and frog. However, amphibians represent a specialized case with their transition from larval aquatic to adult terrestrial life (the clawed frog is secondarily aquatic also as an adult). It would appear more straightforward to compare lobe-finned with ray-finned fish because tetrapods arose from the lobe-finned lineage. Very few fish species of that lineage have survived to the present time, among them lungfish and coelacanths.

Initial analyses of lobe-finned fish found many characteristics previously thought to potentially represent adaptations to terrestrial lifestyle, such as the segregation of the olfactory sensory surface into main and accessory subsystems (lungfish[23]), and the presence of large, rapidly evolving olfactory and gustatory receptor gene families (coelacanth[24−26]). Interestingly, the segregation of the main and accessory olfactory systems could be the evolutionarily older feature because it has already been reported in jawless fish,[27,28] a much more ancestral lineage than either lobe- or ray-finned fish (Figure 1). However, the morphology of the accessory olfactory organs appears very different between lamprey, lungfish, and amphibians/mammals, so an independent evolutionary origin of these structures should also be considered. In this case, the propensity to segregate into olfactory subsystems emerges as the ancient feature, which was lost or not realized in ray-finned fish (but see elsewhere[29]).

EVOLUTIONARY DYNAMICS OF TELEOST FISH OLFACTORY RECEPTOR GENE REPERTOIRES ARE DISTINCTLY DIFFERENT FROM THOSE OF TETRAPODS

The OR Gene Family of Ray-Finned Fish is Smaller but More Divergent than That of Tetrapods

The genomes of many teleost fish species have been scrutinized for OR genes, among them zebrafish, stickleback, medaka, fugu, tetraodon, salmon, and several cichlid species.[8,30−33] Furthermore, transcriptomes of olfactory organs have been analyzed for some species without a published genome (e.g., for goldfish and eel).[34,35] OR family sizes range between 50 and 150 genes, with zebrafish and tilapia, a cichlid, at the upper range and pufferfish at the lower range.[30,32] In comparison, tetrapod OR repertoires contain between a few hundred and a few thousand genes.[36]

Teleost OR gene families are highly divergent and encompass members of four subfamilies not present in mammals — δ, ε, ζ, and η — plus another subfamily (β), which is shared with mammals.[8] For explanation of nomenclature see Ref. 8. Subfamily η is a sister clade to all other ORs, but appears to contain bona fide olfactory receptors: expression in the olfactory organ has been shown for a zebrafish OR (OR137-7) and several *Xenopus* ORs from this subfamily.[30,37]

Subfamily γ, which contains the lion's share of mammalian ORs, is only present with a single member in an early diverging teleost, zebrafish, and subfamily α is absent from teleost genomes altogether.[8] Alpha and γ correspond to the mammalian classes I and II, respectively. These families were also called fish-like and mammalian-like receptors. The latter designation does not reflect newer phylogenies and preferably should not be used any longer. The majority of fish ORs belong to subfamilies δ to η, only very few fish OR genes belong to subfamily β, the sister group of α, and no fish genes belong to subfamily α itself.[8]

High Evolutionary Dynamics in Teleost *Taar* Gene Repertoires

TAARs evolved from aminergic receptors and represent one of four GPCR families that serve as olfactory receptors in teleosts. Cartilaginous fish are the earliest diverging species that possesses TAARs as defined by phylogenetic grouping and the presence of the characteristic TAAR fingerprint motif.[38] A sister gene group in lamprey[10] lacks this motif and phylogenetic analysis suggests an independent origin, albeit also from aminergic receptors.[11]

Two to three ancestral *Taar* genes were identified,[11] with nearly no gene duplications in amphibians and birds and no major radiation in mammals. In contrast, massive gene expansions are seen in several teleost fish species, particularly in stickleback and zebrafish.[11,39] Well over 100 *Taar* genes have been identified in the zebrafish genome, nearly rivaling the number of ORs in that species. Thus of the four major GPCR families of olfactory receptors, the TAAR family is the only one which shows larger evolutionary dynamics in the ray-finned lineage. Correspondingly, an analysis of synonymous to nonsynonymous mutation rate suggested pronounced positive selection in some teleost *Taar* clades, but (nearly) no positive selection in tetrapod *Taar* clades.[11]

A single *Taar* gene, *Taar1*, has not acquired an olfactory function and does not contribute to the evolutionary dynamics of the TAAR family. *Taar1* is highly conserved and direct orthologs are present in species from shark to zebrafish to mouse.[11]

Several rodent TAARs (see Chapter 4 for an in-depth discussion of mammalian TAARs) and one zebrafish TAAR[40] have been shown to be highly specific amine receptors. All possess the so-called aminergic ligand motif, a strong predictor of amine ligands.[41] However, three-fourths of the teleost fish TAARs have lost this motif, consistent with the hypothesis that the large majority of fish TAARs do not recognize amines but rather an as-yet unknown set of ligands. Positive selection is strongest in this group of TAARs, suggesting a highly dynamic evolution.[11]

The V1R-Related *Ora* Gene Family Is Highly Conserved in Teleosts

Mammalian *Vmn1r* genes show a rather dynamic evolution,[42] in striking contrast to the highly conserved orthologous *ora* gene family.[43] Six ancestral *ora* genes can be identified in many teleost species, with rare gene losses or duplications.[33,43,44] The origin of all six ancestral genes can be placed at least in the MRCA of lobe-finned and ray-finned fish because direct orthologs are found for all six genes in coelacanths.[24,26] The family itself is much older, as discussed previously.

A nearly constant size of the receptor repertoire is extremely unusual for olfactory receptor families and is accompanied by generally high sequence conservation and absence of positive selection in *ora* genes.[43] In special cases, such as the rapid speciation observed in East African cichlids, evidence for positive selection has been obtained[45] (see also previous work[46]). Some ORAs might serve as reproductive pheromone receptors,[47] and rapid evolution of such receptors could assist speciation.

A pairwise configuration in the phylogenetic tree suggests the existence of three ORA subclades, one of which—the *ora1/ora2* gene pair—encompasses all mammalian *Vmn1rs* (the other two gene pairs were lost in mammals). This is reminiscent of the situation for ORs, where fish exhibit smaller gene numbers but a much larger diversity in their gene repertoire. These phylogenetic pairs are reflected in the genomic arrangement: *ora1/ora2* and *ora3/ora4* are arranged in closely linked gene pairs across all fish species studied. The gene pairs are asymmetrical, head-to-head for *ora1/ora2* and tail-to-tail for *ora3/ora4*. Such genomic arrangement may be relevant for gene regulation; notably, it is not conserved in coelacanths, which show several gene duplications in the *ora1/ora2* clade.[26]

Another unexpected feature is the presence of introns in *ora3* and *ora4*, with intron borders conserved in all teleost fish analyzed.[43] These introns were gained early within the ray-finned lineage.[26] Nevertheless, all six *ora* genes are expressed in sparse OSNs,[43,48] as expected for bona fide olfactory receptors.

The V2R-Related *olfC* Gene Family Shows Some Species-Specific Expansions

Tetrapod V2R families show extreme differences in size up to several 100 genes in an amphibian and down to a complete loss of this family in some species.[49] The corresponding teleost OlfC families have considerably fewer gene birth and death events, although several species-specific expansions are present. Zebrafish, bitterling, and a cichlid are at the high end of the range with 50−60 *olfC* genes,[50−53] not much different from the repertoire size in several mammalian species. Salmon, stickleback, and fugu have about 30 genes,[54] and medaka and *Tetraodon* between 10 and 20 genes.[51,52]

Two *olfC* genes are highly conserved and their orthologs can be found in all teleost species analyzed. One is OlfCc1, which in phylogenetic trees reliably lies basal to the main OlfC family (see previous work[53]) and thus may be considered the ancestral gene from which all but one *olfC* gene originated. Interestingly, this gene has a direct mammalian ortholog, V2R2, in contrast to all genes from the main OlfC family. The other gene is OlfCa1, originally called 5.24, which also has tetrapod orthologs (e.g., GPRC6a).[55] OlfCa1 is not monophyletic with the other OlfC genes because of the calcium-sensing receptor intervening between OlfCa1 and OlfCc1 in the phylogenetic tree.[50]

Both OlfCa1 and OlfCc1 are expressed broadly[56,57] and are coexpressed with members of the main OlfC family. A knockdown of OlfCc1 has shown this gene to be involved in trafficking and/or function of other OlfC receptors.[57] It is conceivable that at least OlfCc1 and possibly also OlfCa1 form heterodimers with OlfCs from the main family, similar to the heterodimers found for the closely related type 1 taste receptor family[58] (see Chapter 13).

BREAKOUT BOX

TASTE RECEPTOR FAMILIES T1R AND T2R ARE CLOSELY RELATED TO V2R AND V1R, RESPECTIVELY

The sense of taste is very different from olfaction in terms of physiological function. Taste lumps stimuli together in very few categories, whereas olfaction is much more of an analytical sense. Presumably this is valid for aquatic vertebrates as well, even though olfaction and taste make use of the same medium in this case.

Interestingly, two families of taste receptors have as their closest homologs two families of olfactory receptors: for type 1 taste receptors (T1Rs), these are the V2Rs/OlfCs, whereas for type 2 taste receptors (T2Rs), the olfactory sister group is the V1R/ORA gene family. Phylogenetic considerations suggest that the gustatory receptors are evolutionarily younger and have derived from the respective olfactory receptor families.[12]

The T1R family consists of three genes in tetrapods,[129] two of which are present as single genes also in fish. The third gene, T1R2, has expanded to 2–10 genes in different teleost fish species.[25,60,130,131] Thus the T1R family shows a higher evolutionary dynamic in teleosts.

In contrast, the T2R family undergoes very dynamic evolution in tetrapods, with as few as two genes in some avian species and more than 50 genes in an amphibian.[132] The T2R repertoires of ray-finned fish tend to be very small, from three to six genes in a species.[133,134]

It has been argued that terrestrial lifestyles might require larger T2R repertoires. However, the recent identification of a huge T2R repertoire of 80 genes in a lobe-finned fish[26] shows large repertoires and dynamic evolution as characteristic properties of the lobe-finned lineage, but not as a difference between aquatic and land-living species.

Mammalian, avian, and amphibian T2Rs detect bitter substances, which are assumed to signal potential toxicity (see Chapter 13). A single teleost fish T2R has been deorphanized so far and was found to recognize classical bitter substances.[135] Herbivores generally have larger T2R repertoires than carnivores, possibly to avoid a multitude of toxic plant metabolites.[136] Coelacanths prey on various fish and cephalopods,[137] and some cephalopod species are indeed known to harbor potent toxins.[138] One could speculate that the avoidance of such toxins might require a sophisticated detection system with a large number of receptors.

Detection of amino acids by T1R1:T1R3 heterodimers appears to be conserved between aquatic and terrestrial vertebrates, where the taste quality is referred to as umami.[58,135] Interestingly, fish T1R2:T1R3 heterodimers also recognize amino acids[135]; this contrasts with mammals, where this receptor responds to sugars.[58]

OLFACTORY RECEPTOR GENE REPERTOIRES OF A LOBE-FINNED FISH COMBINE TETRAPOD AND TELEOST CHARACTERISTICS

Because tetrapods arose from the lobe-finned lineage, lobe-finned fish are a suitable point of reference to evaluate previously identified differences between mammalian and teleost olfactory systems (Figure 1). Are such differences present because of an aquatic versus terrestrial lifestyle, or do they reflect differences between lobe-finned and ray-finned lineages?

Recently, the coelacanth genome became available,[24,59] and several studies have analyzed the chemosensory receptor repertoires of this lobe-finned fish. As expected, all four olfactory receptor gene families are present, but there are notable differences as compared with both ray-finned fish and tetrapods.

Coelacanths possess a substantial OR repertoire of close to 300 genes,[24] which is much larger than family sizes reported for ray-finned fish and similar in size to an avian repertoire.[8] The coelacanth OR family is very diverse; it contains the tetrapod-specific subfamilies α and γ in addition to all subfamilies found in ray-finned fish.[24] By far the largest expansions are in the fish-like receptor subfamilies, whereas subfamilies α and γ show only moderate gene expansion.[24] Thus, the tendency toward massive gene duplication is more reminiscent of tetrapod OR repertoires, but the substrate of these gene expansions are mostly the fish-like, not the tetrapod-like receptors of coelacanths.

An initial report shows a small TAAR repertoire of only four genes,[25] which is reminiscent of the minute repertoires found in frog and chicken.

The V1R repertoire of coelacanths consists of 20 genes, among them orthologs for all six *ora* genes of teleosts. In addition, several small gene expansions are found within the *ora1/ora2* clade, reminiscent of such expansions in the tetrapod lineage.[24,25] Thus, the V1R repertoire of coelacanths combines characteristics of the ray-finned lineage (the six ancestral genes *Vmn1r1-6*) and of tetrapods (gene expansions in the *Vmn1r1/Vmn1r2* clade).

Coelacanths possess about 60 V2R/OlfC receptors, among them about 10 fish-like genes and one large expansion grouping with tetrapod V2Rs. The only two genes conserved between ray-finned fish (*OlfCa1* and *OlfCc1*)[50] and tetrapods (*GPCR6a*[60] and *Vmn2r1*[61]) are also present.[25]

Taken together, the olfactory receptor gene families of coelacanth show both features of the tetrapod repertoires and of the ray-finned fish repertoires. The former can be assumed to be lineage-specific, whereas the latter may be necessitated by the aquatic habitat. Thus, many features of tetrapod repertoires previously thought to be an adaptation to the water-to-land transition are preformed already in the aquatic stage. It will be a challenging endeavor to understand their evolutionary benefits in the aquatic environment.

AMPHIBIAN OLFACTION IN TRANSITION FROM AQUEOUS TO TERRESTRIAL ENVIRONMENT

Amphibians have been used as model organisms for developmental studies for a long time[62] and are also interesting in evolutionary terms as earlier diverging vertebrates compared to mammals (Figure 1). In the context of aquatic versus terrestrial olfaction they occupy a transition point, as their tadpoles are fully aquatic, whereas postmetamorphotic frogs are either terrestrial or secondarily aquatic.

Larval amphibians possess an accessory olfactory system, the vomeronasal organ (VNO), which does not seem to change much during metamorphosis,[63] as well as a main olfactory organ. The main olfactory organ is water-filled during larval stages, but metamorphoses into the air-filled, so-called principal cavity of adult frogs. This change is accompanied by massive apoptotic cell death, suggesting that presumably all larval OSNs are lost post metamorphosis.[63] Furthermore, during metamorphosis a third organ forms de novo: the so-called middle cavity, which is water-filled and contains ciliated and microvillous OSNs in the case of secondarily aquatic amphibians such as the clawed frog, *Xenopus laevis*.[64] The principal cavity possesses only ciliated OSNs, whereas the tadpole olfactory organ contains both ciliated and microvillous neurons, resembling in this respect the middle cavity and the single olfactory organ of teleost fish.

The genome of *Xenopus tropicalis*, a close relative of *Xenopus laevis*, has recently become available, and the olfactory receptor gene families have been studied in some detail. *Xenopus* OR genes number around 1000, which is several-fold larger than many primate OR repertoires.[8] *Xenopus* possesses six of the seven OR subfamilies, which include all subfamilies shared between fish and mammals, the mammalian-specific one, and all but one of the fish-specific subfamilies (ζ is missing). This high divergence appears to be close to the ancestral situation in the lobe-finned lineage as reflected in the coelacanth OR repertoire, which consists of all seven subfamilies. In contrast to coelacanths (see the previous section), the main gene expansion occurred in the γ subfamily (i.e., class II), similar to the situation in mammals. Thus, the *Xenopus* OR repertoire is in transition between aquatic and terrestrial olfaction: the divergence of the repertoire is large, as seen in aquatic species, but the family size and the main expansion are similar to that observed in terrestrial species.

The *Xenopus* TAAR family, on the other hand, has a single, paltry gene expansion for one of the two ancestral genes,[11] resulting in a family size of five genes.[65] The other ancestral gene, *Taar1*, is present but is not expressed in the olfactory organ, as is true for other species.[65] The spatial distribution of TAAR-expressing cells in the olfactory organ is very close to that of amine-responsive cells, suggesting that *Xenopus* TAARs may mediate at least some of the amine responses.[65]

The V1R family of *Xenopus* has lost three of the six ancestral genes—V1R4, 5, 6 of coelacanths,[24,26] corresponding to ORA4, 5, and 6 of teleost fish.[43] On the other hand, within the *Ora1/Ora2* clade, *Xenopus* has gained about a dozen genes in a single gene expansion, and it has gained a single gene, *Xt_ora14*, which is ancestral to all mammalian genes.[26,43] In other words, mammals have lost all six ancestral genes present in coelacanths and teleosts and have evolved their sometimes impressively large gene repertoires from a single gene first occurring in amphibians.

Although *Xenopus* already possess a VNO, its *Vmn1r* genes are expressed in the main olfactory epithelium (MOE),[66] like teleost *ora* genes, and unlike mammalian V1Rs, which are expressed in the VNO. Thus, the expression of *Vmn1r* genes migrated during evolution of terrestrial olfaction to the accessory olfactory organ *after* the organ formed.

The V2R family of *Xenopus* is the largest ever reported, with close to 350 members.[67] Recently, a very unusual expression pattern was observed for this family. As expected, later diverging genes of subfamily A (i.e., more "modern" ones) are expressed in the VNO like their mammalian counterparts.[68,69] In contrast, earlier diverging genes of subfamily A as well as the single member of subfamily C are expressed exclusively in the MOE[69] (like the corresponding teleost fish *olfC* genes; see earlier discussion). Conceivably, this spatial segregation of early and late-diverging genes reflects some functional divergence between these two groups. In fact, V2Rs expressed in the MOE probably detect amino acids,[69] like their teleost counterparts,[70–72] whereas at least some V2Rs expressed in the VNO may detect sulfated steroids,[73] like some mammalian vomeronasal receptors.[74]

THE LAND-TO-WATER TRANSITION IN SECONDARILY AQUATIC VERTEBRATES LEADS TO LARGE-SCALE PSEUDOGENIZATION OF OLFACTORY RECEPTOR REPERTOIRES

During vertebrate evolution, several lineages have gone back from the terrestrial lifestyle to facultative or fully aquatic life. Among mammals this includes the cetaceans (whales and dolphins) and sirenians (manatees and dugongs) as fully aquatic species, and pinnipeds (seals and sea lions), water shrews, and otters as semiaquatic species (Figure 1).

Generally, semiaquatic species tend to have an olfactory system close to that of their land-living cousins, both with respect to morphology and molecular characteristics. For some of these species, the OR repertoire has been compared with that of their closest terrestrial relatives, in particular the percentage of pseudogenes as an indicator of gene loss. For animals that live in both habitats, such as the sea lion, the comparison with a terrestrial carnivore, dog, showed no significant difference in percentage of pseudogenes found.[75]

In contrast, olfactory organs are reduced or completely absent in fully aquatic species and the large percentage of olfactory receptor pseudogenes shows that most olfactory receptor genes have become obsolete in the new aquatic environment. In the cetacean lineage the percentage of OR pseudogenes ranges between 70% and 100%,[6,75,76] and is thus much higher than that of cow, which as an ungulate is among the closest terrestrial relatives.[77]

Toothed whales show more severe pseudogenization of ORs than baleen whales,[75,76] which retain a reduced olfactory apparatus, whereas toothed whales have lost these structures entirely.[78,79] It has been argued that baleen whales, which are filter feeders, require some residual olfactory capabilities, whereas toothed whales rely on sophisticated echolocation to locate prey. In any case it is clear that the process of pseudogenization has started before the divergence of the whale lineage into toothed and baleen whales.[75]

Interestingly, a gene that is essential in vomeronasal signal transduction, *Trpc2* (transient receptor potential channel c2), is also pseudogenized in several aquatic mammals, including baleen and toothed whales.[80] Thus, the deterioration of olfactory capabilities in obligate aquatic mammals most likely extends to both the main and accessory olfactory systems.

As an aside, some semiaquatic mammals may have evolved a behavioral strategy to make use of airborne olfaction while underwater. They exhale air toward the underwater object to be investigated and breathe back in the resulting air bubble, now enriched with object odors.[81]

AQUATIC ODORS

An odor is defined as such through its detectability by an olfactory receptor. For an aquatic organism, that means that the odor molecule has to be soluble in water (i.e., either charged, or at least polar, or small). There is a wealth of information on substances that fish and amphibians can smell, as evaluated by olfactory-dependent behavioral changes, electroolfactogram of the olfactory organ, and electrophysiological responses of OSNs.[82–84] Commonly accepted as main odor classes in the aquatic environment are amino acids, bile acids, nucleotides, steroids, and prostaglandins.[82,85]

Amino acids serve as a signal to indicate a food source, and fish are generally attracted to this odor.[84] The same function is assumed for nucleotides. Both groups of odors are detected in fish by microvillous OSNs projecting to the lateral olfactory bulb.[70,71,86]

Bile odors are assumed to serve in social communication. Bile components have been most extensively studied in lamprey, among them a species-specific bile acid that is a major component of lamprey migratory pheromone.[87] This odor mix is released by stream-dwelling larval lamprey; adults are guided by the odor to come back to the same streams.[87] Odorous compounds in bile consist not only of bile acids, but also bile alcohols.[88] Teleost bile acid detection may be somewhat less species-specific, as has been shown by a comparison of olfactory sensibilities of eel, goldfish, and tilapia to bile and isolated bile acids from conspecifics and heterospecifics.[89] Bile acids appear to be detected by ciliated OSNs, which in fish project to the medial olfactory bulb.[86]

Steroids and prostaglandins smelled by fish are either reproductive hormones themselves or metabolites thereof, and serve to synchronize male and female reproductive behavior.[90] They appear to be detected by highly specific,[86] but so far unidentified, olfactory receptors and show a considerable degree of species specificity.[91]

Aquatic olfaction is not limited to the four main odor classes discussed previously. Teleost fish and amphibians can smell amines[40,65,92,93] and several behavioral reactions to amines have been reported.[40,92]

Another class of odors for aquatic organisms (and rodents, see Chapters 1 and 10) are peptides. Stickleback smell major histocompatibility complex (MHC) peptides, and female stickleback prefer mates with somewhat dissimilar MHC peptides.[94] Zebrafish prefer to stay with kin and this behavior is presumably mediated via olfactory detection of endogenous MHC peptides.[95] Also, many amphibian pheromones are peptides.[96,97]

Furthermore, there are isolated reports that teleost fish can smell several metabolites,[98] neurotransmitters,[99] and even cations,[100,101] as well as some synthetic psychotropic drugs.[102] Several sea lamprey pheromones and species-specific odors have been isolated and found to be chemically very diverse, among them fatty acid derivatives with a tetrahydrofuran ring[103] and a phenanthrene derivative.[104]

A behaviorally important group of odors are alarm pheromones that signal predation risk to conspecifics and related species.[105] Specialized club cells in the skin of many fish species contain an alarm substance that is released upon injury and produces species-specific avoidance behaviors.[106] Hypoxanthine-3(N)-oxide, a purine derivative, and several pterin derivatives elicit the alarm response.[107,108] On the other hand, chondroitin fragments[109] have been suggested as components of alarm substance. Further research will be needed to unequivocally solve the nature of the alarm substance—or *Schreckstoff*, as it was originally called.[110]

NOT MUCH IS KNOWN ABOUT THE LIGAND SPECTRA OF AQUATIC OLFACTORY RECEPTORS

As discussed previously, there is substantial information on olfactory receptor repertoires, olfactory ligands, and ligand responses in many aquatic species. However, very few receptors from an aquatic organism have been deorphanized so far: in total, four receptors from zebrafish and an ortholog of one of them from another cyprinid fish, goldfish.[56] In addition, two *Xenopus laevis* ORs have been examined in a preliminary investigation.[111] Interestingly, one of them responded to an amino acid mix, suggesting that amphibian ciliated neurons carry some amino acid responses, consistent with the conclusions from a later study.[69]

Two of the four zebrafish receptors belong to the V2R-related OlfC family and respond to amino acids: OlfCa1, formerly called 5.24, and OlfCc1. The best ligands for the goldfish OlfCa1 are the basic amino acids arginine and lysine, with a half maximal effective concentration (EC_{50}) of roughly 1 µM, whereas the zebrafish ortholog has about 10-fold lower affinity to its best ligand, the acidic amino acid glutamate.[112] OlfCc1 responds to high concentrations of several aliphatic neutral amino acids with an EC_{50} of 0.2—0.3 mM,[57] probably too low to carry physiological responses to these amino acids.[70] The affinities of OlfCa1 would be in the expected range for a fish amino acid receptor, but this receptor, like the OlfCc1, is broadly expressed[56] (i.e., does not show the typical sparse expression pattern of olfactory receptors). OlfCa1 may be a receptor cofactor similar to OlfCc1, which enables function and/or intracellular trafficking of other OlfCs.[57] None of the sparsely expressed OlfC receptors have been deorphanized so far.

A single receptor from the TAAR family, TAAR13c, has been deorphanized as cadaverine receptor with an EC_{50} of about 20 µM.[40] Cadaverine is a death-associated odor resulting from bacterial decarboxylation of lysine, which is strongly aversive for humans and zebrafish.[40] TAAR13c appears to be the highest affinity cadaverine receptor in the zebrafish olfactory system, and activation of TAAR13c may be enough to generate the observed avoidance behavior.[40]

One receptor from the V1R-related *ora* gene family, ORA1, has been deorphanized in somewhat of a detective story.[47] Initial experiments showed the activation of ORA1 by tyrosine, but soon it became clear that a contaminant of tyrosine, not the tyrosine itself, was responsible. This compound was isolated and identified as 4-hydroxyphenylacetic acid, which activates ORA1 with an EC_{50} of about 2 µM. ORA1 is narrowly tuned, and even small modifications of the ligand reduce the affinity 10- to 100-fold.[47] Unexpectedly, 4-hydroxyphenylacetic acid was found to increase oviposition frequency in zebrafish mating pairs, suggesting a potential pheromone function and, in extension, ORA1 as putative pheromone receptor. ORA1 is ancestral to all mammalian V1Rs, which are also believed to be pheromone receptors (for a detailed description, see Chapter 10).

Comparing the deorphanized receptors with the known odor classes for fish and amphibians shows glaring gaps in our understanding of aquatic olfactory receptors. The best candidates for amino acid receptors are members of the V2R/OlfC family, which are expressed in microvillous neurons.[70,71] Additionally, there is indirect evidence for OR receptors as amino acid receptors because both ciliated and microvillous neurons respond to amino acids.[19,69] However, so far, no olfactory receptors for bile acids, nucleotides, steroids, and prostaglandins have been found; the sparsely expressed amino acid receptors have not been identified; and olfactory receptors for peptides have not yet been detected, although they should exist.[95]

Signal Transduction by Aquatic Olfactory Receptors

All olfactory receptors described so far in fish have been GPCRs, with the exception of a recently discovered membrane-bound guanylate cyclase, *gucy2f*.[113] Unexpectedly, Saraiva et al. also found a soluble guanylate cyclase, *gucy1b2*, expressed in a sparse population of OSNs.[113] So far, it is not clear whether *gucy1b2* is a signal transduction component downstream of an unidentified receptor or serves itself as nitric oxide (NO) receptor. NO is set free as part of metabolic processes and can diffuse through membranes, so an intracellular receptor could detect an NO signal originating from, for example, a conspecific. Interestingly, *gucy1b2* is also expressed in a small population of mouse OSNs[114] (see Chapter 8).

Many studies have sought to identify the G proteins through which the four olfactory GPCR families signal. Immunohistochemical studies in several teleost and one shark species have shown considerable species differences concerning G protein allocation to OSN populations.[19,72,115–117] However, a firm conclusion would require additional methods of identification such as in situ hybridization, which has been performed in only a few cases.[118] Expression patterns for zebrafish OSNs are similar to the mammalian situation: $G\alpha_{olf}$ is found in ciliated OSNs,[118] crypt neurons express $G\alpha_{i1b}$,[119] and microvillous neurons $G\alpha_o$.[118] However, $G\alpha_{i1b}$ is not orthologous to $G\alpha_{i2}$, the inhibitory G protein coexpressed with mammalian V1Rs. The other two G proteins, $G\alpha_{olf}$ and $G\alpha_o$, do correspond to their mammalian counterparts in ciliated and microvillous neurons, respectively.

For zebrafish OSNs, a pharmacological study has suggested the presence of three different signaling pathways.[93] Evidence was found for the presence of the canonical cyclic adenosine monophosphate (cAMP) pathway, which was activated by bile acids, and for the involvement of phospholipase C (PLC) in a second pathway activated by amino acids.[93] Responses to polyamines seemed to use neither pathway, suggesting the possible existence of a third signaling pathway. This interesting possibility appears not to have been pursued. Comparing response patterns to amino acids and bile acids in the olfactory bulb[86] with known localization of ciliated and microvillous neuron terminals[71] suggests that ciliated OSNs signal via cAMP and microvillous OSNs signal via PLC, which corresponds to the mammalian situation.

In amphibians, an immunohistochemical study has shown $G\alpha_i$ and $G\alpha_o$ in microvillous OSNs and $G\alpha_{olf}$ in ciliated OSNs, which corresponds to the mammalian situation.[120] In the amphibian MOE both cAMP-dependent and -independent pathways have been described in several studies, summarized elsewhere.[121] Recently, the cAMP-independent pathway was investigated pharmacologically and shown to involve PLC, diacylglycerol, and TRPC2.[122] Unexpectedly, this study also pointed to the existence of a third signal transduction pathway. Responses to amino acid odors fell into three categories: PLC-dependent, cAMP-dependent, and neither. It will be interesting to see whether the amphibian and teleost third pathways have anything in common. For a detailed discussion of the diversity of olfactory signal transduction pathways in olfaction, see Chapter 12.

FOUR OLFACTORY SENSORY NEURON TYPES EXPRESSING AQUATIC OLFACTORY RECEPTORS

Fish possess two main populations of OSNs, ciliated and microvillous cells that are very similar to those found in the tetrapod main and accessory olfactory systems. There are minor differences, however. For example, fish microvillous OSNs do not express olfactory marker protein (OMP; see Chapter 7), such that OMP is a specific marker for ciliated OSNs in teleost fish. By contrast, OMP labels ciliated as well as microvillous OSNs in mammals. Interestingly, larval *Xenopus* parallels the situation in fish, with OMP only expressed in ciliated neurons,[69] reflecting the transitional status of this species in evolution from aquatic to terrestrial lifestyle.

In addition, fish possess two minor populations of OSNs, crypt neurons and kappe neurons, both named for their characteristic shape.[21,123] These two cell types differ in soma shape, preferred positions within the sensory surface, and molecular markers both from each other and from ciliated and microvillous OSNs.[21] Crypt neurons express a single receptor, Ora4,[119] and project to a single glomerulus[124] (there is no mirror map in the fish olfactory bulb, in contrast to the situation in rodents). Kappe neurons also project to a single glomerulus,[125] suggesting that they may also express a single receptor. Thus, these two cell types differ from microvillous and ciliated OSNs, which each express large families of olfactory receptors, V2R/OlfC and OR, respectively. The coding principle employed by crypt neurons has been described as the 'one cell type—one receptor—one glomerulus' rule.[124]

Kappe neurons have been described only very recently, and their evolutionary origin is not yet known. Crypt neurons were detected earlier[123] and have been found in many teleost fish[126] and even in some shark species,[127] suggesting that their origin might predate the divergence of bony and cartilaginous fish. Crypt-like cells have not been identified in mammals so far, although the cells of the Grueneberg ganglion bear some morphological similarity[128] (see Chapter 8). Unfortunately, the molecular markers known for both crypt and kappe neurons are antibodies against S100-ir, TrkA-ir (crypt neurons), and $G\alpha_o$-ir (kappe neurons), all of which exploit a cross-reactivity to so far unknown antigens, making a molecular analysis of these cell types difficult. Very recently, a minor TRPC2-expressing cell type, different from both crypt and kappe neurons, has been described in the rodent MOE.[114] Thus, the existence of additional, minor cell types of OSNs appears to be an evolutionarily conserved feature in aquatic and terrestrial olfaction, even though the cell types themselves may differ.

References

1. Churcher AM, Taylor JS. Amphioxus (*Branchiostoma floridae*) has orthologs of vertebrate odorant receptors. *BMC Evol Biol*. 2009;9:242.
2. Niimura Y. On the origin and evolution of vertebrate olfactory receptor genes: comparative genome analysis among 23 chordate species. *Genome Biol Evol*. 2009;1:34—44.
3. Satoh G. Characterization of novel GPCR gene coding locus in amphioxus genome: gene structure, expression, and phylogenetic analysis with implications for its involvement in chemoreception. *Genesis*. 2005;41(2):47—57.
4. Satoh G. Exploring developmental, functional, and evolutionary aspects of amphioxus sensory cells. *Int J Biol Sci*. 2006;2(3):142—148.
5. Niimura Y, Nei M. Evolutionary dynamics of olfactory receptor genes in fishes and tetrapods. *Proc Natl Acad Sci USA*. 2005;102(17):6039—6044.
6. Freitag J, Ludwig G, Andreini I, Rossler P, Breer H. Olfactory receptors in aquatic and terrestrial vertebrates. *J Comp Physiol A*. 1998;183(5):635—650.

7. Libants S, Carr K, Wu H, et al. The sea lamprey *Petromyzon marinus* genome reveals the early origin of several chemosensory receptor families in the vertebrate lineage. *BMC Evol Biol*. 2009;9:180.
8. Niimura Y. Olfactory receptor multigene family in vertebrates: from the viewpoint of evolutionary genomics. *Curr Genomics*. 2012;13(2):103–114.
9. Berghard A, Dryer L. A novel family of ancient vertebrate odorant receptors. *J Neurobiol*. 1998;37(3):383–392.
10. Hashiguchi Y, Nishida M. Evolution of trace amine associated receptor (TAAR) gene family in vertebrates: lineage-specific expansions and degradations of a second class of vertebrate chemosensory receptors expressed in the olfactory epithelium. *Mol Biol Evol*. 2007;24(9):2099–2107.
11. Hussain A, Saraiva LR, Korsching SI. Positive Darwinian selection and the birth of an olfactory receptor clade in teleosts. *Proc Natl Acad Sci USA*. 2009;106(11):4313–4318.
12. Grus WE, Zhang J. Origin of the genetic components of the vomeronasal system in the common ancestor of all extant vertebrates. *Mol Biol Evol*. 2009;26(2):407–419.
13. Venkatesh B, Lee AP, Ravi V, et al. Elephant shark genome provides unique insights into gnathostome evolution. *Nature*. 2014;505(7482):174–179.
14. Nordstrom KJ, Fredriksson R, Schioth HB. The amphioxus (*Branchiostoma floridae*) genome contains a highly diversified set of G protein-coupled receptors. *BMC Evol Biol*. 2008;8:9.
15. Korsching S. The molecular evolution of teleost olfactory receptor gene families. *Results Probl Cell Differ*. 2009;47:37–55.
16. Yoshihara Y. Molecular genetic dissection of the zebrafish olfactory system. *Results Probl Cell Differ*. 2009;47:97–120.
17. Buck LB. Unraveling smell. *Harvey Lect*. 2005;101:117–134.
18. Munger SD, Leinders-Zufall T, Zufall F. Subsystem organization of the mammalian sense of smell. *Annu Rev Physiol*. 2009;71:115–140.
19. Hansen A, Rolen SH, Anderson K, Morita Y, Caprio J, Finger TE. Correlation between olfactory receptor cell type and function in the channel catfish. *J Neurosci*. 2003;23(28):9328–9339.
20. Hamdani el H, Doving KB. The functional organization of the fish olfactory system. *Prog Neurobiol*. 2007;82(2):80–86.
21. Ahuja G, Nia SB, Zapilko V, et al. Kappe neurons, a novel population of olfactory sensory neurons. *Sci Rep*. 2014;4:4037.
22. Broughton RE, Betancur RR, Li C, Arratia G, Orti G. Multi-locus phylogenetic analysis reveals the pattern and tempo of bony fish evolution. *PLoS Curr*. 2013;5.
23. Gonzalez A, Morona R, Lopez JM, Moreno N, Northcutt RG. Lungfishes, like tetrapods, possess a vomeronasal system. *Front Neuroanat*. 2010;4.
24. Nikaido M, Noguchi H, Nishihara H, et al. Coelacanth genomes reveal signatures for evolutionary transition from water to land. *Genome Res*. 2013;23(10):1740–1748.
25. Picone B, Hesse U, Panji S, Van Heusden P, Jonas M, Christoffels A. Taste and odorant receptors of the coelacanth—a gene repertoire in transition. *J Exp Zool B Mol Dev Evol*. 2014;322(6):403–414.
26. Syed AS, Korsching SI. Positive Darwinian selection in the singularly large taste receptor gene family of an 'ancient' fish, *Latimeria chalumnae*. *BMC Genomics*. 2014;15:650.
27. Ren X, Chang S, Laframboise A, Green W, Dubuc R, Zielinski B. Projections from the accessory olfactory organ into the medial region of the olfactory bulb in the sea lamprey (*Petromyzon marinus*): a novel vertebrate sensory structure? *J Comp Neurol*. 2009;516(2):105–116.
28. Chang S, Chung-Davidson YW, Libants SV, et al. The sea lamprey has a primordial accessory olfactory system. *BMC Evol Biol*. 2013;13:172.
29. Taniguchi K, Taniguchi K. Phylogenic studies on the olfactory system in vertebrates. *J Vet Med Sci*. 2014;76(6):781–788.
30. Alioto TS, Ngai J. The odorant receptor repertoire of teleost fish. *BMC Genomics*. 2005;6:173.
31. Hashiguchi Y, Furuta Y, Nishida M. Evolutionary patterns and selective pressures of odorant/pheromone receptor gene families in teleost fishes. *PLoS One*. 2008;3(12):e4083.
32. Azzouzi N, Barloy-Hubler F, Galibert F. Inventory of the cichlid olfactory receptor gene repertoires: identification of olfactory genes with more than one coding exon. *BMC Genomics*. 2014;15:586.
33. Johnstone KA, Lubieniecki KP, Koop BF, Davidson WS. Identification of olfactory receptor genes in Atlantic salmon *Salmo salar*. *J Fish Biol*. 2012;81(2):559–575.
34. Kolmakov NN, Kube M, Reinhardt R, Canario AV. Analysis of the goldfish *Carassius auratus* olfactory epithelium transcriptome reveals the presence of numerous non-olfactory GPCR and putative receptors for progestin pheromones. *BMC Genomics*. 2008;9:429.

II. OLFACTORY TRANSDUCTION

35. Churcher AM, Hubbard PC, Marques JP, Canario AV, Huertas M. Deep sequencing of the olfactory epithelium reveals specific chemosensory receptors are expressed at sexual maturity in the European eel *Anguilla anguilla*. *Mol Ecol*. 2015;24(4):822−834.

36. Niimura Y, Matsui A, Touhara K. Extreme expansion of the olfactory receptor gene repertoire in African elephants and evolutionary dynamics of orthologous gene groups in 13 placental mammals. *Genome Res*. 2014;24(9):1485−1496.

37. Amano T, Gascuel J. Expression of odorant receptor family, type 2 OR in the aquatic olfactory cavity of amphibian frog *Xenopus tropicalis*. *PLoS One*. 2012;7(4):e33922.

38. Lindemann L, Hoener MC. A renaissance in trace amines inspired by a novel GPCR family. *Trends Pharmacol Sci*. 2005;26(5):274−281.

39. Azzouzi N, Barloy-Hubler F, Galibert F. Identification and characterization of cichlid TAAR genes and comparison with other teleost TAAR repertoires. *BMC Genomics*. 2015;16:335.

40. Hussain A, Saraiva LR, Ferrero DM, et al. High-affinity olfactory receptor for the death-associated odor cadaverine. *Proc Natl Acad Sci USA*. 2013;110(48):19579−19584.

41. Huang ES. Construction of a sequence motif characteristic of aminergic G protein-coupled receptors. *Protein Sci*. 2003;12(7):1360−1367.

42. Grus WE, Shi P, Zhang YP, Zhang J. Dramatic variation of the vomeronasal pheromone receptor gene repertoire among five orders of placental and marsupial mammals. *Proc Natl Acad Sci USA*. 2005;102(16):5767−5772.

43. Saraiva LR, Korsching SI. A novel olfactory receptor gene family in teleost fish. *Genome Res*. 2007;17(10):1448−1457.

44. Ota T, Nikaido M, Suzuki H, Hagino-Yamagishi K, Okada N. Characterization of V1R receptor (ora) genes in Lake Victoria cichlids. *Gene*. 2012;499(2):273−279.

45. Nikaido M, Ota T, Hirata T, et al. Multiple episodic evolution events in V1R receptor genes of East-African cichlids. *Genome Biol Evol*. 2014;6(5):1135−1144.

46. Pfister P, Rodriguez I. Olfactory expression of a single and highly variable V1r pheromone receptor-like gene in fish species. *Proc Natl Acad Sci USA*. 2005;102(15):5489−5494.

47. Behrens M, Frank O, Rawel H, et al. ORA1, a zebrafish olfactory receptor ancestral to all mammalian V1R genes, recognizes 4-hydroxyphenylacetic acid, a putative reproductive pheromone. *J Biol Chem*. 2014;289(28):19778−19788.

48. Saraiva JL, Martins RS, Hubbard PC, Canario AV. Lack of evidence for a role of olfaction on first maturation in farmed sea bass *Dicentrarchus labrax*. *Gen Comp Endocrinol*. 2015. http://dx.doi.org/10.1016/j.ygcen.2015.02.017 [Epub ahead of print].

49. Shi P, Zhang J. Extraordinary diversity of chemosensory receptor gene repertoires among vertebrates. *Results Probl Cell Differ*. 2009;47:1−23.

50. Alioto TS, Ngai J. The repertoire of olfactory C family G protein-coupled receptors in zebrafish: candidate chemosensory receptors for amino acids. *BMC Genomics*. 2006;7:309.

51. Hashiguchi Y, Nishida M. Screening the V2R-type putative odorant receptor gene repertoire in bitterling *Tanakia lanceolata*. *Gene*. 2009;441(1−2):74−79.

52. Hashiguchi Y, Nishida M. Evolution and origin of vomeronasal-type odorant receptor gene repertoire in fishes. *BMC Evol Biol*. 2006;6:76.

53. Nikaido M, Suzuki H, Toyoda A, et al. Lineage-specific expansion of vomeronasal type 2 receptor-like (OlfC) genes in cichlids may contribute to diversification of amino acid detection systems. *Genome Biol Evol*. 2013;5(4):711−722.

54. Johnstone KA, Ciborowski KL, Lubieniecki KP, et al. Genomic organization and evolution of the vomeronasal type 2 receptor-like (OlfC) gene clusters in Atlantic salmon, *Salmo salar*. *Mol Biol Evol*. 2009;26(5):1117−1125.

55. Kuang D, Yao Y, Lam J, Tsushima RG, Hampson DR. Cloning and characterization of a family C orphan G-protein coupled receptor. *J Neurochem*. 2005;93(2):383−391.

56. Speca DJ, Lin DM, Sorensen PW, Isacoff EY, Ngai J, Dittman AH. Functional identification of a goldfish odorant receptor. *Neuron*. 1999;23(3):487−498.

57. DeMaria S, Berke AP, Van Name E, Heravian A, Ferreira T, Ngai J. Role of a ubiquitously expressed receptor in the vertebrate olfactory system. *J Neurosci*. 2013;33(38):15235−15247.

58. Hoon MA, Adler E, Lindemeier J, Battey JF, Ryba NJ, Zuker CS. Putative mammalian taste receptors: a class of taste-specific GPCRs with distinct topographic selectivity. *Cell*. 1999;96(4):541−551.

59. Amemiya CT, Alfoldi J, Lee AP, et al. The African coelacanth genome provides insights into tetrapod evolution. *Nature*. 2013;496(7445):311−316.

60. Bjarnadottir TK, Fredriksson R, Schioth HB. The gene repertoire and the common evolutionary history of glutamate, pheromone (V2R), taste(1) and other related G protein-coupled receptors. *Gene*. 2005;362:70−84.

II. OLFACTORY TRANSDUCTION

61. Martini S, Silvotti L, Shirazi A, Ryba NJ, Tirindelli R. Co-expression of putative pheromone receptors in the sensory neurons of the vomeronasal organ. *J Neurosci.* 2001;21(3):843—848.
62. Callery EM. There's more than one frog in the pond: a survey of the Amphibia and their contributions to developmental biology. *Semin Cell Dev Biol.* 2006;17(1):80—92.
63. Dittrich K, Kuttler J, Hassenklover T, Manzini I. Metamorphic remodeling of the olfactory organ of the African clawed frog, *Xenopus laevis. J Comp Neurol.* 2015. http://dx.doi.org/10.1002/cne.23887 [Epub ahead of print].
64. Hansen A, Reiss JO, Gentry CL, Burd GD. Ultrastructure of the olfactory organ in the clawed frog, *Xenopus laevis*, during larval development and metamorphosis. *J Comp Neurol.* 1998;398(2):273—288.
65. Syed AS, Sansone A, Roner S, Bozorg Nia S, Manzini I, Korsching SI. Different expression domains for two closely related amphibian TAARs generate a bimodal distribution similar to neuronal responses to amine odors. *Sci Rep.* 2015;5:13935.
66. Date-Ito A, Ohara H, Ichikawa M, Mori Y, Hagino-Yamagishi K. Xenopus V1R vomeronasal receptor family is expressed in the main olfactory system. *Chem Senses.* 2008;33(4):339—346.
67. Ji Y, Zhang Z, Hu Y. The repertoire of G-protein-coupled receptors in *Xenopus tropicalis. BMC Genomics.* 2009;10:263.
68. Hagino-Yamagishi K, Moriya K, Kubo H, et al. Expression of vomeronasal receptor genes in *Xenopus laevis. J Comp Neurol.* 2004;472(2):246—256.
69. Syed AS, Sansone A, Nadler W, Manzini I, Korsching SI. Ancestral amphibian v2rs are expressed in the main olfactory epithelium. *Proc Natl Acad Sci USA.* 2013;110(19):7714—7719.
70. Friedrich RW, Korsching SI. Combinatorial and chemotopic odorant coding in the zebrafish olfactory bulb visualized by optical imaging. *Neuron.* 1997;18(5):737—752.
71. Sato Y, Miyasaka N, Yoshihara Y. Mutually exclusive glomerular innervation by two distinct types of olfactory sensory neurons revealed in transgenic zebrafish. *J Neurosci.* 2005;25(20):4889—4897.
72. Koide T, Miyasaka N, Morimoto K, et al. Olfactory neural circuitry for attraction to amino acids revealed by transposon-mediated gene trap approach in zebrafish. *Proc Natl Acad Sci USA.* 2009;106(24):9884—9889.
73. Sansone A, Hassenklöver T, Offner T, Fu X, Holy TE, Manzini I. Dual processing of sulfated steroids in the olfactory system of an anuran amphibian. *Front Cell Neurosci.* 2015;9:373 [Epub ahead of print].
74. Nodari F, Hsu FF, Fu X, et al. Sulfated steroids as natural ligands of mouse pheromone-sensing neurons. *J Neurosci.* 2008;28(25):6407—6418.
75. Kishida T, Kubota S, Shirayama Y, Fukami H. The olfactory receptor gene repertoires in secondary-adapted marine vertebrates: evidence for reduction of the functional proportions in cetaceans. *Biol Lett.* 2007;3(4):428—430.
76. McGowen MR, Clark C, Gatesy J. The vestigial olfactory receptor subgenome of odontocete whales: phylogenetic congruence between gene-tree reconciliation and supermatrix methods. *Syst Biol.* 2008;57(4):574—590.
77. Shimamura M, Yasue H, Ohshima K, et al. Molecular evidence from retroposons that whales form a clade within even-toed ungulates. *Nature.* 1997;388(6643):666—670.
78. Breathnach AS. The cetacean central nervous system. *Biol Rev.* 1960;35:187—230.
79. Oelschläger HA. Development of the olfactory and terminalis systems in whales and dolphins. In: Doty RL, Müller-Schwarze D, eds. *Chemical Signals in Vertebrates VI.* Vol. VI. New York: Plenum Press; 1992:141—147.
80. Yu L, Jin W, Wang JX, et al. Characterization of TRPC2, an essential genetic component of VNS chemoreception, provides insights into the evolution of pheromonal olfaction in secondary-adapted marine mammals. *Mol Biol Evol.* 2010;27(7):1467—1477.
81. Marriott S, Cowan E, Cohen J, Hallock RM. Somatosensation, echolocation, and underwater sniffing: adaptations allow mammals without traditional olfactory capabilities to forage for food underwater. *Zool Sci.* 2013;30(2):69—75.
82. Hara TJ. Olfaction and gustation in fish: an overview. *Acta Physiol Scand.* 1994;152(2):207—217.
83. Laberge F, Hara TJ. Neurobiology of fish olfaction: a review. *Brain Res Brain Res Rev.* 2001;36(1):46—59.
84. Whitlock KE. The sense of scents: olfactory behaviors in the zebrafish. *Zebrafish.* 2006;3(2):203—213.
85. Buchinger TJ, Li W, Johnson NS. Bile salts as semiochemicals in fish. *Chem Senses.* 2014;39(8):647—654.
86. Friedrich RW, Korsching SI. Chemotopic, combinatorial, and noncombinatorial odorant representations in the olfactory bulb revealed using a voltage-sensitive axon tracer. *J Neurosci.* 1998;18(23):9977—9988.
87. Sorensen PW, Fine JM, Dvornikovs V, et al. Mixture of new sulfated steroids functions as a migratory pheromone in the sea lamprey. *Nat Chem Biol.* 2005;1(6):324—328.
88. Li K, Brant CO, Siefkes MJ, Kruckman HG, Li W. Characterization of a novel bile alcohol sulfate released by sexually mature male sea lamprey (*Petromyzon marinus*). *PLoS One.* 2013;8(7):e68157.

89. Huertas M, Hagey L, Hofmann AF, Cerda J, Canario AV, Hubbard PC. Olfactory sensitivity to bile fluid and bile salts in the European eel (*Anguilla anguilla*), goldfish (*Carassius auratus*) and Mozambique tilapia (*Oreochromis mossambicus*) suggests a 'broad range' sensitivity not confined to those produced by conspecifics alone. *J Exp Biol*. 2010;213(2):308−317.

90. Stacey N, Chojnacki A, Narayanan A, Cole T, Murphy C. Hormonally derived sex pheromones in fish: exogenous cues and signals from gonad to brain. *Can J Physiol Pharmacol*. 2003;81(4):329−341.

91. Stacey NE, Van Der Kraak GJ, Olsen KH. Male primer endocrine responses to preovulatory female cyprinids under natural conditions in Sweden. *J Fish Biol*. 2012;80(1):147−165.

92. Rolen SH, Sorensen PW, Mattson D, Caprio J. Polyamines as olfactory stimuli in the goldfish *Carassius auratus*. *J Exp Biol*. 2003;206(Pt 10):1683−1696.

93. Michel WC, Sanderson MJ, Olson JK, Lipschitz DL. Evidence of a novel transduction pathway mediating detection of polyamines by the zebrafish olfactory system. *J Exp Biol*. 2003;206(Pt 10):1697−1706.

94. Milinski M, Griffiths S, Wegner KM, Reusch TB, Haas-Assenbaum A, Boehm T. Mate choice decisions of stickleback females predictably modified by MHC peptide ligands. *Proc Natl Acad Sci USA*. 2005;102(12):4414−4418.

95. Hinz C, Namekawa I, Behrmann-Godel J, et al. Olfactory imprinting is triggered by MHC peptide ligands. *Sci Rep*. 2013;3:2800.

96. Wyatt TD. Pheromones: convergence and contrasts in insects and vertebrates. In: Mason RT, Müller-Schwarze D, eds. *Chemical Signals in Vertebrates*. Vol. 10. New York: Springer Press; 2005:7−20.

97. Fraker ME, Hu F, Cuddapah V, et al. Characterization of an alarm pheromone secreted by amphibian tadpoles that induces behavioral inhibition and suppression of the neuroendocrine stress axis. *Horm Behav*. 2009;55(4):520−529.

98. Valdes J, Olivares J, Ponce D, Schmachtenberg O. Analysis of olfactory sensitivity in rainbow trout (*Oncorhynchus mykiss*) reveals their ability to detect lactic acid, pyruvic acid and four B vitamins. *Fish Physiol Biochem*. 2015;41(4):879−885.

99. Hubbard PC, Barata EN, Canario AV. Olfactory sensitivity to catecholamines and their metabolites in the goldfish. *Chem Senses*. 2003;28(3):207−218.

100. Hubbard PC, Canario AV. Evidence that olfactory sensitivities to calcium and sodium are mediated by different mechanisms in the goldfish *Carassius auratus*. *Neurosci Lett*. 2007;414(1):90−93.

101. Dew WA, Pyle GG. Smelling salt: calcium as an odourant for fathead minnows. *Comp Biochem Physiol A Mol Integr Physiol*. 2014;169:1−6.

102. Abreu MS, Giacomini AC, Gusso D, et al. Acute exposure to waterborne psychoactive drugs attract zebrafish. *Comp Biochem Physiol C Toxicol Pharmacol*. 2015;179:37−43.

103. Li K, Huertas M, Brant C, et al. (+)- and (−)-petromyroxols: antipodal tetrahydrofurandiols from larval sea lamprey (*Petromyzon marinus* L.) that elicit enantioselective olfactory responses. *Org Lett*. 2015;17(2):286−289.

104. Li K, Brant CO, Huertas M, Hur SK, Li W. Petromyzonin, a hexahydrophenanthrene sulfate isolated from the larval sea lamprey (*Petromyzon marinus* L.). *Org Lett*. 2013;15(23):5924−5927.

105. Wisenden BD. Olfactory assessment of predation risk in the aquatic environment. *Philos Trans R Soc Lond B Biol Sci*. 2000;355(1401):1205−1208.

106. Doving KB, Lastein S. The alarm reaction in fishes—odorants, modulations of responses, neural pathways. *Ann N Y Acad Sci*. 2009;1170:413−423.

107. Pfeiffer W. Heterocyclic compounds as releasers of the fright reaction in the giant danio *Danio malabaricus* (Jerdon) (Cyprinidae, Ostariophysi, Pisces). *J Chem Ecol*. 1978;4(6):665−673.

108. Pfeiffer W, Riegelbauer G, Meier G, Scheibler B. Effect of hypoxanthine-3(N)-oxide and hypoxanthine-1(N)-oxide on central nervous excitation of the black tetraGymnocorymbus ternetzi (Characidae, Ostariophysi, Pisces) indicated by dorsal light response. *J Chem Ecol*. 1985;11(4):507−523.

109. Mathuru AS, Kibat C, Cheong WF, et al. Chondroitin fragments are odorants that trigger fear behavior in fish. *Curr Biol*. 2012;22(6):538−544.

110. von Frisch K. Zur Psychologie des Fisch-Schwarmes. *Naturwissenschaften*. 1938;26(37):601−606.

111. Mezler M, Fleischer J, Breer H. Characteristic features and ligand specificity of the two olfactory receptor classes from *Xenopus laevis*. *J Exp Biol*. 2001;204(Pt 17):2987−2997.

112. Luu P, Acher F, Bertrand HO, Fan J, Ngai J. Molecular determinants of ligand selectivity in a vertebrate odorant receptor. *J Neurosci*. 2004;24(45):10128−10137.

113. Saraiva LR, Ahuja G, Ivandic I, et al. Molecular and neuronal homology between the olfactory systems of zebrafish and mouse. *Sci Rep*. 2015;5:11487.

114. Omura M, Mombaerts P. Trpc2-expressing sensory neurons in the mouse main olfactory epithelium of type B express the soluble guanylate cyclase Gucy1b2. *Mol Cell Neurosci.* 2015;65:114–124.

115. Ferrando S, Gambardella C, Ravera S, et al. Immunolocalization of G-protein alpha subunits in the olfactory system of the cartilaginous fish *Scyliorhinus canicula. Anat Rec (Hoboken).* 2009;292(11):1771–1779.

116. Hansen A, Anderson KT, Finger TE. Differential distribution of olfactory receptor neurons in goldfish: structural and molecular correlates. *J Comp Neurol.* 2004;477(4):347–359.

117. Hansen A, Zielinski BS. Diversity in the olfactory epithelium of bony fishes: development, lamellar arrangement, sensory neuron cell types and transduction components. *J Neurocytol.* 2005;34(3–5):183–208.

118. Oka Y, Korsching SI. Shared and unique G alpha proteins in the zebrafish versus mammalian senses of taste and smell. *Chem Senses.* 2011;36(4):357–365.

119. Oka Y, Saraiva LR, Korsching SI. Crypt neurons express a single V1R-related ora gene. *Chem Senses.* 2012;37(3):219–227.

120. Gliem S, Syed AS, Sansone A, et al. Bimodal processing of olfactory information in an amphibian nose: odor responses segregate into a medial and a lateral stream. *Cell Mol Life Sci.* 2013;70(11):1965–1984.

121. Manzini I, Schild D. Olfactory coding in larvae of the African clawed frog *Xenopus laevis.* In: Menini A, ed. *The Neurobiology of Olfaction.* Boca Raton (FL); 2010.

122. Sansone A, Hassenklover T, Syed AS, Korsching SI, Manzini I. Phospholipase C and diacylglycerol mediate olfactory responses to amino acids in the main olfactory epithelium of an amphibian. *PLoS One.* 2014;9(1):e87721.

123. Hansen A, Zeiske E. The peripheral olfactory organ of the zebrafish, *Danio rerio*: an ultrastructural study. *Chem Senses.* 1998;23(1):39–48.

124. Ahuja G, Ivandic I, Salturk M, Oka Y, Nadler W, Korsching SI. Zebrafish crypt neurons project to a single, identified mediodorsal glomerulus. *Sci Rep.* 2013;3:2063.

125. Braubach OR, Fine A, Croll RP. Distribution and functional organization of glomeruli in the olfactory bulbs of zebrafish (*Danio rerio*). *J Comp Neurol.* 2012;520(11):2317–2339. Spc2311.

126. Hansen A, Finger TE. Phyletic distribution of crypt-type olfactory receptor neurons in fishes. *Brain Behav Evol.* 2000;55(2):100–110.

127. Ferrando S, Bottaro M, Gallus L, Girosi L, Vacchi M, Tagliafierro G. Observations of crypt neuron-like cells in the olfactory epithelium of a cartilaginous fish. *Neurosci Lett.* 2006;403(3):280–282.

128. Brechbuhl J, Klaey M, Broillet MC. Grueneberg ganglion cells mediate alarm pheromone detection in mice. *Science.* 2008;321(5892):1092–1095.

129. Shi P, Zhang J. Contrasting modes of evolution between vertebrate sweet/umami receptor genes and bitter receptor genes. *Mol Biol Evol.* 2006;23(2):292–300.

130. Ishimaru Y, Okada S, Naito H, et al. Two families of candidate taste receptors in fishes. *Mech Dev.* 2005;122(12):1310–1321.

131. Hashiguchi Y, Furuta Y, Kawahara R, Nishida M. Diversification and adaptive evolution of putative sweet taste receptors in threespine stickleback. *Gene.* 2007;396(1):170–179.

132. Behrens M, Korsching SI, Meyerhof W. Tuning properties of avian and frog bitter taste receptors dynamically fit gene repertoire sizes. *Mol Biol Evol.* 2014;31(12):3216–3227.

133. Go Y, Investigators ST-NY. Proceedings of the SMBE Tri-National Young Investigators' Workshop 2005. Lineage-specific expansions and contractions of the bitter taste receptor gene repertoire in vertebrates. *Mol Biol Evol.* 2006;23(5):964–972.

134. Dong D, Jones G, Zhang S. Dynamic evolution of bitter taste receptor genes in vertebrates. *BMC Evol Biol.* 2009;9:12.

135. Oike H, Nagai T, Furuyama A, et al. Characterization of ligands for fish taste receptors. *J Neurosci.* 2007;27(21):5584–5592.

136. Li D, Zhang J. Diet shapes the evolution of the vertebrate bitter taste receptor gene repertoire. *Mol Biol Evol.* 2014;31(2):303–309.

137. Fricke H, Hissmann K, Schauer J, Erdmann M, Moosa MK, Plante R. Biogeography of the Indonesian coelacanths. *Nature.* 2000;403(6765):38.

138. Fry BG, Roelants K, Norman JA. Tentacles of venom: toxic protein convergence in the kingdom Animalia. *J Mol Evol.* 2009;68(4):311–321.

Insect Olfactory Receptors: An Interface between Chemistry and Biology

Gregory M. Pask[1], Anandasankar Ray[1,2]

[1]Department of Entomology, University of California Riverside, Riverside, CA, USA;
[2]Institute for Integrative Genome Biology, University of California Riverside, Riverside, CA, USA

O U T L I N E

INTRODUCTION

In a complex chemical environment, insects rely on the sensory modality of olfaction to drive several key behaviors. These can include navigation toward a food source, location of a potential mate, or the identification of a suitable egg-laying site. Odorants that signal such behaviors are termed semiochemicals, or chemicals that convey a message to organisms. With such a wide variety of volatile compounds present in nature, insects rely on families of olfactory receptors to detect semiochemicals of interest.

In insects, the primary olfactory appendages are the antenna and maxillary palps. These appendages are covered in sensory hairs, or sensilla, and olfactory sensilla can exhibit a variety of morphological subtypes. The cuticular wall of an olfactory sensillum is covered with many pores, which allow volatiles to enter the sensillum. It has been shown that antennal grooming behavior is important for insects to keep the cuticular pores of olfactory sensilla clean, which helps to maintain olfactory acuity.[1]

Once an odorant molecule passes through a pore, it encounters the aqueous lymph inside the sensillum (see Figure 1). Many odorants are hydrophobic in nature, but the sensillar lymph contains high amounts of odorant-binding proteins (OBPs) that are believed to aid in the solubilization of odorants.[2] Also inside each sensillum are the dendrites of one or more olfactory receptor neurons (ORNs, also referred to as olfactory sensory neurons, or OSNs). A given ORN typically expresses a single type of olfactory receptor on its dendritic

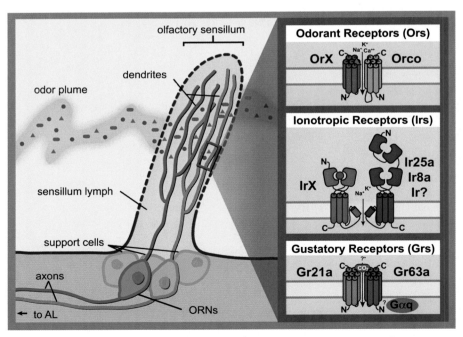

FIGURE 1   **An insect olfactory sensillum and the olfactory receptor families.** ORN, olfactory receptor neuron; AL, antennal lobe; OrX and IrX, generic terms for odorant-binding *Ors* and *Irs*.

membrane, and the receptor imparts sensitivity to a specific subset of odorants.[3−6] The axons of ORNs expressing the same olfactory receptor project to the same spherical region in the antennal lobe of the insect brain called a glomerulus.[7,8] A glomerulus consists of several ORN axons synapsing with the dendrites of projection neurons (PNs). PNs typically innervate a single glomerulus and carry the olfactory signal from the antennal lobe to higher brain centers, such as the mushroom body or the lateral horn of the protocerebrum.[9,10]

Olfactory receptors represent an interface between chemistry and biology, where the receptor converts the recognition of an odorant molecule into an electrical signal in the peripheral olfactory neurons. To detect a wide array of volatile compounds, the insect olfactory system uses a diverse family of olfactory receptors. The first family of receptors identified was termed the odorant receptor family (*Ors*) and is the best studied. A second receptor family was identified and named the ionotropic receptors (*Irs*), which are related to ionotropic glutamate receptors and found in arthropods, nematodes, and mollusks. Additionally, the volatile olfactory cue carbon dioxide (CO_2) is detected by receptors from the gustatory receptor (*Gr*) family expressed in olfactory appendages.

ODORANT RECEPTORS

The first *Or* genes were identified in *Drosophila melanogaster* in 1999 during the completion of the genome project.[11−13] Although olfactory receptors in mammals and nematodes had already been discovered, researchers had difficulty identifying insect orthologs through sequence similarity searches.[14,15] Instead, the *D. melanogaster Or* genes were identified using a novel protein-structure search algorithm and a differential screening technique.[11−13] Completion of additional insect genomes and refinements in sequence analysis have led to annotations of *Or* families across several insects, and the number of *Or* genes is quite variable. In *D. melanogaster,* there are 62 *Ors*, whereas the carpenter ant *Camponotus floridanus* has 352 *Ors*.[16,17] The majority of *Ors* are expressed in either the antennae or the maxillary palps of the adult stage, with a smaller fraction being expressed in the larval stages.[3,8,11,12,18−21] Additionally, insect *Ors* have also been found in other tissues, such as the proboscis and testes.[22,23] With the advent of transcriptomics there has been an improvement in the annotations and expression profiles of *Or* genes as well as other olfactory genes in several insect tissues.[17,24−27]

Evolution of the *Or* Family

Phylogenetic analyses have suggested that the *Or* gene family has evolved from a single lineage within the *Gr* family.[16] The *Or* family is unique to insects and extremely divergent. One member of the family, however, is well conserved across many insect species and is called *Orco* for *Or* coreceptor (previously named *Or83b* in *D. melanogaster*); it is often the first *Or* identified as a new insect genome or transcriptome is assembled.[28−33] *Orco* is distantly related to the *Gr* family, also suggesting that the evolution of the *Or* family began with *Orco*.[16] A recent study investigated the evolution of *Orco* in diverse insect lineages and found that a basal insect *Lepismachilis y-signata* (Archaeognatha) did not have *Orco* or any other *Ors* present in the antennae.[34] In a less basal insect order, Zygentoma, the firebrat *Thermobia domestica* seems to have three paralogs of *Orco* with antennal-specific expression, but no other

Ors.[34] The possible functions of three *Orco* paralogs in this insect is intriguing, given that other insects studied thus far that possess *Ors* only have one *Orco* gene.

Orco is expressed in the majority of the ORNs in *D. melanogaster* and other species tested, suggesting a different role from the other *Or* genes that are expressed in small numbers of neurons.[7,12,30,35,36] *Orco*-mutant *Drosophila* display electrophysiological and behavioral deficits to a broad range of odorants.[35] Subsequent analyses demonstrated that the Orco protein functions as a coreceptor, and is necessary for localization of other *Ors* to the dendritic membrane of ORNs.[35] Olfactory responses of *Orco*-mutant flies can be rescued by expression of *Orco* from mosquitoes or moths, demonstrating a strong functional conservation in insects.[36]

Although *Orco* is required for the proper function of insect *Ors*, it is the other canonical members of the *Or* family that determines the odorant specificity of the receptor complex. The canonical *Ors* across insect taxa are quite divergent. For instance, analysis of ant *Or* families show rapid gains and losses in their evolution.[17] Additionally, *Ors* in the yellow fever mosquito *Aedes aegypti* have less than 20% sequence identity with the *Ors* from the malaria vector mosquito, *Anopheles gambiae*.[21,37] It is believed that differences in behavior and environmental niches between insect species leads to the dynamic evolution and varied expression of unique *Or* repertoires.[38] This has been observed best in *Drosophila sechellia*, a specialist fly species that lays its eggs only on the morinda fruit.[39] The antennae of *D. sechellia* have an increased number of ab3 type sensilla, which express *Ors* highly sensitive to morinda fruit volatiles.[40]

Odorants and Other Ligands of *Ors*

Odorant ligands for several insect *Ors* have been identified. Figure 2 gives an overview of methods used in the functional characterization of insect olfactory receptors. *Or43a* from *D. melanogaster* was the first to be functionally characterized by expression in *Xenopus laevis* oocytes and ectopic expression in *Drosophila* ORNs.[41,42] However, *Or43a* was expressed in antennal ORNs containing an endogenous *Or* that may contribute to the odorant response.[42] In order to test the odor responses of *Ors* more accurately, a novel expression system was developed in *Drosophila*, termed the "empty neuron."[4] A fly line mutant for *Or22a* and *Or22b* resulted in the ab3A neuron losing responses to all odorants tested. Using the *Or22a* promoter and the GAL4-UAS system, an *Or* of interest was expressed in the ab3A neurons in the mutant fly.[4] In this system, expression of *Or47a* in the ab3A neuron resulted in an odorant response profile that mimicked that of the ab5B neuron, the native neuron of *Or47a*.[4]

The empty neuron system was then used to construct a receptor-to-ORN map for the *Drosophila* antenna.[5] By comparing the odor response profiles of antennal sensillar types to those of the deorphanized *Ors*, it could be inferred which *Or* was native to each ORN.[5] These pioneering experiments revealed that several properties of ORNs are dependent upon the *Or*.[5] Not only does the *Or* impart sensitivity to specific odorants, but it also determines the signaling mode (excitatory or inhibitory) and termination kinetics.[5] It was also found that the *Or* determines the spontaneous firing rate of the ORN in the absence of odorants.[5] This activity presumably arises from the spontaneous receptor activity in the dendrites, a feature later confirmed with patch clamp studies of heterologously expressed insect *Ors*.[5,43–45]

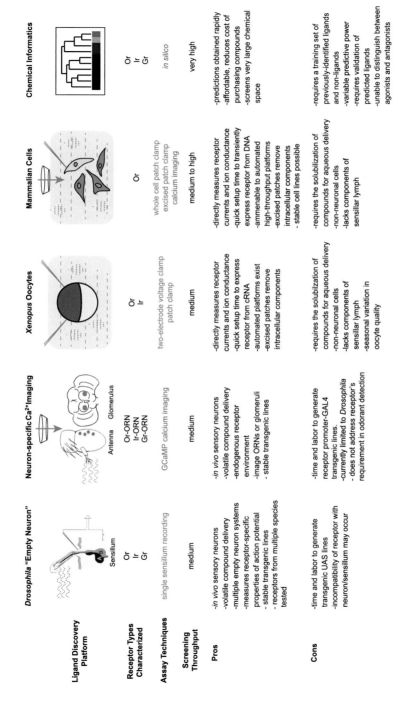

FIGURE 2   Methods used in the functional characterization of insect olfactory receptors.

With the empty neuron system established as a decoder for *Drosophila Ors*, researchers then used this system to deorphanize *Ors* from other insects.[46,47] *An. gambiae Ors* (*AgOrs*) were expressed in the "empty neuron" system and screened against a panel of odorants.[48] In parallel, these *AgOrs* were also expressed and screened in *Xenopus* oocytes and the response profiles were very similar to the *Drosophila* system, validating the use of both methods in insect *Or* deorphanization.[48,49] This study led to a comparison of the odor receptive ranges of *D. melanogaster* and *An. gambiae*.[48] The *D. melanogaster Or* family was found to be most strongly tuned toward esters commonly found in fruits, whereas the *An. gambiae Ors* were tuned more to aromatics and heterocyclics, compound classes typical of human skin emanations.[48]

Using high-throughput techniques, researchers identified a small molecule called VUAA1 (Vanderbilt University Allosteric Agonist 1) that appeared to be a broad-spectrum activator of insect *Ors*.[44,50] Upon further study, VUAA1 was demonstrated to be an *Orco* agonist that could open both homomeric and heteromeric *Or* channels.[44] It was also shown that VUAA1 could activate *Orco*-expressing ORNs in vivo through diffusion into the sensillar lymph.[44] Understanding the structure—activity relationship of VUAA1 and *Orco* has led to the identification of both antagonists and more potent agonists of *Orco*.[47,51−53] Additionally, amiloride and its derivatives have been found to be effective channel blockers of insect *Or* complexes.[45,54] These small molecules provide research tools for furthering our understanding of insect *Or* function.

Or-Mediated Behaviors

A few olfactory-driven behaviors in insects have been linked to specific *Ors* and their ligands. In *Drosophila*, *Or56a* and *Or19a* have been implicated in the detection of harmful food microbes and the preference of citrus for oviposition, respectively.[55,56] Also, specific glomeruli have been identified along with their receptors that are involved in attraction to vinegar (*Or42b*) and aversion to it at high concentrations (*Or85a*).[57] In mosquitoes, the identification of ligands for *Ors* led to their behavioral testing as attractant lures in traps.[58,59] Information from deorphanization of *An. gambiae* receptors has also been integrated with expression data to identify candidate semiochemicals involved in olfactory-driven behaviors.[25,60] Interestingly, the coreceptor *Orco*-mutant *Aedes aegypti*, which lacks the function of all *AaOr* family receptors, is still able to find human skin odor source.[61] This suggests that attraction to complex skin odor may rely on redundant pathways and involve non-*Or* receptor classes.

Or-Mediated Detection of Pheromones

In order to communicate between individuals of the same species, insects emit and detect volatile pheromones, many of which are detected by members of the *Or* family. In general, pheromone-responsive *Ors* are narrowly tuned and elicit robust behaviors when activated. Pheromone-detecting *Ors* were first identified in the silk moth *Bombyx mori*.[62,63] Bombykol and bombykal, pheromones of *B. mori*, respond to *BmOr1* and *BmOr3*, respectively.[62,63] In *D. melanogaster* the males emit *cis*-vaccenyl acetate (cVA), a pheromone involved in aggregation, aggression, and courtship behaviors that is detected by *Or67d* and *Or65a* in both males and females.[64-66] *Or67d*-mutant flies exhibit defects in sexual behaviors, where males attempt to court other males, and females have reduced receptiveness to courting males.[65] Additionally, *Or47b* and *Or88a* in *D. melanogaster* are tuned to other components of fly extracts and may play a role in mate recognition.[66] In the honey bee *Apis mellifera*, *AmOr11* is highly sensitive to 9-oxo-2-decenoic acid, a queen-specific pheromone responsible for inhibiting ovary development of workers.[67] More recently, studies in the cerambycid beetle *Megacyllene caryae* led to annotation and functional characterization of three McOrs that responded to components of a pheromone blend.[26] Despite the distinct ecological roles for pheromones and general odorants, insects utilize different members of the *Or* family for their detection.

Structure/Function of *Ors*

Although the molecular and cellular organization of the olfactory system is similar between both insects and vertebrates, the insect *Ors* are quite different from those in vertebrates. The requirement of the conserved coreceptor *Orco* is unique to insects, as vertebrate ORs function as G-protein–coupled receptors (GPCRs; see Chapter 4).[68] The protein sequences of insect *Ors* lack the DRY and NPxxY amino acid motifs that are common to GPCRs. Additionally, topology prediction algorithms and experimental data strongly suggest that *Orco* and other insect *Ors* have an inverted topology relative to GPCRs with an intracellular amino-terminus and an extracellular carboxyl-terminus.[33,69-71]

Despite the lack of homology to traditional GPCRs, disruption of several proteins involved in the G–protein-coupled signaling cascade resulted in reduced olfactory responses.[62,72-75] Moreover, an increase in the levels of the second messenger inositol trisphosphate was observed in cockroach antennae immediately following pheromone exposure.[76] These studies, however, did not directly address roles of these molecules in insect *Or*-mediated signaling. In 2008, two groups independently came to the conclusion that insect *Ors* function as ligand-gated ion channels, although they proposed different mechanistic models.[43,77] One model suggested that odorant binding triggers both ionotropic and metabotropic signaling at varying speeds and durations.[77] The other model proposed that *Orco* and *Or* subunits form a heteromeric ion channel of unknown stoichiometry.[43] Additional experiments led to a model in which insect *Or* complexes are indeed ligand-gated ion channels but they may be modulated indirectly by G-protein signaling.[44,71,78] These *Or* channels function as nonselective cation channels, with high relative permeability to K^+.[43,77,79] Although the typical ion concentration gradient of neurons promotes outward flux of K^+ ions, the sensillar lymph that bathes ORNs was found to be very high in K^+ ($\sim$200 mM).[80,81] These data suggest that K^+ may be the major ion in the depolarization of ORNs in vivo.

Further study of insect *Or* function has identified roles for *Orco* and *Or* subunits, as well as some of their conserved residues. Although a heteromeric channel is necessary for odorant responses, it has been shown that *Orco* subunits alone are also able to form functional homomeric channels.[44,45,47,51,82] This is similar to other ion channels where the homomeric channels are possible, but the inclusion of subunits provides expanded function. Both *Orco* and *Or* have also been shown to contribute to specific properties of the ion-channel pore, such as ion permeability and susceptibility to channel blockers.[43,45,79,83,84] A simple model stemming from these findings suggests that the channel pore is made up of both *Orco* and *Or* subunits. Point mutations in *Orco* have identified specific residues that are important for channel gating, activation kinetics, cation selectivity, and posttranslational modification.[77,78,84–86] Moreover, mutational studies in a handful of Ors have identified the third and fourth transmembrane domains and the second extracellular loop as being important in odorant binding.[84,87–90] Computational analysis of Or sequences have also identified amino acid motifs in the N-terminal half that can predict ligand sensitivity, and conserved domains in the C-terminal half that may be necessarily for common functions in Ors, such as subunit oligomerization and channel gating.[91]

Having been identified first, the Or family continues to be the best studied among the insect chemosensory receptors. That said, we still know relatively little about the structure and function of insect Ors compared to other ligand-gated ion channels. Obtaining a three-dimensional structure could further our understanding of Ors. Structural information could expand our knowledge of ligand-binding regions as well as the subunit stoichiometry of Or channels. Complementation assays using N-terminal YFP fragments have suggested that at least two of each subunit, Orco and a ligand-binding Or, are present in a functional Or complex, but the absolute number of each subunit is unknown.[69] With an increasing number of heterologous expression methods and pharmacological tools, greater insight into Or structure/function should be on the horizon.

IONOTROPIC RECEPTORS

The coeloconic sensillum type in the *Drosophila* antennae houses several ORNs that do not express Ors or Grs but respond to acids, amines, aldehydes, and humidity, suggesting the existence of another chemoreceptor family.[92] In 2009, a new chemosensory family of receptors, the ionotropic receptors (Irs), was identified from a bioinformatics screen in *D. melanogaster*.[93] The Ir family, consisting of 66 genes in *D. melanogaster*, is an expansion of the ionotropic glutamate receptor (iGluR) family where most genes have lost conserved glutamate-binding residues.[93] Analysis of antennal-specific Irs showed expression in coeloconic sensilla, as well as in the arista and sacculus, thus identifying the receptor family responsible for olfactory responses in these ORNs.[93]

Ir Evolution

The Ir family is an ancient chemoreceptor family that is found across the Protostomia branch of the animal kingdom, with Irs found in arthropods, nematodes, mollusks, and annelids.[93,94] In the spiny lobster, Irs have been identified as the lone receptor family in most of the chemosensory tissues.[95] In the bristletail *Lepismachilis y-signata*, which lacks any Orco or other Or orthologs, it was shown that Irs are expressed in the majority of the antennal ORNs.[34] It is noteworthy that some members of the Ir family in *D. melanogaster* appear to function outside of the olfactory system and have been implicated in the detection of low salt taste and mating pheromones in the gustatory system.[94,96–98]

Functional Characterization of Irs

The assembly of subunits that comprise a functional Ir channel is more complex than in Or channels, but there are some similarities. Two Irs, Ir25a and Ir8a, are the most closely related to the iGluRs and are broadly expressed in *Ir*-expressing ORNs.[93,99] Additionally, *Ir25a* and *Ir8a* are required for odorant-evoked responses to several coeloconic ORNs and have been described as *Irco* receptors.[93,99] Another receptor, *Ir76b*, also shows broad expression in coeloconic sensilla and may act as a third coreceptor.[93,99] There is evidence that some *Irs* may function without these three known coreceptors, since knockout or knockdown of *Ir25a*, *Ir8a*, and *Ir76b* did not reduce ammonia responses of *Ir92a+* ORNs.[100] However, ectopic expression of *Ir92a* in *Or*-expressing neurons does not confer ammonia sensitivity, suggesting other coreceptor subunits may be required for proper function.[100]

Initial characterization studies have identified several ligands for odorant-binding *Irs*. By mapping individual *Irs* to specific coeloconic sensilla on the *Drosophila* antennae, it was confirmed that the *Irs* are responsible for the previously identified odorant-evoked responses in these sensilla.[93,101] Additionally, misexpression of specific *Irs* in other coeloconic ORNs conferred odorant response profiles similar to its endogenous neuron.[93] After screening an expanded panel of odorants, it was shown that *Ir*-expressing ORNs in *D. melanogaster* are primarily tuned to acids, amines, and aldehydes.[101] The *Or* repertoire in *Drosophila* is not tuned to these classes of ecologically important odorants, suggesting an important role of the *Ir* family in flies to cover this chemical space.

Ectopic expression of *Irs* has identified the subunit combinations that are sufficient to confer specific odorant responses in *Drosophila*. When expressed in an *Or*-expressing ORN, *Ir8a* and *Ir84a* are sufficient for a response to phenylacetaldehyde.[99] Interestingly, when either *Ir* is expressed alone, it is unable to localize to the dendrites of the ORN or respond to the odorant.[99] *Ir8a* and *Ir84a* have also been functionally characterized in *Xenopus* oocytes, where odorant responses were validated in the absence of other olfactory proteins, such as *Ors*, that were present in the fly expression paradigm.[99] Two other pairs of *Irs*, *Ir8a* + *Ir75a* and *Ir8a* + *Ir64a*, have also been characterized in oocytes and mirror the odorant-evoked responses observed in in vivo studies.[99,102]

Ir Structure and Function

The subunit stoichiometry of the *Ir8a* + *Ir84a* complex was examined using single-molecule imaging.[99] The data suggest that *Irs* assemble as heterotetramers, like AMPA and kainate iGluRs, and the *Ir8a* + *Ir84a* channel contains two subunits of each.[99] It appears that *Ir* complexes containing *Ir8a* may require only one odorant-binding *Ir* to function, while channels with *Ir25a* appear to be different.[99] When ectopically expressed, proper dendritic localization and odorant-evoked responses are only observed when three *Irs*—*Ir25a*, *Ir76a*, and *Ir76b*—are expressed together.[99] It appears the *Ir* family exhibits a complex oligomeric organization that is specific to the *Ir* subunits involved.

Irs, like *Ors*, also function as odorant-gated ion channels. Investigation of the cation permeability in *Ir* pairs expressed in oocytes has shown that these currents are mostly carried by the monovalent cations, Na^+ and K^+.[99] However, the *Ir8a* + *Ir84a* complexes did flux a relatively small amount of Ca^{2+} ions.[99] *Ir84a* retains a glutamine residue in the pore selectivity filter that has been associated with Ca^{2+} permeability in iGluRs, and indeed a mutation at this residue abolished Ca^{2+} currents.[99] Moreover, truncations and point mutations in the amino terminus and ligand-binding domain have been shown to affect dendritic localization as well as odorant responsiveness.[99]

Ir-Mediated Behavior

In flies and other insects, several *Irs* have been linked to specific behaviors. *Ir64a*, which is expressed in the sacculus of the *Drosophila* antenna with *Ir8a*, is tuned to acids and is shown to contribute to acid aversion behavior.[103] Another *D. melanogaster Ir*, *Ir84a*, has been shown to contribute to male courtship behavior.[104] The strong ligands of *Ir84a* are common to food sources, where mating and oviposition also occur.[104] It is believed that this *Ir*-mediated sensory pathway is responsible for the coordination of reproductive behaviors with suitable feeding and oviposition sites.[104]

Ammonia and other amines are attractive to drosophilids and mosquitoes, and it is believed that *Irs* mediate this behavior. *Ir92a* in *D. melanogaster* has been identified as an amine and ammonia receptor, and knockdown of this receptor results in decreased attraction.[100] Ammonia is a component of human sweat and an attractant for host-seeking mosquitoes.[105,106] However, a clear ortholog of *Ir92a* is not present in mosquito genomes. In *An. gambiae* larvae, knockdown of *AgIr76b* altered the butylamine-evoked behavior, suggesting that this receptor is necessary for amine detection in mosquitoes.[107]

A conserved *Ir* has been implicated in the detection and avoidance of DEET.[108] In *D. melanogaster*, *Ir40a*-expressing neurons in the sacculus were found to respond to DEET and several other repellents.[108] Furthermore, RNAi knockdown of *Ir40a* abolished DEET repellency in behavioral assays.[108] It has been proposed that DEET exerts its behavioral effect through multiple receptors and sensory modalities.[20,108–110] This is consistent with the observation that *Orco*-mutant *Aedes aegypti* continue to avoid a DEET-treated heat source, suggesting the involvement of another family of receptors.[111] Since DEET is a synthetic compound, the natural ligand for *Ir40a*-expressing ORNs is not known.

CO_2-SENSITIVE GUSTATORY RECEPTORS

After the *D. melanogaster* gustatory receptor *(Gr)* family was identified (see Chapter 14), a handful of *Grs* were identified that showed consistent expression in the antennae.[112,113] It was shown that the ORNs expressing one of these, *Gr21a*, send their axonal projections to a CO_2-sensitive glomerulus in the antennal lobe, and synaptic disruption of *Gr21a*-expressing ORNs impaired CO_2 avoidance behavior.[113,114] Another member of this family, *Gr63a*, is coexpressed with Gr21a, and a mutant in this gene lacked electrophysiological and behavioral responses to CO_2.[115,116] Indeed, ectopic coexpression of *Gr21a* and *Gr63a* confers CO_2 responses in the "empty neuron," establishing these Grs as bona fide olfactory receptors of CO_2.[115,116] Although these GRs respond to CO_2, other odorants have been shown to affect these receptors as well. In *D. melanogaster*, volatiles from ripening fruits are able to inhibit CO_2-sensitive ORNs in the antennae.[117] Heterologous expression of Gr21 and Gr63a showed that the inhibition was receptor-dependent, and the inhibitory odorants were able to abolish CO_2 avoidance.[117]

CO_2 Receptor Evolution and Conservation

CO_2 is a common sensory cue, and its detection is important for several insect species. While the *Gr63a/Gr21a* receptor is conserved in many insect species, the behavioral valence of CO$_2$ can be quite different. While *D. melanogaster* avoid CO$_2$ in a T-maze, hematophagous insects such as mosquitoes, sand flies, and kissing bugs are attracted to it and use plumes of CO$_2$ to navigate to host during host-seeking behavior.[118,119] Foraging hawkmoths also prefer flowers with higher levels of CO$_2$, as it often correlates with a greater amount of nectar.[120] Homing desert ants have been shown to track plumes of CO$_2$ emitted from underground nests.[121] Although the drosophilids have two CO$_2$-sensitive *Grs*, an additional paralog of *Gr21a* is present in mosquitoes, moths, and beetles.[122] However, expression of only *AgGr22* and *AgGr24*, the respective orthologs of *Gr21a* and *Gr63a* in *An. gambiae*, is able to confer CO$_2$ responses in flies.[123] Surprisingly, no orthologs of these *Grs* are found within ants, bees, and wasps, suggesting that the detection of CO$_2$ in Hymenopterans is mediated by other olfactory receptors.[122]

Structure/Function of CO$_2$ Receptors

The molecular mechanism of CO$_2$ receptor function has yet to be fully understood. In *D. melanogaster*, mutants of metabotropic pathways such as *Gα$_q$* and *Gγ30A* have reduced CO$_2$ responses, and the inclusion of *Gα$_q$* with ectopically expressed *Gr21a* and *Gr63a* greatly enhanced CO$_2$ sensitivity.[124] However, CO$_2$ receptors may also exhibit ionotropic function as another member of the *Gr* family, a fructose-sensitive *Gr* from *Bombyx mori*, exhibited cation channel properties in patch clamp experiments similar to the related *Or* family.[125] Further investigation is needed to elucidate the precise signaling mechanism of the CO$_2$ receptors.

Manipulation of CO$_2$ Detection in Mosquitoes

The importance of CO$_2$ as an attractive cue for mosquitoes and other hematophagous disease vectors has led to the identification of compounds that can alter host-seeking behavior. Moreover, several odorants from human skin were also found to activate the CO$_2$-sensitive ORN, cpA, of both *An. gambiae* and *Ae. aegypti* mosquitoes, making it a dual-detector of exhaled CO$_2$ and skin odors.[61,123] Electrophysiological and in silico chemical screens of the cpA neurons have led to the identification of a variety of ligands, including activators, inhibitors, and prolonged activators, which are capable of stimulating the ORN for over 5 min after application.[61,118,126]

The odorant activators of the CO$_2$ receptor could potentially be used as convenient evaporating lures in mosquito traps for surveillance or control. Further analysis of CO$_2$ receptor activators led to an in silico screen to predict novel, safe, and affordable ligands, and several candidate cpA activators and inhibitors were confirmed in vivo and displayed behavioral phenotypes.[61] A cpA activator, cyclopentanone, is an effective lure to attract and trap female *Culex quinquefasciatus* mosquitoes when tested in a greenhouse.[61] Conversely, a predicted cpA inhibitor, ethyl pyruvate, was found to inhibit skin-odor responses in *Ae. aegypti* and reduced their attraction to skin.[61]

Prolonged activation of cpA also has a strong behavioral phenotype, where both wind-tunnel attraction to CO$_2$ and house-entry behaviors were impaired.[126] Additionally, a prolonged inhibitor of the cpA neuron, butyryl chloride, has also been identified and exposure to it reduces CO$_2$ responses for up to 24 h.[61] Moreover, upwind attraction behavior toward human foot odor was greatly reduced in *Ae. aegypti* with an inactivated cpA neuron.[61] Currently, the mechanism of prolonged activation or inhibition of the cpA neuron remains unclear, but it is an effective method to disrupt CO_2-related behaviors in mosquitoes.[61,126]

Genetic manipulation of CO_2 receptors in mosquitoes has also been shown to disrupt host-seeking behavior. A mutant mosquito lacking a functional *AaGr3* was generated and electrophysiological responses to CO_2 were abolished.[127] In behavioral assays, *AaGr3* mutants showed severe defects in CO_2-evoked activation and attraction to both heat and lactic acid.[127] However, in assays using live animals, the host-seeking behavior of *AaGr3* mutants was found to be dependent on the spatial scale of the assay arena.[127] The *AaGr3* mutant mosquitoes had difficulty in finding mice in a large cage; however, this defect was not observed in small cages that pose less of a host-seeking challenge. Similarly, in greenhouse experiments *AaGr3* mutants exhibited only a minor reduction in attraction to human hosts.[127] Taken together, these observations suggest that in the absence of the CO_2 receptor pathway, mosquitoes can utilize other cues such as other skin-odor detecting receptors and heat to locate hosts.

BREAKOUT BOX

PROCESSING OF ODOR MIXTURES IN INSECTS

Although the majority of studies in insect olfaction utilize single-odorant stimuli, insects must be able to discriminate relevant volatile cues within a complex chemical environment. At the periphery, components of an odor mixture are expected to differentially affect ORNs based on the properties of olfactory receptor pharmacology and ephaptic coupling of ORNs.[128,129] The activation pattern in the periphery is conveyed to glomeruli within the antennal lobe (AL), where the combinatorial activity is integrated and relayed to higher brain regions using projection neurons (PNs).

ORN axons synapse with PNs in the glomeruli of the AL, but another group of neurons called local interneurons (LNs) allow for cross-talk between glomeruli and aid in the processing of complex odors. The lateral connections mediated by LNs are bimodal, excitatory, or inhibitory. One study found that after removing the direct synaptic input of an ORN, the corresponding PN could be excited by activation of other olfactory circuits.[130] By recruiting other PNs, it is believed that excitatory LNs amplify an odorant signal and increase sensitivity.[130,131] Additionally, LNs can also be inhibitory and have been shown to release GABA around ORN axon terminals to reduce synaptic transmission.[132] The role of LNs is believed to contribute to the gain control of olfactory processing, where excitatory and inhibitory LNs maintain a dynamic range of detection in both high and low odorant concentrations.[131,132]

In some insects, pheromonal cues are emitted as a complex of volatiles in specific ratios, and it appears that PNs can detect specific compound ratios within the AL. In the *Manduca sexta* sphinx moth, the natural ratio of its sex pheromone complex is 2:1, and deviation from this ratio results in reduced mating behaviors.[133] It was found that the natural ratio of the pheromone complex resulted in the most synchronous PN firing patterns.[133] This synchrony observed in the AL has also been suggested to correlate with floral attraction behaviors in *M. sexta*, and is believed to play a role in the specific detection of complex odor stimuli in the environment.[134]

OTHER OLFACTORY PROTEINS AT THE PERIPHERY

To facilitate the solubilization of hydrophobic volatiles into the aqueous sensillar lymph, it has been proposed that the lymph contains high amounts of odorant-binding proteins (OBPs) expressed in the support cells and secreted into the sensillar lymph.[135] In vitro biochemical experiments show that OBPs can bind to several odorant molecules.[136,137] Also, the addition of OBPs in heterologous expression systems has been shown to increase odorant sensitivity.[138,139]

An OBP in *Drosophila*, LUSH, binds to the pheromone cVA in the at1 sensillum.[140] Initial studies found that *lush* mutants were unable to detect cVA and spontaneous activity of the ORN was abolished.[140] Additionally, a single-point mutant of LUSH that closely resembles the cVA-bound state was able to activate at1 neurons in the absence of pheromone, leading to a model where the cVA-bound LUSH protein is the true ligand of Or67d.[141] However, a subsequent study has contradicted this model and demonstrated that high concentrations of cVA are able to evoke responses in *lush* mutants, supporting a model where cVA binds directly to *Or67d* to elicit ORN activity.[142] Additionally, *Or67d* expressed in the empty neuron conferred responses to cVA in the absence of LUSH, albeit at 1000-fold lower sensitivity than native ORNs.[66] With the exception of LUSH, genetic studies and in vivo analyses are lacking for OBPs. However, it is possible that OBPs could contribute to olfactory sensitivity at lower concentrations of pheromones and general odorants.

Another class of accessory proteins named the odorant-degrading enzymes (ODEs) is also secreted into the sensillar lymph, where these proteins have been proposed to break down accumulated odorant molecules. The best understood ODE, ApolPDE, is an esterase found in sensillar lymph and responsible for rapidly degrading sex pheromone in the giant silk moth *Antheraea polyphemus*.[143,144] Additional candidate ODEs have also been proposed, such as aldehyde oxidases and cytochrome P450s, due to antennal-specific expression and the ability to degrade both pheromones and general odorants.[145,146]

A family of scavenger receptors named sensory neuron membrane proteins (SNMPs) have a role in the detection of pheromones at the periphery.[147] In *D. melanogaster*, SNMP colocalizes with *Or67d* in the dendrites of cVA-sensitive ORNs, and a knockout of SNMP results in defects in cVA detection as well as highly elevated spontaneous activity in the ORN.[148] *Snmp* mutants still respond to high concentrations of cVA, but the kinetics of activation and deactivation were much slower than in the wild type.[149] These kinetic differences were also found in *Xenopus* oocytes expressing *BmOrco + BmOr1*, where coexpression of SNMP resulted in faster activation and deactivation.[149] These results suggest that in addition to the effect on odorant sensitivity, SNMP has a role in the temporal kinetics of the response. SNMP has also been implicated in the detection of farnesol, an attractive odorant found in ripe citrus, is detected by *Or83c*-expressing ORNs that also express SNMP.[150] SNMP mutants retain the ability to detect farnesol, but the activation and deactivation kinetics were significantly slower.[150]

Recent studies have found enriched expression of ammonium transporters (AMTs) in the antennae of both *An. gambiae* and *D. melanogaster*. In flies, *Amt* is localized to the support cells surrounding an ammonia-sensitive ORN that expresses *Ir92a*.[151] *Amt* mutants showed reduced electrophysiological responses to ammonia, while other odorant-evoked responses were unaffected.[151] In *Xenopus* oocytes, both ammonium and methylammonium evoked *AgAmt*-dependent inward currents.[152] These ions could be present in sensillar lymph as protonated ammonia or methylamine, or even as possible breakdown products after degradation of nitrogenous odorants. A possible role of AMTs could be one of lymph clearance, where AMTs prevent overaccumulation of ammonium and methylammonium ions in the sensillum lymph to maintain olfactory acuity.

METHODS FOR FUNCTIONAL CHARACTERIZATION OF INSECT OLFACTORY RECEPTORS

Expression in *Drosophila* ORNs

The function of *Ors*, *Irs*, and CO_2-sensitive *Grs* from several insect species has been examined through expression in selected ORNs. The "empty neuron" system has proven to be robust in the functional characterization of insect *Ors*.[4] In this line, a deletion has removed the coding regions of *Or22a* and *Or22b*, rendering the ab3A neuron unresponsive to odorants.[4] Using the ab3A promoter of *Or22a-GAL4*, one can drive expression of any *Or* from a *UAS-Or* transgene. This system allows for characterization of an *Or* in the absence of another *Or* using single-sensillum electrophysiological recording.[4] *Or* expression in the empty neuron confers a unique odorant response profile and spontaneous firing frequency that resembles that of its endogenous ORN and has been useful for mapping *Or* expression to particular ORN types.[5] Various aspects of the receptor properties are faithfully maintained, such as the odor response profile, odor response levels, spontaneous activity, and termination kinetics, indicating that the system is suitable for reliable deorphanization of most *Or* receptors. The empty neuron system has since been used to screen several members of the *Or* family in *D. melanogaster* and *An. gambiae*, as well as a handful of other olfactory receptors from various insects.[5,6,48,116,138]

Using a very similar GAL4-UAS approach, a second empty neuron system was developed in the trichoid sensilla at1 neuron, which normally expresses the cVA receptor, *Or67d*.[65] Some *Ors* work better in this trichoid neuron system. Expression of *Or83c* does not confer farnesol responses in the traditional ab3 empty neuron, but farnesol responses mimicked those of the endogenous ORN when expressed in the empty at1 system.[150] Together these two in vivo platforms provide a remarkably powerful *Or* deorphanization method for insect *Ors*. However, characterizing *Ors* from evolutionarily distant insect species should be interpreted with caution because these *Ors* could respond differently when complexed with a nonnative Orco from *Drosophila*.

Activity Imaging in Vivo from Neurons

Odorant ligands for specific ORNs have been identified in *Drosophila* by expressing a Ca^{2+}-sensitive protein from a UAS-GCaMP3 transgene using a receptor-specific promoter-*Gal4* transgene in the same fly.[108,153] This methodology is particularly suitable for ligand identification for *Ors* that are present in ORNs that are difficult to access with an electrode. With the advent of improved varieties of Ca^{2+}- and voltage-sensitive proteins, this method can provide even greater utility.

Expression in Cell Systems

The function of insect olfactory receptors has also been studied in vitro using the classic ion channel expression system, the *Xenopus laevis* oocyte. Once an olfactory receptor is cloned, complementary RNA (cRNA) is synthesized by in vitro transcription and then injected into mature oocytes. After allowing time for the translation and membrane localization of the receptor, two-electrode voltage clamp electrophysiology (TEVC) is used to measure currents during the aqueous perfusion of odorant. It is also possible to perform patch clamp electrophysiology on excised membranes, and high-throughput automated platforms exist for both cRNA injection and TEVC. To date, receptors from both the *Or* and *Ir* families have been characterized in oocytes, and the response profiles have been similar when compared to those observed in their native ORNs.[41,99,102,123]

Functional insect *Ors* can also be expressed in cultured mammalian cells. Expression vectors containing *Orco* and an *OrX* can be transfected in a variety of cultured cells, with human embryonic kidney (HEK) and Chinese hamster ovary (CHO) cells being the most popular. Expression of the *Ors* can be transient or the transgenes can be integrated into the genome, resulting in stable, and sometimes inducible, *Or*-expressing cell lines. *Or* function can be assayed using patch clamp electrophysiology (whole cell or excised membrane patches) or imaging using Ca^{2+}-sensitive dyes, such as Fura-2 and Fluo4. Cultured cells are also very amenable to 96- and 384-well plating formats and have been used for high-throughput screening of *Ors*.[44,50] Only receptors in the *Or* family have functioned in mammalian cell systems thus far.

In both of these in vitro expression systems, odorants and other molecules are delivered in an aqueous phase. To help solubilize these often-hydrophobic molecules, a solvent such as DMSO is added to the assay buffer. Recombinant OBPs can also be directly added to the assay buffer, and this method has been shown to increase the sensitivity of moth pheromone receptors in oocytes.[139] In these cell systems, it is also possible that some important factors of in vivo ORNs are missing and may be required for certain response characteristics. As was the case for SNMP, these factors can be coexpressed with olfactory receptors and assayed for their role in olfactory signaling.[149]

In Silico Screening

Recent advances in chemical informatics have led to development of a computational screening pipeline to discover new ligands for *Or*, *Irs*, and *Grs*. The computer first analyzes three-dimensional structures of the known ligands to identify shared structural features of actives. Using this structural information, the computer next screens large chemical structure databases to identify ligands that are close in structure to known actives.[61,108,154] The success rates reported so far are >70% for predictions that are validated by electrophysiology.[154]

With the continued development of new screening platforms for insect olfactory receptors, such as in silico screens and cell-free systems, a plethora of options are becoming available to advance our understanding of olfactory receptor function and ultimately how insects navigate a chemical environment.[154,155]

References

1. Böröczky K, Wada-Katsumata A, Batchelor D, Zhukovskaya M, Schal C. Insects groom their antennae to enhance olfactory acuity. *Proc Natl Acad Sci*. 2013;110(9):3615−3620. http://dx.doi.org/10.1073/pnas.1212466110.
2. Steinbrecht RA. Odorant-binding proteins: expression and function. *Ann NY Acad Sci*. 1998;855:323−332.
3. Goldman AL, van Naters WVDG, Lessing D, Warr CG, Carlson JR. Coexpression of two functional odor receptors in one neuron. *Neuron*. 2005;45(5):661−666. http://dx.doi.org/10.1016/j.neuron.2005.01.025.
4. Dobritsa AA, van der Goes van Naters W, Warr CG, Steinbrecht RA, Carlson JR. Integrating the molecular and cellular basis of odor coding in the *Drosophila* antenna. *Neuron*. 2003;37(5):827−841.
5. Hallem EA, Ho MG, Carlson JR. The molecular basis of odor coding in the *Drosophila* antenna. *Cell*. 2004;117(7):965−979. http://dx.doi.org/10.1016/j.cell.2004.05.012.
6. Hallem EA, Carlson JR. Coding of odors by a receptor repertoire. *Cell*. 2006;125(1):143−160. http://dx.doi.org/10.1016/j.cell.2006.01.050.
7. Vosshall LB, Wong AM, Axel R. An olfactory sensory map in the fly brain. *Cell*. 2000;102(2):147−159.
8. Couto A, Alenius M, Dickson BJ. Molecular, anatomical, and functional organization of the *Drosophila* olfactory system. *Curr Biol*. 2005;15(17):1535−1547. http://dx.doi.org/10.1016/j.cub.2005.07.034.
9. Marin EC, Jefferis GSXE, Komiyama T, Zhu H, Luo L. Representation of the glomerular olfactory map in the *Drosophila* brain. *Cell*. 2002;109(2):243−255.
10. Wong AM, Wang JW, Axel R. Spatial representation of the glomerular map in the *Drosophila* protocerebrum. *Cell*. 2002;109(2):229−241.
11. Clyne PJ, Warr CG, Freeman MR, Lessing D, Kim J, Carlson JR. A novel family of divergent seven-transmembrane proteins: candidate odorant receptors in *Drosophila*. *Neuron*. 1999;22(2):327−338.
12. Vosshall LB, Amrein H, Morozov PS, Rzhetsky A, Axel R. A spatial map of olfactory receptor expression in the *Drosophila* antenna. *Cell*. 1999;96(5):725−736.
13. Gao Q, Chess A. Identification of candidate *Drosophila* olfactory receptors from genomic DNA sequence. *Genomics*. 1999;60(1):31−39. http://dx.doi.org/10.1006/geno.1999.5894.
14. Buck L, Axel R. A novel multigene family may encode odorant receptors: a molecular basis for odor recognition. *Cell*. 1991;65(1):175−187.
15. Troemel ER, Chou JH, Dwyer ND, Colbert HA, Bargmann CI. Divergent seven transmembrane receptors are candidate chemosensory receptors in *C. elegans*. *Cell*. 1995;83(2):207−218.
16. Robertson HM, Warr CG, Carlson JR. Molecular evolution of the insect chemoreceptor gene superfamily in *Drosophila melanogaster*. *Proc Natl Acad Sci USA*. 2003;100(Suppl 2):14537−14542. http://dx.doi.org/10.1073/pnas.2335847100.
17. Zhou X, Slone JD, Rokas A, et al. Phylogenetic and transcriptomic analysis of chemosensory receptors in a pair of divergent ant species reveals sex-specific signatures of odor coding. *PLoS Genet*. 2012;8(8):e1002930. http://dx.doi.org/10.1371/journal.pgen.1002930.
18. Kreher SA, Kwon JY, Carlson JR. The molecular basis of odor coding in the *Drosophila* larva. *Neuron*. 2005;46(3):445−456. http://dx.doi.org/10.1016/j.neuron.2005.04.007.
19. Fishilevich E, Domingos AI, Asahina K, Naef F, Vosshall LB, Louis M. Chemotaxis behavior mediated by single larval olfactory neurons in *Drosophila*. *Curr Biol*. 2005;15(23):2086−2096. http://dx.doi.org/10.1016/j.cub.2005.11.016.
20. Xia Y, Wang G, Buscariollo D, Pitts RJ, Wenger H, Zwiebel LJ. The molecular and cellular basis of olfactory-driven behavior in *Anopheles gambiae* larvae. *Proc Natl Acad Sci*. 2008;105(17):6433−6438. http://dx.doi.org/10.1073/pnas.0801007105.
21. Bohbot J, Pitts RJ, Kwon H-W, Rützler M, Robertson HM, Zwiebel LJ. Molecular characterization of the *Aedes aegypti* odorant receptor gene family. *Insect Mol Biol*. 2007;16(5):525−537. http://dx.doi.org/10.1111/j.1365-2583.2007.00748.x.
22. Kwon H-W, Lu T, Rützler M, Zwiebel LJ. Olfactory responses in a gustatory organ of the malaria vector mosquito *Anopheles gambiae*. *Proc Natl Acad Sci USA*. 2006;103(36):13526−13531. http://dx.doi.org/10.1073/pnas.0601107103.
23. Pitts RJ, Liu C, Zhou X, Malpartida JC, Zwiebel LJ. Odorant receptor-mediated sperm activation in disease vector mosquitoes. *Proc Natl Acad Sci*. 2014. http://dx.doi.org/10.1073/pnas.1322923111.
24. Pitts RJ, Rinker DC, Jones PL, Rokas A, Zwiebel LJ. Transcriptome profiling of chemosensory appendages in the malaria vector *Anopheles gambiae* reveals tissue- and sex-specific signatures of odor coding. *BMC Genomics*. 2011;12(1):271. http://dx.doi.org/10.1186/1471-2164-12-271.

25. Rinker DC, Pitts RJ, Zhou X, Suh E, Rokas A, Zwiebel LJ. Blood meal-induced changes to antennal transcriptome profiles reveal shifts in odor sensitivities in *Anopheles gambiae*. *Proc Natl Acad Sci*. 2013. http://dx.doi.org/10.1073/pnas.1302562110.

26. Mitchell RF, Hughes DT, Luetje CW, et al. Sequencing and characterizing odorant receptors of the cerambycid beetle *Megacyllene caryae*. *Insect Biochem Mol Biol*. 2012. http://dx.doi.org/10.1016/j.ibmb.2012.03.007.

27. Andersson MN, Grosse-Wilde E, Keeling CI, et al. Antennal transcriptome analysis of the chemosensory gene families in the tree killing bark beetles, Ips typographus and *Dendroctonus ponderosae* (Coleoptera: Curculionidae: Scolytinae). *BMC Genomics*. 2013;14:198. http://dx.doi.org/10.1186/1471-2164-14-198.

28. Hill CA, Fox AN, Pitts RJ, et al. G protein-coupled receptors in *Anopheles gambiae*. *Science*. 2002;298(5591):176–178. http://dx.doi.org/10.1126/science.1076196.

29. Krieger J, Klink O, Mohl C, Raming K, Breer H. A candidate olfactory receptor subtype highly conserved across different insect orders. *J Comp Physiol A*. 2003;189(7):519–526. http://dx.doi.org/10.1007/s00359-003-0427-x.

30. Pitts RJ, Fox AN, Zwiebel LJ. A highly conserved candidate chemoreceptor expressed in both olfactory and gustatory tissues in the malaria vector *Anopheles gambiae*. *Proc Natl Acad Sci USA*. 2004;101(14):5058–5063. http://dx.doi.org/10.1073/pnas.0308146101.

31. Xia Y, Zwiebel LJ. Identification and characterization of an odorant receptor from the West Nile virus mosquito, *Culex quinquefasciatus*. *Insect Biochem Mol Biol*. 2006;36(3):169–176. http://dx.doi.org/10.1016/j.ibmb.2005.12.003.

32. Yang Y, Krieger J, Zhang L, Breer H. The olfactory co-receptor Orco from the migratory locust (*Locusta migratoria*) and the desert locust (*Schistocerca gregaria*): identification and expression pattern. *Int J Biol Sci*. 2012;8(2):159–170.

33. Hull JJ, Hoffmann EJ, Perera OP, Snodgrass GL. Identification of the western tarnished plant bug (*Lygus hesperus*) olfactory co-receptor orco: expression profile and confirmation of atypical membrane topology. *Arch Insect Biochem Physiol*. 2012;81(4):179–198. http://dx.doi.org/10.1002/arch.21042.

34. Mißbach C, Dweck HK, Vogel H, et al. Evolution of insect olfactory receptors. *eLife*. 2014;3:e02115. http://dx.doi.org/10.7554/eLife.02115.

35. Larsson MC, Domingos AI, Jones WD, Chiappe ME, Amrein H, Vosshall LB. Or83b encodes a broadly expressed odorant receptor essential for *Drosophila* olfaction. *Neuron*. 2004;43(5):703–714. http://dx.doi.org/10.1016/j.neuron.2004.08.019.

36. Jones WD, Nguyen T-AT, Kloss B, Lee KJ, Vosshall LB. Functional conservation of an insect odorant receptor gene across 250 million years of evolution. *Curr Biol*. 2005;15(4):R119–R121. http://dx.doi.org/10.1016/j.cub.2005.02.007.

37. Bohbot JD, Jones PL, Wang G, Pitts RJ, Pask GM, Zwiebel LJ. Conservation of indole responsive odorant receptors in mosquitoes reveals an ancient olfactory trait. *Chem Senses*. 2011;36(2):149–160. http://dx.doi.org/10.1093/chemse/bjq105.

38. de Bruyne M, Smart R, Zammit E, Warr CG. Functional and molecular evolution of olfactory neurons and receptors for aliphatic esters across the *Drosophila* genus. *J Comp Physiol A Neuroethol Sens Neural Behav Physiol*. 2010;196(2):97–109. http://dx.doi.org/10.1007/s00359-009-0496-6.

39. Mcbride CS, Arguello JR, O'meara BC. Five *Drosophila* genomes reveal nonneutral evolution and the signature of host specialization in the chemoreceptor superfamily. *Genetics*. 2007;177(3):1395–1416. http://dx.doi.org/10.1534/genetics.107.078683.

40. Dekker T, Ibba I, Siju KP, Stensmyr MC, Hansson BS. Olfactory shifts parallel superspecialism for toxic fruit in *Drosophila melanogaster* sibling, *D. sechellia*. *Curr Biol*. 2006;16(1):101–109. http://dx.doi.org/10.1016/j.cub.2005.11.075.

41. Wetzel CH, Behrendt HJ, Gisselmann G, Störtkuhl KF, Hovemann B, Hatt H. Functional expression and characterization of a *Drosophila* odorant receptor in a heterologous cell system. *Proc Natl Acad Sci USA*. 2001;98(16):9377–9380. http://dx.doi.org/10.1073/pnas.151103998.

42. Störtkuhl KF, Kettler R. Functional analysis of an olfactory receptor in *Drosophila melanogaster*. *Proc Natl Acad Sci USA*. 2001;98(16):9381–9385. http://dx.doi.org/10.1073/pnas.151105698.

43. Sato K, Pellegrino M, Nakagawa T, Nakagawa T, Vosshall LB, Touhara K. Insect olfactory receptors are heteromeric ligand-gated ion channels. *Nature*. 2008;452(7190):1002–1006. http://dx.doi.org/10.1038/nature06850.

44. Jones PL, Pask GM, Rinker DC, Zwiebel LJ. Functional agonism of insect odorant receptor ion channels. *Proc Natl Acad Sci*. 2011;108(21):8821–8825. http://dx.doi.org/10.1073/pnas.1102425108.

45. Pask GM, Bobkov YV, Corey EA, Ache BW, Zwiebel LJ. Blockade of insect odorant receptor currents by amiloride derivatives. *Chem Senses*. 2013. http://dx.doi.org/10.1093/chemse/bjs100.

46. Hallem EA, Fox AN, Zwiebel LJ, Carlson JR. Olfaction: mosquito receptor for human-sweat odorant. *Nature.* 2004;427(6971):212—213. http://dx.doi.org/10.1038/427212a.

47. Jones PL, Pask GM, Romaine IM, et al. Allosteric antagonism of insect odorant receptor ion channels. *PLoS One.* 2012;7(1):e30304. http://dx.doi.org/10.1371/journal.pone.0030304.

48. Carey AF, Wang G, Su C-Y, Zwiebel LJ, Carlson JR. Odorant reception in the malaria mosquito *Anopheles gambiae. Nature.* 2010;464(7285):66—71. http://dx.doi.org/10.1038/nature08834.

49. Wang G, Carey AF, Carlson JR, Zwiebel LJ. Molecular basis of odor coding in the malaria vector mosquito *Anopheles gambiae. Proc Natl Acad Sci.* 2010;107(9):4418—4423. http://dx.doi.org/10.1073/pnas.0913392107.

50. Rinker DC, Jones PL, Pitts RJ, et al. Novel high-throughput screens of *Anopheles gambiae* odorant receptors reveal candidate behaviour-modifying chemicals for mosquitoes. *Physiol Entomol.* 2012;37(1):33—41. http://dx.doi.org/10.1111/j.1365-3032.2011.00821.x.

51. Chen S, Luetje CW. Identification of new agonists and antagonists of the insect odorant receptor co-receptor subunit. *PLoS One.* 2012;7(5):e36784. http://dx.doi.org/10.1371/journal.pone.0036784.

52. Taylor RW, Romaine IM, Liu C, et al. Structure-activity relationship of a broad-spectrum insect odorant receptor agonist. *ACS Chem Biol.* 2012. http://dx.doi.org/10.1021/cb300331z.

53. Romaine IM, Taylor RW, Saidu SP, et al. Narrow SAR in odorant sensing Orco receptor agonists. *Bioorg Med Chem Lett.* 2014;24(12):2613—2616. http://dx.doi.org/10.1016/j.bmcl.2014.04.081.

54. Röllecke K, Werner M, Ziemba PM, Neuhaus EM, Hatt H, Gisselmann G. Amiloride derivatives are effective blockers of insect odorant receptors. *Chem Senses.* 2013. http://dx.doi.org/10.1093/chemse/bjs140.

55. Stensmyr MC, Dweck HKM, Farhan A, et al. A conserved dedicated olfactory circuit for detecting harmful microbes in *Drosophila. Cell.* 2012;151(6):1345—1357. http://dx.doi.org/10.1016/j.cell.2012.09.046.

56. Dweck HKM, Ebrahim SAM, Kromann S, et al. Olfactory preference for egg laying on citrus substrates in *Drosophila. Curr Biol.* 2013. http://dx.doi.org/10.1016/j.cub.2013.10.047.

57. Semmelhack JL, Wang JW. Select *Drosophila* glomeruli mediate innate olfactory attraction and aversion. *Nature.* 2009;459(7244):218—223. http://dx.doi.org/10.1038/nature07983.

58. Okumu FO, Killeen GF, Ogoma S, et al. Development and field evaluation of a synthetic mosquito lure that is more attractive than humans. *PLoS One.* 2010;5(1):e8951. http://dx.doi.org/10.1371/journal.pone.0008951.

59. Mukabana WR, Mweresa CK, Otieno B, et al. A novel synthetic odorant blend for trapping of malaria and other African mosquito species. *J Chem Ecol.* 2012. http://dx.doi.org/10.1007/s10886-012-0088-8.

60. Rinker DC, Zhou X, Pitts RJ, AGC Consortium, Rokas A, Zwiebel LJ. Antennal transcriptome profiles of anopheline mosquitoes reveal human host olfactory specialization in *Anopheles gambiae. BMC Genomics.* 2013;14:749. http://dx.doi.org/10.1186/1471-2164-14-749.

61. Tauxe GM, Macwilliam D, Boyle SM, Guda T, Ray A. Targeting a dual detector of skin and CO_2 to modify mosquito host seeking. *Cell.* 2013;155(6):1365—1379. http://dx.doi.org/10.1016/j.cell.2013.11.013.

62. Sakurai T, Nakagawa T, Mitsuno H, et al. Identification and functional characterization of a sex pheromone receptor in the silkmoth *Bombyx mori. Proc Natl Acad Sci USA.* 2004;101(47):16653—16658. http://dx.doi.org/10.1073/pnas.0407596101.

63. Nakagawa T, Sakurai T, Nishioka T, Touhara K. Insect sex-pheromone signals mediated by specific combinations of olfactory receptors. *Science.* 2005;307(5715):1638—1642. http://dx.doi.org/10.1126/science.1106267.

64. Ha TS, Smith DP. A pheromone receptor mediates 11-cis-vaccenyl acetate-induced responses in *Drosophila. J Neurosci.* 2006;26(34):8727—8733. http://dx.doi.org/10.1523/JNEUROSCI.0876-06.2006.

65. Kurtovic A, Widmer A, Dickson BJ. A single class of olfactory neurons mediates behavioural responses to a *Drosophila* sex pheromone. *Nature.* 2007;446(7135):542—546. http://dx.doi.org/10.1038/nature05672.

66. van der Goes van Naters W, Carlson JR. Receptors and neurons for fly odors in *Drosophila. Curr Biol.* 2007;17(7):606—612. http://dx.doi.org/10.1016/j.cub.2007.02.043.

67. Wanner KW, Nichols AS, Walden KKO, Brockmann A, Luetje CW, Robertson HM. A honey bee odorant receptor for the queen substance 9-oxo-2-decenoic acid. *Proc Natl Acad Sci USA.* 2007;104(36):14383—14388. http://dx.doi.org/10.1073/pnas.0705459104.

68. Kaupp UB. Olfactory signalling in vertebrates and insects: differences and commonalities. *Nat Rev Neurosci.* 2010;11(3):188—200. http://dx.doi.org/10.1038/nrn2789.

69. Benton R, Sachse S, Michnick SW, Vosshall LB. Atypical membrane topology and heteromeric function of *Drosophila* odorant receptors in vivo. *PLoS Biol.* 2006;4(2):e20. http://dx.doi.org/10.1371/journal.pbio.0040020.

70. Lundin C, Käll L, Kreher SA, et al. Membrane topology of the *Drosophila* OR83b odorant receptor. *FEBS Lett.* 2007;581(29):5601—5604. http://dx.doi.org/10.1016/j.febslet.2007.11.007.

71. Smart R, Kiely A, Beale M, et al. *Drosophila* odorant receptors are novel seven transmembrane domain proteins that can signal independently of heterotrimeric G proteins. *Insect Biochem Mol Biol.* 2008;38(8):770—780. http://dx.doi.org/10.1016/j.ibmb.2008.05.002.

72. Zufall F, Hatt H. A calcium-activated nonspecific cation channel from olfactory receptor neurones of the silkmoth *Antheraea polyphemus. J Exp Biol.* 1991;161:455—468.

73. Zufall F, Hatt H. Dual activation of a sex pheromone-dependent ion channel from insect olfactory dendrites by protein kinase C activators and cyclic GMP. *Proc Natl Acad Sci USA.* 1991;88(19):8520—8524.

74. Walker WB, Smith EM, Jan T, Zwiebel LJ. A functional role for *Anopheles gambiae* Arrestin1 in olfactory signal transduction. *J Insect Physiol.* 2008;54(4):680—690. http://dx.doi.org/10.1016/j.jinsphys.2008.01.007.

75. Kain P, Chakraborty TS, Sundaram S, Siddiqi O, Rodrigues V, Hasan G. Reduced odor responses from antennal neurons of Gq{alpha}, phospholipase C{beta}, and rdgA mutants in *Drosophila* support a role for a phospholipid intermediate in insect olfactory transduction. *J Neurosci.* 2008;28(18):4745—4755. http://dx.doi.org/10.1523/JNEUROSCI.5306-07.2008.

76. Boekhoff I, Raming K, Breer H. Pheromone-induced stimulation of inositol-trisphosphate formation in insect antennae is mediated by G-proteins. *J Comp Physiol B.* 1990;160(1):99—103.

77. Wicher D, Schäfer R, Bauernfeind R, et al. *Drosophila* odorant receptors are both ligand-gated and cyclic-nucleotide-activated cation channels. *Nature.* 2008;452(7190):1007—1011. http://dx.doi.org/10.1038/nature06861.

78. Sargsyan V, Getahun MN, Llanos SL, Olsson SB, Hansson BS, Wicher D. Phosphorylation via PKC regulates the function of the *Drosophila* odorant co-receptor. *Front Cell Neurosci.* 2011;5:5. http://dx.doi.org/10.3389/fncel.2011.00005.

79. Pask GM, Jones PL, Rützler M, Rinker DC, Zwiebel LJ. Heteromeric Anopheline odorant receptors exhibit distinct channel properties. *PLoS One.* 2011;6(12):e28774. http://dx.doi.org/10.1371/journal.pone.0028774.

80. Zufall F, Stengl M, Franke C, Hildebrand JG, Hatt H. Ionic currents of cultured olfactory receptor neurons from antennae of male *Manduca sexta. J Neurosci.* 1991;11(4):956—965.

81. Kaissling K-E. Single unit and electroantennogram recordings in insect olfactory organs. In: *Experimental Cell Biology of Taste and Olfaction.* 1995:361—377.

82. Neuhaus EM, Gisselmann G, Zhang W, Dooley R, Störtkuhl K, Hatt H. Odorant receptor heterodimerization in the olfactory system of *Drosophila melanogaster. Nat Neurosci.* 2005;8(1):15—17. http://dx.doi.org/10.1038/nn1371.

83. Nichols AS, Chen S, Luetje CW. Subunit contributions to insect olfactory receptor function: channel block and odorant recognition. *Chem Senses.* 2011;36(9):781—790. http://dx.doi.org/10.1093/chemse/bjr053.

84. Nakagawa T, Pellegrino M, Sato K, Vosshall LB, Touhara K. Amino acid residues contributing to function of the heteromeric insect olfactory receptor complex. *PLoS One.* 2012;7(3):e32372. http://dx.doi.org/10.1371/journal.pone.0032372.

85. Kumar BN, Taylor RW, Pask GM, Zwiebel LJ, Newcomb RD, Christie DL. A conserved aspartic acid is important for agonist (VUAA1) and odorant/tuning receptor-dependent activation of the insect odorant co-receptor (Orco). *PLoS One.* 2013;8(7):e70218. http://dx.doi.org/10.1371/journal.pone.0070218.

86. Turner RM, Derryberry SL, Kumar BN, et al. Mutational analysis of cysteine residues of the insect odorant co-receptor (Orco) from *Drosophila melanogaster* reveals differential effects on agonist- and odorant/tuning receptor-dependent activation. *J Biol Chem.* 2014. http://dx.doi.org/10.1074/jbc.M114.603993.

87. Nichols AS, Luetje CW. Transmembrane segment 3 of *Drosophila melanogaster* odorant receptor subunit 85B contributes to ligand-receptor interactions. *J Biol Chem.* 2010:1—26. http://dx.doi.org/10.1074/jbc.M109.058321.

88. Leary GP, Allen JE, Bunger PL, et al. Single mutation to a sex pheromone receptor provides adaptive specificity between closely related moth species. *Proc Natl Acad Sci.* 2012;109(35):14081—14086. http://dx.doi.org/10.1073/pnas.1204661109.

89. Hughes DT, Wang G, Zwiebel LJ, Luetje CW. A determinant of odorant specificity is located at the extracellular loop 2-transmembrane domain 4 interface of an *Anopheles gambiae* odorant receptor subunit. *Chem Senses.* 2014. http://dx.doi.org/10.1093/chemse/bju048.

90. Xu P, Leal WS. Probing insect odorant receptors with their cognate ligands: insights into structural features. *Biochem Biophys Res Comm.* 2013;435(3):477—482. http://dx.doi.org/10.1016/j.bbrc.2013.05.015.

91. Ray A, van Naters WG, Carlson JR. Molecular determinants of odorant receptor function in insects. *J Biosci.* 2014;39(4):555—563.

92. Yao CA, Ignell R, Carlson JR. Chemosensory coding by neurons in the coeloconic sensilla of the *Drosophila* antenna. *J Neurosci.* 2005;25(37):8359—8367. http://dx.doi.org/10.1523/JNEUROSCI.2432-05.2005.

93. Benton R, Vannice KS, Gomez-Diaz C, Vosshall LB. Variant ionotropic glutamate receptors as chemosensory receptors in *Drosophila*. *Cell*. 2009;136(1):149–162. http://dx.doi.org/10.1016/j.cell.2008.12.001.

94. Croset V, Rytz R, Cummins SF, et al. Ancient protostome origin of chemosensory ionotropic glutamate receptors and the evolution of insect taste and olfaction. *PLoS Genet*. 2010;6(8):e1001064. http://dx.doi.org/10.1371/journal.pgen.1001064.

95. Corey EA, Bobkov Y, Ukhanov K, Ache BW. Ionotropic Crustacean olfactory receptors. *PLoS One*. 2013;8(4):e60551. http://dx.doi.org/10.1371/journal.pone.0060551.

96. Voets T, Prenen J, Vriens J, et al. Molecular determinants of permeation through the cation channel TRPV4. *J Biol Chem*. 2002;277(37):33704.

97. Zhang YV, Ni J, Montell C. The molecular basis for attractive salt-taste coding in *Drosophila*. *Science*. 2013;340(6138):1334–1338. http://dx.doi.org/10.1126/science.1234133.

98. Koh T-W, He Z, Gorur-Shandilya S, et al. The *Drosophila* IR20a clade of ionotropic receptors are candidate taste and pheromone receptors. *Neuron*. 2014;83(4):850–865. http://dx.doi.org/10.1016/j.neuron.2014.07.012.

99. Abuin L, Bargeton B, Ulbrich MH, Isacoff EY, Kellenberger S, Benton R. Functional architecture of olfactory ionotropic glutamate receptors. *Neuron*. 2011;69(1):44–60. http://dx.doi.org/10.1016/j.neuron.2010.11.042.

100. Min S, Ai M, Shin SA, Suh GSB. Dedicated olfactory neurons mediating attraction behavior to ammonia and amines in *Drosophila*. *Proc Natl Acad Sci*. 2013. http://dx.doi.org/10.1073/pnas.1215680110.

101. Silbering AF, Rytz R, Grosjean Y, et al. Complementary function and integrated wiring of the evolutionarily distinct *Drosophila* olfactory subsystems. *J Neurosci*. 2011;31(38):13357–13375. http://dx.doi.org/10.1523/JNEUROSCI.2360-11.2011.

102. Ai M, Blais S, Park J-Y, Min S, Neubert TA, Suh GSB. Ionotropic glutamate receptors IR64a and IR8a form a functional odorant receptor complex in vivo in *Drosophila*. *J Neurosci*. 2013;33(26):10741–10749. http://dx.doi.org/10.1523/JNEUROSCI.5419-12.2013.

103. Ai M, Min S, Grosjean Y, et al. Acid sensing by the *Drosophila* olfactory system. *Nature*. 2010;468(7324):691–695. http://dx.doi.org/10.1038/nature09537.

104. Grosjean Y, Rytz R, Farine J-P, et al. An olfactory receptor for food-derived odours promotes male courtship in *Drosophila*. *Nature*. 2011;478(7368):236–240. http://dx.doi.org/10.1038/nature10428.

105. Cork A, Park KC. Identification of electrophysiologically-active compounds for the malaria mosquito, *Anopheles gambiae*, in human sweat extracts. *Med Vet Entomol*. 1996;10(3):269–276.

106. Smallegange RC, Qiu YT, van Loon JJA, Takken W. Synergism between ammonia, lactic acid and carboxylic acids as kairomones in the host-seeking behaviour of the malaria mosquito *Anopheles gambiae* sensu stricto (Diptera: Culicidae). *Chem Senses*. 2005;30(2):145–152. http://dx.doi.org/10.1093/chemse/bji010.

107. Liu C, Pitts RJ, Bohbot JD, Jones PL, Wang G, Zwiebel LJ. Distinct olfactory signaling mechanisms in the malaria vector mosquito *Anopheles gambiae*. *PLoS Biol*. 2010;8(8). http://dx.doi.org/10.1371/journal.pbio.1000467.

108. Kain P, Boyle SM, Tharadra SK, et al. Odour receptors and neurons for DEET and new insect repellents. *Nature*. 2013. http://dx.doi.org/10.1038/nature12594.

109. Degennaro M, McBride CS, Seeholzer L, et al. Orco mutant mosquitoes lose strong preference for humans and are not repelled by volatile DEET. *Nature*. 2013. http://dx.doi.org/10.1038/nature12206.

110. Xu P, Choo Y-M, La Rosa De A, Leal WS. Mosquito odorant receptor for DEET and methyl jasmonate. *Proc Natl Acad Sci*. 2014. http://dx.doi.org/10.1073/pnas.1417244111.

111. Guda T, Kain P, Sharma KR, Pham CK, Ray A. *Repellent Compound with Larger Protective Zone than DEET Identified through Activity-Screening of Ir40a Neurons, Does not Require or Function*. 2015. http://dx.doi.org/10.1101/017145.

112. Clyne PJ, Warr CG, Carlson JR. Candidate taste receptors in *Drosophila*. *Science*. 2000;287(5459):1830–1834.

113. Scott K, Brady R, Cravchik A, et al. A chemosensory gene family encoding candidate gustatory and olfactory receptors in *Drosophila*. *Cell*. 2001;104(5):661–673.

114. Suh GSB, Wong AM, Hergarden AC, et al. A single population of olfactory sensory neurons mediates an innate avoidance behaviour in *Drosophila*. *Nature*. 2004;431(7010):854–859. http://dx.doi.org/10.1038/nature02980.

115. Jones WD, Cayirlioglu P, Kadow IG, Vosshall LB. Two chemosensory receptors together mediate carbon dioxide detection in *Drosophila*. *Nature*. 2007;445(7123):86–90. http://dx.doi.org/10.1038/nature05466.

116. Kwon JY, Dahanukar A, Weiss LA, Carlson JR. The molecular basis of CO_2 reception in *Drosophila*. *Proc Natl Acad Sci USA*. 2007;104(9):3574–3578. http://dx.doi.org/10.1073/pnas.0700079104.

117. Turner SL, Ray A. Modification of CO_2 avoidance behaviour in *Drosophila* by inhibitory odorants. *Nature*. 2009;461(7261):277–281. http://dx.doi.org/10.1038/nature08295.

118. Dekker T, Geier M, Cardé RT. Carbon dioxide instantly sensitizes female yellow fever mosquitoes to human skin odours. *J Exp Biol*. 2005;208(Pt 15):2963–2972. http://dx.doi.org/10.1242/jeb.01736.
119. Taneja J, Guerin PM. Oriented responses of the triatomine bugs *Rhodnius prolixus* and *Triatoma infestans* to vertebrate odours on a servosphere. *J Comp Physiol A*. 1995;176(4):455–464.
120. Thom C, Guerenstein PG, Mechaber WL, Hildebrand JG. Floral CO_2 reveals flower profitability to moths. *J Chem Ecol*. 2004;30(6):1285–1288.
121. Buehlmann C, Hansson BS, Knaden M. Path integration controls nest-plume following in desert ants. *Curr Biol*. 2012;22(7):645–649. http://dx.doi.org/10.1016/j.cub.2012.02.029.
122. Robertson HM, Kent LB. Evolution of the gene lineage encoding the carbon dioxide receptor in insects. *J Insect Sci*. 2009;9:19. http://dx.doi.org/10.1673/031.009.1901.
123. Lu T, Qiu YT, Wang G, et al. Odor coding in the maxillary palp of the malaria vector mosquito *Anopheles gambiae*. *Curr Biol*. 2007;17(18):1533–1544. http://dx.doi.org/10.1016/j.cub.2007.07.062.
124. Yao CA, Carlson JR. Role of G-proteins in odor-sensing and CO_2-sensing neurons in *Drosophila*. *J Neurosci*. 2010;30(13):4562–4572. http://dx.doi.org/10.1523/JNEUROSCI.6357-09.2010.
125. Sato K, Tanaka K, Touhara K. Sugar-regulated cation channel formed by an insect gustatory receptor. *Proc Natl Acad Sci*. 2011;108(28):11680–11685. http://dx.doi.org/10.1073/pnas.1019622108.
126. Turner SL, Li N, Guda T, Githure J, Cardé RT, Ray A. Ultra-prolonged activation of CO_2-sensing neurons disorients mosquitoes. *Nature*. 2011;474(7349):87–91. http://dx.doi.org/10.1038/nature10081.
127. McMeniman CJ, Corfas RA, Matthews BJ, Ritchie SA, Vosshall LB. Multimodal integration of carbon dioxide and other sensory cues drives mosquito attraction to humans. *Cell*. 2014;156(5):1060–1071. http://dx.doi.org/10.1016/j.cell.2013.12.044.
128. Su C-Y, Martelli C, Emonet T, Carlson JR. Temporal coding of odor mixtures in an olfactory receptor neuron. *Proc Natl Acad Sci USA*. 2011. http://dx.doi.org/10.1073/pnas.1100369108.
129. Su C-Y, Menuz K, Reisert J, Carlson JR. Non-synaptic inhibition between grouped neurons in an olfactory circuit. *Nature*. 2012. http://dx.doi.org/10.1038/nature11712.
130. Olsen SR, Bhandawat V, Wilson RI. Excitatory interactions between olfactory processing channels in the *Drosophila* antennal lobe. *Neuron*. 2007;54(1):89–103. http://dx.doi.org/10.1016/j.neuron.2007.03.010.
131. Assisi C, Stopfer M, Bazhenov M. Excitatory local interneurons enhance tuning of sensory information. *PLoS Comput Biol*. 2012;8(7):e1002563. http://dx.doi.org/10.1371/journal.pcbi.1002563.
132. Olsen SR, Wilson RI. Lateral presynaptic inhibition mediates gain control in an olfactory circuit. *Nature*. 2008;452(7190):956–960. http://dx.doi.org/10.1038/nature06864.
133. Martin JP, Lei H, Riffell JA, Hildebrand JG. Synchronous firing of antennal-lobe projection neurons encodes the behaviorally effective ratio of sex-pheromone components in male *Manduca sexta*. *J Comp Physiol A Neuroethol Sens Neural Behav Physiol*. 2013;199(11):963–979. http://dx.doi.org/10.1007/s00359-013-0849-z.
134. Riffell JA, Lei H, Hildebrand JG. Neural correlates of behavior in the moth *Manduca sexta* in response to complex odors. *Proc Natl Acad Sci*. 2009;106(46):19219–19226. http://dx.doi.org/10.1073/pnas.0910592106.
135. Leal WS. Odorant reception in insects: roles of receptors, binding proteins, and degrading enzymes. *Annu Rev Entomol*. 2012. http://dx.doi.org/10.1146/annurev-ento-120811-153635.
136. Wojtasek H, Leal WS. Conformational change in the pheromone-binding protein from *Bombyx mori* induced by pH and by interaction with membranes. *J Biol Chem*. 1999;274(43):30950–30956.
137. Leal WS, Barbosa RMR, Xu W, et al. Reverse and conventional chemical ecology approaches for the development of oviposition attractants for Culex mosquitoes. *PLoS One*. 2008;3(8):e3045. http://dx.doi.org/10.1371/journal.pone.0003045.
138. Syed Z, Ishida Y, Taylor K, Kimbrell DA, Leal WS. Pheromone reception in fruit flies expressing a moth's odorant receptor. *Proc Natl Acad Sci USA*. 2006;103(44):16538–16543. http://dx.doi.org/10.1073/pnas.0607874103.
139. Sun M, Liu Y, Walker WB, et al. Identification and characterization of pheromone receptors and interplay between receptors and pheromone binding proteins in the diamondback moth, *Plutella xyllostella*. *PLoS One*. 2013;8(4):e62098. http://dx.doi.org/10.1371/journal.pone.0062098.
140. Xu P, Atkinson R, Jones DNM, Smith DP. *Drosophila* OBP LUSH is required for activity of pheromone-sensitive neurons. *Neuron*. 2005;45(2):193–200. http://dx.doi.org/10.1016/j.neuron.2004.12.031.
141. Laughlin J, Ha T, Jones D, Smith D. Activation of pheromone-sensitive neurons is mediated by conformational activation of pheromone-binding protein. *Cell*. 2008;133(7):1255–1265. http://dx.doi.org/10.1016/j.cell.2008.04.046.

142. Gomez-Diaz C, Reina JH, Cambillau C, Benton R. Ligands for pheromone-sensing neurons are not conformationally activated odorant binding proteins. *PLoS Biol.* 2013;11(4):e1001546. http://dx.doi.org/10.1371/journal.pbio.1001546.
143. Vogt RG, Riddiford LM, Prestwich GD. Kinetic properties of a sex pheromone-degrading enzyme: the sensillar esterase of *Antheraea polyphemus*. *Proc Natl Acad Sci USA.* 1985;82(24):8827–8831.
144. Ishida Y, Leal WS. Rapid inactivation of a moth pheromone. *Proc Natl Acad Sci USA.* 2005;102(39):14075–14079. http://dx.doi.org/10.1073/pnas.0505340102.
145. Choo Y-M, Pelletier J, Atungulu E, Leal WS. Identification and characterization of an antennae-specific aldehyde oxidase from the navel orangeworm. *PLoS One.* 2013;8(6):e67794. http://dx.doi.org/10.1371/journal.pone.0067794.
146. Keeling CI, Henderson H, Li M, Dullat HK, Ohnishi T, Bohlmann J. CYP345E2, an antenna-specific cytochrome P450 from the mountain pine beetle, *Dendroctonus ponderosae* Hopkins, catalyses the oxidation of pine host monoterpene volatiles. *Insect Biochem Mol Biol.* 2013;43(12):1142–1151. http://dx.doi.org/10.1016/j.ibmb.2013.10.001.
147. Rogers ME, Krieger J, Vogt RG. Antennal SNMPs (sensory neuron membrane proteins) of Lepidoptera define a unique family of invertebrate CD36-like proteins. *J Neurobiol.* 2001;49(1):47–61.
148. Benton R, Vannice KS, Vosshall LB. An essential role for a CD36-related receptor in pheromone detection in *Drosophila*. *Nature.* 2007;450(7167):289–293. http://dx.doi.org/10.1038/nature06328.
149. Li Z, Ni JD, Huang J, Montell C. Requirement for *Drosophila* SNMP1 for rapid activation and termination of pheromone-induced activity. *PLoS Genet.* 2014;10(9):e1004600. http://dx.doi.org/10.1371/journal.pgen.1004600.
150. Ronderos DS, Lin C-C, Potter CJ, Smith DP. Farnesol-detecting olfactory neurons in *Drosophila*. *J Neurosci.* 2014;34(11):3959–3968. http://dx.doi.org/10.1523/JNEUROSCI.4582-13.2014.
151. Menuz K, Larter NK, Park J, Carlson JR. An RNA-seq screen of the *Drosophila* antenna identifies a transporter necessary for ammonia detection. *PLoS Genet.* 2014;10(11):e1004810. http://dx.doi.org/10.1371/journal.pgen.1004810.
152. Pitts RJ, Derryberry Jr SL, Pulous FE, Zwiebel LJ. Antennal-expressed ammonium transporters in the malaria vector mosquito *Anopheles gambiae*. *PLoS One.* 2014;9(10):e111858. http://dx.doi.org/10.1371/journal.pone.0111858.
153. Pelz D, Roeske T, Syed Z, de Bruyne M, Galizia CG. The molecular receptive range of an olfactory receptor in vivo (*Drosophila melanogaster* Or22a). *J Neurobiol.* 2006;66(14):1544–1563. http://dx.doi.org/10.1002/neu.20333.
154. Boyle SM, McInally S, Ray A, Luo L. Expanding the olfactory code by in silico decoding of odor-receptor chemical space. *eLife.* 2013;2. http://dx.doi.org/10.7554/eLife.01120.
155. Tegler LT, Corin K, Hillger J, Wassie B, Yu Y, Zhang S. Cell-free expression, purification, and ligand-binding analysis of *Drosophila melanogaster* olfactory receptors DmOR67a, DmOR85b and DmORCO. *Sci Rep.* 2015;5:7867. http://dx.doi.org/10.1038/srep07867.

Cyclic AMP Signaling in the Main Olfactory Epithelium

Christopher H. Ferguson, Haiqing Zhao

Department of Biology, The Johns Hopkins University, Baltimore, MD, USA

INTRODUCTION

Cyclic adenosine monophosphate (cAMP) is a cyclic nucleotide that functions as a key mediator of extracellular detection in a variety of signaling pathways, including glycogen metabolism, gene regulation, and olfactory sensory transduction. Since its discovery by Earl Sutherland in the late 1950s,[1] the importance of cAMP as a second messenger for diverse cellular functions has garnered enormous investigative effort, and we have gained substantial amounts of knowledge about the cAMP signaling system.[2]

Chemosensory Transduction
http://dx.doi.org/10.1016/B978-0-12-801694-7.00007-X

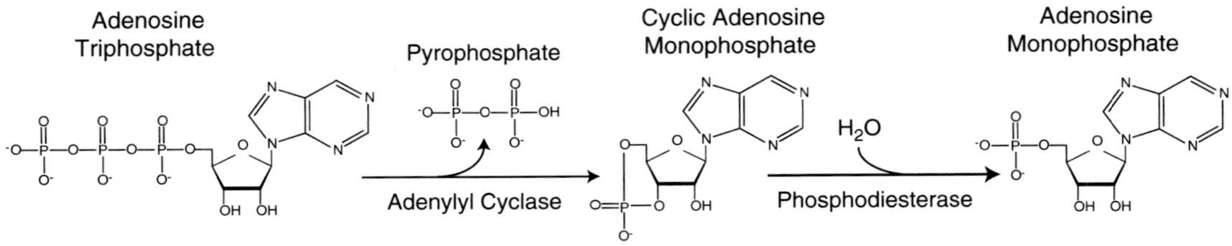

FIGURE 1 **Cyclic AMP synthesis and degradation.** Cyclic AMP (cAMP) is synthesized by adenylyl cyclase (AC) enzymes from molecules of adenosine triphosphate (ATP). The degradation of cAMP is catalyzed by phosphodiesterase (PDE) enzymes through a hydrolysis reaction to form adenosine monophosphate (AMP). In olfactory sensory neurons of the main olfactory epithelium, AC3 and PDE1C are responsible for cAMP synthesis and degradation in the cilia, where olfactory transduction occurs.

    In the cell, cAMP is generated by the activity of adenylyl cyclases (ACs) that convert adenosine triphosphate (ATP) into the reactive cyclical adenosine monophosphate (AMP) nucleotide. cAMP is degraded primarily by the activity of cyclic nucleotide phosphodiesterases (PDEs), which cleave the cyclical phosphodiester bond to convert a molecule of cAMP into a molecule of AMP (Figure 1). There are 10 AC genes encoded in the mammalian genome,[3] and all act as a source of significant amplification in various signaling pathways, as a single cyclase enzyme is capable of generating more than a thousand molecules of cAMP per second upon activation.[4] The family of cyclic nucleotide PDEs is made up of 21 genes, many of which preferentially degrade cAMP.[2] ATP is abundant in the cell with concentrations in the low millimolar range.[5,6] Reported cAMP concentrations have been more variable, being anywhere from a low of 10 nM up to 50 μM.[7] Cyclic AMP mediates cellular function by acting on effector molecules including cAMP-dependent protein kinases (PKAs),[8,9] the guanine exchange factor EPAC,[10,11] and cyclic nucleotide-gated (CNG) ion channels.[12,13]

    Given that small changes in cAMP levels may elicit large shifts in signaling dynamics, the regulation of cAMP transients is critically important for cellular function. These transients can be regulated at the level of production, through modulation of AC activity, or at the level of degradation through modulation of PDE activity. Commonly, alterations in the production and the degradation of cAMP are employed simultaneously.[2]

    The importance of cAMP is particularly apparent in the main olfactory epithelium (MOE) of the olfactory system. The MOE is a specialized tissue that lines the nasal cavity, and within the MOE are the olfactory sensory neurons (OSNs). OSNs convert odorous chemicals in the environment into neural electrical signals that are conducted along OSN axons to the brain and ultimately give rise to the perception of smell. The vast majority of OSNs in the MOE transduce odorants into membrane depolarization through a cAMP-mediated signaling cascade.[14,15] Without the generation of cAMP transients, the OSNs are incapable of responding to odorant stimuli through the canonical signaling pathway, rendering the organism largely anosmic.[16–18] In addition, cAMP also regulates many other cellular processes, the most notable of which is the targeting of OSN axons (see Breakout Box 1).[19]

BREAKOUT BOX 1

CYCLIC AMP REGULATES AXON TARGETING OF OLFACTORY SENSORY NEURONS

Membrane depolarization in olfactory cilia is relayed to the brain through OSN axons that synapse with second-order neurons in the main olfactory bulb. OSNs expressing the same OR converge their axons to the same location within the olfactory bulb in a neuropil structure known as a glomerulus. Targeting of OSN axons to the bulb depends, at least in part, on signaling through cAMP.[19,123]

OSN axons first project to the approximate location of their final targets, defining a coarse map, after which they sort into distinct glomeruli. Coarse targeting involves patterning along the anteroposterior (A-P) axis as well as the dorsal-ventral (D-V) axis. A-P patterning has been shown to depend on levels of cAMP generated through spontaneous OR activity, whereas D-V patterning appears to function independent of cAMP generation and is instead determined by the location of OSNs within epithelial zones of the MOE.[124,125] In A-P patterning, ORs that are more spontaneously active generate higher basal levels of cAMPwhich activates cAMP-dependent protein kinase (PKA) and

in turn phosphorylates the transcription factor CREB, ultimately promoting more posterior targeting of an axon.[126] Conversely, OSNs expressing ORs with low spontaneous activity generate less basal cAMP, leading to more anterior targeting. As with A-P patterning, the local sorting of OSN axons is also dependent on the expressed OR and levels of cAMP generation. OSNs expressing the same OR sort together through differential expression of homophilic attractive and heterophilic repulsive molecules, the expression of which is driven by different levels of neuronal activity triggered by ligand-induced cAMP generation.[127]

Many questions still remain to be answered in regards to the process of OSN axon targeting including where the cAMP is generated and how the site of cAMP generation relates to the site of influence, how cAMP-dependent neuronal activity is converted into changes in targeting factors especially with regard to local sorting, and what is the identity of the PDE responsible for regulating the cAMP levels associated with axon targeting.

THE EARLY HISTORY OF cAMP IN OLFACTORY TRANSDUCTION

The pivotal role played by cAMP in olfactory signal transduction was only truly appreciated after several hallmark discoveries. Pioneering studies in the early 1980s demonstrated that the transduction machinery presides in the cilia of the OSNs (Figure 2). Flushing of the frog nasal cavity with a mild detergent dislodged OSN cilia and eliminated electrical responses to odorant stimulation without disrupting the integrity of the remainder of the olfactory epithelium.[20,21] Determining that the transduction cascade was confined to the

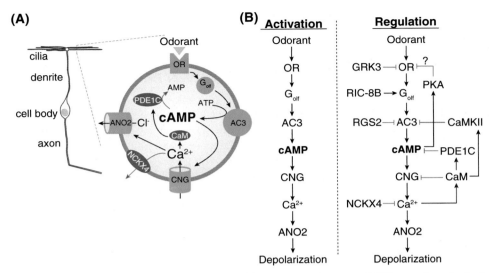

FIGURE 2   **Vertebrate olfactory transduction.** Each olfactory sensory neuron (OSN) extends a single dendritic process to the surface of the olfactory epithelium. This dendritic process ends in a dendritic knob from which 10–20 cilia emanate out into the mucus overlying the epithelium. These cilia are the sites of olfactory transduction. The transduction process converts binding of an odorant molecule with its cognate receptor into membrane depolarization through a cAMP-mediated transduction cascade, and is the first step in generating the perception of scent. Following activation, the system is restored to a resting state through numerous regulatory mechanisms. The ultimate realization of olfaction is a dynamic balance between activation and regulation. (A) Diagrams of an OSN (left) and a cross-section of a cilium showing the main components of olfactory transduction cascade (right). (B) Activation cascade for olfactory transduction (left) and regulatory mechanisms in olfactory transduction process (right). Negative regulatory interactions are indicated by red lines. AC3, adenylyl cyclase 3; ANO2, anoctamin 2, Ca^{2+}-activated chloride channel; CaM; calmodulin; CaMKII, Ca^{2+}/CaM-dependent protein kinase II; cAMP, cyclic adenosine monophosphate; CNG, olfactory cyclic nucleotide-gated cation channel; $G\alpha_{olf}$, olfactory G protein; GRK3, G protein-receptor kinase 3; NCKX4, the potassium-dependent Na^+/Ca^{2+} exchanger 4; OR, odorant receptor; PDE1C, phosphodiesterase 1C; PKA, cAMP-dependent protein kinase A; RGS2, regulator of G protein signaling 2.

olfactory cilia prompted implementation of a cilium isolation technique for enrichment of olfactory transduction components that could then be used for biochemical analysis. This cilium preparation method, originally developed for the isolation of flagella from unicellular ciliates such as *Tetrahymena*, involves treatment of isolated MOE with a high concentration of Ca^{2+} ions, causing detachment of cilia from the dendritic knob.[22,23]

In 1985, a milestone publication by the Lancet group, which used cilia prepared from frog olfactory epithelium, provided strong evidence that cAMP was involved in olfactory transduction.[23] The authors demonstrated enrichment of an odorant-sensitive AC in the OSN cilia. In this and following studies,[23] they also identified a stimulatory G protein in the OSN cilia that was likely responsible for activating the resident AC. These seminal findings suggested that odorant receptors (ORs; see Chapter 3) were G protein–coupled receptors (GPCRs) and that they signaled through a stimulatory Gα protein variant to activate an AC and spur the production of cAMP. Shortly thereafter, Nakamura and Gold demonstrated, using the patch-clamp technique on plasma membranes directly excised from toad OSN cilia, the presence of a CNG ion channel that was the probable effector of cAMP.[12] The involvement of the CNG channel was further strengthened by a series of electrophysiological investigations.[24–27] The opening of the CNG channel leads to influx of Na^+ and Ca^{2+} and consequently membrane depolarization. To further reinforce the role of cAMP, the Breer group used a fast-quench system to monitor levels of cAMP in cilium preparations and showed that cAMP levels peaked within 50 ms of odorant presentation, a speed that is sufficiently fast to mediate olfactory transduction.[28] Together, these electrophysiological and biochemical investigations laid the groundwork for rapid expansion in the understanding of cAMP-mediated olfactory transduction that still continues today.

MOLECULAR IDENTIFICATION OF OLFACTORY TRANSDUCTION COMPONENTS

Following the strong evidence that olfactory transduction involves a cAMP-mediated signaling cascade, researchers were prompted to identify and clone the genes that encode the transduction components (Figure 2). The first wave of breakthrough was achieved between the late 1980s and early 1990s. By screening a complementary DNA (cDNA) library prepared from the rat olfactory epithelium, the Reed group cloned the stimulatory G protein alpha subunit $G\alpha_{olf}$ and became the first to successfully clone an olfactory transduction component.[29] Similar cDNA library screening efforts led the Reed group to identify the AC in the OSN cilia as adenylyl cyclase 3 (AC3)[30] and led the Reed and Yau groups to clone the gene encoding the principle subunit of the olfactory CNG channel, CNGA2.[31] Subsequent genetic knockout work confirmed the key roles played by each of these proteins as mice lacking any of these transduction components were rendered broadly anosmic.[16–18]

Around the same time, significant effort was aimed at determining the molecular identity of the odorant receptors. In 1991, Buck and Axel published their groundbreaking discovery on the cloning of OR genes through the use of the polymerase chain reaction (PCR).[32] Their success stemmed from the assumptions that ORs were coupled to a G protein cascade mediated by cAMP and that ORs constituted a multigene family. These assumptions led them to design degenerate primers for the PCR based on conserved residues of known GPCRs and guided the analysis of the PCR products.[32] Subsequent studies have shown that OSNs follow a one-neuron, one-receptor rule, whereby each OSN expresses only one OR and that only one of the two alleles for a given OR is expressed.[33–37] Notably, even though canonical OSNs can express any one of over a thousand ORs (at least in rodents), they all share the same downstream transduction pathway.

Efforts quickly followed to identify the additional molecular constituents of the heterotetrameric CNG channel and the heterotrimeric G_{olf} complex. When heterologously expressed, the CNGA2 subunit is capable of forming a homomeric channel gated by cAMP; however, this homomeric channel exhibited channel properties different from those of the native olfactory channel.[31,38] Further screening discovered two additional subunits that contribute to the native CNG channel: the CNGA4[38,39] subunit and the CNGB1b[40,41] subunit. Although

CNGA4 and CNGB1b are individually incapable of forming a homomeric channel, when coexpressed with the CNGA2, the resulting heteromeric channel exhibits properties similar to the native olfactory channel.[40,42] Further studies suggested that the native CNG channel is likely formed with a stoichiometry of two CNGA2 subunits to one CNGA4 subunit to one CNGB1b subunit.[42,43] In the case of the heterotrimeric G_{olf} complex, the $G\beta$ subunit is likely to be the $G\beta1$,[44] and the $G\gamma$ subunit has been identified as $G\gamma13$.[45]

The olfactory transduction cascade possesses a unique feature in which, downstream of cAMP activation of the CNG channel and Ca^{2+} influx, a Ca^{2+}-activated efflux of chloride ions leads to additional depolarization of the OSN cilia. Early electrophysiology suggested that this Cl^- current is a source of amplification in the transduction cascade, accounting for approximately 50—90% of OSN depolarization.[46—50] Recent mass spectrometry—based proteomic analysis of mouse cilium preparations helped identify anoctamin 2 (ANO2) as the olfactory Ca^{2+}-activated Cl^- channel.[51—53] Analysis of mutant mice lacking ANO2 supported its role in current generation in OSNs.[54] However, behavioral analysis suggests that ANO2 may be dispensable for proper olfaction in mice,[54] and it is unclear how ANO2 contributes to human olfaction.[55]

Concurrent with the exploration of components involved in the generation of cAMP and subsequent electrical responses, the PDE(s) that degrade cAMP were also actively pursued. Early work determined that PDE activity was present in the olfactory cilia and that this activity was dependent on Ca^{2+} and calmodulin (CaM).[56,57] Through screening of a rat olfactory cDNA library, the Beavo group identified the PDE responsible for degrading cAMP in the olfactory cilia as PDE1C2, a PDE that is activated by Ca^{2+}/CaM.[58,59]

Akin to the importance of the PDE for controlling cilial cAMP levels, the Ca^{2+} transporter(s) responsible for controlling cilial Ca^{2+} levels is critically important for olfactory transduction. Mass spectrometry—based proteomic analysis of mouse cilium preparations also helped to identify the potassium-dependent Na^+-Ca^{2+} exchanger 4 (NCKX4) as the Ca^{2+} transporter responsible for the extrusion of Ca^{2+} from the cilia.[60]

REGULATION OF THE cAMP SIGNALING CASCADE

One characteristic trait of sensory receptor cells, including OSNs, is the ability to alter response sensitivity according to the intensity and duration of environmental stimuli. OSNs exhibit reduced sensitivity upon sustained or repeated odor exposure—a phenomenon known as olfactory adaptation. Adaptation at the cellular level is believed to contribute to changes in the perception of the odorous environment, including the fading of conscious sensation of smell over time, and to aid animals in discriminating between odors in a complex odor environment. Regulation of the cAMP-mediated olfactory transduction cascade forms the basis of olfactory adaptation.

Mechanisms regulating cAMP signaling roughly fall into two categories: those that regulate the cAMP transient and those that affect the impact of cAMP in the system. In olfactory transduction, Ca^{2+}-dependent negative feedback mechanisms play a major role both in regulating the cAMP transient and in affecting the impact of cAMP (Figure 2).[15,61]

The importance of Ca^{2+} in regulating the cAMP transient is intuitively appreciable as the activity of both AC3, for cAMP production, and PDE1C, for cAMP degradation, can be regulated by Ca^{2+}. AC3 activity is inhibited by elevated Ca^{2+} through phosphorylation by Ca^{2+}/CaM-dependent protein kinase II (CaMKII).[62,63] The phosphorylation occurs at a single amino acid of AC3, serine1076, and mutation of this serine to alanine abolishes Ca^{2+}-induced inhibition of AC3 activity when that protein is expressed in cultured cells.[63] In OSNs, CaMKII inhibitors blocked the phosphorylation of AC3 at serine1076 and prolonged odorant-evoked cAMP transients.[64] These findings support a model in which regulation of cAMP transients in the olfactory transduction cascade could occur through Ca^{2+}-induced phosphorylation of AC3 by CaMKII. Therefore, Ca^{2+} influx could cause reduced cAMP production following odorant stimulation and thus reduced response sensitivity. Consistent with this model, use of CaMKII inhibitors perturbs the electrophysiological response of OSNs after sustained odorant stimuli.[65] However, mice lacking the serine1076 phosphorylation site display no deficits in olfactory response.[66] Thus, whether and to what extent Ca^{2+}-induced AC3 phosphorylation may contribute to regulation of the cAMP transient in OSNs requires further investigation. In addition to Ca^{2+}-dependent inhibitory phosphorylation, AC3 can also be inhibited by regulator of G protein signaling 2 (RGS2) in OSNs.[67] This unusual RGS function could also serve as a means to regulate cAMP transients in OSNs.

PDE1C is a Ca^{2+}/CaM-activated PDE with similar affinities for both cAMP and cGMP (cyclic guanosine monophosphate) in the low micromolar range.[58] Additionally, PDE1C2, the isoform specific to OSNs, has greater affinity for Ca^{2+} than does PDE1C1.[59] Biochemical and electrophysiological studies using PDE inhibitors determined that PDE activity would likely be critical for the termination of OSN responses to odorants and, as such, it was speculated that a loss of PDE1C would slow response termination.[24,68] Interestingly, mice lacking PDE1C display more rapid response termination, rather than the slowed termination that was predicted, as well as reduced response amplitude.[69] One potential reason for the lack of prolonged response termination is that rapid diffusion of cAMP out of the cilia and into the dendritic knob could allow for cAMP degradation by another PDE.[70] Thus far, two PDEs are known to be expressed in canonical OSNs: PDE1C and PDE4A.[71,72] PDE1C and PDE4A are present in distinct compartments within OSNs, with PDE1C being localized exclusively to the cilia and PDE4A being localized to the dendritic knob and the rest of the cell.[72] When PDE4A is knocked out alone, mutant OSNs display no deficit in olfactory responses, suggesting that PDE1C is sufficient for degradation of cilial cAMP. However, when both PDE4A and PDE1C were knocked out in the same animal, mutant OSNs displayed significantly prolonged response termination.[69] The phenotype observed in the double mutants confirms the importance of cAMP degradation for the termination of OSN responses.

The phenotype of reduced response amplitude in PDE1C mutant mice indicated that PDE1C could also have a role in setting the resting response sensitivity. A large portion of OSNs possesses spontaneous activity, which results, at least in part, from the spontaneous activation of ORs in the absence of ligand binding.[73] PDE1C may act to degrade the spontaneously produced cAMP, thus preventing the opening of CNG channels and subsequent Ca^{2+} influx in the absence of the stimulus. In PDE1C mutant OSNs, lack of cAMP degradation in the cilia would increase cilial cAMP and, subsequently, cilial Ca^{2+} levels. Elevated Ca^{2+} could then cause reduced response sensitivity.

In addition to the direct modulation of AC3 and PDE1C activity, regulatory events acting upstream of AC3, namely on $G\alpha_{olf}$ and ORs, could play a role in regulating the cAMP transient. Ric-8b, a putative GTP exchange factor (GEF) found in OSNs, has been suggested to affect the cAMP transient by acting on $G\alpha_{olf}$.[74] Ric-8b has been shown to associate with the members of the G_{olf} protein complex[74,75] and has also been implicated in transport of ORs to the cell surface in culture.[76] Coexpression of Ric-8b with ORs in cell culture showed that Ric-8b can enhance odorant-elicited cAMP production, possibly through acting as a GEF on $G\alpha_{olf}$ to amplify odorant-induced responses.[74]

In various GPCR signaling pathways, receptor phosphorylation is a common mechanism for attenuating receptor signaling.[77] In olfactory transduction, GPCR kinase 3 (GRK3) was reported to regulate OR activity and thus the cAMP transient.[78–80] Inhibition of GRK activity in olfactory cilium preparations prevented odorant-elicited increases in cAMP from recovering to basal levels, and this effect was attributed to GRK3 through the use of antibodies specific to GRK3.[79] In line with this, cilium preparations from mice lacking GRK3 displayed a prolonged, yet small, elevation of cAMP levels in response to odorant stimulation.[80] Subsequent research, however, found that OSNs lacking GRK3 did not show electrophysiological deficits in odorant responses, arguing against a significant role for GRK3 in attenuating OR responsiveness.[81] Early research also noted an increase in radiolabeled phosphate incorporation upon stimulation of isolated olfactory cilium preparations with odorant mixtures. This phosphate incorporation was reduced upon inhibition of either cAMP-mediated protein kinase A (PKA) or the Ca^{2+}-activated protein kinase C (PKC), suggesting a role for these two kinases in regulating olfactory transduction.[82] Of particular interest was the presence of a cluster of phosphorylated proteins around the 50-kD region of the gel, which falls around the predicted size of ORs[82] and implicates a role for OR phosphorylation in olfactory regulation. At this time, the influence of PKA- and PKC-mediated phosphorylation in olfactory transduction remains to be explored.

In olfactory transduction, mechanisms that affect the impact of cAMP are mainly ascribed to the Ca^{2+}-mediated desensitization of the CNG channel to cAMP. Ca^{2+} that enters the cilium through the CNG channel during a response can reduce the sensitivity of the channel to cAMP[83] through binding to CaM associated directly with the CNG channel.[84]

The modulatory subunits of the olfactory CNG channel, CNGB1b and CNGA4, not only greatly increase the apparent affinity of the heteromeric channel to cAMP,[38–41,85–87] but also regulate Ca^{2+}/CaM-mediated channel desensitization.[85–87] Both CNGB1b and CNGA4 subunits are required for rapid desensitization kinetics.[85–87] OSNs from CNGA4 gene knockout mice display defects in rapid adaptation,[85] and these knockout mice show elevated odor detection threshold and impaired odor adaptation in behavioral tests.[88] This Ca^{2+}/CaM-mediated CNG channel desensitization was suggested to act as a major mechanism for rapid olfactory adaptation[65,85,89,90]; however, a recent study suggested that this desensitization is not necessary for rapid adaptation.[91] Each of the olfactory CNG channel subunits contains a CaM-binding site,[84,92] but only the site in the CNGB1b subunit is necessary and sufficient for Ca^{2+}/CaM-mediated desensitization under the native channel configuration.[84,93] Genetic deletion of the CaM-binding site of the CNGB1b subunit in mice does not impair OSN responses to repeated odor stimulation,[91] suggesting that the Ca^{2+}/CaM-mediated CNG channel desensitization is dispensable for rapid olfactory adaptation. Deletion of the CaM-binding site in CNGB1b does cause impaired attenuation of the response to sustained odor stimulation as well as a slightly prolonged OSN response to a brief stimulation.[91]

Given the importance of Ca^{2+} in regulating both the generation of the cAMP transient and the impact of cAMP within the system, the proper regulation of Ca^{2+} levels within the OSN cilia is also critically important for normal olfaction. The regulation of resting Ca^{2+} levels in OSN cilia is believed to be the result of Ca^{2+} extrusion by Na^+/Ca^{2+} exchangers and ATP-dependent Ca^{2+} pumps.[15,94] NCKX4 was identified as the principal Na^+/Ca^{2+} exchanger relevant to olfactory transduction. OSNs in mutant mice lacking NCKX4 show substantially prolonged odorant responses, demonstrating a key role for NCKX4 in regulating the kinetics of the cAMP-mediated olfactory response through control of Ca^{2+} levels.[60] The role for the plasma membrane Ca^{2+}-ATPase (PMCA) in regulating olfactory transduction is less clear. Early studies using pharmacological inhibitors indicated that PMCA-based efflux could be important for response termination; however, recent work with more specific inhibitors suggests that the impact on response termination may have been an indirect effect of the inhibitors on Na^+/K^+ ATPase activity.[95,96]

FUTURE DIRECTIONS

Although the olfactory field has a strong grasp on the core players of cAMP-mediated olfactory transduction cascade, many details of the molecular mechanism underlying the regulation of olfaction remain unclear. Although only a small fraction of many potential mechanisms has been explored, the predominant limitation is an inability to assess the impact of individual regulatory mechanisms in real time and in intact tissues.

Because levels of cAMP and Ca^{2+} are directly indicative of changes in an olfactory response, understanding cAMP and Ca^{2+} dynamics is critical for understanding olfaction (Figure 3). Ca^{2+} dynamics in OSN cilia has been investigated using Ca^{2+} indicator dyes.[97–99] These imaging studies represent the initial efforts in understanding the Ca^{2+} dynamics of olfactory responses and have provided insightful information about the regulation of transduction. However, there is still is a lack of effective approaches for *in vivo*, or even *ex vivo*, measurement of cAMP in the cilia of OSNs. To date, cAMP dynamics has been studied by making use of either biochemical assays performed on isolated cilia[80,100] or electrophysiological recordings of current flow through the CNG channel.[26,65] However, biochemical measurements do not provide individual cell or single cilium resolution and suffer from altered physiological conditions resulting from cilial rupture. Additionally, electrophysiological recordings of CNG channel activity are indirect because channel activity is a product of cAMP activation of the channel as well as regulation of the channel by other factors, including Ca^{2+}/CaM-dependent desensitization.

Recently, there has been the exciting development of genetically encoded indicators for cAMP.[101,102] These genetically encoded indicators are either ratiometric Förster resonance energy transfer–based indicators, such as indicator of cAMP using Epac 3 (ICUE3),[103] or

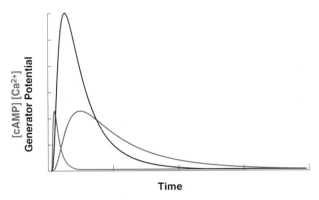

Time

FIGURE 3   **Olfactory response dynamics.** Hypothetical changes in cAMP concentration (blue), Ca²⁺ concentration (red), and the generator potential (black) in response to a brief odorant pulse. An accurate determination of these
molecular dynamics will represent a major advancement toward an understanding of olfactory transduction. cAMP,
cyclic adenosine monophosphate.

single fluorescent protein-based indicators, such as Flamindo 2.[104] These indicators have been
applied in numerous systems, including β2 adrenergic receptor signaling and insulin release
by pancreatic beta cells, and allow researchers to measure spatiotemporal changes of cAMP
in specific subcellular compartments.[101,105] A major benefit of these indicators is that they can
be incorporated into the genome of an animal and targeted to specific cellular compartments,
including cilia, through the inclusion of a localization tag. Although this innovative technique
is begging to be implemented in the study of olfactory transduction, there are several potential limitations that will need to be overcome. First, there is the need for targeting, ideally specifically to the OSN cilia, which could reduce confounding background signals from the cell
bodies and dendritic knob. At present, there are several proposed cilia targeting motifs,[106,107]
and although some have seen successful application in primary cilia,[108,109] none has been
tested for their ability to specifically target proteins to OSN cilia *in vivo*. Second, indicator kinetics are critically important and must be sufficiently quick to rise and fall with changes in
cAMP because these changes occur on the millisecond timescale.[110] Although current studies
have focused on a timescale of seconds to minutes, development of these indicators has progressed rapidly in an effort to improve all aspects of their function.[101] Additionally, implementation of cAMP imaging may be more challenging because of the spatial restriction of
the cilia (mammalian olfactory cilia range from 5 to 50 μm in length with diameters of
300 nm at the proximal segment and 100 nm at the distal segment, yielding an approximate
volume of 1 fl). The miniscule volume of olfactory cilia may limit measurement of cAMP dynamics because small changes in the pool of available cAMP could affect measurements, and
indicator presence in the cilia could buffer cAMP levels and negatively impact signaling.
Nevertheless, imaging cAMP dynamics with a tool that is capable of accurately and reliably
representing changes of cAMP in the transduction cascade would be critically important for
understanding signaling in the olfactory epithelium. Combining this indicator with the available mutant mice containing genetic disruptions of key olfactory transduction components

would be the ideal way to determine the exact impact these olfactory mutations have on regulation.

Regulation of AC3 and PDE1C is an area of study that still requires substantial investigation. How AC3 is regulated *in vivo* is still unclear,[66] and in the case of PDE1C little is understood about how PDE1C is regulated except for the stimulation of PDE activity by Ca^{2+}/CaM. Additionally, we still do not know the membrane densities or cytosolic concentrations of AC3 or PDE1C in the olfactory cilia. Gaining insight into the relative abundance of these two proteins in a given olfactory cilium as well as *in vivo* activity levels at rest, during and after odorant stimulation, will be important for understanding the role of cAMP in olfaction.

Few studies have looked at the potential role for PKA, the most common effector of cAMP, in the olfactory transduction despite the fact that PKA is a highly studied signaling protein in other systems.[111,112] The evidence of a role for PKA in olfactory transduction is limited to studies of PKA inhibition decreasing total phosphate incorporation in olfactory cilia[82]; however, the precise proteins that are regulated by this phosphorylation and whether PKA-mediated phosphorylation plays a role in regulating cAMP generation and olfactory transduction are still unknown.

Another area that needs to be explored further is the interplay between cAMP and Ca^{2+} in the transduction cascade. Although it is known that cAMP is responsible for increases in intracellular Ca^{2+} and that this Ca^{2+} in turn inhibits the production of more cAMP, the temporal nature of this relationship as well as the weight of each in the signal cascade has yet to be explored. The ability to simultaneously measure the dynamics of cAMP and Ca^{2+} should help to gain better understanding of the relationship between cAMP and Ca^{2+} transients.

Computational modeling efforts have started to address many of these gray areas of olfactory transduction. With recent increases in computational power, greater effort has been made to model the olfactory cascade in an attempt to understand the underlying principles with a focus on negative regulation.[113,114] One limitation of current modeling efforts is that all of the enzymatic rates and ionic concentrations on which the models are built are estimated from single-cell recordings. Although incredibly useful for understanding general roles of molecular components in the olfactory transduction cascade, ultimately it is impossible to take these measured rates too literally because the environment for the dissociated OSN is not necessarily the same as the environment within the intact olfactory epithelium. However, with each iteration, the published models of olfactory transduction become more complex and, despite the existing limitations, more closely represent the physiological responses.

Last but not least, our current understanding of how human olfactory disorders relate to changes in the cAMP-mediated transduction process is relatively poor,[115,116] although several genetic variants have been linked with impaired olfaction in humans.[117–121] A recent study has identified a genetic mutation in the CNGA2 channel subunit responsible for a case of isolated congenital anosmia.[122] Taking advantage of the rapid advances in high-throughput sequencing should facilitate the characterization of changes in gene expression resulting from aging and disease states as well as the identification of genetic variations, particularly those related to the peripheral transduction process that are responsible for human smell disorders.

BREAKOUT BOX 2

OLFACTORY MARKER PROTEIN

Frank Margolis discovered olfactory marker protein (OMP) in 1972.[128] Since then, OMP has been integral for our understanding of vertebrate olfaction. OMP is a 19-kD cytoplasmic protein and is encoded by a single-exon gene in rodents.[129] It is highly conserved between vertebrate species, but absent in invertebrates.

In the MOE, OMP has been used to label mature OSNs because of its specific and abundant expression. OMP is not detected in other cell types in the MOE, including basal cells, sustentacular cells, immature OSNs,[130] and a small subset of mature OSNs that transduce odorant information through a noncanonical, cGMP-based signaling cascade.[131] In the peripheral olfactory system, OMP is also detected in the receptor cells of the vomeronasal organ, the septal organ, and the Grueneberg ganglion, suggesting a role in general olfactory functionality. The specific and robust expression of OMP in olfactory sensors has led to widespread application of the *Omp* gene as a driver for olfactory system—specific mutations critical for physiological and behavioral analysis of olfactory function.[132]

Recently, OMP as an indicator of olfactory signaling has also been applied to locate olfactory signaling components in tissues outside of the olfactory system.[133]

Gene knockout studies suggest a modulatory role for OMP in olfactory transduction. OSNs lacking OMP display reduced response amplitudes as well as slowed onset and termination kinetics.[134] Consistently, mice lacking OMP display impaired threshold detection and altered odorant quality perception[135,136] and do not show preference for their biological mothers when nursing.[137] Despite decades of work, however, the molecular mechanism by which OMP modulates olfactory signaling is still a mystery. It has been suggested that OMP acts as a modulator of olfactory signaling upstream of cAMP production,[138] as a modulator of Ca^{2+} homeostasis,[139] and a critical player for the functional maturation of OSNs.[137] Structural studies[140,141] have shown that OMP may participate in protein—protein interactions and that these interactions with protein partners, such as the Bex family of proteins, may ultimately facilitate the modulatory role played by OMP.[142]

SUMMARY

Decades of research has uncovered the extensive roles of cAMP in the MOE. The field has gone from knowing relatively little about the process of olfactory transduction to having a firm grasp on the key players of the cAMP-mediated transduction cascade as well as many of the regulatory mechanisms involved. Many questions still need to be answered regarding the role of cAMP in the main olfactory epithelium, and researchers continue to adapt new methods in order to expose these scientific mysteries.

Acknowledgments

This work was supported by National Institutes of Health grant DC007395.

References

1. Sutherland EW, Rall TW. Fractionation and characterization of a cyclic adenine ribonucleotide formed by tissue particles. *J Biol Chem.* 1958;232(2):1077–1091.
2. Beavo JA, Brunton LL. Cyclic nucleotide research — still expanding after half a century. *Nat Rev Mol Cell Biol.* 2002;3(9):710–718.
3. Hanoune J, Defer N. Regulation and role of adenylyl cyclase isoforms. *Annu Rev Pharmacol Toxicol.* 2001;41:145–174.
4. Sunahara RK, Dessauer CW, Gilman AG. Complexity and diversity of mammalian adenylyl cyclases. *Annu Rev Pharmacol Toxicol.* 1996;36:461–480.
5. Beis I, Newsholme EA. The contents of adenine nucleotides, phosphagens and some glycolytic intermediates in resting muscles from vertebrates and invertebrates. *Biochem J.* 1975;152(1):23–32.
6. Traut TW. Physiological concentrations of purines and pyrimidines. *Mol Cell Biochem.* 1994;140(1):1–22.
7. Conti M, Mika D, Richter W. Cyclic AMP compartments and signaling specificity: role of cyclic nucleotide phosphodiesterases. *J Gen Physiol.* 2014;143(1):29–38.
8. Walsh DA, Perkins JP, Krebs EG. An adenosine 3′,5′-monophosphate-dependant protein kinase from rabbit skeletal muscle. *J Biol Chem.* 1968;243(13):3763–3765.
9. Tao M, Salas ML, Lipmann F. Mechanism of activation by adenosine 3′:5′-cyclic monophosphate of a protein phosphokinase from rabbit reticulocytes. *Proc Natl Acad Sci USA.* 1970;67(1):408–414.
10. de Rooij J, Zwartkruis FJ, Verheijen MH, et al. Epac is a Rap1 guanine-nucleotide-exchange factor directly activated by cyclic AMP. *Nature.* 1998;396(6710):474–477.
11. Kawasaki H, Springett GM, Mochizuki N, et al. A family of cAMP-binding proteins that directly activate Rap1. *Science.* 1998;282(5397):2275–2279.
12. Nakamura T, Gold GH. A cyclic nucleotide-gated conductance in olfactory receptor cilia. *Nature.* 1987;325(6103):442–444.
13. Kaupp UB, Seifert R. Cyclic nucleotide-gated ion channels. *Physiol Rev.* 2002;82(3):769–824.
14. Firestein S. How the olfactory system makes sense of scents. *Nature.* 2001;413(6852):211–218.
15. Kleene SJ. The electrochemical basis of odor transduction in vertebrate olfactory cilia. *Chem Senses.* 2008;33(9):839–859.
16. Brunet LJ, Gold GH, Ngai J. General anosmia caused by a targeted disruption of the mouse olfactory cyclic nucleotide-gated cation channel. *Neuron.* 1996;17(4):681–693.
17. Belluscio L, Gold GH, Nemes A, Axel R. Mice deficient in G(olf) are anosmic. *Neuron.* 1998;20(1):69–81.
18. Wong ST, Trinh K, Hacker B, et al. Disruption of the type III adenylyl cyclase gene leads to peripheral and behavioral anosmia in transgenic mice. *Neuron.* 2000;27(3):487–497.
19. Sakano H. Neural map formation in the mouse olfactory system. *Neuron.* 2010;67(4):530–542.
20. Mair RG, Gesteland RC, Blank DL. Changes in morphology and physiology of olfactory receptor cilia during development. *Neuroscience.* 1982;7(12):3091–3103.
21. Adamek GD, Gesteland RC, Mair RG, Oakley B. Transduction physiology of olfactory receptor cilia. *Brain Res.* 1984;310(1):87–97.
22. Anholt RR, Aebi U, Snyder SH. A partially purified preparation of isolated chemosensory cilia from the olfactory epithelium of the bullfrog, *Rana catesbeiana.* *J Neurosci.* 1986;6(7):1962–1969.
23. Chen Z, Pace U, Heldman J, Shapira A, Lancet D. Isolated frog olfactory cilia: a preparation of dendritic membranes from chemosensory neurons. *J Neurosci.* 1986;6(8):2146–2154.
24. Firestein S, Darrow B, Shepherd GM. Activation of the sensory current in salamander olfactory receptor neurons depends on a G protein-mediated cAMP second messenger system. *Neuron.* 1991;6(5):825–835.
25. Firestein S, Zufall F, Shepherd GM. Single odor-sensitive channels in olfactory receptor neurons are also gated by cyclic nucleotides. *J Neurosci.* 1991;11(11):3565–3572.

26. Zufall F, Firestein S, Shepherd GM. Analysis of single cyclic nucleotide-gated channels in olfactory receptor cells. *J Neurosci.* 1991;11(11):3573–3580.

27. Lowe G, Gold GH. Contribution of the ciliary cyclic nucleotide-gated conductance to olfactory transduction in the salamander. *J Physiol.* 1993;462:175–196.

28. Breer H, Boekhoff I, Tareilus E. Rapid kinetics of second messenger formation in olfactory transduction. *Nature.* 1990;345(6270):65–68.

29. Jones DT, Reed RR. Golf: an olfactory neuron specific-G protein involved in odorant signal transduction. *Science.* 1989;244(4906):790–795.

30. Bakalyar HA, Reed RR. Identification of a specialized adenylyl cyclase that may mediate odorant detection. *Science.* 1990;250(4986):1403–1406.

31. Dhallan RS, Yau KW, Schrader KA, Reed RR. Primary structure and functional expression of a cyclic nucleotide-activated channel from olfactory neurons. *Nature.* 1990;347(6289):184–187.

32. Buck L, Axel R. A novel multigene family may encode odorant receptors: a molecular basis for odor recognition. *Cell.* 1991;65(1):175–187.

33. Chess A, Simon I, Cedar H, Axel R. Allelic inactivation regulates olfactory receptor gene expression. *Cell.* 1994;78(5):823–834.

34. Malnic B, Hirono J, Sato T, Buck LB. Combinatorial receptor codes for odors. *Cell.* 1999;96(5):713–723.

35. Serizawa S, Miyamichi K, Nakatani H, et al. Negative feedback regulation ensures the one receptor-one olfactory neuron rule in mouse. *Science.* 2003;302(5653):2088–2094.

36. Mombaerts P. Odorant receptor gene choice in olfactory sensory neurons: the one receptor-one neuron hypothesis revisited. *Curr Opin Neurobiol.* 2004;14(1):31–36.

37. Markenscoff-Papadimitriou E, Allen WE, Colquitt BM, et al. Enhancer interaction networks as a means for singular olfactory receptor expression. *Cell.* 2014;159(3):543–557.

38. Bradley J, Li J, Davidson N, Lester HA, Zinn K. Heteromeric olfactory cyclic nucleotide-gated channels: a subunit that confers increased sensitivity to cAMP. *Proc Natl Acad Sci USA.* 1994;91(19):8890–8894.

39. Liman ER, Buck LB. A second subunit of the olfactory cyclic nucleotide-gated channel confers high sensitivity to cAMP. *Neuron.* 1994;13(3):611–621.

40. Bonigk W, Bradley J, Muller F, et al. The native rat olfactory cyclic nucleotide-gated channel is composed of three distinct subunits. *J Neurosci.* 1999;19(13):5332–5347.

41. Sautter A, Zong X, Hofmann F, Biel M. An isoform of the rod photoreceptor cyclic nucleotide-gated channel beta subunit expressed in olfactory neurons. *Proc Natl Acad Sci USA.* 1998;95(8):4696–4701.

42. Zheng J, Zagotta WN. Stoichiometry and assembly of olfactory cyclic nucleotide-gated channels. *Neuron.* 2004;42(3):411–421.

43. Zhong H, Lai J, Yau KW. Selective heteromeric assembly of cyclic nucleotide-gated channels. *Proc Natl Acad Sci USA.* 2003;100(9):5509–5513.

44. Sathyanesan A, Feijoo AA, Mehta ST, Nimarko AF, Lin W. Expression profile of G-protein betagamma subunit gene transcripts in the mouse olfactory sensory epithelia. *Front Cell Neurosci.* 2013;7:84.

45. Li F, Ponissery-Saidu S, Yee KK, et al. Heterotrimeric G protein subunit Ggamma13 is critical to olfaction. *J Neurosci.* 2013;33(18):7975–7984.

46. Kleene SJ, Gesteland RC. Calcium-activated chloride conductance in frog olfactory cilia. *J Neurosci.* 1991;11(11):3624–3629.

47. Kleene SJ. Origin of the chloride current in olfactory transduction. *Neuron.* 1993;11(1):123–132.

48. Lowe G, Gold GH. Nonlinear amplification by calcium-dependent chloride channels in olfactory receptor cells. *Nature.* 1993;366(6452):283–286.

49. Kurahashi T, Yau KW. Co-existence of cationic and chloride components in odorant-induced current of vertebrate olfactory receptor cells. *Nature.* 1993;363(6424):71–74.

50. Kleene SJ. High-gain, low-noise amplification in olfactory transduction. *Biophys J.* 1997;73(2):1110–1117.

51. Stephan AB, Shum EY, Hirsh S, Cygnar KD, Reisert J, Zhao H. ANO2 is the cilial calcium-activated chloride channel that may mediate olfactory amplification. *Proc Natl Acad Sci USA.* 2009;106(28):11776–11781.

52. Rasche S, Toetter B, Adler J, et al. Tmem16b is specifically expressed in the cilia of olfactory sensory neurons. *Chem Senses.* 2010;35(3):239–245.

53. Pifferi S, Dibattista M, Menini A. TMEM16B induces chloride currents activated by calcium in mammalian cells. *Pflugers Arch.* 2009;458(6):1023–1038.

54. Billig GM, Pal B, Fidzinski P, Jentsch TJ. Ca^{2+}-activated Cl$^-$ currents are dispensable for olfaction. *Nat Neurosci.* 2011;14(6):763—769.

55. Cenedese V, Mezzavilla M, Morgan A, et al. Assessment of the olfactory function in Italian patients with type 3 von Willebrand disease caused by a homozygous 253 Kb deletion involving VWF and TMEM16B/ANO2. *PLoS One.* 2015;10(1):e0116483.

56. Anholt RR, Rivers AM. Olfactory transduction: cross-talk between second-messenger systems. *Biochemistry.* 1990;29(17):4049—4054.

57. Borisy FF, Ronnett GV, Cunningham AM, Juilfs D, Beavo J, Snyder SH. Calcium/calmodulin-activated phosphodiesterase expressed in olfactory receptor neurons. *J Neurosci.* 1992;12(3):915—923.

58. Yan C, Zhao AZ, Bentley JK, Loughney K, Ferguson K, Beavo JA. Molecular cloning and characterization of a calmodulin-dependent phosphodiesterase enriched in olfactory sensory neurons. *Proc Natl Acad Sci USA.* 1995;92(21):9677—9681.

59. Yan C, Zhao AZ, Bentley JK, Beavo JA. The calmodulin-dependent phosphodiesterase gene PDE1C encodes several functionally different splice variants in a tissue-specific manner. *J Biol Chem.* 1996;271(41):25699—25706.

60. Stephan AB, Tobochnik S, Dibattista M, Wall CM, Reisert J, Zhao H. The Na(+)/Ca(2+) exchanger NCKX4 governs termination and adaptation of the mammalian olfactory response. *Nat Neurosci.* 2012;15(1):131—137.

61. Zufall F, Leinders-Zufall T. The cellular and molecular basis of odor adaptation. *Chem Senses.* 2000;25(4):473—481.

62. Wayman GA, Impey S, Storm DR. Ca^{2+} inhibition of type III adenylyl cyclase in vivo. *J Biol Chem.* 1995;270(37):21480—21486.

63. Wei J, Wayman G, Storm DR. Phosphorylation and inhibition of type III adenylyl cyclase by calmodulin-dependent protein kinase II in vivo. *J Biol Chem.* 1996;271(39):24231—24235.

64. Wei J, Zhao AZ, Chan GC, et al. Phosphorylation and inhibition of olfactory adenylyl cyclase by CaM kinase II in neurons: a mechanism for attenuation of olfactory signals. *Neuron.* 1998;21(3):495—504.

65. Leinders-Zufall T, Ma M, Zufall F. Impaired odor adaptation in olfactory receptor neurons after inhibition of Ca^{2+}/calmodulin kinase II. *J Neurosci.* 1999;19(14):RC19.

66. Cygnar KD, Collins SE, Ferguson CH, Bodkin-Clarke C, Zhao H. Phosphorylation of adenylyl cyclase III at serine1076 does not attenuate olfactory response in mice. *J Neurosci.* 2012;32(42):14557—14562.

67. Sinnarajah S, Dessauer CW, Srikumar D, et al. RGS2 regulates signal transduction in olfactory neurons by attenuating activation of adenylyl cyclase III. *Nature.* 2001;409(6823):1051—1055.

68. Boekhoff I, Breer H. Termination of second messenger signaling in olfaction. *Proc Natl Acad Sci USA.* 1992;89(2):471—474.

69. Cygnar KD, Zhao H. Phosphodiesterase 1C is dispensable for rapid response termination of olfactory sensory neurons. *Nat Neurosci.* 2009;12(4):454—462.

70. Chen C, Nakamura T, Koutalos Y. Cyclic AMP diffusion coefficient in frog olfactory cilia. *Biophysical J.* 1999;76(5):2861—2867.

71. Cherry JA, Davis RL. A mouse homolog of dunce, a gene important for learning and memory in *Drosophila*, is preferentially expressed in olfactory receptor neurons. *J Neurobiol.* 1995;28(1):102—113.

72. Juilfs DM, Fulle HJ, Zhao AZ, Houslay MD, Garbers DL, Beavo JA. A subset of olfactory neurons that selectively express cGMP-stimulated phosphodiesterase (PDE2) and guanylyl cyclase-D define a unique olfactory signal transduction pathway. *Proc Natl Acad Sci USA.* 1997;94(7):3388—3395.

73. Reisert J. Origin of basal activity in mammalian olfactory receptor neurons. *J Gen Physiol.* 2010;136(5):529—540.

74. Von Dannecker LE, Mercadante AF, Malnic B. Ric-8B, an olfactory putative GTP exchange factor, amplifies signal transduction through the olfactory-specific G-protein Galphaolf. *J Neurosci.* 2005;25(15):3793—3800.

75. Kerr DS, Von Dannecker LE, Davalos M, Michaloski JS, Malnic B. Ric-8B interacts with G alpha olf and G gamma 13 and co-localizes with G alpha olf, G beta 1 and G gamma 13 in the cilia of olfactory sensory neurons. *Mol Cell Neurosci.* 2008;38(3):341—348.

76. Zhuang H, Matsunami H. Synergism of accessory factors in functional expression of mammalian odorant receptors. *J Biol Chem.* 2007;282(20):15284—15293.

77. Reiter E, Lefkowitz RJ. GRKs and beta-arrestins: roles in receptor silencing, trafficking and signaling. *Trends Endocrinol Metab.* 2006;17(4):159—165.

78. Dawson TM, Arriza JL, Jaworsky DE, et al. Beta-adrenergic receptor kinase-2 and beta-arrestin-2 as mediators of odorant-induced desensitization. *Science.* 1993;259(5096):825—829.

79. Schleicher S, Boekhoff I, Arriza J, Lefkowitz RJ, Breer H. A beta-adrenergic receptor kinase-like enzyme is involved in olfactory signal termination. *Proc Natl Acad Sci USA*. 1993;90(4):1420−1424.

80. Peppel K, Boekhoff I, McDonald P, Breer H, Caron MG, Lefkowitz RJ. G protein-coupled receptor kinase 3 (GRK3) gene disruption leads to loss of odorant receptor desensitization. *J Biol Chem*. 1997;272(41):25425−25428.

81. Kato A, Reisert J, Ihara S, Yoshikawa K, Touhara K. Evaluation of the role of g protein-coupled receptor kinase 3 in desensitization of mouse odorant receptors in a mammalian cell line and in olfactory sensory neurons. *Chem Senses*. 2014;39(9):771−780.

82. Boekhoff I, Schleicher S, Strotmann J, Breer H. Odor-induced phosphorylation of olfactory cilia proteins. *Proc Natl Acad Sci USA*. 1992;89(24):11983−11987.

83. Chen TY, Yau KW. Direct modulation by Ca(2+)-calmodulin of cyclic nucleotide-activated channel of rat olfactory receptor neurons. *Nature*. 1994;368(6471):545−548.

84. Bradley J, Bonigk W, Yau KW, Frings S. Calmodulin permanently associates with rat olfactory CNG channels under native conditions. *Nat Neurosci*. 2004;7(7):705−710.

85. Munger SD, Lane AP, Zhong H, et al. Central role of the CNGA4 channel subunit in Ca^{2+}-calmodulin-dependent odor adaptation. *Science*. 2001;294(5549):2172−2175.

86. Bradley J, Reuter D, Frings S. Facilitation of calmodulin-mediated odor adaptation by cAMP-gated channel subunits. *Science*. 2001;294(5549):2176−2178.

87. Michalakis S, Reisert J, Geiger H, et al. Loss of CNGB1 protein leads to olfactory dysfunction and subciliary cyclic nucleotide-gated channel trapping. *J Biol Chem*. 2006;281(46):35156−35166.

88. Kelliher KR, Ziesmann J, Munger SD, Reed RR, Zufall F. Importance of the CNGA4 channel gene for odor discrimination and adaptation in behaving mice. *Proc Natl Acad Sci USA*. 2003;100(7):4299−4304.

89. Kurahashi T, Menini A. Mechanism of odorant adaptation in the olfactory receptor cell. *Nature*. 1997;385(6618):725−729.

90. Boccaccio A, Lagostena L, Hagen V, Menini A. Fast adaptation in mouse olfactory sensory neurons does not require the activity of phosphodiesterase. *J Gen Physiol*. 2006;128(2):171−184.

91. Song Y, Cygnar KD, Sagdullaev B, et al. Olfactory CNG channel desensitization by Ca^{2+}/CaM via the B1b subunit affects response termination but not sensitivity to recurring stimulation. *Neuron*. 2008;58(3):374−386.

92. Liu M, Chen TY, Ahamed B, Li J, Yau KW. Calcium-calmodulin modulation of the olfactory cyclic nucleotide-gated cation channel. *Science*. 1994;266(5189):1348−1354.

93. Waldeck C, Vocke K, Ungerer N, Frings S, Mohrlen F. Activation and desensitization of the olfactory cAMP-gated transduction channel: identification of functional modules. *J Gen Physiol*. 2009;134(5):397−408.

94. Pyrski M, Koo JH, Polumuri SK, et al. Sodium/calcium exchanger expression in the mouse and rat olfactory systems. *J Comp Neurol*. 2007;501(6):944−958.

95. Kleene SJ. Limits of calcium clearance by plasma membrane calcium ATPase in olfactory cilia. *PLoS One*. 2009;4(4):e5266.

96. Griff ER, Kleene NK, Kleene SJ. A selective PMCA inhibitor does not prolong the electroolfactogram in mouse. *PLoS One*. 2012;7(5):e37148.

97. Leinders-Zufall T, Greer CA, Shepherd GM, Zufall F. Imaging odor-induced calcium transients in single olfactory cilia: specificity of activation and role in transduction. *J Neurosci*. 1998;18(15):5630−5639.

98. Castillo K, Restrepo D, Bacigalupo J. Cellular and molecular Ca^{2+} microdomains in olfactory cilia support low signaling amplification of odor transduction. *Eur J Neurosci*. 2010;32(6):932−938.

99. Lopez F, Delgado R, Lopez R, Bacigalupo J, Restrepo D. Transduction for pheromones in the main olfactory epithelium is mediated by the Ca^{2+}-activated channel TRPM5. *J Neurosci*. 2014;34(9):3268−3278.

100. Breer H, Boekhoff I. Second messenger signalling in olfaction. *Curr Opin Neurobiol*. 1992;2(4):439−443.

101. DiPilato LM, Zhang J. Fluorescent protein-based biosensors: resolving spatiotemporal dynamics of signaling. *Curr Opin Chem Biol*. 2010;14(1):37−42.

102. Gorshkov K, Zhang J. Visualization of cyclic nucleotide dynamics in neurons. *Front Cell Neurosci*. 2014;8:395.

103. DiPilato LM, Zhang J. The role of membrane microdomains in shaping beta2-adrenergic receptor-mediated cAMP dynamics. *Mol Biosyst*. 2009;5(8):832−837.

104. Odaka H, Arai S, Inoue T, Kitaguchi T. Genetically-encoded yellow fluorescent cAMP indicator with an expanded dynamic range for dual-color imaging. *PLoS One*. 2014;9(6):e100252.

105. Dyachok O, Isakov Y, Sagetorp J, Tengholm A. Oscillations of cyclic AMP in hormone-stimulated insulin-secreting beta-cells. *Nature*. 2006;439(7074):349−352.

106. Berbari NF, Johnson AD, Lewis JS, Askwith CC, Mykytyn K. Identification of ciliary localization sequences within the third intracellular loop of G protein-coupled receptors. *Mol Biol Cell*. 2008;19(4):1540−1547.

107. Follit JA, Li L, Vucica Y, Pazour GJ. The cytoplasmic tail of fibrocystin contains a ciliary targeting sequence. *J Cell Biol*. 2010;188(1):21−28.

108. Su S, Phua SC, DeRose R, et al. Genetically encoded calcium indicator illuminates calcium dynamics in primary cilia. *Nat Methods*. 2013;10(11):1105−1107.

109. Delling M, DeCaen PG, Doerner JF, Febvay S, Clapham DE. Primary cilia are specialized calcium signalling organelles. *Nature*. 2013;504(7479):311−314.

110. Boekhoff I, Tareilus E, Strotmann J, Breer H. Rapid activation of alternative second messenger pathways in olfactory cilia from rats by different odorants. *EMBO J*. 1990;9(8):2453−2458.

111. Johnson DA, Akamine P, Radzio-Andzelm E, Madhusudan M, Taylor SS. Dynamics of cAMP-dependent protein kinase. *Chem Rev*. 2001;101(8):2243−2270.

112. Kandel ER. The molecular biology of memory: cAMP, PKA, CRE, CREB-1, CREB-2, and CPEB. *Mol Brain*. 2012;5:14.

113. Dougherty DP, Wright GA, Yew AC. Computational model of the cAMP-mediated sensory response and calcium-dependent adaptation in vertebrate olfactory receptor neurons. *Proc Natl Acad Sci USA*. 2005;102(30):10415−10420.

114. De Palo G, Boccaccio A, Miri A, Menini A, Altafini C. A dynamical feedback model for adaptation in the olfactory transduction pathway. *Biophys J*. 2012;102(12):2677−2686.

115. Feldmesser E, Bercovich D, Avidan N, et al. Mutations in olfactory signal transduction genes are not a major cause of human congenital general anosmia. *Chem Senses*. 2007;32(1):21−30.

116. Menashe I, Abaffy T, Hasin Y, et al. Genetic elucidation of human hyperosmia to isovaleric acid. *PLoS Biol*. 2007;5(11):e284.

117. Franco B, Guioli S, Pragliola A, et al. A gene deleted in Kallmann's syndrome shares homology with neural cell adhesion and axonal path-finding molecules. *Nature*. 1991;353(6344):529−536.

118. Legouis R, Hardelin JP, Levilliers J, et al. The candidate gene for the X-linked Kallmann syndrome encodes a protein related to adhesion molecules. *Cell*. 1991;67(2):423−435.

119. Dode C, Levilliers J, Dupont JM, et al. Loss-of-function mutations in FGFR1 cause autosomal dominant Kallmann syndrome. *Nat Genet*. 2003;33(4):463−465.

120. Weiss J, Pyrski M, Jacobi E, et al. Loss-of-function mutations in sodium channel Nav1.7 cause anosmia. *Nature*. 2011;472(7342):186−190.

121. Keydar I, Ben-Asher E, Feldmesser E, et al. General olfactory sensitivity database (GOSdb): candidate genes and their genomic variations. *Hum Mutat*. 2013;34(1):32−41.

122. Karstensen HG, Mang Y, Fark T, Hummel T, Tommerup N. The first mutation in CNGA2 in two brothers with anosmia. *Clin Genet*. 2015;88(3):293−296.

123. Chesler AT, Zou DJ, Le Pichon CE, et al. A G protein/cAMP signal cascade is required for axonal convergence into olfactory glomeruli. *Proc Natl Acad Sci USA*. 2007;104(3):1039−1044.

124. Miyamichi K, Serizawa S, Kimura HM, Sakano H. Continuous and overlapping expression domains of odorant receptor genes in the olfactory epithelium determine the dorsal/ventral positioning of glomeruli in the olfactory bulb. *J Neurosci*. 2005;25(14):3586−3592.

125. Imai T, Suzuki M, Sakano H. Odorant receptor-derived cAMP signals direct axonal targeting. *Science*. 2006;314(5799):657−661.

126. Nakashima A, Takeuchi H, Imai T, et al. Agonist-independent GPCR activity regulates anterior-posterior targeting of olfactory sensory neurons. *Cell*. 2013;154(6):1314−1325.

127. Serizawa S, Miyamichi K, Takeuchi H, Yamagishi Y, Suzuki M, Sakano H. A neuronal identity code for the odorant receptor-specific and activity-dependent axon sorting. *Cell*. 2006;127(5):1057−1069.

128. Margolis FL. A brain protein unique to the olfactory bulb. *Proc Natl Acad Sci USA*. 1972;69(5):1221−1224.

129. Rogers KE, Dasgupta P, Gubler U, Grillo M, Khew-Goodall YS, Margolis FL. Molecular cloning and sequencing of a cDNA for olfactory marker protein. *Proc Natl Acad Sci USA*. 1987;84(6):1704−1708.

130. Graziadei GA, Stanley RS, Graziadei PP. The olfactory marker protein in the olfactory system of the mouse during development. *Neuroscience*. 1980;5(7):1239−1252.

131. Cockerham RE, Puche AC, Munger SD. Heterogeneous sensory innervation and extensive intrabulbar connections of olfactory necklace glomeruli. *PLoS One*. 2009;4(2):e4657.

132. Li J, Ishii T, Feinstein P, Mombaerts P. Odorant receptor gene choice is reset by nuclear transfer from mouse olfactory sensory neurons. *Nature*. 2004;428(6981):393–399.

133. Kang N, Kim H, Jae Y, et al. Olfactory marker protein expression is an indicator of olfactory receptor-associated events in non-olfactory tissues. *PLoS One*. 2015;10(1):e0116097.

134. Buiakova OI, Baker H, Scott JW, et al. Olfactory marker protein (OMP) gene deletion causes altered physiological activity of olfactory sensory neurons. *Proc Natl Acad Sci USA*. 1996;93(18):9858–9863.

135. Youngentob SL, Margolis FL. OMP gene deletion causes an elevation in behavioral threshold sensitivity. *Neuroreport*. 1999;10(1):15–19.

136. Youngentob SL, Margolis FL, Youngentob LM. OMP gene deletion results in an alteration in odorant quality perception. *Behav Neurosci*. 2001;115(3):626–631.

137. Lee AC, He J, Ma M. Olfactory marker protein is critical for functional maturation of olfactory sensory neurons and development of mother preference. *J Neurosci*. 2011;31(8):2974–2982.

138. Reisert J, Yau KW, Margolis FL. Olfactory marker protein modulates the cAMP kinetics of the odour-induced response in cilia of mouse olfactory receptor neurons. *J Physiol*. 2007;585(Pt 3):731–740.

139. Kwon HJ, Koo JH, Zufall F, Leinders-Zufall T, Margolis FL. Ca extrusion by NCX is compromised in olfactory sensory neurons of OMP mice. *PLoS One*. 2009;4(1):e4260.

140. Baldisseri DM, Margolis JW, Weber DJ, Koo JH, Margolis FL. Olfactory marker protein (OMP) exhibits a beta-clam fold in solution: implications for target peptide interaction and olfactory signal transduction. *J Mol Biol*. 2002;319(3):823–837.

141. Smith PC, Firestein S, Hunt JF. The crystal structure of the olfactory marker protein at 2.3 A resolution. *J Mol Biol*. 2002;319(3):807–821.

142. Behrens M, Margolis JW, Margolis FL. Identification of members of the Bex gene family as olfactory marker protein (OMP) binding partners. *J Neurochem*. 2003;86(5):1289–1296.

Cyclic GMP Signaling in Olfactory Sensory Neurons

Trese Leinders-Zufall, Pablo Chamero

Department of Physiology, Center for Integrative Physiology and Molecular Medicine,
University of Saarland School of Medicine, Homburg, Germany

INTRODUCTION

The intracellular second messenger cyclic guanosine-3′,5′-monophosphate (cGMP) affects and modulates various physiological processes through the regulation of different signaling cascades. There are two types of guanylyl cyclases—soluble (sGC) and particulate guanylyl cyclases (pGC; a.k.a., receptor guanylyl cyclases)—both of which synthesize cGMP.[1] Both types, but not all isoforms, have been described in olfactory tissue.[2–7] The sGCs are ubiquitous cytosolic receptors that are activated by gaseous messengers. It seems that the presence of sGC activators, particularly the enzymes that produce the gaseous messengers, are differentially regulated in olfaction. The pGCs are a family of seven membrane-bound proteins, two of which (GC-D and GC-G) are found in some olfactory sensory neurons (OSNs).

The degradation of cGMP occurs exclusively via cyclic nucleotide phosphodiesterases (PDEs). In this process, cGMP can act as a regulator of some PDE activity, thus influencing the intracellular concentration of the other cyclic nucleotide, cyclic adenosine monophosphate (cAMP).[8] This kind of cross-talk between the cyclic nucleotides, along with the presence of several cGMP synthesizing and degrading components and the presence of distinct target molecules like cyclic nucleotide-gated (CNG) channels, may have profound effects on the physiological function of OSNs. In this review, we focus on the current knowledge regarding how cGMP is synthesized and degraded to exert its influence on target molecules (e.g., CNG channels) in OSNs.

SOLUBLE GUANYLYL CYCLASE

sGC (Figure 1(A)) is an intracellular receptor that facilitates the formation of cGMP.[9–13] It is a heterodimer composed of one α and one heme-binding β subunit. Two isoforms of each subunit have been identified, with the $\alpha 1 \beta 1$ protein being the most common form.[13] Because of its heme-binding domain, which evolved from a family of bacterial proteins called heme-nitric oxide/oxygen domain proteins, sGC can function as gas sensor.[14,15] The domain containing the heme group is fused to a catalytic domain that shares similarities with adenylyl cyclase,[16] but in this case converts guanosine triphosphate (GTP) to cGMP. sGC binds molecules like nitric oxide (NO) and carbon monoxide (CO), which are known as gaseous messengers or gaseous mediator molecules. Of these two gaseous mediators, NO is the more potent activator of sGC.[17] The half-life of the gas-sGC complex ranges between 4 min and 3 h at room temperature.[18,19] Another activator of sGC is protoporphyrin IX (a precursor of heme), which operates independently of either heme or NO.[20] To support sGC's catalytic activity and thus the production of cGMP, divalent cations like Mg^{2+} and Mn^{2+}, which complex with GTP, are required. Interestingly, the production of cGMP by NO/CO-sensing sGCs can be modulated by adenosine triphosphate through its influence on NO binding and dissociation[21,22] (additional references can be found in the review by Derbyshire and Marletta[13]).

In worms, flies, and mammals, the presence of predators, prey, mates, and food quality can be sensed through the environmental content of respiratory gases detected by chemosensory neurons.[23] The respiratory gases are mainly oxygen (O_2) and carbon dioxide (CO_2), with a small amount of NO. The most abundant form of sGC, the $\alpha 1 / \beta 1$ heterodimer,[24,25] is unable to form a stable complex with the main respiratory gas, O_2.[15] Yet, changes in O_2 can be detected by so-called atypical sGCs in worms and flies. These atypical sGCs, however, have not yet been reported to exist in mammals. *Caenorhabditis elegans* responds to changes in O_2 to balance its environmental needs: low O_2 decreases energy production and high O_2 induces oxidative stress. The atypical sGCs *gcy-35/36* and *gcy-31/33*, expressed in two different sensory neurons, are required to select a preferred environment and work by responding to O_2 increases or decreases, respectively.[26–28] Similarly, O_2-sensitive sGCs in insects mediate feeding behavior to both a hyperoxic and hypoxic

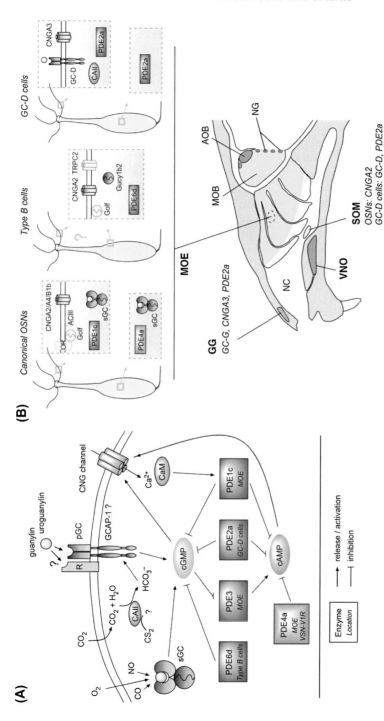

FIGURE 1 (A) Schematic representation of mammalian olfactory cyclic guanosine-3′,5′-monophosphate (cGMP) signaling, which depends on the balance between cGMP synthesis (violet) and degradation (green) to influence the function of target molecules. cGMP is generated by either soluble (sGC) or membrane-bound (particulate, pGC) guanylyl cyclases (GCs). The distribution of GCs in OSNs depends on their isoform. The expression of different GCs within an OSN could thus cause localized variations in cGMP concentration. Similarly, the presence of the cGMP degrading enzymes, the phosphodiesterases (PDEs), could modulate cGMP levels. The presence of multiple PDEs in separate cellular compartments of the sensory neuron, each with its distinct regulatory properties, could form negative feedback loops influencing cGMP signaling. It is, however, not yet clear, which combination of cGMP-modulating molecules exist in any given OSN. (B) Organization of the rodent olfactory system (lateral view) and known expression of cGMP elements in subpopulations of olfactory neurons, in particular in the three MOE neurons: the canonical OSNs, the type B and the GC-D cells. The sensory neurons in the various subsystems express both specific and identical cGMP-dependent molecules. Not all molecules mentioned in the main text and known to exist in the peripheral olfactory system are indicated, because for these molecules it is unknown which OSNs express them. AOB, accessory olfactory bulb; GG, Grueneberg ganglion; MOB, main olfactory bulb; MOE, main olfactory epithelium; NC, nasal cavity; NG, necklace glomeruli; SOM, septal organ of Masera; VNO, vomeronasal organ; VSN-V1R, subpopulation of VSN expressing the vomeronasal receptor subtype 1 (V1R).

environment.[23,29] Thus, both worms and flies use two classes of atypical sGCs to sense O_2 levels above and below a particular setpoint. If homeostasis is disturbed, the chemosensory neurons, independent of which atypical sGC they express, respond by increasing cGMP, which activates a CNG channel inducing a depolarization and resulting in the animal to avoid the source of changed O_2. Interestingly, an sGC β2 subunit, *Gucy1b2*, has recently been detected in a subpopulation of murine OSNs located in the main olfactory epithelium (MOE),[6] which contains several olfactory subsystems serving specific chemosensory roles (Figure 1(B)). It is not clear if *Gucy1b2* belongs to the atypical sGCs, but *Gucy1b2* is speculated to respond to changes in either NO or O_2. In a heterologous system this β2 subunit can produce cGMP in the absence of an α subunit when stimulated with NO, but requires the presence of the divalent cation manganese.[30] Olfactory detection of the second major respiratory gas, CO_2, has mainly been attributed to the initiation of pGC activity through modulation of bicarbonate ions (see the following section). It is therefore generally accepted that sGC acts as an NO or CO sensor in vertebrate OSNs.

Two different roles for sGC-induced cGMP signaling in olfaction have been proposed, based on the expression of the enzymes necessary to produce the sGC activators NO or CO (i.e., nitric oxide synthase (NOS), nicotinamide adenine dinucleotide phosphate diaphorase, or heme oxygenase) and measurements of either cGMP levels or the activity of CNG channels.[4,7,9,31] The function of sGC-induced cGMP signaling depends strongly on the expression of either NO- or CO-producing enzymes, which change during OSN maturation. In the first role, which pertains to odor detection in mature canonical OSNs, cGMP modulates the primary transduction pathway at the level of the CNG channels. Briefly, the primary response of a canonical vertebrate OSN to odor stimulation is a rapid rise in cAMP, which activates CNG channels, causing a receptor potential that, depending on its size, will lead to the generation of an action potential (see Chapter 7). The CNG channel in canonical OSNs is a heteromultimeric protein composed of three distinct subunits (CNGA2, CNGA4, and CNGB1b). An interesting feature of this channel is that it can be gated not only by cAMP but also by cGMP,[32−35] for which the channel actually displays a higher affinity than for cAMP.[36−38] Odor stimulation also leads to an elevation of cGMP, but this rise in cGMP occurs on a much slower time scale.[7,31,39,40] The increase in cellular cGMP concentration is produced by sGC as a result of the odor-induced activity of the CO-producing enzyme heme oxygenase-2. The persistent cGMP-induced activity of the calcium-permeable CNG channel causes a prolonged calcium elevation in the OSN cilia, which in turn promotes various calcium-dependent processes that reduce odor responsiveness.[41] The mechanism by which olfactory receptor activation is coupled to CO synthesis remains unclear. Considering the timing of the cGMP response, its effect may act as a feedback messenger in odor adaptation, affecting CNG channel open probability.[4,42] Another long-term adaptive response initiated by sGC-induced cGMP signaling may involve amplification of adenylyl cyclase activity[7] or modulation of Epac (cyclic AMP-activated guanine nucleotide exchange factors for Ras-like GTPases).[43] These actions may impinge on long-term events related to transcriptional regulation, thus enhancing the survival of OSNs.[44]

The second proposed role for sGC-induced cGMP is in the proliferation or maturation of MOE neurons resulting from transient expression of NOS and heme oxygenase-1, the other CO-producing enzyme.[31,45,46] However, NOS is present in mature bovine MOE neurons,[47] putting into question the possibility of NOS-mediated cGMP effects on development and regeneration in this species.

The distribution and isoforms of sGC in the vertebrate olfactory system remain largely uninvestigated. Although sGC is present in the ciliary layer of the MOE,[47−49] the sensory epithelium is currently better known for its different subtypes of OSNs. Some OSN subtypes express pGCs (see the following section), which would make cGMP signaling here fascinating, especially if both types of GCs are present in these neurons.

MEMBRANE-BOUND GUANYLYL CYCLASE

Cyclic GMP levels can also be modulated by membrane-bound or particulate guanylyl cyclases, pGCs (Figure 1(A)).[1,2,50] These molecules, also known as receptor GCs, are usually homodimers in which each polypeptide consists of an extracellular ligand-binding domain, a transmembrane segment, an intracellular protein kinase-like domain, and a C-terminal catalytic domain. Mammals can express up to seven pGC isoforms: GC-A to GC-G. Of these, GC-D and GC-G have been identified in two subpopulations of OSNs[5,51,52]: GC-D−expressing OSNs are found within the MOE, whereas GC-G is found in neurons of a separate chemosensory organ, the Grueneberg ganglion (Figure 1(B)). pGCs were initially classified as receptors for extracellular ligands or cellular proteins, but it now appears that pGCs can be regulated by both extra- and intracellular signaling pathways.[1,50]

Natriuretic peptides activate GC-A and GC-B.[53] Bacterial enterotoxin as well as the guanylin-family peptides guanylin and uroguanylin stimulate GC-C.[54] The first role ascribed to GC-D−expressing OSNs in mammals was that of CO_2 sensors.[55] However, GC-D can be activated by both extra- and intracellular ligands, namely guanylin, uroguanylin, and bicarbonate ions[56−59] (Figure 1(A)). CO_2 is readily converted in solution to bicarbonate, a reaction that is accelerated by carbonic anhydrase, which is present in the same olfactory sensory neurons.[55,58,59] Interestingly, carbonic anhydrase may also alter the metabolism of carbon disulfide (CS_2).[60−62] Guanylin, uroguanylin, and CS_2 each promote food-related social learning, a function that is dependent on the presence of GC-D in the MOE.[59,63,64] From the previously mentioned findings, it is conceivable that in addition to guanylin and uroguanylin, the bicarbonate ion acts as a catalyst for GC-D activity. GC-E and GC-F, which are expressed in mammalian photoreceptors, are without a known ligand for the receptor domain. Their protein kinase-like domain associates with intracellular, Ca^{2+}-binding, GC-activating proteins GCAP-1 and GCAP-2 to stimulate GC activity. Although GC-E and GC-F have not been identified in OSNs, the ciliary layer of the MOE expresses retinal GCAP-1.[48] The molecular partner and thus the function of GCAP-1 in OSNs has not yet been elucidated. Other small molecules, like p21-activated kinase, can also directly activate pGCs,[65] including GC-D.[51] GC-G, the other pGC found in olfactory tissues,[52] contains an extracellular domain that is structurally similar to GC-A and GC-B. Still, this receptor cannot be activated extracellularly by natriuretic peptides. GC-G is sensitive to intracellular bicarbonate ions, as demonstrated in GC-G—expressing HEK-293 human kidney cells and neuronal-like NG108 cells.[66] GC-G might also act as a sensor for cool temperatures and has therefore been proposed as the thermosensory protein in the Grueneberg ganglion,[67] whose sensory neurons respond to decreases in temperature.[68] The mechanism by which GC-G activity is triggered by lowered temperatures remains unclear. Likewise, the pGC subtype *gcy-12* from *C. elegans* can be modulated by cool temperatures in a heterologous system.[69] *Gcy-12*, however, seems not to play a role in thermosensation in vivo.[70] In this case, thermotaxis in *C. elegans* depends on the presence of pGCs (*gcy-8/18/23*), causing activation of CNG channels and the resulting change in membrane potential.[70] The thermoreceptor in this case was hypothesized to be either a G-protein—coupled receptor or the three pGCs *gcy-8/18/23*. Another study in *C. elegans* proposed a signaling cascade involving *gcy-8* (not *gcy-18/23*) and the phosphodiesterase type 2A (PDE2A), which regulates cGMP levels in a temperature-dependent manner.[71] Regardless of the mechanism, it appears that thermosensation in *C. elegans* or in the Grueneberg ganglion of mice changes the levels of intracellular cGMP.

The distribution of pGCs in the vertebrate peripheral olfactory system is currently thought to be limited to the two subpopulations of OSNs: the GC-D—expressing OSNs[51,72] and the group of Grueneberg ganglion neurons expressing GC-G.[52] GC-D—expressing OSNs reside mainly within the MOE and are restricted to a broad but defined zone of cells on the mouse olfactory turbinates that are clustered in hotspots at the dorsal rim of each endoturbinate.[3,51] The GC-D protein appears to be expressed at the highest concentrations in the olfactory cilia.[72] These GC-D—expressing OSNs are discernable from the canonical OSNs because of their difference in signal transduction cascade. In contrast to the canonical OSNs that use cAMP to transduce odor stimuli (see previous chapters and Chapter 7), GC-D—expressing OSNs use a cGMP-mediated signaling cascade, including the cGMP-sensitive CNG channel CNGA3 and PDE2A transduce chemosensory signals.[56,72] GC-D—expressing OSNs project to a ring of glomeruli in the caudal region (neck) of the olfactory bulb, which are thus termed the necklace glomeruli[73,74] (see Breakout Box).

The GC-G—expressing neurons are a subpopulation of rodent Grueneberg ganglion neurons,[52] which can be found at the dorsal tip of the nasal cavity. Like the GC-D—expressing OSNs in the MOE, these neurons express the cGMP-sensitive CNG channel CNGA3 and phosphodiesterase PDE2A.[52,68,75,76] They innervate an area in the olfactory bulb located near the necklace glomeruli and the accessory olfactory bulb[55,77] (see Breakout Box). At the moment, it remains unclear whether the GC-G expression depends on the maturity of the neurons or on the developmental stage of the animal (postnatal day 0—10). For instance, a substantial portion of Grueneberg ganglion neurons express trace amine-associated receptors (TAARs; see Chapter 4) during the late embryonic and neonatal stages, but not during adulthood.[78] Similarly, Grueneberg ganglion neurons from neonatal mice respond broadly to chemosensory stimulation, in contrast to adult neurons,[79,80] suggesting a large change in the functional role of these sensory neurons during development. Given that all Grueneberg ganglion neurons seem to sense temperature changes and express cGMP-dependent CNGA3 channels, multiple thermosensors, as in *C. elegans*, may exist in this vertebrate organ.

BREAKOUT BOX

THE NECKLACE SYSTEM

Olfactory sensory information is initially processed in the MOB. Axonal projections into specific areas of the MOB depend on the type of OSN. They terminate in characteristic spheroid shapes, called glomeruli, where they form synapses with dendrites from projection neurons (mitral and tufted cells) and periglomerular interneurons. The necklace system contains glomeruli that primarily receive input from the GC-D-positive/CNGA3-positive OSNs located in the MOE. These glomeruli form a ring around the stalk (neck) of the posterior MOB.[73,74] They consist of ∼20 glomeruli/bulb, which have been identified using immunocytochemistry against PDE2A and genetic markers through gene targeting.[56,72,77] Individual necklace glomeruli receive input from both GC-D—positive/PDE2A-positive and GC-D—negative/PDE2A-negative OSNs and appear therefore to be a heterogenous population.[77,112,113] Axons of a subpopulation of sensory neurons, the Grueneberg ganglion neurons, which express GC-G, project to the dorsal aspect of the caudal MOB, near the AOB, innervating ∼10 glomeruli.[73] In addition to GC-G, the Grueneberg ganglion neurons express—like the GC-D neurons—the cGMP-dependent channel CNGA3 and the cGMP-degrading enzyme PDE2A.[68,75,76] However, they do not express GC-D or carbonic anhydrase II.[52] As both types of neurons are positive for PDE2A,[52,72,76] it is difficult to distinguish the two types of projections that innervate the same region.[114] In one genetic model, GC-D—positive OSNs are also found to innervate numerous small glomeruli anterior to the necklace region,[55,77] suggesting some degree of functional segregation. The functional complexity of the necklace system remains unclear. It is presumed to play a role in odor-induced, food-related social learning as this behavior is abolished in animals in which GC-D—expressing OSNs cannot respond to guanylin, uroguanylin, or CS_2.[59,63] Experiments using either electrophysiological recordings in the region of necklace glomeruli or immunodetection of the immediate early gene c-fos to identify neural activity in the necklace system have also indicated that innate avoidance behavior in mice is regulated by this cGMP-dependent system. This behavior can be induced by both CO_2[55] or by particular predator odors.[79,80] In these latter studies, the predator odor 2-propyl thietane (2PT)—induced activity in the necklace system depends entirely on Grueneberg ganglion neuron activation and a functional CNGA3 channel. By contrast, activity of the necklace system induced by the predator odor 2,4,5-trimethylthiazoline (TMT) is independent of CNGA3 expression and is only slightly reduced by severing the axonal connection from the Grueneberg ganglion.[79] Thus, it appears that the necklace system receives additional sensory input from neurons located outside of the Grueneberg ganglion, most likely from neurons located in the MOE.

What is the role of pGC-induced cGMP signaling in these two subpopulations of OSNs? Stimulating GC-D—expressing OSNs with urine, guanylin, uroguanylin, CS_2, or CO_2 leads to membrane depolarization and an increase in action potential frequency that depend on the expression of both GC-D and CNGA3.[55,56,59] In a heterologous system, uroguanylin-induced GC-D activity increases intracellular cGMP concentration.[57] Similar to the GC-D—expressing OSNs, the GC-G—expressing neurons of the Grueneberg ganglion express cGMP-related signaling proteins.[52,68,75,76] Increasing the intracellular cGMP level in these Grueneberg ganglion neurons similarly induces an excitatory response and an increase in intracellular calcium.[68,81] Therefore, it seems that for the primary transduction of chemosensory signals both subpopulations of olfactory neurons use pGC-induced cGMP signaling.

CYCLIC NUCLEOTIDE PHOSPHODIESTERASES

Cyclic nucleotide PDEs (Figure 1(A)) are enzymes that hydrolyze the phosphodiester bond in either cAMP or cGMP. Structurally, all mammalian PDEs share a common organization, with a regulatory domain at the N-terminus and a conserved catalytic domain at the C-terminus. PDEs regulate the localization, duration, and amplitude of cGMP signaling as a result of having various interaction sites for phosphorylation, binding sites for small ligands and protein—protein interactions, and dimerization domains.[82–84] PDEs may therefore play an essential role in signaling specificity and compartmentalization. The 11 PDE families can be classified according to their substrate specificity: some are selective for cAMP (PDE4, 7—8), some for cGMP (PDE5, 6, and 9), and others for both of the cyclic nucleotides (PDE1—3 and 10—11).[82–84] Each PDE family can be subdivided based on its splice variants. In mammalian OSNs, three PDEs were initially identified: the cGMP/cAMP-sensitive isoforms PDE1C and PDE2A, and the cAMP-selective PDE4A.[6,55,56,72,85–89] However, deep sequencing of the murine olfactory neuron transcriptome has revealed the expression of all 11 PDE gene families in the olfactory epithelium.[90] The most abundant cGMP-specific PDE in MOE neurons is PDE6D. Other cGMP-sensitive PDEs include PDE3A, 3B, 5A, 6G, 9A, 10A, and 11A.[90] The cellular localization of these cGMP-sensitive PDEs has not been clarified, nor has their function in olfactory detection and processing. Knowledge gained from other systems indicates that, depending on the PDE isoform, an increase in cGMP can affect cAMP levels. For example, activated PDE2 decreases cAMP concentrations[91] and inhibition of PDE3 by cGMP can cause an increase in cAMP[92] (Figure 1(A)). This kind of cross-talk could dramatically influence odor responsiveness through the modulation of cyclic nucleotide concentration, therefore controlling ion channel open probability and membrane depolarization. CNG channels in canonical MOE neurons, which are gated by both cGMP and cAMP, could be especially affected by this kind of cross-talk.

cGMP/cAMP-sensitive PDE1c[93] is localized in the cilia of canonical OSNs.[72,85] It was expected that PDE1c would be critical for the rapid termination of the sensory response through hydrolysis of odor-induced cAMP; however, the odor-induced generator potential was not prolonged in *Pde1c*-deficient mice, as seen through electroolfactogram recordings.[85] In contrast, neurons deficient in PDE1C displayed reduced sensitivity to repeated odor stimulation, whereas a deficiency in PDE4A had no noticeable effect. Only the elimination of both PDE1C and PDE4A prolonged the termination of the sensory response.[85] In light of the newly identified PDEs in OSNs,[90] it is possible that upregulation of another PDE following the elimination of PDE1c hampered the outcome and prevented the expected change in the kinetics of the sensory response termination. In addition, one aspect of cGMP (as well as cAMP) signaling has not yet been thoroughly addressed and raises a number of physiological questions related to the kinetics and energetics of the cyclic nucleotide signal. Like calcium ions, cyclic nucleotides have numerous targets that must be differentially regulated and which are localized within different parts of the cell. Furthermore, the signaling specificity of cyclic nucleotides for their molecular targets (which could, for example, trigger distinct PDE activities) remains relatively unclear for olfactory sensory neurons. The high energy requirement for the synthesis of cyclic nucleotides suggests that it would be beneficial for a cell to localize the production of these molecules. It would therefore be quite useful for researchers to have a method for detecting localized intracellular cyclic nucleotide signals.

TECHNIQUES FOR MONITORING CHANGES IN cGMP

Obtaining accurate, reliable measurements of cGMP in most cells and tissues has proven challenging, with OSNs being no exception. Classical biochemical methods for measuring cGMP are based on either radioimmunoassays using radioactive-labeled cGMP or enzyme-linked immunoassays. These methods, despite being quite sensitive and specific, are not well suited for real-time cGMP measurements in individual cells due to a lack of temporal and spatial resolution. The need for high spatial resolution at the subcellular level is particularly critical given that cGMP signaling may not be uniformly distributed throughout the cytosol, but is likely organized within subcellular microdomains.[94] Using CNG channels as cGMP sensors partially resolves this issue.[95–97] Depending on their subunit composition, CNG channels can bind both cGMP and cAMP, and are activated in the presence of high levels of cyclic nucleotides.[36–38,98] This method, however, is indirect: cGMP and cAMP themselves are not monitored, but rather the cation currents or intracellular calcium flux, using electrophysiological or calcium imaging methods, respectively. New indicators based on fluorescence Förster-resonance energy transfer (FRET) have improved real-time cGMP detection. The general strategy for creating these genetically encoded indicators is to sandwich a cGMP-sensitive domain between two mutants of fluorescent proteins—often cyan fluorescent protein and yellow fluorescent protein. Binding of cGMP induces a conformational shift within the sensor, modulating the energy transfer between the fluorescent proteins. The sensors vary in terms of dynamics, binding affinity, and cyclic-nucleotide specificity, mostly depending on the type of cGMP-responsive protein being used. Many of these sensors—termed "cygnets"—contain a partially truncated cGMP kinase Iα (cGKIα) with cGMP binding domains.[99] Using a cygnet biosensor, odor-induced cGMP synthesis in cultured olfactory sensory neurons was not only shown to depend on the activation of sGC via gaseous messengers, but seems to require cAMP.[43] A variety of other FRET sensors that use shorter (cGi500) or longer (CGY-Del1) cGKIα domains, or that show cGMP binding in the GAF domains of PDE2 and PDE5, have been developed.[100–103] Use of the FRET technique in *C. elegans* chemosensory neurons revealed that cGMP dynamics differ between cellular compartments. These dynamics are shaped by competing negative feedback loops triggered by increased levels of both sustained Ca^{2+} and cGMP, which themselves are induced by O_2-sensitive atypical GC and coordinated by different PDEs.[104] The wide repertoire of FRET sensors offers varied sensitivities, kinetic properties, and selectivities, permitting detailed investigation of intracellular cGMP signaling. Furthermore, these sensors are genetically encoded, thus enabling the targeting of specific subcellular compartments and cell types, and making it possible to image multiple second messengers (such as calcium and cAMP) simultaneously.[105]

The disadvantages of most FRET-based biosensors reside in their low dynamic range, low temporal resolution, and poor performance in deep tissue during in vivo recordings. Therefore, novel non-FRET sensors have also been developed (e.g., circularly permuted green fluorescent protein fused to cGMP binding domains).[106] This technology, originally used to create fluorescent calcium sensors,[107] monitors increases in cpGFP fluorescence intensity after a cGMP binding—induced conformational change. Known as FlincGs (fluorescent indicator of cGMP), these sensors use different truncated cGKI variants as cGMP-binding domains, leading to a number of dynamic ranges, dissociation constants, and selectivities for cGMP.[106,108]

Both FRET and FlincGs sensors are fluorescent sensors that require an external light source. High levels of autofluorescence or investigations in combination with other light-sensitive systems may limit such fluorescence-based approaches. Bioluminescence provides a good alternative to fluorescence sensors due to a good signal-to-noise ratio. This has led to the development of sensors that use bioluminescence resonance energy transfer (BRET) to monitor cyclic nucleotides. Similar to the working mechanism of FRET, BRET makes use of the energy derived from a donor enzyme (e.g., through a luciferase reaction) to excite a fluorescent protein in close proximity to the donor enzyme. BRET sensors for cGMP signaling use cGMP binding domains, such as the GAF domain from PDE5 that is sandwiched between GFP and luciferase.[109]

As is apparent from this short overview, a variety of biosensors have been developed that allow direct visualization of cGMP signaling in intact living cells (for a more complete overview, see Sprenger et al.[110]). Furthermore, new sensors targeted to specific microdomains are currently under development. These new cGMP biosensors may aid in the investigation of cGMP dynamics in different compartments of olfactory sensory neurons following chemosensory stimuli, as they may offer unsurpassed spatial and temporal resolution, thus resolving the issues regarding cGMP's subcellular heterogeneity in intact olfactory tissue.

FUTURE PERSPECTIVES

Many outstanding questions and exciting challenges for the future still exist. There are multiple regulatory mechanisms controlling cGMP levels in OSNs, which could significantly influence local cGMP dynamics. It remains unclear, however, where many of these regulators are located within the OSN. Not only is their location within the neurons and in the various subsystems of the olfactory system important to elucidate, but it will also be of value to examine the precise cellular components regulating cGMP levels, namely the GCs, the numerous PDEs, and the undiscussed protein kinases and phosphatases.[111] Furthermore, the dialog between the cyclic nucleotide regulatory systems deserves special attention, as it allows the fine-tuning of odorant-induced activation and adaptation of sensory responses as well as long-term cellular responses like transcriptional activation and neuronal development, in OSNs.

References

1. Lucas KA, Pitari GM, Kazerounian S, et al. Guanylyl cyclases and signaling by cyclic GMP. *Pharmacol Rev.* 2000;52(3):375–414.
2. Gibson AD, Garbers DL. Guanylyl cyclases as a family of putative odorant receptors. *Annu Rev Neurosci.* 2000;23:417–439.
3. Munger SD, Leinders-Zufall T, Zufall F. Subsystem organization of the mammalian sense of smell. *Annu Rev Physiol.* 2009;71:115–140.
4. Zufall F, Leinders-Zufall T. The cellular and molecular basis of odor adaptation. *Chem Senses.* 2000;25(4):473–481.
5. Zufall F, Munger SD. Receptor guanylyl cyclases in mammalian olfactory function. *Mol Cell Biochem.* 2010;334(1–2):191–197.
6. Omura M, Mombaerts P. Trpc2-expressing sensory neurons in the mouse main olfactory epithelium of type B express the soluble guanylate cyclase Gucy1b2. *Mol Cell Neurosci.* 2015;65.
7. Moon C, Simpson PJ, Tu Y, Cho H, Ronnett GV. Regulation of intracellular cyclic GMP levels in olfactory sensory neurons. *J Neurochem.* 2005;95(1):200–209.
8. Zaccolo M, Movsesian MA. cAMP and cGMP signaling cross-talk: role of phosphodiesterases and implications for cardiac pathophysiology. *Circ Res.* 2007;100(11):1569–1578.
9. Verma A, Hirsch DJ, Glatt CE, Ronnett GV, Snyder SH. Carbon monoxide: a putative neural messenger. *Science.* 1993;259(5093):381–384.
10. Garthwaite J, Garthwaite G, Palmer RM, Moncada S. NMDA receptor activation induces nitric oxide synthesis from arginine in rat brain slices. *Eur J Pharmacol.* 1989;172(4–5):413–416.
11. Bredt DS, Snyder SH. Nitric oxide mediates glutamate-linked enhancement of cGMP levels in the cerebellum. *Proc Natl Acad Sci USA.* 1989;86(22):9030–9033.
12. Ignarro LJ. Signal transduction mechanisms involving nitric oxide. *Biochem Pharmacol.* 1991;41(4):485–490.
13. Derbyshire ER, Marletta MA. Structure and regulation of soluble guanylate cyclase. *Annu Rev Biochem.* 2012;81:533–559.
14. Iyer LM, Anantharaman V, Aravind L. Ancient conserved domains shared by animal soluble guanylyl cyclases and bacterial signaling proteins. *BMC Genomics.* 2003;4(1):5.
15. Derbyshire ER, Marletta MA. Biochemistry of soluble guanylate cyclase. *Handb Exp Pharmacol.* 2009;(191):17–31.
16. Sunahara RK, Beuve A, Tesmer JJ, Sprang SR, Garbers DL, Gilman AG. Exchange of substrate and inhibitor specificities between adenylyl and guanylyl cyclases. *J Biol Chem.* 1998;273(26):16332–16338.
17. Stone JR, Marletta MA. Soluble guanylate cyclase from bovine lung: activation with nitric oxide and carbon monoxide and spectral characterization of the ferrous and ferric states. *Biochemistry.* 1994;33(18):5636–5640.
18. Sharma VS, Ranney HM. The dissociation of NO from nitrosylhemoglobin. *J Biol Chem.* 1978;253(18):6467–6472.
19. Hille R, Olson JS, Palmer G. Spectral transitions of nitrosyl hemes during ligand binding to hemoglobin. *J Biol Chem.* 1979;254(23):12110–12120.
20. Ignarro LJ, Wood KS, Wolin MS. Activation of purified soluble guanylate cyclase by protoporphyrin IX. *Proc Natl Acad Sci USA.* 1982;79(9):2870–2873.
21. Cary SP, Winger JA, Marletta MA. Tonic and acute nitric oxide signaling through soluble guanylate cyclase is mediated by nonheme nitric oxide, ATP, and GTP. *Proc Natl Acad Sci USA.* 2005;102(37):13064–13069.
22. Russwurm M, Koesling D. NO activation of guanylyl cyclase. *EMBO J.* 2004;23(22):4443–4450.
23. Scott K. Out of thin air: sensory detection of oxygen and carbon dioxide. *Neuron.* 2011;69(2):194–202.
24. Harteneck C, Koesling D, Soling A, Schultz G, Bohme E. Expression of soluble guanylyl cyclase. Catalytic activity requires two enzyme subunits. *FEBS Lett.* 1990;272(1–2):221–223.
25. Buechler WA, Nakane M, Murad F. Expression of soluble guanylate cyclase activity requires both enzyme subunits. *Biochem Biophys Res Commun.* 1991;174(1):351–357.

26. Zimmer M, Gray JM, Pokala N, et al. Neurons detect increases and decreases in oxygen levels using distinct guanylate cyclases. *Neuron.* 2009;61(6):865–879.

27. Cheung BH, Cohen M, Rogers C, Albayram O, de Bono M. Experience-dependent modulation of *C. elegans* behavior by ambient oxygen. *Curr Biol.* 2005;15(10):905–917.

28. Gray JM, Karow DS, Lu H, et al. Oxygen sensation and social feeding mediated by a *C. elegans* guanylate cyclase homologue. *Nature.* 2004;430(6997):317–322.

29. Vermehren A, Langlais KK, Morton DB. Oxygen-sensitive guanylyl cyclases in insects and their potential roles in oxygen detection and in feeding behaviors. *J Insect Physiol.* 2006;52(4):340–348.

30. Koglin M, Vehse K, Budaeus L, Scholz H, Behrends S. Nitric oxide activates the β_2 subunit of soluble guanylyl cyclase in the absence of a second subunit. *J Biol Chem.* 2001;276(33):30737–30743.

31. Chen J, Tu Y, Moon C, Nagata E, Ronnett GV. Heme oxygenase-1 and heme oxygenase-2 have distinct roles in the proliferation and survival of olfactory receptor neurons mediated by cGMP and bilirubin, respectively. *J Neurochem.* 2003;85(5):1247–1261.

32. Pifferi S, Boccaccio A, Menini A. Cyclic nucleotide-gated ion channels in sensory transduction. *FEBS Lett.* 2006;580(12):2853–2859.

33. Biel M, Michalakis S. Cyclic nucleotide-gated channels. *Handb Exp Pharmacol.* 2009;(191):111–136.

34. Kaupp UB, Seifert R. Cyclic nucleotide-gated ion channels. *Physiol Rev.* 2002;82(3):769–824.

35. Zufall F, Firestein S, Shepherd GM. Cyclic nucleotide-gated ion channels and sensory transduction in olfactory receptor neurons. *Annu Rev Biophys Biomol Struct.* 1994;23:577–607.

36. Nakamura T, Gold GH. A cyclic nucleotide-gated conductance in olfactory receptor cilia. *Nature.* 1987;325(6103):442–444.

37. Frings S, Lynch JW, Lindemann B. Properties of cyclic nucleotide-gated channels mediating olfactory transduction. Activation, selectivity, and blockage. *J Gen Physiol.* 1992;100(1):45–67.

38. Firestein S, Zufall F, Shepherd GM. Single odor-sensitive channels in olfactory receptor neurons are also gated by cyclic nucleotides. *J Neurosci.* 1991;11(11):3565–3572.

39. Leinders-Zufall T, Shepherd GM, Zufall F. Modulation by cyclic GMP of the odour sensitivity of vertebrate olfactory receptor cells. *Proc R Soc Lond B Biol Sci.* 1996;263(1371):803–811.

40. Zufall F, Leinders-Zufall T. Identification of a long-lasting form of odor adaptation that depends on the carbon monoxide/cGMP second-messenger system. *J Neurosci.* 1997;17(8):2703–2712.

41. Leinders-Zufall T, Greer CA, Shepherd GM, Zufall F. Imaging odor-induced calcium transients in single olfactory cilia: specificity of activation and role in transduction. *J Neurosci.* 1998;18(15):5630–5639.

42. Breer H, Shepherd GM. Implications of the NO/cGMP system for olfaction. *Trends Neurosci.* 1993;16(1):5–9.

43. Pietrobon M, Zamparo I, Maritan M, Franchi SA, Pozzan T, Lodovichi C. Interplay among cGMP, cAMP, and Ca^{2+} in living olfactory sensory neurons in vitro and in vivo. *J Neurosci.* 2011;31(23):8395–8405.

44. Watt WC, Sakano H, Lee ZY, Reusch JE, Trinh K, Storm DR. Odorant stimulation enhances survival of olfactory sensory neurons via MAPK and CREB. *Neuron.* 2004;41(6):955–967.

45. Chen J, Tu Y, Moon C, Matarazzo V, Palmer AM, Ronnett GV. The localization of neuronal nitric oxide synthase may influence its role in neuronal precursor proliferation and synaptic maintenance. *Dev Biol.* 2004;269(1):165–182.

46. Zufall F, Shepherd GM, Barnstable CJ. Cyclic nucleotide gated channels as regulators of CNS development and plasticity. *Curr Opin Neurobiol.* 1997;7(3):404–412.

47. Wenisch S, Arnhold S. NADPH-diaphorase activity and NO synthase expression in the olfactory epithelium of the bovine. *Anat Histol Embryol.* 2010;39(3):201–206.

48. Moon C, Jaberi P, Otto-Bruc A, Baehr W, Palczewski K, Ronnett GV. Calcium-sensitive particulate guanylyl cyclase as a modulator of cAMP in olfactory receptor neurons. *J Neurosci.* 1998;18(9):3195–3205.

49. Ingi T, Ronnett GV. Direct demonstration of a physiological role for carbon monoxide in olfactory receptor neurons. *J Neurosci.* 1995;15(12):8214–8222.

50. Kuhn M. Function and dysfunction of mammalian membrane guanylyl cyclase receptors: lessons from genetic mouse models and implications for human diseases. *Handb Exp Pharmacol.* 2009;(191):47–69.

51. Fulle HJ, Vassar R, Foster DC, Yang RB, Axel R, Garbers DL. A receptor guanylyl cyclase expressed specifically in olfactory sensory neurons. *Proc Natl Acad Sci USA.* 1995;92(8):3571–3575.

52. Fleischer J, Mamasuew K, Breer H. Expression of cGMP signaling elements in the Grueneberg ganglion. *Histochem Cell Biol.* 2009;131(1):75–88.

53. Potter LR, Hunter T. Guanylyl cyclase-linked natriuretic peptide receptors: structure and regulation. *J Biol Chem.* 2001;276(9):6057−6060.

54. Forte Jr LR. Uroguanylin and guanylin peptides: pharmacology and experimental therapeutics. *Pharmacol Ther.* 2004;104(2):137−162.

55. Hu J, Zhong C, Ding C, et al. Detection of near-atmospheric concentrations of CO_2 by an olfactory subsystem in the mouse. *Science.* 2007;317(5840):953−957.

56. Leinders-Zufall T, Cockerham RE, Michalakis S, et al. Contribution of the receptor guanylyl cyclase GC-D to chemosensory function in the olfactory epithelium. *Proc Natl Acad Sci USA.* 2007;104(36):14507−14512.

57. Duda T, Sharma RK. ONE-GC membrane guanylate cyclase, a trimodal odorant signal transducer. *Biochem Biophys Res Commun.* 2008;367(2):440−445.

58. Sun L, Wang H, Hu J, Han J, Matsunami H, Luo M. Guanylyl cyclase-D in the olfactory CO_2 neurons is activated by bicarbonate. *Proc Natl Acad Sci USA.* 2009;106(6):2041−2046.

59. Munger SD, Leinders-Zufall T, McDougall LM, et al. An olfactory subsystem that detects carbon disulfide and mediates food-related social learning. *Curr Biol.* 2010;20(16):1438−1444.

60. Beauchamp Jr RO, Bus JS, Popp JA, Boreiko CJ, Goldberg L. A critical review of the literature on carbon disulfide toxicity. *Crit Rev Toxicol.* 1983;11(3):169−278.

61. Haritos VS, Dojchinov G. Carbonic anhydrase metabolism is a key factor in the toxicity of CO_2 and COS but not CS_2 toward the flour beetle *Tribolium castaneum* [Coleoptera: Tenebrionidae]. *Comp Biochem Physiol C Toxicol Pharmacol.* 2005;140(1):139−147.

62. Dalvi RR, Neal RA. Metabolism in vivo of carbon disulfide to carbonyl sulfide and carbon dioxide in the rat. *Biochem Pharmacol.* 1978;27(11):1608−1609.

63. Arakawa H, Kelliher KR, Zufall F, Munger SD. The receptor guanylyl cyclase type D (GC-D) ligand uroguanylin promotes the acquisition of food preferences in mice. *Chem Senses.* 2013;38(5):391−397.

64. Kelliher KR, Munger SD. Chemostimuli for guanylyl cyclase-D-expressing olfactory sensory neurons promote the acquisition of preferences for foods adulterated with the rodenticide warfarin. *Front Neurosci.* 2015;9:262.

65. Guo D, Zhang JJ, Huang XY. A new Rac/PAK/GC/cGMP signaling pathway. *Mol Cell Biochem.* 2010;334(1−2):99−103.

66. Chao YC, Cheng CJ, Hsieh HT, Lin CC, Chen CC, Yang RB. Guanylate cyclase-G, expressed in the Grueneberg ganglion olfactory subsystem, is activated by bicarbonate. *Biochem J.* 2010;432(2):267−273.

67. Chao YC, Chen CC, Lin YC, Breer H, Fleischer J, Yang RB. Receptor guanylyl cyclase-G is a novel thermosensory protein activated by cool temperatures. *EMBO J.* 2015;34(3):294−306.

68. Schmid A, Pyrski M, Biel M, Leinders-Zufall T, Zufall F. Grueneberg ganglion neurons are finely tuned cold sensors. *J Neurosci.* 2010;30(22):7563−7568.

69. Yu S, Avery L, Baude E, Garbers DL. Guanylyl cyclase expression in specific sensory neurons: a new family of chemosensory receptors. *Proc Natl Acad Sci USA.* 1997;94(7):3384−3387.

70. Inada H, Ito H, Satterlee J, Sengupta P, Matsumoto K, Mori I. Identification of guanylyl cyclases that function in thermosensory neurons of *Caenorhabditis elegans*. *Genetics.* 2006;172(4):2239−2252.

71. Wang D, O'Halloran D, Goodman MB. GCY-8, PDE-2, and NCS-1 are critical elements of the cGMP-dependent thermotransduction cascade in the AFD neurons responsible for *C. elegans* thermotaxis. *J Gen Physiol.* 2013;142(4):437−449.

72. Juilfs DM, Fulle HJ, Zhao AZ, Houslay MD, Garbers DL, Beavo JA. A subset of olfactory neurons that selectively express cGMP-stimulated phosphodiesterase (PDE2) and guanylyl cyclase-D define a unique olfactory signal transduction pathway. *Proc Natl Acad Sci USA.* 1997;94(7):3388−3395.

73. Fuss SH, Omura M, Mombaerts P. The Grueneberg ganglion of the mouse projects axons to glomeruli in the olfactory bulb. *Eur J Neurosci.* 2005;22(10):2649−2654.

74. Shinoda K, Shiotani Y, Osawa Y. "Necklace olfactory glomeruli" form unique components of the rat primary olfactory system. *J Comp Neurol.* 1989;284(3):362−373.

75. Mamasuew K, Hofmann N, Kretzschmann V, et al. Chemo- and thermosensory responsiveness of Grueneberg ganglion neurons relies on cyclic guanosine monophosphate signaling elements. *Neurosignals.* 2011;19(4):198−209.

76. Liu CY, Fraser SE, Koos DS. Grueneberg ganglion olfactory subsystem employs a cGMP signaling pathway. *J Comp Neurol.* 2009;516(1):36−48.

77. Walz A, Feinstein P, Khan M, Mombaerts P. Axonal wiring of guanylate cyclase-D-expressing olfactory neurons is dependent on neuropilin 2 and semaphorin 3F. *Development*. 2007;134(22):4063–4072.

78. Fleischer J, Schwarzenbacher K, Breer H. Expression of trace amine-associated receptors in the Grueneberg ganglion. *Chem Senses*. 2007;32(6):623–631.

79. Perez-Gomez A, Bleymehl K, Stein B, et al. Innate predator odor aversion driven by parallel olfactory subsystems that converge in the ventromedial hypothalamus. *Curr Biol*. 2015;25.

80. Brechbuhl J, Moine F, Klaey M, et al. Mouse alarm pheromone shares structural similarity with predator scents. *Proc Natl Acad Sci USA*. 2013;110(12):4762–4767.

81. Liu CY, Xiao C, Fraser SE, Lester HA, Koos DS. Electrophysiological characterization of Grueneberg ganglion olfactory neurons: spontaneous firing, sodium conductance, and hyperpolarization-activated currents. *J Neurophysiol*. 2012;108(5):1318–1334.

82. Bender AT, Beavo JA. Cyclic nucleotide phosphodiesterases: molecular regulation to clinical use. *Pharmacol Rev*. 2006;58(3):488–520.

83. Mehats C, Andersen CB, Filopanti M, Jin SL, Conti M. Cyclic nucleotide phosphodiesterases and their role in endocrine cell signaling. *Trends Endocrinol Metab*. 2002;13(1):29–35.

84. Maurice DH, Ke H, Ahmad F, Wang Y, Chung J, Manganiello VC. Advances in targeting cyclic nucleotide phosphodiesterases. *Nat Rev Drug Discov*. 2014;13(4):290–314.

85. Cygnar KD, Zhao H. Phosphodiesterase 1C is dispensable for rapid response termination of olfactory sensory neurons. *Nat Neurosci*. 2009;12(4):454–462.

86. Yan C, Zhao AZ, Bentley JK, Beavo JA. The calmodulin-dependent phosphodiesterase gene PDE1C encodes several functionally different splice variants in a tissue-specific manner. *J Biol Chem*. 1996;271(41):25699–25706.

87. Cherry JA, Davis RL. A mouse homolog of dunce, a gene important for learning and memory in *Drosophila*, is preferentially expressed in olfactory receptor neurons. *J Neurobiol*. 1995;28(1):102–113.

88. Lau YE, Cherry JA. Distribution of PDE4A and G(o) alpha immunoreactivity in the accessory olfactory system of the mouse. *Neuroreport*. 2000;11(1):27–32.

89. Leinders-Zufall T, Brennan P, Widmayer P, et al. MHC class I peptides as chemosensory signals in the vomeronasal organ. *Science*. 2004;306(5698):1033–1037.

90. Kanageswaran N, Demond M, Nagel M, et al. Deep sequencing of the murine olfactory receptor neuron transcriptome. *PLoS One*. 2015;10(1):e0113170.

91. Nikolaev VO, Gambaryan S, Engelhardt S, Walter U, Lohse MJ. Real-time monitoring of the PDE2 activity of live cells: hormone-stimulated cAMP hydrolysis is faster than hormone-stimulated cAMP synthesis. *J Biol Chem*. 2005;280(3):1716–1719.

92. Degerman E, Belfrage P, Manganiello VC. Structure, localization, and regulation of cGMP-inhibited phosphodiesterase (PDE3). *J Biol Chem*. 1997;272(11):6823–6826.

93. Loughney K, Martins TJ, Harris EA, et al. Isolation and characterization of cDNAs corresponding to two human calcium, calmodulin-regulated, 3′,5′-cyclic nucleotide phosphodiesterases. *J Biol Chem*. 1996;271(2):796–806.

94. Fischmeister R, Castro LR, Abi-Gerges A, et al. Compartmentation of cyclic nucleotide signaling in the heart: the role of cyclic nucleotide phosphodiesterases. *Circ Res*. 2006;99(8):816–828.

95. Leinders-Zufall T, Shepherd GM, Zufall F. Regulation of cyclic nucleotide-gated channels and membrane excitability in olfactory receptor cells by carbon monoxide. *J Neurophysiol*. 1995;74(4):1498–1508.

96. Kramer RH. Patch-cram detection of cyclic GMP in intact cells. In: Walz A, Boulton AA, Baker GB, eds. *Patch-Clamp Analysis: Advanced Techniques*. Vol. 35. Totowa, NJ: Humana Press; 2002:245–264.

97. Rich TC, Fagan KA, Nakata H, Schaack J, Cooper DM, Karpen JW. Cyclic nucleotide-gated channels colocalize with adenylyl cyclase in regions of restricted cAMP diffusion. *J Gen Physiol*. 2000;116(2):147–161.

98. Craven KB, Zagotta WN. CNG and HCN channels: two peas, one pod. *Annu Rev Physiol*. 2006;68:375–401.

99. Honda A, Adams SR, Sawyer CL, Lev-Ram V, Tsien RY, Dostmann WR. Spatiotemporal dynamics of guanosine 3′,5′-cyclic monophosphate revealed by a genetically encoded, fluorescent indicator. *Proc Natl Acad Sci USA*. 2001;98(5):2437–2442.

100. Russwurm M, Mullershausen F, Friebe A, Jager R, Russwurm C, Koesling D. Design of fluorescence resonance energy transfer (FRET)-based cGMP indicators: a systematic approach. *Biochem J*. 2007;407(1):69–77.

101. Sato M, Hida N, Ozawa T, Umezawa Y. Fluorescent indicators for cyclic GMP based on cyclic GMP-dependent protein kinase Ialpha and green fluorescent proteins. *Anal Chem*. 2000;72(24):5918–5924.

102. Nikolaev VO, Gambaryan S, Lohse MJ. Fluorescent sensors for rapid monitoring of intracellular cGMP. *Nat Methods.* 2006;3(1):23–25.
103. Niino Y, Hotta K, Oka K. Simultaneous live cell imaging using dual FRET sensors with a single excitation light. *PLoS One.* 2009;4(6):e6036.
104. Couto A, Oda S, Nikolaev VO, Soltesz Z, de Bono M. In vivo genetic dissection of O_2-evoked cGMP dynamics in a *Caenorhabditis elegans* gas sensor. *Proc Natl Acad Sci USA.* 2013;110(35):E3301–E3310.
105. Landa Jr LR, Harbeck M, Kaihara K, et al. Interplay of Ca^{2+} and cAMP signaling in the insulin-secreting MIN6 β-cell line. *J Biol Chem.* 2005;280(35):31294–31302.
106. Nausch LW, Ledoux J, Bonev AD, Nelson MT, Dostmann WR. Differential patterning of cGMP in vascular smooth muscle cells revealed by single GFP-linked biosensors. *Proc Natl Acad Sci USA.* 2008;105(1):365–370.
107. Nagai T, Sawano A, Park ES, Miyawaki A. Circularly permuted green fluorescent proteins engineered to sense Ca^{2+}. *Proc Natl Acad Sci USA.* 2001;98(6):3197–3202.
108. Bhargava Y, Hampden-Smith K, Chachlaki K, et al. Improved genetically-encoded, FlincG-type fluorescent biosensors for neural cGMP imaging. *Front Mol Neurosci.* 2013;6:26.
109. Biswas KH, Sopory S, Visweswariah SS. The GAF domain of the cGMP-binding, cGMP-specific phosphodiesterase (PDE5) is a sensor and a sink for cGMP. *Biochemistry.* 2008;47(11):3534–3543.
110. Sprenger JU, Nikolaev VO. Biophysical techniques for detection of cAMP and cGMP in living cells. *Int J Mol Sci.* 2013;14(4):8025–8046.
111. Francis SH, Busch JL, Corbin JD, Sibley D. cGMP-dependent protein kinases and cGMP phosphodiesterases in nitric oxide and cGMP action. *Pharmacol Rev.* 2010;62(3):525–563.
112. Cockerham RE, Leinders-Zufall T, Munger SD, Zufall F. Functional analysis of the guanylyl cyclase type D signaling system in the olfactory epithelium. *Ann NY Acad Sci.* 2009;1170:173–176.
113. Cockerham RE, Puche AC, Munger SD. Heterogeneous sensory innervation and extensive intrabulbar connections of olfactory necklace glomeruli. *PLoS One.* 2009;4(2):e4657.
114. Matsuo T, Rossier DA, Kan C, Rodriguez I. The wiring of Grueneberg ganglion axons is dependent on neuropilin 1. *Development.* 2012;139(15):2783–2791.

Ciliary Trafficking of Transduction Molecules

Jeremy C. McIntyre, Jeffrey R. Martens

Department of Pharmacology and Therapeutics, Center for Smell and Taste, University of Florida, College of Medicine, Gainesville, FL, USA

INTRODUCTION

The olfactory system provides an awareness of our environment by detecting volatile chemicals in the air. The olfactory system is necessary for detecting odors that inform our behaviors, is crucial for our appreciation of flavor, and plays important roles in our quality of life, health, and safety. Olfactory dysfunction, either from an impaired sense of smell (dysosmia) or a complete loss of the ability to smell (anosmia), can have dramatic effects on quality of life, can prevent us from detecting signs of danger such as smoke or spoiled food, and can have significant emotional consequences.[1,2] Impaired olfactory function is thought to affect several million Americans, and more than 50% of individuals are older than age 65 years.[3] However, this may be a gross underestimate given that olfactory dysfunction frequently goes unreported.[4] Although in some cases we understand the cause of olfactory dysfunction, in at least 20% of cases the underlying etiology remains unknown.[5] With ongoing research, the role of genetics and gene mutations in olfactory loss is becoming more clear. Within the past 10 years, olfactory dysfunction has emerged as a symptom of a class of genetic disorders known as ciliopathies. Ciliopathies are pleiotropic disorders involving the malformation or malfunction of cilia that affects the function of numerous organ systems. This chapter will focus on the role of cilia in olfaction, how proteins are trafficked into olfactory cilia, and olfactory disorders caused by mutations in cilia-related proteins.

Inhalation of odorants into the nasal cavity activates a signaling pathway that is localized to the cilia of canonical odorant receptor expressing olfactory sensory neurons (OSNs) found in the main olfactory epithelium (MOE) (Figure 1(A)). Although OSNs comprise the majority of chemical detecting cells in the MOE, several types of microvillar cells are present as well (see Breakout Box). This first process involves binding of an odor molecule to an odorant receptor (OR) (see Chapter 3) that initiates a stimulatory heterotrimeric G-protein to activate adenylyl cyclase type 3 (AC3) (Figure 1(B) and (C); see Chapter 7). The subsequent increase in ciliary cyclic adenosine monophosphate (cAMP) concentrations opens the olfactory cyclic nucleotide–gated (CNG) channel leading to an influx of calcium ions (Ca^{2+}) into the cilium (Figure 1(D)). The olfactory CNG channel is a heterotetrameric structure responsible for the cAMP-mediated calcium current that initiates OSN depolarization.[6–8] The channel comprises CNGA2,

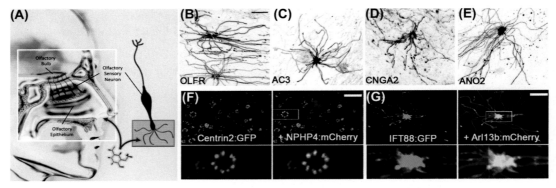

FIGURE 1　**En face confocal microscopy reveals.** (A) Olfactory sensory neurons (OSNs) located in the main olfactory epithelium (MOE) project cilia along the surface of the tissue. These cilia are the sites of odorant binding and the initiation of the olfactory signaling pathway. (B–E) Fluorescent fusion proteins ectopically expressed in OSNs show olfactory signaling proteins, including (B) the odorant receptors (OLFR); (C) AC3; (D) CNGA2; and (E) ANO2 localize to the full length of olfactory cilia. (F) En face imaging of CETN2-green fluorescent protein (GFP; green) mice shows that each cilium possesses a basal body, and the transition zone protein, nephronophthisis 4 (NPHP4)-mCherry (red), just distal to the basal body. (G) Ectopic expression of IFT88-GFP (green) in olfactory cilia shows that the IFT components are found in the cilium, labeled with Arl13b-mCherry (red) in a punctate pattern. (*Figures (F) and (G), modified from Williams et al.[31]*) Scale bars 10 μm.

CNGB1b, and CNGA4 in a 2:1:1 stoichiometry.[9,10] Although CNGA2 is capable of forming functional homotetramers, CNGA4 and CNGB1b modulate properties of the channel,[11,12] and all three subunits are required for the proper functioning of the channel in OSNs.[13,14] This depolarization can be further amplified through Ca^{2+}-activated Cl^- channels (Figure 1(E)). All of these components are localized to the cilia and play essential roles in olfactory signal transduction. However, because cilia lack the necessary protein translation machinery, each signaling protein is produced in the cell body and must be transported into the cilium. Here we will discuss what is known about the mechanisms responsible for this trafficking and how perturbation of trafficking or cilia formation can causes impaired olfactory function.

CILIA STRUCTURE

The cilium is a highly evolutionarily conserved organelle necessary for the function of many single-cell organisms and other lower eukaryotes. In vertebrates, including humans, most cell types in the body are capable of forming a cilium.[15] Both the number of cilia and cilia structure can vary widely across the diverse types of mammalian cells. Although many cells possess only a single primary cilium, a few neurons project 2-3 primary cilia,[16] and a subset of cells, such as those lining the respiratory epithelia and brain ventricles, can possess hundreds of motile cilia.[17] The OSN is a highly specialized cell type: upward of 40 cilia will project from their dendritic knobs along the surface of the MOE (Figure 2). The number and length of these organelles significantly enhance the surface area on which an OSN is capable of detecting odors that enter the nasal cavity.

Typically, cilia are categorized based on their motility and axonemal structure. The ciliary axoneme is most often composed of nine doublets of microtubules arranged symmetrically around a central core that either contains the (9 + 2) configuration or lacks the (9 + 0) configuration, a central pair of singlet microtubules. Cilia of the (9 + 2) configuration are often termed motile, whereas cilia of the (9 + 0) configuration have been classically defined as nonmotile, or primary, cilia. However, these distinctions are not exact, as rare motile (9 + 0) cilia present in the embryonic node allow the development of proper left-right asymmetry in the body.[18] Conversely, nonmotile (9 + 2) cilia are found in sensory organs such as the inner ear and MOE of many vertebrates.[19,20] The immobility of mammalian olfactory cilia results from their lack of dynein arms connecting the inner and outer microtubules.[20] In contrast, some nonmammalian vertebrates such as fish and amphibians[21,22] do possess these linkages; this allows for motile olfactory cilia, possibly to function in odorant clearance.[23,24]

The ciliary axoneme is supported by microtubules comprising polymers of alpha and beta tubulin.[25,26] These tubulin subunits possess several different posttranslational modifications. These modifications occur to both the alpha and beta tubulin subunits, and include acetylation (α), polyglutamylation ($\alpha + \beta$), polyglycylation ($\alpha + \beta$), and detyrosination (α). Tubulin

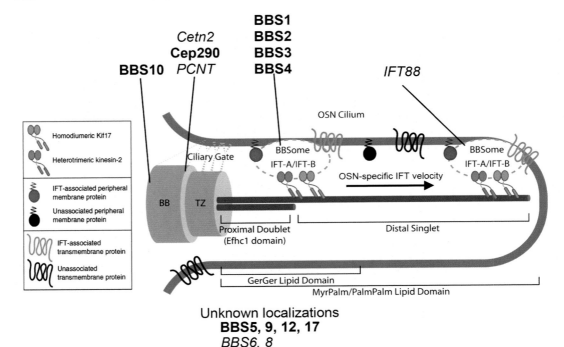

FIGURE 2 **Mutations known to affect olfactory cilia structure and function.** Model depicting the organization of a single olfactory cilium. Shown are the basal body, transition zone, and microtubule structure that supports the cilium. The microtubules provide a highway that the kinesin motors can move along, in complex with intraflagellar transport (IFT) proteins and Bardet-Biedl syndrome (BBS) proteins. These complexes can then move olfactory signaling proteins within the cilium. Highlighted are cilia proteins that are known to be necessary for olfactory function and where they are found within olfactory cilia. Mutations in the proteins in bold have been identified in patients with olfactory function, whereas those in italics are known from mouse models. BBSome, proteins associated with the pleiotropic disorder BBS; OSN, olfactory sensory neurons. *Figure modified from Williams et al.*[31]

modifications play roles in both the regulation of cargo transport[27] and ciliary maintenance in the case of zebrafish olfactory cilia.[28,29] However, little is known regarding the function of these tubulin modifications in mammalian olfactory cilia.

It is believed that the projection of cilia serves to create a large receptive field in which the OSN can be activated by odorants entering the nasal cavity. Cilia from a single OSN project in all directions, extending to lengths of 50–60 µm, increasing the sensory surface of the MOE by more than 40 times.[30,31] Calcium imaging of individual olfactory cilia has revealed that each cilium acts as an independent signaling compartment.[32,33] Importantly, the calcium transient needed to depolarize the OSN begins in the cilium before moving into the dendritic knob and soma.[33,34] Much of our early knowledge of the structure of olfactory cilia came from electron microscopy studies.[22,35–39] These studies elucidated a thicker proximal region corresponding to the 9 + 2 microtubules within the first 2—3 μm of a projecting cilium. The distal region of cilia then projects up to 60 μm as it becomes thinner and the axonemal structure tapers to only one to four singlet microtubules, usually consisting of just one pair. New technologies have verified and extended these findings through the localization of fluorescently labeled proteins within individual olfactory cilia.[31,40] These studies have been particularly insightful in highlighting the distinct protein distribution between the proximal and distal segments.

The proximal and distal segments of olfactory cilia likely represent distinct compartments because signaling proteins are differentially distributed between these two regions.[41] In developing, nascent cilia, distribution of signaling proteins is more uniform across the segments; however, these proteins appear to preferentially localize to the distal segments in mature cilia.[41—43] Until recently, however, a detailed analysis of these distinct domains in olfactory cilia has been lacking. This is likely from the distal segments of the olfactory cilia being oriented parallel to the epithelial surface, with significant overlap with the cilia from nearby OSNs. This intertwined mat makes it nearly impossible to segregate individual cilia using immunohistochemical methods. Through expression of fluorescently tagged proteins known to localize to specific regions of a cilium, Williams and colleagues used confocal microscopy of live OSNs to delineate olfactory cilia structure.[31] Transgenic expression of centrin2 (CETN2) fused to green fluorescent protein (GFP) revealed a basal body at the base of each cilium that project from the dendritic knob. Just distal to this point sits the transition zone, identified by the presence of the protein nephronophthisis 4, (NPHP4) (Figure 1(F)). Finally, the proximal segments of olfactory cilia are revealed through localization of the protein EF-hand domain (C-terminal) containing 1, which interacts with doublet microtubules, a region limited to approximately 2.5 μm in length.

LIPID COMPOSITION

The cilium, like the rest of the cell body, is enclosed by a lipid bilayer. However, the lipid composition (i.e., the lipid ratio and organization) at the base of the cilium and within the cilium itself likely differs from the rest of the cell. For example, the base of the primary cilia in cultured Madin—Darby Canine Kidney cells is composed of an annulus of distinct lipids.[44] Additionally, others have reported that there is an enrichment of cholesterol in the shaft, but not the necklace region, of epithelial cilia that extends during ciliogenesis.[45] The presence of lipid differences suggests that the lipid composition of olfactory cilia may be important for supporting olfactory signal transduction. Although there is little information regarding the precise lipid composition of this important membrane structure, recent data demonstrate that the lipid composition of olfactory cilia is distinct from the rest of the cell.[31]

When expressed as a soluble protein, GFP enriches poorly in olfactory cilia. Taking advantage of this fact, Williams and colleagues expressed GFP proteins that had been modified to allow the addition of lipid anchors.[31] GFP proteins that carried acylation signals localized throughout the cilia. GFP that was prenylated, however, only localized to the proximal regions of the cilia. These data demonstrate the lipid composition of olfactory cilia is distinct and, given the localization of the acylated GFP, is likely composed of highly organized (raft-like) lipids. This may have important implications for the organization and function of canonical olfactory signaling proteins. Interestingly, multiple protein components of the odor detection pathway have been localized to lipid rafts.[46–50] The olfactory cilia are likely the subcellular site where multiple pathways coalesce, raising the possibility that mutations that affect lipid-based signaling could underlie some cases of olfactory dysfunction. Additionally, pharmacological agents that affect membrane organization or lower total lipid content in the body (i.e., statins) could cause olfactory dysfunction in patients.[51–53]

MOVEMENT OF PROTEINS WITHIN THE CILIUM

Perhaps one of the most widely recognized processes that occur in cilia is that of intraflagellar transport (IFT). IFT is evolutionarily conserved and was first identified in the green algae *Chlamydomonas reinhardtii*.[54–56] Because cilia lack components necessary for local protein synthesis, cargo, such as microtubule components and signaling proteins, must be synthesized in the cell and transported into the cilia. The ciliary microtubules provide the structure upon which molecular motors move with the plus end at the distal tip of the cilium. This orientation ensures that individual microtubule subunits are added or removed from the cilia tip. Movement toward the plus end, termed anterograde, is accomplished by kinesin motors,[54] whereas minus end–directed movement, termed retrograde, is carried out by cytoplasmic dynein motors.[57] This process, which uses two types of molecular motor proteins in complex with transport molecules termed IFT particles,[25,26,58] is necessary for the building and maintenance of all cilia. Although the fundamental mechanisms of IFT are generally conserved between both species and cilia types, differences are clearly present in the precise processes of cilia formation and maintenance. Although the majority of mammalian cilia comprise mostly doublet microtubules, olfactory cilia have only a short region of doublets, with the remainder being singlet microtubules. Given the differences in axoneme structure, the movement of IFT complexes in olfactory cilia remains largely unknown. Comparative analysis with mammalian olfactory cilia elucidates some of these differences.

In *Caenorhabditis elegans* chemosensory neurons, two kinesin-2 motors, the heterotrimeric kinesin-II and the homodimeric OSM-3, coordinate to both form and maintain the cilium.[59] However, differences in the necessity of these cargo-carrying motors have been identified between specialized cilia types in invertebrates and mammals.[60,61] For example, although loss of function of the mammalian OSM-3 homolog, kinesin family member 17 (KIF17), impaired ciliary trafficking of the olfactory CNG channel, it had no effect on cilia length in mammalian cells.[60,61] In addition to these differences in motor protein function, there are differences in the underlying microtubule structure on which these motors move. In *C. elegans*, kinesin-II is restricted to movement on doublet microtubules.[61] By contrast, recent evidence shows that the mammalian heterotrimeric kinesin motor is not restricted to doublet microtubules and can move on the distal microtubule singlets of olfactory cilia. These data were provided by direct visualization of GFP-tagged kinesin family member 3a, a component of the heterotrimeric kinesin motor, and KIF17 in mammalian olfactory cilia, providing the first real-time evidence of the IFT process in cilia of native cells.[31] Expression of a dominant-negative KIF17 had no detrimental effects on cilia structure, suggesting that it is dispensable for maintaining olfactory cilia.[31] These data provide new evidence of kinesin movement occurring throughout olfactory cilia and highlight the similarities and differences in the formation and maintenance of chemosensory cilia in mammals and lower organisms.

In the olfactory system, IFT particles were first visualized through ultrastructure analysis of frog olfactory cilia as electron dense complexes along the ciliary axoneme.[22,62] It is now known from work in other systems that two distinct complexes of proteins named for their molecular weight comprise the IFT particles. These two complexes, complex A (consisting of IFT144, 140, 139, 122, and possibly 43) and complex B (consisting of IFT 172, 88, 81, 80, 74/72, 57/55, 52, 46, 27, and 20) are extremely well conserved across species.[62] Precise roles of the IFT complexes in mammalian olfactory cilia transport remain undefined, although IFT88 is necessary for the formation and maintenance of olfactory cilia.[40] In addition to movement of cargo proteins within the cilium, IFT proteins share homology with Golgi-localized clathrin trafficking machinery and may help with the localization of some proteins, such as ORs, to the cilia.[63,64]

IFT has now been visualized in mammalian olfactory cilia using total internal reflection fluorescence (TIRF) microscopy with fluorescently tagged IFT proteins, IFT122 and IFT88 (Figure 1(G)).[31] Using live-cell, dual-color TIRF imaging, it could be seen that the two proteins moved together in complex throughout olfactory cilia. However, the speed of IFT particle movement was different in the anterograde and retrograde direction. Interestingly, IFT speeds were more similar within the different cilia of the same neuron than they were to speeds within other OSNs, indicating stochastic variation of IFT speeds between OSNs. This difference may reflect neuron age, activation state, or possibly the specific OR expressed by the neurons. Further testing is needed to determine the mechanism underlying these differences and their importance to olfactory function.

It has been largely assumed that IFT trains carry cargo proteins such as tubulin and signaling proteins. Recent data have also suggested that IFT complexes can move transmembrane proteins such as G protein—coupled receptors (GPCRs) as well as accessory proteins such as those associated with a complex of proteins associated with the pleiotropic disorder Bardet-Biedl syndrome (BBS) called BBSome. Using dual-color TIRF imaging, the BBSome protein BBS1, BBS2, BBS3, and BBS4 and adenosine biphosphate—ribosylation factor-like 13B (Arl13b) were all visualized moving in native mammalian olfactory cilia.[31] Furthermore, when both BBS4 and IFT88 were expressed in the same OSN, the two proteins were always seen moving together. These data indicate that core BBS proteins are necessary components of IFT in native mammalian olfactory cilia.[31,65,66]

Olfactory signaling proteins, such as AC3, CNGA2, and the odorant receptor OLFR78 have also been definitively identified as cargo for IFT in olfactory cilia.[31] Although largely nonmobile at steady state, each signaling protein was capable of moving in IFT-like fashion. Coexpression of fluorescently tagged proteins, in combination with BBS4, provided further evidence that the movement was occurring through IFT. Together, these data suggest that olfactory signaling proteins are largely stationary in olfactory cilia, likely only undergoing passive diffusion while residing in the cilia membrane. However, olfactory signaling proteins are competent to move via IFT. It is unknown whether this is indicative of normal delivery of the protein, a product of stochastic movement, or is associated with internalization or removal of nonfunctional proteins.

MECHANISMS REGULATING THE SELECTIVE CILIARY ENRICHMENT OF OLFACTORY SIGNALING PROTEINS

The ciliary compartment is a selective organelle that restricts entry and maintains a specialized protein pool. The underlying mechanisms that permit this specialization remain largely unknown. That only a subset of cellular proteins is able to gain access to the cilium suggests the presence of a barrier to diffusion that restricts entry into the cilium.[67] This barrier is thought to occur at the basal body through interactions with a large complex of proteins.[25,26,68,69] Olfactory cilia basal bodies are located in the dendritic knobs of OSNs, with each cilium possessing its own basal body.[31] Several components of the basal bodies have been identified that regulate the entry of selective olfactory signaling proteins into cilia.

Mutation or deletion of proteins associated with the basal bodies of olfactory cilia can have distinct effects on olfactory cilia formation, maintenance and cargo trafficking. For example, the basal body protein pericentrin is required for olfactory cilia formation, although the formation of other cilia types is unaffected.[70] Another basal body protein, centrosomal protein 290 (CEP290) is important in the ciliary localization of a subset of olfactory signaling molecules. Analysis of a mouse model with hypomorphic deletion of CEP290 showed that the olfactory G proteins, $G\alpha_{olf}$ and $G\gamma_{13}$, did not localize to the cilia.[71] Interestingly, this mutation in CEP290 did not affect olfactory cilia structure or the localization of several other olfactory signaling molecules. Genetic deletion of another basal body protein, CETN2, results in a similar mislocalization of a subset of signaling proteins in the olfactory epithelium. Ciliary trafficking of AC3 and CNGA2 is affected in mice lacking CETN2, leading to a loss of these proteins in the cilia during early development.[72] The ciliary localization of other olfactory signaling proteins such as ORs and G proteins was unaffected at early developmental ages. However, this analysis is complicated given that olfactory cilia are also lost as these mice age. Finally, transmembrane protein 67 (TMEM67, also known as MKS3) also localizes to the dendritic knobs of OSNs.[73] In a rat model harboring a mutation in MKS3, olfactory cilia are malformed with altered microtubule organization. Localization of both $G\alpha_{olf}$ and AC3 to cilia is reduced in this mutant model, although localization of ORs appears normal within individual OSNs. Together, these studies provide compelling evidence that basal body proteins differentially regulate cilia maintenance and cargo entry into olfactory cilia and that these processes are distinct for different signaling proteins.

Protein–protein interactions also regulate ciliary localization of olfactory signaling proteins. For example, ciliary trafficking of the olfactory CNG channel complex requires the CNGB1b subunit.[60,74] The CNG complex must also interact with KIF17 to gain access to the cilium. Coexpression of the CNG subunits with a dominant-negative KIF17 prevented the ciliary localization of the channel.[60] These data highlight the importance of protein—protein interactions in ciliary localization.

Individual cargo proteins also possess amino acid sequences necessary for cilia entry. Several of these "targeting" sequences have been identified in proteins that localize to olfactory cilia, including the "RVxP" motif originally identified in polycystin-2,[75] the AX[S/A]XQ motif found in some GPCRs[76] and a nuclear localization signal.[77] For olfactory signaling proteins, the RVxP motif is necessary for the ciliary delivery of the olfactory CNG channel.[60] Interestingly, the RVxP motif is present in CNGB1b, but not CNGA2. Mutation of this motif in CNGB1b blocks the localization of the CNG channel to cilia. However, insertion of the RVxP motif into the C-terminus of CNGA2 is not sufficient for ciliary targeting in the absence CNGB1b. Thus, the RVxP motif is necessary for the CNG channel to localize to cilia, but is not sufficient for the CNGA2 subunit to enter the cilia. Although these motifs control ciliary localization, the precise mechanisms by which they do so remain unknown. That they are present in only a subset of olfactory ciliary proteins provides further indication that there are likely multiple ciliary targeting motifs acting through distinct mechanisms.

Posttranslational modifications of olfactory cargo proteins also regulate entry into the cilium. Phosphorylation of the CNG channel by the enzyme PACS-1 is necessary for its localization to cilia.[78] Inhibition of this phosphorylation through pharmacological inhibitors or dominant-negative protein expression prevents both delivery of the channel to the cilia and olfactory function. Other forms of posttranslational modification also play a role in ciliary trafficking. SUMOylation is a modification in which the small ubiquitin-like protein (SUMO) is covalently attached to other proteins. In addition to its role regulating nuclear trafficking, SUMOylation has roles in membrane trafficking and regulation of transmembrane proteins. In olfactory cilia, SUMO modification of AC3 is necessary for its trafficking into the cilia.[79] Thus it is likely that more yet-unidentified posttranslational modifications exist that regulates trafficking of olfactory signaling proteins into the cilium.

CILIARY LOCALIZATION OF ODORANT RECEPTORS

The binding of environmental odors to ORs, the largest class of GPCRs,[80] initiates the olfactory signaling cascade. The observation that the absence of olfactory cilia is correlated with olfactory dysfunction supports the idea that ciliary localization of ORs is necessary for odor detection.[40,65,66] Even so, relatively little is known about the mechanisms that regulate this process. Some clues to OR cilia trafficking may be derived from other GPCRs that localize to cilia in different tissues. Sequence motifs such as VxPx and AX[S/A]XQ regulate trafficking of GPCRs including rhodopsin, somatostatin receptor 3 (SSTR3), and serotonin receptor 6 (HTR6).[76,81–83] Although a few ORs contain the AX[S/A]XQ targeting motif,[76] the vast majority do not, suggesting that it is unlikely to be critical for ciliary trafficking of ORs. It remains possible that ORs contain as-yet unidentified sequence motifs that target these receptors to the cilia. However, for many receptors targeting sequences act in concert with other mechanisms.

Ciliary trafficking for GPCRs, such as SSTR3, HTR6, and rhodopsin, also depends on interactions with other proteins such as those that make up the BBSome and the Tubby (TUB) family of proteins.[84–88] In mice lacking either BBS2 or BBS4, many GPCRs fail to localize to cilia present on primary neurons in the brain.[85] However, the role that BBS proteins play in OR trafficking remains unclear because olfactory cilia are absent in BBS null mice. Furthermore, analysis of the MOE of TUB knockout mice revealed that ciliary localization of at least one OR was indistinguishable from wild-type mice, although related proteins also expressed in the olfactory epithelium may compensate for the loss of TUB.[84,89] Future work aimed at identifying the mechanisms by which ORs are enriched in the cilia will aid in understanding olfactory function.

BREAKOUT BOX

MICROVILLI AND CHEMICAL DETECTION

Cilia are not the only apical cellular projections that contribute to the function of the olfactory system and the detection of chemical signals. Several cell types in the nasal cavity contain apical microvilli that are important for cellular function.[30,112,113] These organelles are distinct from cilia in that they are actin based. Additionally, although kinesin and dynein motors transport proteins in cilia, myosin motors move along the actin network in microvilli. In the main olfactory epithelium (MOE), several distinct cell types possess microvilli. The sustentacular cells, supporting cells that have the most apical cell bodies in the MOE but that extend processes to the basal lamina, project apical microvilli directly beneath the cilia of OSNs. Several other microvillar cells, which are distinct from either sustentacular cells or OSNs, appear to lack axons or projections to the base of the MOE. Sparse trigeminal innervation of these cells has been seen, suggesting that a few of these may be a type of solitary chemosensory cell.[114,115] One potential role for these cells is the detection of odorous irritants.

Microvilli are also important for the function of the vomeronasal organ (VNO)

(see Chapter 11), which is present in many mammals. Vomeronasal sensory neurons (VSNs) detect a variety of semiochemicals, and the microvilli contain the receptors and sensory transduction cascades used in that detection. Interestingly, VSN microvilli are distinct from those on sustentacular cells in the VNO. For example, VSN microvilli contain the calcium-sensitive actin bundling protein villin, whereas VNO sustentacular cells lack this protein.[116] Nothing is known about how vomeronasal receptors, which are G protein–coupled receptors, or other signaling proteins are trafficked to microvilli.

What might be the role of cilia in these microvillar cells? The presence of one organelle does not preclude a cell from possessing the other. For example, the well-studied Madin–Darby Canine Kidney cells possess both a primary cilium and many microvilli. It is possible that the sustentacular cells that form the apical surface of the MOE possess both cilia and microvilli. Given the limits of standard immunohistochemical approaches in revealing organelle structure in the MOE, genetic models may be needed to understand these differences.

CILIOPATHIES AND OLFACTORY FUNCTION

It is estimated that several million individuals suffer from general or clinical anosmia in the United States, although this may be an underestimate because of patient underreporting.[4] Although acute traumas and infections are common causes of anosmia, within the past ~10 years it has become apparent that patients with genetic mutations or disorders affecting cilia can also exhibit olfactory dysfunction (Figure 2).[65,71,90] Currently, olfactory deficits are known comorbidities with two different ciliopathies, BBS and Leber congenital amaurosis (LCA).

BBS is a pleiotropic disorder leading to numerous phenotypes including olfactory dysfunction.[91–98] Mouse models with genetic deletions of BBS1, BBS2, BBS4, BBS6, and BBS8 exhibit severely impaired olfactory function.[65,66,90,91,99–101] However, deletions of different BBS proteins do not appear to share common underlying mechanisms of olfactory dysfunction, potentially because of differential localization and/or functions of individual BBS proteins (Table 1; Figure 2).[102,103] Using fluorescently tagged, ectopically expressed proteins, differential localization of BBS proteins in the dendritic knobs and cilia of OSNs is apparent.[31] Localization differences and differences in interactions with the IFT complex may underlie the diversity in phenotypes between tissues and loss of different BBS genes. For example, olfactory cilia are lost in BBS4 null mice, whereas cilia are present on other cell types.[90] This difference may be due to the fact that in olfactory cilia, BBS1 and BBS4 move in a 1:1 stoichiometry with IFT proteins suggesting a strict coupling in their function, which may not be as robust in other tissues.[31] The penetrance of olfactory phenotypes with the loss of each of the other BBS proteins remains to be determined. Because cilia assembly is often unaffected in multiple cell types in other BBS null mouse models, olfactory cilia may be especially sensitive to loss of BBS proteins.[65,90] Nevertheless, it is eminently clear that mutations in BBS proteins affect cilia function and lead to olfactory deficits, which may be a significant contributor to congenital anosmias.

Mutations in the centrosomal/basal body protein, CEP290, can cause a form of retinal dystrophy known as LCA.[104,105] LCA patients with identified mutations in CEP290 are olfactory impaired.[71] Together, these studies suggest that olfactory dysfunction from ciliary defects can occur by two separate mechanisms: (1) a complete loss of olfactory cilia and (2) a defect in protein trafficking leading to a loss in olfactory signaling.

TABLE 1 List of Olfactory Cilia-Related Genes

Gene name	Gene symbol	Function	Localization
McKusick-Kaufman syndrome	Mkks/BBS6	BBS assembly complex	Unknown
Bardet-Biedl syndrome 10	BBS10	BBS assembly complex	Basal body
Bardet-Biedl syndrome 12	BBS12	BBS assembly complex	Unknown
Bardet-Biedl syndrome 1	BBS1	BBSome	Full length
Bardet-Biedl syndrome 2	BBS2	BBSome	Full length

(*Continued*)

TABLE 1 List of Olfactory Cilia-Related Genes—cont'd

Gene name	Gene symbol	Function	Localization
Bardet-Biedl syndrome 4	BBS4	BBSome	Full length
Bardet-Biedl syndrome 5	BBS5	BBSome	Unknown
Tetratricopeptide repeat domain 8	Ttc8/BBS8	BBSome	Unknown
Bardet-Biedl syndrome 9	BBS9	BBSome	Unknown
Leucine zipper transcription factor-like 1	LZTFL1/BBS17	BBSome trafficking regulator	Unknown
Centrin 2	CETN2	Centriole function	Basal body
Pericentrin	PECTN	Centriole function	Basal body
Transmembrane protein 67	TMEM67/MKS3	Centriole function	Basal body
ADP-ribosylation factor-like 6	ARL6/BBS3	GTPase/trafficking	Full length
ADP-ribosylation factor-like 13B	Arl13b	GTPase/trafficking	Full length
Intraflagellar transport protein 122	IFT122	Intraflagellar transport	Full length
Intraflagellar transport protein 88	IFT88	Intraflagellar transport	Full length
Kinesin 17	Kif17	Intraflagellar transport	Full length
Kinesin 3a	Kif3a	Intraflagellar transport	Full length
EF-hand domain (C-terminal) containing 1	EFHC1	Microtubule interactions	Proximal doublets
Nephronophthisis 4	NPHP4	Regulation of ciliary trafficking	Transition zone
Centrosomal protein 290	CEP290	Regulation of ciliary trafficking	Basal body
Adenylyl cyclase 3	AC3	Signal transduction	Full length
Cyclic nucleotide gated channel subunit A2	CNGA2	Signal transduction	Full length
Cyclic nucleotide gated channel subunit A4	CNGA4	Signal transduction	Full length
Cyclic nucleotide–gated channel subunit B1b	CNGB1b	Signal transduction	Full length
Anoctamin 2	Ano2	Signal transduction	Full length

Names and symbols of genes presented in this chapter and their basic function and localization, if known, are provided.
BBS, Bardet-Biedl syndrome; BBSome, a complex of proteins associated with BBS.

POTENTIAL TREATMENTS FOR CILIOPATHY-INDUCED ANOSMIA

Currently, there are no curative treatments for olfactory dysfunctions, particularly congenital loss. With the growing catalog of mutations that lead to cilia defects, the list of congenital anosmias that are associated with these disorders is likely to grow. Gene therapy approaches, however, represent a viable option to potentially correct anosmias resulting from genetic mutations.[40,106]

Understanding the effects of mutations on cilia formation and function will have significant implications for the treatment of olfactory dysfunction resulting from ciliopathies. Many of the mouse models of cilia disorders exhibit a loss of olfactory cilia.[40,65,66,73,99] This suggests that an important consideration of treatment is the ability to restore the cilia structure to OSNs. Proof of principle for this has recently been accomplished.[40] The role of IFT is well established in cilia formation and maintenance, making mutant models of IFT dysfunction useful for studying restoration of cilia. Mice with a hypomorphic mutation in the gene *Ift88*, termed the Oak Ridge Polycystic Kidney (ORPK) mouse model, have been used for nearly two decades to understand the role of cilia in physiology and function of numerous organ systems.[107,108] Mutation in *Ift88* leads to a loss of olfactory cilia and a loss of olfactory function.[40] To test if restoration of cilia structures is possible, ORPK mice were treated with an adenovirus vector encoding the wild-type IFT88 protein.[40] Adenoviral expression of IFT88 lead to the reestablishment of cilia on transduced OSNs. These data demonstrate the ability of mature differentiated cells to build cilia when the necessary components are present. Additionally, the reexpression of IFT88 restored function to transduced OSNs, as seen by animals treated with gene therapy being able to respond to odors. These experiments have provided critical information to the importance of cilia for olfactory function and also offer evidence that the ability to restore normal olfaction in ciliopathy patients is possible.

SUMMARY

Olfactory cilia are necessary organelles for the detection of odorants. The cilia represent a privileged subcellular compartment optimized for transducing information from the surrounding environment. Understanding the mechanisms for the selective enrichment of the membrane and protein components of cilia is therefore important for our understanding of olfactory function. Similarly, understanding the mechanisms by which mutations cause disruption of cilia function is critical for understanding the etiology of olfactory loss. Continued study of the penetrance of ciliopathy phenotypes in the olfactory system will help elucidate the contribution of congenital anosmias to the broader picture of chemosensory loss in the general population. Because OSNs are continually produced throughout life, the process of ciliogenesis and protein transport is critical for olfactory function. Disruption of olfactory cilia through other means such as trauma, inflammation, or infection could also contribute to the etiology of olfactory disorders. Moving beyond our understanding of the pathogenesis of chemosensory loss, future work should focus on therapeutic strategies to restore olfactory function. With regard to congenital anosmia and ciliopathies, the olfactory system lends itself to the use of gene therapy techniques because OSNs are uniquely situated in the nasal cavity, exposed to the external environment. Thus drugs or therapeutic agents injected into the nasal cavity can directly act on the neurons. Importantly, OSNs are also highly amenable to transduction via adenovirus vectors, making this an attractive approach for treatment.[109–111]

References

1. Toller SV. Assessing the impact of anosmia: review of a questionnaire's findings. *Chem Senses*. 1999;24(6):705–712.
2. Keller A, Malaspina D. Hidden consequences of olfactory dysfunction: a patient report series. *BMC Ear Nose Throat Disord*. 2013;13(1):8.
3. Murphy C, Schubert CR, Cruickshanks KJ, Klein BE, Klein R, Nondahl DM. Prevalence of olfactory impairment in older adults. *Jama*. 2002;288(18):2307–2312.
4. Nguyen-Khoa BA, Goehring Jr EL, Vendiola RM, Pezzullo JC, Jones JK. Epidemiologic study of smell disturbance in 2 medical insurance claims populations. *Arch Otolaryngol Head Neck Surg*. 2007;133(8):748–757.
5. Jafek BW, Murrow B, Linschoten M. Evaluation and treatment of anosmia. *Curr Opin Otolaryngol Head Neck Surg*. 2000;8:5.
6. Fesenko EE, Novoselov VI, Novikov JV. Molecular mechanisms of olfactory reception. VI. Kinetic characteristics of camphor interaction with binding sites of rat olfactory epithelium. *Biochim Biophys Acta*. 1985;839(3):268–275.
7. Firestein S, Werblin F. Odor-induced membrane currents in vertebrate-olfactory receptor neurons. *Science*. 1989;244(4900):79–82.
8. Nakamura T, Gold GH. A cyclic nucleotide-gated conductance in olfactory receptor cilia. *Nature*. 1987;325(6103):442–444.
9. Bonigk W, Bradley J, Muller F, et al. The native rat olfactory cyclic nucleotide-gated channel is composed of three distinct subunits. *J Neurosci*. 1999;19(13):5332–5347.
10. Zheng J, Zagotta WN. Stoichiometry and assembly of olfactory cyclic nucleotide-gated channels. *Neuron*. 2004;42(3):411–421.
11. Nache V, Zimmer T, Wongsamitkul N, et al. Differential regulation by cyclic nucleotides of the CNGA4 and CNGB1b subunits in olfactory cyclic nucleotide-gated channels. *Sci Signal*. 2012;5(232):ra48.
12. Kaupp UB, Seifert R. Cyclic nucleotide-gated ion channels. *Physiol Rev*. 2002;82(3):769–824.
13. Munger SD, Lane AP, Zhong H, et al. Central role of the CNGA4 channel subunit in Ca^{2+}-calmodulin-dependent odor adaptation. *Science*. 2001;294(5549):2172–2175.
14. Kelliher KR, Ziesmann J, Munger SD, Reed RR, Zufall F. Importance of the CNGA4 channel gene for odor discrimination and adaptation in behaving mice. *Proc Natl Acad Sci USA*. 2003;100(7):4299–4304.
15. Davenport JR, Yoder BK. An incredible decade for the primary cilium: a look at a once-forgotten organelle. *Am J Physiol Renal Physiol*. 2005;289(6):F1159–F1169.
16. Koemeter-Cox AI, Sherwood TW, Green JA, et al. Primary cilia enhance kisspeptin receptor signaling on gonadotropin-releasing hormone neurons. *Proc Natl Acad Sci USA*. 2014;111(28):10335–10340.
17. Livraghi A, Randell SH. Cystic fibrosis and other respiratory diseases of impaired mucus clearance. *Toxicol Pathol*. 2007;35(1):116–129.
18. Okada Y, Takeda S, Tanaka Y, Izpisua Belmonte JC, Hirokawa N. Mechanism of nodal flow: a conserved symmetry breaking event in left-right axis determination. *Cell*. 2005;121(4):633–644.
19. Dabdoub A, Kelley MW. Planar cell polarity and a potential role for a Wnt morphogen gradient in stereociliary bundle orientation in the mammalian inner ear. *J Neurobiol*. 2005;64(4):446–457.

20. Menco BP. Ciliated and microvillous structures of rat olfactory and nasal respiratory epithelia. A study using ultra-rapid cryo-fixation followed by freeze-substitution or freeze-etching. *Cell Tissue Res.* 1984;235(2):225–241.

21. Lidow MS, Menco BP. Observations on axonemes and membranes of olfactory and respiratory cilia in frogs and rats using tannic acid-supplemented fixation and photographic rotation. *J Ultrastruct Res.* 1984;86(1):18–30.

22. Reese TS. Olfactory cilia in the frog. *J Cell Biol.* 1965;25(2):209–230.

23. Bronshtein AA, Minor AV. Significance of flagellae and their mobility for olfactory receptor function. *Dokl Akad Nauk SSSR.* 1973;213(4):987–989.

24. Mair RG, Gesteland RC, Blank DL. Changes in morphology and physiology of olfactory receptor cilia during development. *Neuroscience.* 1982;7(12):3091–3103.

25. Rosenbaum JL, Witman GB. Intraflagellar transport. *Nat Rev Mol Cell Biol.* 2002;3(11):813–825.

26. Scholey JM. Intraflagellar transport. *Annu Rev cell Dev Biol.* 2003;19:423–443.

27. Hammond JW, Cai D, Verhey KJ. Tubulin modifications and their cellular functions. *Curr Opin cell Biol.* 2008;20(1):71–76.

28. Pathak N, Obara T, Mangos S, Liu Y, Drummond IA. The zebrafish fleer gene encodes an essential regulator of cilia tubulin polyglutamylation. *Mol Biol Cell.* 2007;18(11):4353–4364.

29. Schwarzenbacher K, Fleischer J, Breer H. Formation and maturation of olfactory cilia monitored by odorant receptor-specific antibodies. *Histochem Cell Biol.* 2005;123(4–5):419–428.

30. Menco BP. Lectins bind differentially to cilia and microvilli of major and minor cell populations in olfactory and nasal respiratory epithelia. *Microsc Res Tech.* 1992;23(2):181–199.

31. Williams CL, McIntyre JC, Norris SR, et al. Direct evidence for BBSome-associated intraflagellar transport reveals distinct properties of native mammalian cilia. *Nat Commun.* 2014;5:5813.

32 Leinders-Zufall T, Rand MN, Shepherd GM, Greer CA, Zufall F. Calcium entry through cyclic nucleotide-gated channels in individual cilia of olfactory receptor cells: spatiotemporal dynamics. *J Neurosci.* 1997;17(11): 4136–4148.

33. Leinders-Zufall T, Greer CA, Shepherd GM, Zufall F. Imaging odor-induced calcium transients in single olfactory cilia: specificity of activation and role in transduction. *J Neurosci.* 1998;18(15):5630–5639.

34. Zufall F, Leinders-Zufall T, Greer CA. Amplification of odor-induced Ca(2+) transients by store-operated Ca(2+) release and its role in olfactory signal transduction. *J Neurophysiol.* 2000;83(1):501–512.

35. Cuschieri A, Bannister LH. The development of the olfactory mucosa in the mouse: electron microscopy. *J Anat.* 1975;119(Pt 3):471–498.

36. Cuschieri A, Bannister LH. The development of the olfactory mucosa in the mouse: light microscopy. *J Anat.* 1975;119(Pt 2):277–286.

37. Menco BP. Qualitative and quantitative freeze-fracture studies on olfactory and nasal respiratory epithelial surfaces of frog, ox, rat, and dog. II. Cell apices, cilia, and microvilli. *Cell Tissue Res.* 1980;211(1):5–29.

38. Menco M. Qualitative and quantitative freeze-fracture studies on olfactory and respiratory epithelial surfaces of frog, ox, rat, and dog. IV. Ciliogenesis and ciliary necklaces (including high-voltage observations). *Cell Tissue Res.* 1980;212(1):1–16.

39. Menco BP. Qualitative and quantitative freeze-fracture studies on olfactory and nasal respiratory structures of frog, ox, rat, and dog. I. A general survey. *Cell Tissue Res.* 1980;207(2):183–209.

40. McIntyre JC, Davis EE, Joiner A, et al. Gene therapy rescues cilia defects and restores olfactory function in a mammalian ciliopathy model. *Nat Med.* 2012;18(9):1423–1428.

41. Matsuzaki O, Bakin RE, Cai X, Menco BP, Ronnett GV. Localization of the olfactory cyclic nucleotide-gated channel subunit 1 in normal, embryonic and regenerating olfactory epithelium. *Neuroscience.* 1999;94(1):131–140.

42. Menco BP, Cunningham AM, Qasba P, Levy N, Reed RR. Putative odour receptors localize in cilia of olfactory receptor cells in rat and mouse: a freeze-substitution ultrastructural study. *J Neurocytol.* 1997;26(10):691–706.

43. Flannery RJ, French DA, Kleene SJ. Clustering of cyclic-nucleotide-gated channels in olfactory cilia. *Biophys J.* 2006;91(1):179–188.

44. Vieira OV, Gaus K, Verkade P, Fullekrug J, Vaz WL, Simons K. FAPP2, cilium formation, and compartmentalization of the apical membrane in polarized Madin-Darby canine kidney (MDCK) cells. *Proc Natl Acad Sci USA.* 2006;103(49):18556–18561.

45. Chailley B, Boisvieux-Ulrich E, Sandoz D. Evolution of filipin-sterol complexes and intramembrane particle distribution during ciliogenesis. *J Submicrosc Cytol.* 1983;15(1):275–280.

46. Brady JD, Rich ED, Martens JR, Karpen JW, Varnum MD, Brown RL. Interplay between PIP3 and calmodulin regulation of olfactory cyclic nucleotide-gated channels. *Proc Natl Acad Sci USA*. 2006;103(42):15635−15640.
47. Brady JD, Rich TC, Le X, et al. Functional role of lipid raft microdomains in cyclic nucleotide-gated channel activation. *Mol Pharmacol*. 2004;65(3):503−511.
48. Kobayakawa K, Hayashi R, Morita K, et al. Stomatin-related olfactory protein, SRO, specifically expressed in the murine olfactory sensory neurons. *J Neurosci*. 2002;22(14):5931−5937.
49. Schreiber S, Fleischer J, Breer H, Boekhoff I. A possible role for caveolin as a signaling organizer in olfactory sensory membranes. *J Biol Chem*. 2000;275(31):24115−24123.
50. Goldstein BJ, Kulaga HM, Reed RR. Cloning and characterization of SLP3: a novel member of the stomatin family expressed by olfactory receptor neurons. *J Assoc Res Otolaryngol*. 2003;4(1):74−82.
51. Agarwal V, Mishra B. Recent trends in drug delivery systems: intranasal drug delivery. *Indian J Exp Biol*. 1999;37(1):6−16.
52. Illum L. Nasal drug delivery−possibilities, problems and solutions. *J Control Release*. 2003;87(1−3):187−198.
53. Doty RL, Philip S, Reddy K, Kerr KL. Influences of antihypertensive and antihyperlipidemic drugs on the senses of taste and smell: a review. *J Hypertens*. 2003;21(10):1805−1813.
54. Cole DG, Diener DR, Himelblau AL, Beech PL, Fuster JC, Rosenbaum JL. *Chlamydomonas* kinesin-II-dependent intraflagellar transport (IFT): IFT particles contain proteins required for ciliary assembly in *Caenorhabditis elegans* sensory neurons. *J Cell Biol*. 1998;141(4):993−1008.
55. Kozminski KG, Beech PL, Rosenbaum JL. The *Chlamydomonas* kinesin-like protein FLA10 is involved in motility associated with the flagellar membrane. *J Cell Biol*. 1995;131(6 Pt 1):1517−1527.
56. Kozminski KG, Johnson KA, Forscher P, Rosenbaum JL. A motility in the eukaryotic flagellum unrelated to flagellar beating. *Proc Natl Acad Sci USA*. 1993;90(12):5519−5523.
57. Pazour GJ, Wilkerson CG, Witman GB. A dynein light chain is essential for the retrograde particle movement of intraflagellar transport (IFT). *J Cell Biol*. 1998;141(4):979−992.
58. Scholey JM. Intraflagellar transport motors in cilia: moving along the cell's antenna. *J Cell Biol*. 2008;180(1):23−29.
59. Snow JJ, Ou G, Gunnarson AL, et al. Two anterograde intraflagellar transport motors cooperate to build sensory cilia on *C. elegans* neurons. *Nat Cell Biol*. 2004;6(11):1109−1113.
60. Jenkins PM, Hurd TW, Zhang L, et al. Ciliary targeting of olfactory CNG channels requires the CNGB1b subunit and the kinesin-2 motor protein, KIF17. *Curr Biol*. 2006;16(12):1211−1216.
61. Ou G, Blacque OE, Snow JJ, Leroux MR, Scholey JM. Functional coordination of intraflagellar transport motors. *Nature*. 2005;436(7050):583−587.
62. Cole DG. The intraflagellar transport machinery of *Chlamydomonas reinhardtii*. *Traffic*. 2003;4(7):435−442.
63. Avidor-Reiss T, Maer AM, Koundakjian E, et al. Decoding cilia function: defining specialized genes required for compartmentalized cilia biogenesis. *Cell*. 2004;117(4):527−539.
64. Dwyer ND, Adler CE, Crump JG, L'Etoile ND, Bargmann CI. Polarized dendritic transport and the AP-1 mu1 clathrin adaptor UNC-101 localize odorant receptors to olfactory cilia. *Neuron*. 2001;31(2):277−287.
65. Kulaga HM, Leitch CC, Eichers ER, et al. Loss of BBS proteins causes anosmia in humans and defects in olfactory cilia structure and function in the mouse. *Nat Genet*. 2004;36(9):994−998.
66. Tadenev AL, Kulaga HM, May-Simera HL, Kelley MW, Katsanis N, Reed RR. Loss of Bardet-Biedl syndrome protein-8 (BBS8) perturbs olfactory function, protein localization, and axon targeting. *Proc Natl Acad Sci USA*. 2011;108(25):10320−10325.
67. Inglis PN, Boroevich KA, Leroux MR. Piecing together a ciliome. *Trends Genet*. 2006;22(9):491−500.
68. Breslow DK, Nachury MV. Primary cilia: how to keep the riff-raff in the plasma membrane. *Curr Biol*. 2011;21(11):R434−R436.
69. Dishinger JF, Kee HL, Verhey KJ. Analysis of ciliary import. *Methods Enzymol*. 2013;524:75−89.
70. Miyoshi K, Kasahara K, Miyazaki I, et al. Pericentrin, a centrosomal protein related to microcephalic primordial dwarfism, is required for olfactory cilia assembly in mice. *Faseb J*. 2009;23(10):3289−3297.
71. McEwen DP, Koenekoop RK, Khanna H, et al. Hypomorphic CEP290/NPHP6 mutations result in anosmia caused by the selective loss of G proteins in cilia of olfactory sensory neurons. *Proc Natl Acad Sci USA*. 2007;104(40):15917−15922.
72. Ying G, Avasthi P, Irwin M, et al. Centrin 2 is required for mouse olfactory ciliary trafficking and development of ependymal cilia planar polarity. *J Neurosci*. 2014;34(18):6377−6388.

73. Pluznick JL, Rodriguez-Gil DJ, Hull M, et al. Renal cystic disease proteins play critical roles in the organization of the olfactory epithelium. *PLoS One*. 2011;6(5):e19694.

74. Michalakis S, Reisert J, Geiger H, et al. Loss of CNGB1 protein leads to olfactory dysfunction and subciliary cyclic nucleotide-gated channel trapping. *J Biol Chem*. 2006;281(46):35156—35166.

75. Geng L, Okuhara D, Yu Z, et al. Polycystin-2 traffics to cilia independently of polycystin-1 by using an N-terminal RVxP motif. *J Cell Sci*. 2006;119(Pt 7):1383—1395.

76. Berbari NF, Johnson AD, Lewis JS, Askwith CC, Mykytyn K. Identification of ciliary localization sequences within the third intracellular loop of G protein-coupled receptors. *Mol Biol Cell*. 2008;19(4):1540—1547.

77. Dishinger JF, Kee HL, Jenkins PM, et al. Ciliary entry of the kinesin-2 motor KIF17 is regulated by importin-beta2 and RanGTP. *Nat Cell Biol*. 2010;12(7):703—710.

78. Jenkins PM, Zhang L, Thomas G, Martens JR. PACS-1 mediates phosphorylation-dependent ciliary trafficking of the cyclic-nucleotide-gated channel in olfactory sensory neurons. *J Neurosci*. 2009;29(34):10541—10551.

79. McIntyre JC, Joiner AM, Zhang L, Iniguez-Lluhi J, Martens JR. SUMOylation regulates ciliary localization of olfactory signaling proteins. *J Cell Sci*. 2015;128(10):1934—1945.

80. Buck L, Axel R. A novel multigene family may encode odorant receptors: a molecular basis for odor recognition. *Cell*. 1991;65(1):175—187.

81. Deretic D. Post-Golgi trafficking of rhodopsin in retinal photoreceptors. *Eye*. 1998;12(Pt 3b):526—530.

82. Deretic D, Schmerl S, Hargrave PA, Arendt A, McDowell JH. Regulation of sorting and post-Golgi trafficking of rhodopsin by its C-terminal sequence QVS(A)PA. *Proc Natl Acad Sci USA*. 1998;95(18):10620—10625.

83. Mazelova J, Astuto-Gribble L, Inoue H, et al. Ciliary targeting motif VxPx directs assembly of a trafficking module through Arf4. *Embo J*. 2009;28(3):183—192.

84. Sun X, Haley J, Bulgakov OV, Cai X, McGinnis J, Li T. Tubby is required for trafficking G protein-coupled receptors to neuronal cilia. *Cilia*. 2012;1(1):21.

85. Berbari NF, Lewis JS, Bishop GA, Askwith CC, Mykytyn K. Bardet-Biedl syndrome proteins are required for the localization of G protein-coupled receptors to primary cilia. *Proc Natl Acad Sci USA*. 2008;105(11):4242—4246.

86. Loktev AV, Jackson PK. Neuropeptide Y family receptors traffic via the Bardet-Biedl syndrome pathway to signal in neuronal primary cilia. *Cell Rep*. 2013;5(5):1316—1329.

87. Mukhopadhyay S, Wen X, Chih B, et al. TULP3 bridges the IFT-A complex and membrane phosphoinositides to promote trafficking of G protein-coupled receptors into primary cilia. *Genes Dev*. 2010;24(19):2180—2193.

88. Domire JS, Green JA, Lee KG, Johnson AD, Askwith CC, Mykytyn K. Dopamine receptor 1 localizes to neuronal cilia in a dynamic process that requires the Bardet-Biedl syndrome proteins. *Cell Mol Life Sci*. 2011;68(17):2951—2960.

89. Sammeta N, Yu TT, Bose SC, McClintock TS. Mouse olfactory sensory neurons express 10,000 genes. *J Comp Neurol*. 2007;502(6):1138—1156.

90. Iannaccone A, Mykytyn K, Persico AM, et al. Clinical evidence of decreased olfaction in Bardet-Biedl syndrome caused by a deletion in the BBS4 gene. *Am J Med Genet A*. 2005;132(4):343—346.

91. Rachel RA, May-Simera HL, Veleri S, et al. Combining Cep290 and Mkks ciliopathy alleles in mice rescues sensory defects and restores ciliogenesis. *J Clin Invest*. 2012;122(4):1233—1245.

92. Braun JJ, Noblet V, Durand M, et al. Olfaction evaluation and correlation with brain atrophy in Bardet-Biedl syndrome. *Clin Genet*. 2014;86(6):521—529.

93. Hichri H, Stoetzel C, Laurier V, et al. Testing for triallelism: analysis of six BBS genes in a Bardet-Biedl syndrome family cohort. *Eur J Hum Genet*. 2005;13(5):607—616.

94. Mykytyn K, Nishimura DY, Searby CC, et al. Evaluation of complex inheritance involving the most common Bardet-Biedl syndrome locus (BBS1). *Am J Hum Genet*. 2003;72(2):429—437.

95. Biedl A. A pair of siblings with adiposo-genital dystrophy. 1922. *Obes Res*. 1995;3(4):404.

96. Beales PL, Badano JL, Ross AJ, et al. Genetic interaction of BBS1 mutations with alleles at other BBS loci can result in non-Mendelian Bardet-Biedl syndrome. *Am J Hum Genet*. 2003;72(5):1187—1199.

97. Beales PL, Elcioglu N, Woolf AS, Parker D, Flinter FA. New criteria for improved diagnosis of Bardet-Biedl syndrome: results of a population survey. *J Med Genet*. 1999;36(6):437—446.

98. Bardet G. On congenital obesity syndrome with polydactyly and retinitis pigmentosa (a contribution to the study of clinical forms of hypophyseal obesity). 1920. *Obes Res*. 1995;3(4):387—399.

99. Fath MA, Mullins RF, Searby C, et al. Mkks-null mice have a phenotype resembling Bardet-Biedl syndrome. *Hum Mol Genet*. 2005;14(9):1109—1118.

II. OLFACTORY TRANSDUCTION

100. Nishimura DY, Fath M, Mullins RF, et al. Bbs2-null mice have neurosensory deficits, a defect in social dominance, and retinopathy associated with mislocalization of rhodopsin. *Proc Natl Acad Sci USA*. 2004;101(47):16588–16593.

101. Mykytyn K, Mullins RF, Andrews M, et al. Bardet-Biedl syndrome type 4 (BBS4)-null mice implicate Bbs4 in flagella formation but not global cilia assembly. *Proc Natl Acad Sci USA*. 2004;101(23):8664–8669.

102. Klysik M. Ciliary syndromes and treatment. *Pathol Res Pract*. 2008;204(2):77–88.

103. Stoetzel C, Laurier V, Davis EE, et al. BBS10 encodes a vertebrate-specific chaperonin-like protein and is a major BBS locus. *Nat Genet*. 2006;38(5):521–524.

104. Cideciyan AV, Aleman TS, Jacobson SG, et al. Centrosomal-ciliary gene CEP290/NPHP6 mutations result in blindness with unexpected sparing of photoreceptors and visual brain: implications for therapy of Leber congenital amaurosis. *Hum Mutat*. 2007;28(11):1074–1083.

105. den Hollander AI, Koenekoop RK, Yzer S, et al. Mutations in the CEP290 (NPHP6) gene are a frequent cause of Leber congenital amaurosis. *Am J Hum Genet*. 2006;79(3):556–561.

106. McIntyre JC, Williams CL, Martens JR. Smelling the roses and seeing the light: gene therapy for ciliopathies. *Trends Biotechnol*. 2013;31(6):355–363.

107. Berbari NF, O'Connor AK, Haycraft CJ, Yoder BK. The primary cilium as a complex signaling center. *Curr Biol*. 2009;19(13):R526–R535.

108. Lehman JM, Michaud EJ, Schoeb TR, Aydin-Son Y, Miller M, Yoder BK. The oak ridge polycystic kidney mouse: modeling ciliopathies of mice and men. *Dev Dyn*. 2008;237(8):1960–1971.

109. Gau P, Rodriguez S, De Leonardis C, Chen P, Lin DM. Air-assisted intranasal instillation enhances adenoviral delivery to the olfactory epithelium and respiratory tract. *Gene Ther*. 2011;18(5):432–436.

110. Ivic L, Pyrski MM, Margolis JW, Richards LJ, Firestein S, Margolis FL. Adenoviral vector-mediated rescue of the OMP-null phenotype in vivo. *Nat Neurosci*. 2000;3(11):1113–1120.

111. Zhao H, Ivic L, Otaki JM, Hashimoto M, Mikoshiba K, Firestein S. Functional expression of a mammalian odorant receptor. *Science*. 1998;279(5348):237–242.

112. Asan E, Drenckhahn D. Immunocytochemical characterization of two types of microvillar cells in rodent olfactory epithelium. *Histochem Cell Biol*. 2005;123(2):157–168.

113. Elsaesser R, Paysan J. The sense of smell, its signalling pathways, and the dichotomy of cilia and microvilli in olfactory sensory cells. *BMC Neurosci*. 2007;8(suppl 3):S1.

114. Lin W, Ezekwe Jr EA, Zhao Z, Liman ER, Restrepo D. TRPM5-expressing microvillous cells in the main olfactory epithelium. *BMC Neurosci*. 2008;9:114.

115. Lin W, Ogura T, Margolskee RF, Finger TE, Restrepo D. TRPM5-expressing solitary chemosensory cells respond to odorous irritants. *J Neurophysiol*. 2008;99(3):1451–1460.

116. Hofer D, Shin DW, Drenckhahn D. Identification of cytoskeletal markers for the different microvilli and cell types of the rat vomeronasal sensory epithelium. *J Neurocytol*. 2000;29(3):147–156.

Vomeronasal Receptors: V1Rs, V2Rs, and FPRs

Ivan Rodriguez[1, 2]

[1]Department of Genetics and Evolution, University of Geneva, Geneva, Switzerland;
[2]Geneva Neuroscience Center, University of Geneva, Geneva, Switzerland

INTRODUCTION

All life forms, from unicellular beings to cognitively proficient species, use chemosensors. These molecular devices fulfill broad functions that range from the maintenance of homeostasis to associative learning. As a result of the independent evolution of phylogenetic clades and of the number of different stimuli potentially useful to perceive, a remarkable assortment of these sensors is found in today's living species. In terrestrial vertebrates, these molecular probes comprise a myriad of different neurotransmitter receptors in the brain, the sweet and bitter taste receptors on the tongue, and the odorant receptors in the nose, for example.

This chapter is aimed at describing a type of chemoreceptor found in the sensory part of the vomeronasal organ. This structure, also called the Jacobson organ, is a bilateral olfactory probe that is located on the top of the palate in mammals.[1] Its size varies greatly depending on the species. In snakes, for example, it can dwarf the main olfactory sensory epithelium. It can also be vestigial, like in some haplorrhine primate species (in particular humans) or in cetaceans.[2] An opening, located rostrally, allows chemicals to enter the sensory structure. In some species, this process is active while in others it is passive. In some, the nostrils are the entry point for the signaling molecules, and in others it is the mouth. Encapsulated inside the vomeronasal organ are a few hundred thousand sensory neurons. They are organized in a crescent-shaped, pseudostratified epithelium, lined with a fluid-containing lumen. It is via this lumen that chemicals contact the specialized dendritic endings of vomeronasal sensory neurons (VSNs). After recognition of the chemicals by the vomeronasal cells, or more precisely by the receptors expressed by the sensory neurons, a signal is directly transmitted via the neuron's axon, to the accessory olfactory bulb in the brain. This signal eventually reaches key areas of the limbic system, in particular the amygdala and the hypothalamus, that regulate both endocrine responses and innate behaviors.

Vomeronasal receptors (VRs) are special, not in terms of structure but because they represent the door to a remarkable system, the activation of which modulates and even triggers a variety of social behaviors. This system is specialized in perceiving molecules emitted by peers or produced by individuals from other species.[3] These chemical compounds that bear a communicative value are called semiochemicals, a useful term that encompasses various types of molecules including pheromones, allomones, and kairomones, and thus avoids the debate about who benefits from the emission (see Chapters 1 and 2). The behaviors that result from vomeronasal activation by semiochemicals are varied. For example, aggression between males, a flight response from a predator, sick conspecific avoidance, or various sexual behaviors can be elicited.[4-9] Thus, vomeronasal receptors not only represent important tools for survival at the level of the individual, but are at the base of the survival of the species.

To date, in the mouse, three families of vomeronasal receptors have been identified. They are called type 1 and type 2 vomeronasal receptors (V1Rs and V2Rs, respectively) and formyl peptide receptors (FPRs).

TYPE 1 VOMERONASAL RECEPTORS

In 1995, using an elegant single-cell subtractive approach that in today's transcriptomic era looks very old-fashioned, Catherine Dulac and Richard Axel identified the first V1Rs.[10] At the time, these receptors were the only known vomeronasal chemoreceptors. Like the odorant receptors found four years earlier by Linda Buck and Richard Axel,[11] and in fact like most known vertebrate chemosensors, V1Rs are G-protein—coupled, seven-transmembrane domain receptors (GPCRs).[3,12,13] GPCRs are classified into six classes according to their amino acid sequences. The largest of these classes is termed the class A or rhodopsin-like family. V1Rs are today considered as part of this group, although they lack many of the most common sequence motifs that characterize its members. They thus share no sequence homology with odorant receptors (ORs, see Chapter 3), although they are not complete loners since they do exhibit similarities with the bitter taste receptors (T2Rs, see Chapter 13), for

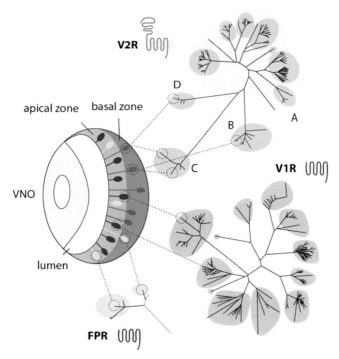

FIGURE 1   Vomeronasal receptor gene families in the mouse. A single V1R or FPR gene is expressed per sensory neuron, while two V2R genes are co-transcribed in each cell. V1Rs and V2Rs are expressed in the apical and basal neuroepithelium, respectively, while FPRs, depending on the gene, are expressed in one layer or the other. *Modified from Rodriguez and Boehm.[121]*

example. A typical V1R is about 300 amino acids long, with short extracellular and intracellular N- and C-termini, respectively.

Based on molecular markers, one can divide the pseudostratified vomeronasal neuroepithelium into two layers.[14–16] Both layers are about the same thickness in the mouse (Figure 1). The first is located at the base of the epithelium and the second lies on top of it, close to the vomeronasal lumen. V1R expression is limited to sensory neurons located in this apical part. V1Rs are present on the membrane of their dendritic endings: that is, on the specialized microvilli that bathe in the mucus. There they are thought to be coupled to a heterotrimeric G-protein containing a $G\alpha_{i2}$ subunit.

V1R Genes

In the mouse genome, V1R genes (termed *Vmn1rs*) are present on most autosomes and are arranged in gene clusters.[17–19] Each V1R gene is separated from the next by 5–50 kb, with no apparent rule in terms of direction of transcription.[17,20] In between V1R genes, one finds an exceptional density of long interspersed nuclear elements (LINEs), whose role, if any, is still unclear.[21] In mammalian species, V1R genes have a very simple structure. They possess a first (and sometimes a second) noncoding exon(s) followed by a final exon that contains the entire

V1R coding sequence. V1R promoters are highly conserved,[19,22] at least between members of the same subfamilies, which is in contrast with the very low sequence identity between promoters of other chemoreceptor gene families, in particular OR genes.[23,24]

V1R genes emerged over 400 million years ago, and are still present in the genome of teleosts, reptiles, and mammals.[25–29] Most of these genes are expressed in the vomeronasal organ, although a few are transcribed in sensory neurons from the main olfactory epithelium.[30–33] A total of 392 different V1R genes are present in the mouse genome, among which 239 have an intact open reading frame (Figure 1).[34–36] Their products have been organized according to their sequence similarities into 12 subfamilies in the mouse (V1Ra-V1Rl), with members inside a given family sharing at least 40% amino acid sequence identity.[34] Such a large V1R gene repertoire is also found in a few other species like the rabbit, in strepsirrhine primates such as the mouse lemur, and also, unexpectedly, the platypus, which leads the vertebrate ranking with 283 apparently functional *Vmn1rs*.[37] Dogs, cows, snakes, and humans are at the lower end of the list, which ends with some species like dolphins and flying foxes that simply lack intact *Vmn1rs*.[27,28,30,35,37–39] Depending on the species, the proportion of potentially functional *Vmn1rs* and pseudogenes is variable. In rats and mice, for example, pseudogenes account for about 50% of the total number of *Vmn1rs*, but reach over 90% in many primates.[40] Not only is the size of the repertoires different between species, but their contents also vary. And this can be quick in evolutionary terms, since a high diversity is even found at a subspecies level.[41] These dynamic *Vmn1r* families result from high rates of gene duplications, gene conversions, expansions or contractions of specific clades, and often pseudogenizations. This is true for most vertebrate clades, except teleosts, whose *Vmn1r* repertoires apparently escaped this instability.[25,26,42,43] Indeed, not only do fish have a very limited number of *Vmn1rs*, but these are conserved between teleosts (likely pointing to V1R agonists that are shared between these species).

Is there a pressure for the remarkable variety in the size and diversity of V1R repertoires? It first appears that those that benefit from a large number of V1Rs possess well-developed vomeronasal systems,[28,35] a correlation that is not surprising. It was suggested that during evolution, in particular after the move of species from water to land, an increase in the number of V1Rs took place, or rather a decrease in the ratio between V2Rs (a second vomeronasal family that will be discussed later) and V1Rs. This idea was laid to rest after the deciphering of the corn snake vomeronasal repertoire, which contains an extremely limited number of V1R genes and a huge V2R repertoire.[27] There is, however, a correlation that seems to resist the sequencing of new species and the identification of their corresponding chemoreceptor repertoires. It is the link between the size of V1R gene repertoires and specific ecological niches or behaviors: the genomes of species that build nests and species that are rather nocturnal each contain large numbers of potentially functional V1R genes.[44] Congruent with the adaptation of V1R repertoires to specific social structures and ecological niches, close analyses of V1R genes show that they are subjected to different types of selective pressures, including positive Darwinian selection, depending on the gene and the species.[21,35,41,43,45–51]

Similar to what is observed for OR genes in the main olfactory system,[52] each VSN "chooses," randomly, a single V1R gene.[53–55] This rule, which is very strict,[56] extends to the allele, which is transcribed, again randomly, either from the paternal or the maternal copy. The mechanisms underlying this singular expression are not understood. A hint may lie in the clustered organization of V1R genes. This likely reflects evolutionary proximity between *Vmn1rs*, which often duplicate locally, but it also may point to the existence of cluster-wide regulatory mechanisms. In fact, there is evidence of cluster-dependent *Vmn1r* transcriptional regulation: a very strong bias is observed against having two V1Rs coexpressed in cis from the same cluster.[56] This situation apparently contradicts the strict V1R monogenic rule mentioned above. However, this expression of a second V1R in a given vomeronasal sensory neuron occurs only if the choice of a first one does not lead to a functional protein.[56] The receptor itself thus appears to play a role in the non-expression of other V1Rs. Recent insights in OR regulation point to a feedback loop that involves a chemoreceptor-induced unfolded protein response.[57,58] Given the many similarities between ORs and VRs in terms of genomic organization and transcriptional characteristics (e.g., the forced expression of an OR in a VSN blocks the expression of V1Rs and therefore results in a negative feedback identical to that induced by V1Rs themselves[59]), it is tempting to speculate that similar mechanisms may be at work to achieve monogenic and monoallelic transcription of both chemosensory gene superfamilies.

V1R Functions

V1R-expressing VSNs project their axons to the dorsal part of the main olfactory bulb (MOB), an area called the accessory olfactory bulb (AOB). It is the first vomeronasal relay in the brain. V1R-expressing VSNs project to the rostral part of this structure. Functionally identical neurons (that is, those expressing the same V1R) coalesce into 10 to 30 neuropil-rich structures called glomeruli.[54,55,60] Each vomeronasal glomerulus is thus innervated by one (or a few) specific population(s) of VSNs. Interestingly, V1Rs appear to play a role in this process, since the deletion of a specific V1R gene leads neurons expressing the nonfunctional allele to project in an apparently disorganized map to the AOB.[54,55] Whether this reflects that VSNs lacking V1Rs are missing key targeting information or that the new projection pattern is the result of VSNs expressing another, randomly selected V1R, is unclear. Whatever the answer to this question, V1Rs appear, directly or indirectly, to be involved in the targeting process. Again, VSN convergence shares similarities with what is observed in the main olfactory system, where like-fibers (that is, those expressing the same ORs) also coalesce into glomeruli. Supporting this parallel and extending it to a mechanistic level, the M71 OR has the striking capacity to rescue the ability of VSN axons to coalesce and converge in the AOB when genetically swapped with the V1Rb2 receptor.[54] Likewise, the inverse experiment, which is the expression of the V1Rb2 receptor instead of the M71 receptor in olfactory sensory neurons (OSNs), rescues the ability of these neurons to find each other, and to coalesce in MOB glomeruli.[61]

V1Rs thus play a role in regulating their transcription and in guiding axons. But a main function of V1Rs is to recognize chemicals, in particular those produced in bodily secretions such as urine, tears, sweat, or feces. V1Rs are highly selective and sensitive to small organic compounds present in natural blends.[8,9,62–68] A remarkable and systematic survey was recently undertaken in the mouse. This study evaluated the response of VSNs to chemical cues derived from conspecifics or from other species.[64] The response of VSNs expressing defined VRs to specific ligands or to natural blends was assessed by the transcription of the immediate early gene *Egr1*. It was found that complex natural cues, like chemosignals from hawk bedding, strongly activate mouse V1R-expressing VSNs. It was also observed that some V1R clades are specialized in detecting particular types of ligands. For example, the V1Rd clade is activated by female-specific cues. Specialization is not limited to subfamilies: V1Rc25, V1Rg8, and V1Rh16 were exclusively activated by chemical signals produced by owls. Surprisingly, some V1R clades showed activation by apparently conflicting signals. For example, members of the V1Re subfamily that are activated by predator, nonpredator, and mouse cues.[64]

Some compounds, in particular sulfated steroids, a class of chemicals present in female urine, are potent V1R agonists: 80% of vomeronasal activation by female urine has been attributed to this type of agonist.[69,70] These sulfated steroids are selectively identified by specific V1Rs.[64,67,71,72] V1Rj2, for example, responds to estradiols and androgens but is not activated by estriols; V1Re6 and V1Re2 selectively detect corticosteroids; and V1Rf3 does not respond to androgens but is activated by estriols and estradiols.[64] V1Rs thus recognize different structural classes of steroids, which may provide the recipient animal with a sense of the physiological status of its peers. Taken together, V1Rs appear to include both specialist and generalist receptors.

As previously mentioned, various behavioral alterations affecting interindividual interactions in mice result from the silencing of the VNO. This silencing was obtained either via the knockout of the gene coding for the diacylglycerol-activated TRPC2 channel, a major downstream component of the vomeronasal transduction cascade (see Chapter 11), or via the physical removal of the organ itself. The resulting behavioral anomalies include a lack of sexual discrimination between males and females, a loss of sexual inhibition toward immature individuals, a loss of predator avoidance, a decrease of inter-male aggression, and a loss of sick conspecific avoidance.[4–7] What specific contributions were made by V1Rs to these behaviors is unclear, because the silencing of the whole organ also silences VSNs expressing V2Rs or FPRs (see below). However, a relatively limited decrease in the size of the V1R gene repertoire leads to significant effects. In 2002, a chromosome-engineered mouse was generated in which a *Vmn1r* cluster that included all 16 expressed members of the V1Ra and V1Rb subfamilies was deleted. The corresponding female mice exhibited low levels of maternal care toward their pups, while the mutant males showed low sexual activity toward females.[18] The anomalies exhibited by these mutant mice may have also affected other behaviors linked to vomeronasal activity, but the genetic deletion was generated and tested prior to the identification of the multiple functions played by the vomeronasal system, in particular predator and sick conspecific avoidance.

TYPE 2 VOMERONASAL RECEPTORS

Two years after the identification of V1Rs, three groups simultaneously and independently reported the existence of a second VR family, the V2Rs.[73–75] Like V1Rs, V2Rs belong to the GPCR superfamily. But they are not classified in the rhodopsin-like group; they are part of the class C, also known as the metabotropic glutamate receptor family. V2Rs share sequence homology with the T1R sweet taste receptors (see Chapter 13), and unlike V1Rs possess a long N-terminus that is thought to contain the primary agonist binding site.[3,12,13] The typical size of a V2R is about 600 amino acids. However, numerous splice variants from a single V2R gene can result in differences in the length of the extracellular N-terminus.

While V1Rs are expressed by VSNs located in the apical zone of the vomeronasal epithelium, V2Rs are only found in VSNs located basally. This segregation between neuronal types is not limited to the chemosensors; it also applies to the transduction machinery (V2Rs are thought to be coupled to a G-protein containing a $G\alpha_o$ subunit), to their projection sites in the AOB (V2R-expressing axons project to the caudal part of the structure[76]), and even applies to higher-order projections into the amygdala and the hypothalamus.[77,78] Like the V1Rs and the vomeronasal marker TRPC2,[79] very few V2Rs are expressed by OSNs in the main olfactory epithelium.[80,81]

V2R Genes

Unlike V1Rs, the coding sequence of V2R genes (called *Vmn2rs*) is split into multiple exons, with a single, large exon encoding the seven-transmembrane part of the receptor. *Vmn2rs* are, like *Vmn1rs*, organized in clusters. Depending on the species, they can be numerous or almost absent.[29,82] Among the poorly equipped species are Old and New World monkeys, dogs, and cows.[83] Well-equipped species include the opossum, some amphibians, snakes, and many rodent species. Like *Vmn1rs*, *Vmn2r* repertoires expanded and contracted significantly. They did so long ago, but also very recently, which today makes V2R orthologous relationships between species difficult to identify.[82–85] In the mouse, whose genome comprises 121 potentially functional V2Rs (Figure 1), V2R gene products are usually classified into four families, termed A to D. This classification is not a mere tool to make a large repertoire easier to describe. First, these families are apparently under different selective pressures. As a result, one observes a nonrandom distribution of V2R variants in different clades when looking at single nucleotide polymorphisms within and between species/subspecies of mice. In particular, V2Rs that have been linked to the detection of sympatric cues (from nonpredatory rodents) differ more than those that detect predators (from snakes, for example) or conspecific cues.[41,64] Second, unlike other olfactory receptors including ORs (Chapter 3), TAARs (Chapter 4), V1Rs, and FPRs (see below), V2Rs do not strictly follow the "monogenic rule." All V2R-expressing VSNs express a member of the V2RC family, plus one from another V2R family.[86–89] The role played by this coexpression is unclear, but it is found in many species, even reptiles.[27] The identity of VSNs from the basal vomeronasal neuroepithelium is further complicated by the expression of nonclassical major histocompatibility complex 1b proteins in specific subsets of very basal V2R-positive neurons, suggesting a further division of the basal, $G\alpha_o$-, V2R-expressing layer.[90–93]

V2R Functions

While V1Rs appear to specialize in the detection of small chemicals, V2Rs recognize large molecules, usually peptides or proteins. These are found in bodily secretions such as urine or tears that are produced either by conspecifics or heterospecifics. As is observed for V1R response profiles, V2R clades appear to be specialized in detecting specific classes of molecules. For example, male mouse bedding activates most members of clade 7 from the A family in mouse, while rat bedding activates most members of clade 8 from the same family.[64] Again like V1Rs, some clades appear more versatile, such as clade 3 from the A family, the members of which respond to both female and male mouse bedding and to predator cues.[64] Single compounds are known to be V2R agonists. The first ones to be identified were major histocompatibility complex (MHC) class I peptides[94–96] that activate V2R-expressing VSN populations at very low concentrations. For example, V2R1b, a receptor encoded by *Vmn2r26*, recognizes MHC peptides at subpicomolar concentrations.[95] This class of ligands is of interest because these peptides, given the polymorphism of MHC molecules between individuals that translates into a corresponding variability of their bound peptides, represent unique and specific molecular signatures. Supporting this view, the well-known selective prevention of pregnancy induced by the scent of unfamiliar males, a response called the Bruce effect,[97] can be recapitulated by adding these peptides to an otherwise neutral urine sample.[94]

Other V2R agonists have been identified, including the exocrine gland-secreting peptides (ESPs) and major urinary proteins (MUPs). ESPs are small peptides produced in exocrine glands and released in the tear fluid or in saliva of some rodent species.[98] In mice and rats, they are encoded by a large gene family composed of 38 members, and the expression of some of them is strain, age, and sex-specific.[99] Their perception can trigger or prevent innate and stereotyped behaviors. For example, ESP22 is a juvenile pheromone the perception of which by VSNs exerts a strong inhibitory effect on adult male mating.[100] Another remarkable example is the first semiochemical-VR-behavior string identified in a rodent, and even in a mammal: the male-produced ESP1 enhances female lordosis (in other words sexual receptivity) by activating VSNs expressing V2Rp5.[64,101] This ESP—V2R interaction is not unique. For example, V2Rp1 recognizes ESP5, and V2Rp2 recognizes both ESP6 and ESP5.[102]

MUPs, which as their name suggests are often found in urine, are hydrophobic lipocalins. They are, like ESPs, encoded by a large gene family. Their expression is gender- and strain-specific in the mouse, thus providing potential individuality signals. Interestingly, MUPs are not only intraspecific pheromonal cues. MUPS produced by cats and other predators are sensed by the vomeronasal system of rodents, where they trigger an innate defensive response.[5,103] Multiple responses are thus induced by the V2R-mediated recognition of MUPs, ESPs, and other compounds, which include aggressive, sexual, and avoidance behaviors.[5,7,101,104—107]

The topographical and molecular dichotomy between V1R- and V2R-expressing neurons suggests different roles played by both receptor families. To test this hypothesis, an obvious approach would require the genetic deletion of either the entire V1R or the entire V2R repertoire followed by the evaluation of potential deficiencies in the corresponding animals. Given that members of both receptor gene families are present on most autosomes in the mouse, the systematic gene targeting of all V1Rs will take some years. As an indirect and faster alternative, two groups took advantage of the molecular dichotomy that characterizes the two VSN populations, more precisely in deleting two key players in the V1R and the V2R transduction cascades, respectively, namely, the $G\alpha_{i2}$ or the $G\alpha_o$ subunits. The results are not as clear-cut as one could have expected: the specific deletion of $G\alpha_o$ in VSNs leads to a decrease in male and female aggressive responses, to an alteration of the female reaction to the estrus-inducing activity of male urine, and to a decrease in lordosis and recognition of familiar male scents in females.[108] The deletion of $G\alpha_{i2}$ leads to a rather milder phenotype (although many of the previously mentioned behaviors were not tested), that consists in decreased male aggressiveness toward intruders and lowered maternal aggressive behavior.[109]

FORMYL PEPTIDE RECEPTORS

FPR Genes

A few years ago, in an attempt to discover potentially unknown vomeronasal receptors, two groups independently identified a third family of chemoreceptors.[110,111] This family, called formyl peptide receptors (FPRs), contains seven members in the mouse (Figure 1).

Two of them (FPR1 and FPR2) are not expressed in VSNs. However, these proteins were previously known to be expressed by immune cells in mammals, where they recognize formylated peptides produced by bacteria and other compounds related to pathogens or pathogenic states. In the mouse and other rodent species, the remaining FPRs (FPR-rs1/3/4/6/7) are, like V1Rs and V2Rs, transcribed in a punctate pattern in the vomeronasal neuroepithelium. In the mouse, VSNs that transcribe FPRs do not coexpress other members of their own family nor members of the V1R or V2R families.[110,111] And like V1R and V2R genes, both immune and vomeronasal FPR genes are clustered in the genome. This olfactory expression of FPRs may be limited to rodent species[112] because although the use of FPRs by immune cells is likely shared by all mammals, no orthologs of vomeronasal FPRs are found outside the rodent clade (which is still significant since about 40% of all mammalian species living today are rodents). Thus, not only are VR repertoires prone to expansions, contractions, and variations, but they appear also to incorporate new chemoreceptors that neofunctionalize.

Vomeronasal FPR expression follows the V1R/V2R dichotomy between apical and basal VSNs. However, depending on their identity, FPRs are expressed either in the upper zone (Fpr-rs3, 4, 6, 7), or in the lower layer of the neuroepithelium (Fpr-rs1).[110,111] In addition, the likeliness of FPR-rs3 choice by each neuron exhibits a rostrocaudal gradient along the length of the vomeronasal organ.[113] The more rostrally located the VSN is, the more likely FPR-rs3 will be chosen. This observation is reminiscent of the specific zones in which each OR or TAAR is expressed in the MOE.

FPR Functions

FPRs are both present on the dendritic endings of VSNs and on their axonal termini in the AOB, where neurons expressing the same receptor converge into multiple glomeruli.[113] Various vomeronasal FPR agonists have been identified, many of them having previously been shown to activate immune FPRs.[110] This is at the origin of a hypothesis, still to be tested, that proposes that FPRs may be involved in the detection of pathogens or of pathogenic states.[110] The list of potential FPR agonists has today been lengthened, with the addition of formylated peptides from mitochondria[114] and another peptide (termed W) for which the FPR-rs1 receptor exhibits stereoselective tuning for a form containing D-amino acids.[115] This last point is interesting because most peptides found in nature are built with L-amino acids, except those from the cell wall of bacteria or those from toxins produced by some fungi.

ODORANT RECEPTORS

Members of a fourth chemoreceptor family are also expressed by vomeronasal sensory neurons. These are ORs, the very ones that are also found in sensory neurons of the main olfactory system. ORs have been reported in VSNs from the apical vomeronasal sensory epithelium.[116] Whether their presence outside the MOE reflects spurious expression or is functionally relevant is unknown. However, the observation that VSNs expressing the ORs Olfr17 or Olfr78 coalesce into glomeruli in the rostral accessory olfactory bulb[116] suggests that they may play a role within the accessory olfactory system.

BREAKOUT BOX

CHEMORECEPTORS, CHEMICAL ISOLATION, AND SPECIATION

The diversity of VR repertoires provides us with a unique footprint of evolution, or at least of the different chemical histories of species. This is because to a given ecological niche, some chemoreceptors are more adapted than other chemoreceptors. Usually, members of a species share the same niche. At some point, this species may face a novel situation, such as the arrival of a new predator or the invasion of a new territory. This novelty, to which some individuals are more adapted than others, may lead to the split of subpopulations and eventually to speciation events. Depending on the novel environment, specific pressures will then act on and shape chemosensory receptor repertoires of the new species. However, these events could take place in a different order, particularly in the case of sympatric speciation (i.e., speciation taking place within the same geographic area). One could consider changes in VR repertoires as a founder event that will modify the abilities of individuals to recognize each other and eventually to interact socially. Thus, changes in VR repertoires result not in the geographic separation but the chemical separation of a given species into different groups that would no longer exchange genetic information. Similar to switching from a diurnal to a nocturnal schedule or to a change of courting songs, these alterations could lead to loss of social communication and therefore to genetic isolation between groups. In fact, evolution of pheromones and of their receptors has been suggested to be instrumental in reproductive isolation in invertebrates, from yeasts to insects.[117–120] In this way, the diversity of vomeronasal receptor toolboxes may well have directed the emergence of novel species.

CODING LINES

Whether vomeronasal-induced behavioral outputs result, as a general rule, from the simple activity of specific coding lines (e.g., a VSN population expressing one or a few VR genes) is unclear. Evidence for single, independent, and nonredundant lines exists, such as the behavioral response of female mice to ESP1 via activation of VSNs expressing V2Rp5.[101] However, in the mouse there is also evidence for multimodal or combinatorial coding that requires the concomitant activation of multiple chemosignaling lines to adequately respond to a given stimulus. For example, the deletion of only 10% of the mouse V1R repertoire was shown to affect the behavior of both males and females (which suggests the necessity to integrate different signals to produce an adequate output). Additionally, the genetic silencing of all V2R-expressing VSNs via the deletion of the $G\alpha_o$ subunit leads to phenotypes that partially overlap with those observed following the silencing of V1R-expressing neurons via the deletion of the $G\alpha_{i2}$ subunit.[108,109,114]

As is often observed with biological systems, the logic on which vomeronasal perception is based is likely to be complex. It certainly involves a mix between single receptors that activate independent, robust, and predetermined circuits, and integrative or even synergistic sensory processes that implicate the activation of various receptor types in different systems. The truth is that we are still at the beginning of the uncovering of the molecular logic by which mammals, via their VRs, extract information from their neighbors. But this early phase will not last, thanks to the development of genetic tools that allow activation or silencing of specific neuronal populations at will, and to easily disrupt gene expression, providing us with very efficient instruments to decipher these rules.

Acknowledgments

I thank all members of my laboratory for comments on the manuscript.

References

1. Doving KB, Trotier D. Structure and function of the vomeronasal organ. *J Exp Biol.* November 1998;201(Pt 21): 2913–2925.
2. Mucignat-Caretta C. The rodent accessory olfactory system. *J Comp Physiol A Neuroethol Sens Neural Behav Physiol.* October 2010;196(10):767–777.
3. Dulac C, Torello AT. Molecular detection of pheromone signals in mammals: from genes to behaviour. *Nat Rev Neurosci.* July 2003;4(7):551–562.
4. Stowers L, Holy TE, Meister M, Dulac C, Koentges G. Loss of sex discrimination and male-male aggression in mice deficient for TRP2. *Science.* February 22, 2002;295(5559):1493–1500.
5. Papes F, Logan DW, Stowers L. The vomeronasal organ mediates interspecies defensive behaviors through detection of protein pheromone homologs. *Cell.* May 14, 2010;141(4):692–703.
6. Boillat M, Challet L, Rossier D, Kan C, Carleton A, Rodriguez I. The vomeronasal system mediates sick conspecific avoidance. *Curr Biol.* January 19, 2015;25(2):251–255.
7. Chamero P, Marton TF, Logan DW, et al. Identification of protein pheromones that promote aggressive behaviour. *Nature.* December 6, 2007;450(7171):899–902.
8. Ben-Shaul Y, Katz LC, Mooney R, Dulac C. In vivo vomeronasal stimulation reveals sensory encoding of conspecific and allospecific cues by the mouse accessory olfactory bulb. *Proc Natl Acad Sci USA.* March 16, 2010;107(11):5172–5177.
9. Holy TE, Dulac C, Meister M. Responses of vomeronasal neurons to natural stimuli. *Science.* 2000;289(5484):1569–1572.
10. Dulac C, Axel R. A novel family of genes encoding putative pheromone receptors in mammals. *Cell.* 1995;83(2):195–206.
11. Buck L, Axel R. A novel multigene family may encode odorant receptors: a molecular basis for odor recognition. *Cell.* 1991;65(1):175–187.
12. Mombaerts P. Genes and ligands for odorant, vomeronasal and taste receptors. *Nat Rev Neurosci.* April 2004;5(4):263–278.
13. Tirindelli R, Mucignat-Caretta C, Ryba NJ. Molecular aspects of pheromonal communication via the vomeronasal organ of mammals. *Trends Neurosci.* 1998;21(11):482–486.
14. Takigami S, Mori Y, Tanioka Y, Ichikawa M. Morphological evidence for two types of mammalian vomeronasal system. *Chem Senses.* May 2004;29(4):301–310.
15. Berghard A, Buck LB. Sensory transduction in vomeronasal neurons: evidence for G alpha o, G alpha i2, and adenylyl cyclase II as major components of a pheromone signaling cascade. *J Neurosci.* February 1, 1996;16(3):909–918.
16. Jia C, Halpern M. Subclasses of vomeronasal receptor neurons: differential expression of G proteins (Gi alpha 2 and G(o alpha)) and segregated projections to the accessory olfactory bulb. *Brain Res.* May 6, 1996;719(1–2):117–128.
17. Del Punta K, Rothman A, Rodriguez I, Mombaerts P. Sequence diversity and genomic organization of vomeronasal receptor genes in the mouse. *Genome Res.* December 2000;10(12):1958–1967.

18. Del Punta K, Leinders-Zufall T, Rodriguez I, et al. Deficient pheromone responses in mice lacking a cluster of vomeronasal receptor genes. *Nature*. September 5, 2002;419(6902):70–74.

19. Lane RP, Cutforth T, Axel R, Hood L, Trask BJ. Sequence analysis of mouse vomeronasal receptor gene clusters reveals common promoter motifs and a history of recent expansion. *Proc Natl Acad Sci USA*. January 8, 2002;99(1):291–296.

20. Zhang X, Rodriguez I, Mombaerts P, Firestein S. Odorant and vomeronasal receptor genes in two mouse genome assemblies. *Genomics*. 2004;83(5):802–811.

21. Young JM, Kambere M, Trask BJ, Lane RP. Divergent V1R repertoires in five species: amplification in rodents, decimation in primates, and a surprisingly small repertoire in dogs. *Genome Res*. February 2005;15(2):231–240.

22. Stewart RLR. V1R promoters are well conserved and exhibit common putative regulatory motifs. *BMC Genomics*. 2007;25:253.

23. Clowney EJMA, Colquitt BM, Pathak N, Lane RP, Lomvardas S. High-throughput mapping of the promoters of the mouse olfactory receptor genes reveals a new type of mammalian promoter and provides insight into olfactory receptor gene regulation. *Genome Res*. 2011;21(8):1249–1259.

24. Vassalli A, Rothman A, Feinstein P, Zapotocky M, Mombaerts P. Minigenes impart odorant receptor-specific axon guidance in the olfactory bulb. *Neuron*. August 15, 2002;35(4):681–696.

25. Pfister P, Rodriguez I. Olfactory expression of a single and highly variable V1r pheromone receptor-like gene in fish species. *Proc Natl Acad Sci USA*. April 4, 2005.

26. Saraiva LR, Korsching SI. A novel olfactory receptor gene family in teleost fish. *Genome Res*. October 2007;17(10):1448–1457.

27. Brykczynska U, Tzika AC, Rodriguez I, Milinkovitch MC. Contrasted evolution of the vomeronasal receptor repertoires in mammals and squamate reptiles. *Genome Biol Evol*. 2013;5(2):389–401.

28. Grus W, Shi P, Zhang Y, Zhang J. Dramatic variation of the vomeronasal pheromone gene repertoire among five orders of placental and marsupial mammals. *PNAS*. 2005;102(16):5767–5772.

29. Shi P, Zhang J. Comparative genomic analysis identifies an evolutionary shift of vomeronasal receptor gene repertoires in the vertebrate transition from water to land. *Genome Res*. 2007;17(2):166–174.

30. Rodriguez I, Greer CA, Mok MY, Mombaerts P. A putative pheromone receptor gene expressed in human olfactory mucosa. *Nat Genet*. September 2000;26(1):18–19.

31. Wakabayashi Y, Ohkura S, Okamura H, Mori Y, Ichikawa M. Expression of a vomeronasal receptor gene (V1r) and G protein alpha subunits in goat, *Capra hircus*, olfactory receptor neurons. *J Comp Neurol*. 2007;503(2):371–380.

32. Wakabayashi Y, Mori Y, Ichikawa M, Yazaki K, Hagino-Yamagishi K. A putative pheromone receptor gene is expressed in two distinct olfactory organs in goats. *Chem Senses*. March 2002;27(3):207–213.

33. Karunadasa DK, Chapman C, Bicknell RJ. Expression of pheromone receptor gene families during olfactory development in the mouse: expression of a V1 receptor in the main olfactory epithelium. *Eur J Neurosci*. May 2006;23(10):2563–2572.

34. Rodriguez I, Del Punta K, Rothman A, Ishii T, Mombaerts P. Multiple new and isolated families within the mouse superfamily of V1r vomeronasal receptors. *Nat Neurosci*. February 2002;5(2):134–140.

35. Young JM, Massa HF, Hsu L, Trask BJ. Extreme variability among mammalian V1R gene families. *Genome Res*. January 2010;20(1):10–18.

36. Zhang X, Firestein S. Comparative genomics of odorant and pheromone receptor genes in rodents. *Genomics*. 2007;89(4):441–450.

37. Grus WE, Shi P, Zhang J. Largest vertebrate vomeronasal type 1 receptor gene repertoire in the semiaquatic platypus. *Mol Biol Evol*. October 2007;24(10):2153–2157.

38. Rodriguez I, Mombaerts P. Novel human vomeronasal receptor-like genes reveal species-specific families. *Curr Biol*. June 25, 2002;12(12):R409–R411.

39. Yoder AD, Larsen PA. The molecular evolutionary dynamics of the vomeronasal receptor (class 1) genes in primates: a gene family on the verge of a functional breakdown. *Front Neuroanat*. 2014;8:153.

40. Nei M, Niimura Y, Nozawa M. The evolution of animal chemosensory receptor gene repertoires: roles of chance and necessity. *Nat Rev Genet*. December 2008;9(12):951–963.

41. Wynn EH, Sánchez-Andrade G, Carss KJ, Logan DW. Genomic variation in the vomeronasal receptor gene repertoires of inbred mice. *BMC Genomics*. 21 August 2012;13:415.

42. Hashiguchi Y, Furuta Y, Nishida M. Evolutionary patterns and selective pressures of odorant/pheromone receptor gene families in teleost fishes. *PLoS One*. 2008;3(12):e4083.

43. Pfister P, Randall J, Montoya-Burgos JI, Rodriguez I. Divergent evolution among teleost V1R receptor genes. *PLoS One.* 2007;2:e379.

44. Wang G, Shi P, Zhu Z, Zhang YP. More functional V1R genes occur in nest-living and nocturnal terricolous mammals. *Genome Biol Evol.* 2010;2:277−283.

45. Shi P, Bielawski JP, Yang H, Zhang YP. Adaptive diversification of vomeronasal receptor 1 genes in rodents. *J Mol Evol.* May 2005;60(5):566−576.

46. Mundy NI, Cook S. Positive selection during the diversification of class I vomeronasal receptor-like (V1RL) genes, putative pheromone receptor genes, in human and primate evolution. *Mol Biol Evol.* November 2003;20(11):1805−1810.

47. Grus WE, Zhang J. Rapid turnover and species-specificity of vomeronasal pheromone receptor genes in mice and rats. *Gene.* October 13, 2004;340(2):303−312.

48. Kurzweil VC, Getman M, NICS Comparative Sequencing Program, Green ED, Lane RP. Dynamic evolution of V1R putative pheromone receptors between *Mus musculus* and *Mus spretus. BMC Genomics.* February 9, 2009;10:74.

49. Lane RP, Young J, Newman T, Trask BJ. Species specificity in rodent pheromone receptor repertoires. *Genome Res.* April 2004;14(4):603−608.

50. Niimura Y, Nei M. Evolutionary dynamics of olfactory and other chemosensory receptor genes in vertebrates. *J Hum Genet.* 2006;51(6):505−517.

51. Nozawa M, Kawahara Y, Nei M. Genomic drift and copy number variation of sensory receptor genes in humans. *Proc Natl Acad Sci USA.* December 18, 2007;104(51):20421−20426.

52. Chess A, Simon I, Cedar H, Axel R. Allelic inactivation regulates olfactory receptor gene expression. *Cell.* 1994;78(5):823−834.

53. Dulac C, Axel R. Expression of candidate pheromone receptor genes in vomeronasal neurons. *Chem Senses.* 1998;23(4):467−475.

54. Rodriguez I, Feinstein P, Mombaerts P. Variable patterns of axonal projections of sensory neurons in the mouse vomeronasal system. *Cell.* April 16, 1999;97(2):199−208.

55. Belluscio L, Koentges G, Axel R, Dulac C. A map of pheromone receptor activation in the mammalian brain. *Cell.* 1999;97(2):209−220.

56. Roppolo D, Vollery S, Kan CD, Luscher C, Broillet MC, Rodriguez I. Gene cluster lock after pheromone receptor gene choice. *EMBO J.* July 25, 2007;26(14):3423−3430.

57. Dalton R, Lyons D, Lomvardas S. Co-opting the unfolded protein response to elicit olfactory receptor feedback. *Cell.* 2013.

58. Rodriguez I. Singular expression of olfactory receptor genes. *Cell.* October 10, 2013;155(2):274−277.

59. Capello L, Roppolo D, Jungo VP, Feinstein P, Rodriguez I. A common gene exclusion mechanism used by two chemosensory systems. *Eur J Neurosci.* 2009;29(4):671−678.

60. Wagner S, Gresser AL, Torello AT, Dulac C. A multireceptor genetic approach uncovers an ordered integration of VNO sensory inputs in the accessory olfactory bulb. *Neuron.* June 1, 2006;50(5):697−709.

61. Feinstein P, Bozza T, Rodriguez I, Vassalli A, Mombaerts P. Axon guidance of mouse olfactory sensory neurons by odorant receptors and the beta2 adrenergic receptor. *Cell.* June 11, 2004;117(6):833−846.

62. Leinders-Zufall T, Lane AP, Puche AC, et al. Ultrasensitive pheromone detection by mammalian vomeronasal neurons. *Nature.* 2000;405(6788):792−796.

63. Boschat C, Pelofi C, Randin O, et al. Pheromone detection mediated by a V1r vomeronasal receptor. *Nat Neurosci.* December 2002;5(12):1261−1262.

64. Isogai Y, Si S, Pont-Lezica L, et al. Molecular organization of vomeronasal chemoreception. *Nature.* October 13, 2011;478(7368):241−245.

65. Trinh K, Storm DR. Vomeronasal organ detects odorants in absence of signaling through main olfactory epithelium. *Nat Neurosci.* May 2003;6(5):519−525.

66. Xu F, Schaefer M, Kida I, et al. Simultaneous activation of mouse main and accessory olfactory bulbs by odors or pheromones. *J Comp Neurol.* September 5, 2005;489(4):491−500.

67. Nodari F, Hsu FF, Fu X, et al. Sulfated steroids as natural ligands of mouse pheromone-sensing neurons. *J Neurosci.* June 18, 2008;28(25):6407−6418.

68. Hammen GF, Turaga D, Holy TE, Meeks JP. Functional organization of glomerular maps in the mouse accessory olfactory bulb. *Nat Neurosci.* July 2014;17(7):953−961.

69. Holekamp TF, Turaga D, Holy TE. Fast three-dimensional fluorescence imaging of activity in neural populations by objective-coupled planar illumination microscopy. *Neuron*. March 13, 2008;57(5):661−672.

70. Turaga D, Holy TE. Organization of vomeronasal sensory coding revealed by fast volumetric calcium imaging. *J Neurosci*. February 1, 2012;32(5):1612−1621.

71. Meeks JP, Arnson HA, Holy TE. Representation and transformation of sensory information in the mouse accessory olfactory system. *Nat Neurosci*. June 2010;13(6):723−730.

72. Hsu FF, Nodari F, Kao LF, et al. Structural characterization of sulfated steroids that activate mouse pheromone-sensing neurons. *Biochemistry*. December 30, 2008;47(52):14009−14019.

73. Herrada G, Dulac C. A novel family of putative pheromone receptors in mammals with a topographically organized and sexually dimorphic distribution. *Cell*. 1997;90(4):763−773.

74. Matsunami H, Buck LB. A multigene family encoding a diverse array of putative pheromone receptors in mammals. *Cell*. 1997;90(4):775−784.

75. Ryba NJ, Tirindelli R. A new multigene family of putative pheromone receptors. *Neuron*. 1997;19(2):371−379.

76. Del Punta K, Puche A, Adams NC, Rodriguez I, Mombaerts P. A divergent pattern of sensory axonal projections is rendered convergent by second-order neurons in the accessory olfactory bulb. *Neuron*. September 12, 2002;35(6):1057−1066.

77. Mohedano-Moriano A, Pro-Sistiaga P, Ubeda-Banon I, de la Rosa-Prieto C, Saiz-Sanchez D, Martinez-Marcos A. V1R and V2R segregated vomeronasal pathways to the hypothalamus. *Neuroreport*. October 29, 2008;19(16):1623−1626.

78. Mohedano-Moriano A, Pro-Sistiaga P, Ubeda-Banon I, Crespo C, Insausti R, Martinez-Marcos A. Segregated pathways to the vomeronasal amygdala: differential projections from the anterior and posterior divisions of the accessory olfactory bulb. *Eur J Neurosci*. April 2007;25(7):2065−2080.

79. Omura M, Mombaerts P. Trpc2-expressing sensory neurons in the main olfactory epithelium of the mouse. *Cell Rep*. July 24, 2014;8(2):583−595.

80. Syed AS, Sansone A, Nadler W, Manzini I, Korsching SI. Ancestral amphibian v2rs are expressed in the main olfactory epithelium. *Proc Natl Acad Sci USA*. May 7, 2013;110(19):7714−7719.

81. Hohenbrink P, Dempewolf S, Zimmermann E, Mundy NI, Radespiel U. Functional promiscuity in a mammalian chemosensory system: extensive expression of vomeronasal receptors in the main olfactory epithelium of mouse lemurs. *Front Neuroanat*. 2014;8:102.

82. Yang H, Shi P, Zhang YP, Zhang J. Composition and evolution of the V2r vomeronasal receptor gene repertoire in mice and rats. *Genomics*. September 2005;86(3):306−315.

83. Young JM, Trask BJ. V2R gene families degenerated in primates, dog and cow, but expanded in opossum. *Trends Genet*. May 2007;23(5):212−215.

84. Hashiguchi Y, Nishida M. Evolution and origin of vomeronasal-type odorant receptor gene repertoire in fishes. *BMC Evol Biol*. 2006;6:76.

85. Hashiguchi Y, Nishida M. Evolution of vomeronasal-type odorant receptor genes in the zebrafish genome. *Gene*. December 5, 2005;362:19−28.

86. Ishii T, Mombaerts P. Coordinated coexpression of two vomeronasal receptor V2R genes per neuron in the mouse. *Mol Cell Neurosci*. February 2011;46(2):397−408.

87. Martini S, Silvotti L, Shirazi A, Ryba NJ, Tirindelli R. Co-expression of putative pheromone receptors in the sensory neurons of the vomeronasal organ. *J Neurosci*. February 1, 2001;21(3):843−848.

88. Silvotti L, Moiani A, Gatti R, Tirindelli R. Combinatorial co-expression of pheromone receptors, V2Rs. *J Neurochem*. December 2007;103(5):1753−1763.

89. Silvotti L, Cavalca E, Gatti R, Percudani R, Tirindelli R. A recent class of chemosensory neurons developed in mouse and rat. *PLoS One*. 2011;6(9):e24462.

90. Ishii T, Hirota J, Mombaerts P. Combinatorial coexpression of neural and immune multigene families in mouse vomeronasal sensory neurons. *Curr Biol*. March 4, 2003;13(5):394−400.

91. Ishii T, Mombaerts P. Expression of nonclassical class I major histocompatibility genes defines a tripartite organization of the mouse vomeronasal system. *J Neurosci*. March 5, 2008;28(10):2332−2341.

92. Leinders-Zufall T, Ishii T, Chamero P, et al. A family of nonclassical class I MHC genes contributes to ultrasensitive chemodetection by mouse vomeronasal sensory neurons. *J Neurosci*. April 9, 2014;34(15):5121−5133.

93. Loconto J, Papes F, Chang E, et al. Functional expression of murine V2R pheromone receptors involves selective association with the M10 and M1 families of MHC class Ib molecules. *Cell.* March 7, 2003;112(5): 607−618.

94. Leinders-Zufall T, Brennan P, Widmayer P, et al. MHC class I peptides as chemosensory signals in the vomeronasal organ. *Science.* November 5, 2004;306(5698):1033−1037.

95. Leinders-Zufall T, Ishii T, Mombaerts P, Zufall F, Boehm T. Structural requirements for the activation of vomeronasal sensory neurons by MHC peptides. *Nat Neurosci.* December 2009;12(12):1551−1558.

96. He J, Ma L, Kim S, Nakai J, Yu CR. Encoding gender and individual information in the mouse vomeronasal organ. *Science.* April 25, 2008;320(5875):535−538.

97. Bruce HM. An exteroceptive block to pregnancy in the mouse. *Nature.* July 11, 1959;184:105.

98. Kimoto H, Haga S, Sato K, Touhara K. Sex-specific peptides from exocrine glands stimulate mouse vomeronasal sensory neurons. *Nature.* October 6, 2005;437(7060):898−901.

99. Kimoto H, Sato K, Nodari F, Haga S, Holy TE, Touhara K. Sex- and strain-specific expression and vomeronasal activity of mouse ESP family peptides. *Curr Biol.* November 6, 2007;17(21):1879−1884.

100. Ferrero DM, Moeller LM, Osakada T, et al. A juvenile mouse pheromone inhibits sexual behaviour through the vomeronasal system. *Nature.* October 17, 2013;502(7471):368−371.

101. Haga S, Hattori T, Sato T, et al. The male mouse pheromone ESP1 enhances female sexual receptive behaviour through a specific vomeronasal receptor. *Nature.* July 1, 2010;466(7302):118−122.

102. Dey S, Matsunami H. Calreticulin chaperones regulate functional expression of vomeronasal type 2 pheromone receptors. *Proc Natl Acad Sci USA.* October 4, 2011;108(40):16651−16656.

103. Rodriguez I. The chemical MUPpeteer. *Cell.* May 14, 2010;141(4):568−570.

104. Kaur AW, Ackels T, Kuo TH, et al. Murine pheromone proteins constitute a context-dependent combinatorial code governing multiple social behaviors. *Cell.* April 24, 2014;157(3):676−688.

105. Hurst JL, Payne CE, Nevison CM, et al. Individual recognition in mice mediated by major urinary proteins. *Nature.* December 6, 2001;414(6864):631−634.

106. Roberts SA, Davidson AJ, McLean L, Beynon RJ, Hurst JL. Pheromonal induction of spatial learning in mice. *Science.* December 14, 2012;338(6113):1462−1465.

107. Roberts SA, Simpson DM, Armstrong SD, et al. Darcin: a male pheromone that stimulates female memory and sexual attraction to an individual male's odour. *BMC Biol.* 2010;8:75.

108. Oboti L, Perez-Gomez A, Keller M, et al. A wide range of pheromone-stimulated sexual and reproductive behaviors in female mice depend on G protein Galphao. *BMC Biol.* 2014;12:31.

109. Norlin EM, Gussing F, Berghard A. Vomeronasal phenotype and behavioral alterations in G alpha i2 mutant mice. *Curr Biol.* July 15, 2003;13(14):1214−1219.

110. Riviere S, Challet L, Fluegge D, Spehr M, Rodriguez I. Formyl peptide receptor-like proteins are a novel family of vomeronasal chemosensors. *Nature.* May 28, 2009;459(7246):574−577.

111. Liberles SD, Horowitz LF, Kuang D, et al. Formyl peptide receptors are candidate chemosensory receptors in the vomeronasal organ. *Proc Natl Acad Sci USA.* June 16, 2009;106(24):9842−9847.

112. Yang H, Shi P. Molecular and evolutionary analyses of formyl peptide receptors suggest the absence of VNO-specific FPRs in primates. *J Genet Genomics.* December 2010;37(12):771−778.

113. Dietschi Q, Assens A, Challet L, Carleton A, Rodriguez I. Convergence of FPR-rs3-expressing neurons in the mouse accessory olfactory bulb. *Mol Cell Neurosci.* September 2013;56:140−147.

114. Chamero P, Katsoulidou V, Hendrix P, et al. G protein G(alpha)o is essential for vomeronasal function and aggressive behavior in mice. *Proc Natl Acad Sci USA.* August 2, 2011;108(31):12898−12903.

115. Bufe B, Schumann T, Zufall F. Formyl peptide receptors from immune and vomeronasal system exhibit distinct agonist properties. *J Biol Chem.* September 28, 2012;287(40):33644−33655.

116. Levai O, Feistel T, Breer H, Strotmann J. Cells in the vomeronasal organ express odorant receptors but project to the accessory olfactory bulb. *J Comp Neurol.* October 1, 2006;498(4):476−490.

117. Seike T, Nakamura T, Shimoda C. Molecular coevolution of a sex pheromone and its receptor triggers reproductive isolation in *Schizosaccharomyces pombe. Proc Natl Acad Sci USA.* April 7, 2015;112(14):4405−4410.

118. Engsontia P, Sangket U, Chotigeat W, Satasook C. Molecular evolution of the odorant and gustatory receptor genes in lepidopteran insects: implications for their adaptation and speciation. *J Mol Evol.* August 2014;79(1−2):21−39.

119. Pillay N, Rymer TL. Behavioural divergence, interfertility and speciation: a review. *Behav Process*. November 2012;91(3):223—235.
120. Gould F, Estock M, Hillier NK, et al. Sexual isolation of male moths explained by a single pheromone response QTL containing four receptor genes. *Proc Natl Acad Sci USA*. May 11, 2010;107(19):8660—8665.
121. Rodriguez I, Boehm U. Pheromone sensing in mice. *Results Probl Cell Differ*. 2009;47:77—96.

Vomeronasal Transduction and Cell Signaling

Marc Spehr

Department of Chemosensation, Institute for Biology II, RWTH Aachen University, Aachen, Germany

INTRODUCTION

More than 200 years ago, in 1813, the Danish anatomist Ludvig L. Jacobson (1783—1843) first described an organ "located in the foremost part of the nasal cavity, in close contact with the nasal septum, on palatal elongations of the intermaxillary bone."[1] Comparative analysis in both domesticated and wild mammalian species led Jacobson to conclude that "the organ exists in all mammals" as a "sensory organ which may be of assistance to the sense of smell."[1] The "organ of Jacobson" was later (re)named "organon vomeronasale (Jacobsoni)" or simply the vomeronasal organ (VNO).

Today, we know that the VNO is the peripheral sensory structure of the accessory olfactory system (AOS) (see Chapters 1, 10, and 12), which exists in most mammalian species.

Substantial morphological diversity, however, renders extrapolation of findings between species quite complicated.[2,3] Therefore, this chapter focuses on the rodent AOS, placing particular emphasis on knowledge that emerged from genetically modified mouse models. Mice are by far the best-studied VNO model and their genetic amenability has allowed multilevel investigations on the molecular, cellular, and systems level.[4]

THE ANATOMY AND CELLULAR COMPOSITION OF THE VNO

The VNO is a paired tubular structure at the base of the anterior nasal septum (Figure 1).[5,6] Enclosed in a cartilaginous capsule, the blind-ended cylindrical organ opens anteriorly—via the vomeronasal or nasopalatine duct—into either the nasal or the oral cavity, respectively. The VNO lumen is filled with lateral gland secretions that add up to $\sim 0.5\,mm^3$ mucus.[7] Medially, a crescent-shaped neuroepithelium harbors three distinct cell types: (1) basal cells, a group of pluripotent neural stem cells important for epithelial regeneration[8]; (2) sustentacular/supporting cells, glia-like cells that serve structural and metabolic functions; and (3) a few hundred thousand short-lived vomeronasal sensory neurons (VSNs), which are continuously replaced from the basal cell reservoir and function as the peripheral sensory "antennae" for the AOS. VSNs are small bipolar neurons that extend a single, unbranched dendrite apically and an unmyelinated axon basally. At the epithelial surface, the apical dendrite terminates in a microvillar knob that is bathed in the lumen's mucus (Figure 1(C)). At the basal lamina, the VSN axons gather into nerve bundles, which project dorsally through openings in

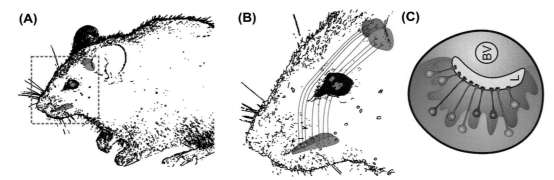

FIGURE 1   **The anatomical basis of vomeronasal signaling.** Schematic drawings of the mouse accessory olfactory system. (A, B) The VNO consists of bilaterally symmetrical blind-ended tubes at the anterior base of the nasal septum. (C) At least two subpopulations of bipolar microvillous sensory neurons reside in either the more apical V1R-positive (red) or the more basal V2R-positive (green) layer of a medial crescent-shaped sensory neuroepithelium. FPR-rs-expressing neurons (blue) are predominantly found in the apical layer, although a few reside in the basal layer. Chemostimuli are sucked into the VNO lumen (L) upon vascular contractions of a large lateral blood vessel (BV). (B) The layer-specific organizational dichotomy is maintained at the accessory olfactory bulb (AOB) level, where vomeronasal neurons make synaptic contacts with mitral cell dendrites. Here, apical VSNs project their axons to the rostral half of the AOB, whereas basal VSNs innervate the caudal half of the AOB. VNO, vomeronasal organ; VSN, vomeronasal sensory neuron. FPR, formyl peptide receptor.

the cribriform plate of the skull and along the medial olfactory bulbs to target the glomerular layer of the accessory olfactory bulb (AOB).[6]

On its lateral side, the VNO is composed of highly vascularized cavernous tissue that harbors a large blood vessel (Figure 1(C)). This characteristic lateral vessel is innervated by sympathetic nerve fibers of the superior cervical ganglion that, upon activation, trigger substantial vasoconstriction.[9,10] In situations of stress, novelty, and/or arousal, this mechanism results in negative intraluminal pressure, effectively constituting a peristaltic pump that mediates luminal fluid entry into the VNO. It is this pumping mechanism that allows uptake of relatively nonvolatile stimuli from urine, saliva, vaginal fluid, and other gland secretions.[11,12]

VOMERONASAL CHEMORECEPTOR FUNCTION

To understand VSN signaling, insight into vomeronasal chemoreceptor function is critical. In Chapter 10, a detailed account of VNO receptor biology has been presented. Thus, the following paragraphs will only provide a summary of those aspects of vomeronasal receptor protein function that are crucial to the concepts of VSN signal transduction.

In most mammalian species, the AOS shows a structural and functional dichotomy[13–15] (Figure 1). Largely based on marker protein expression, the VNO neuroepithelium is topographically segregated into at least two neuronal subpopulations. VSNs expressing both the G protein α-subunit $G\alpha_{i2}$, the phosphodiesterase isoform PDE4A and a single member of the *Vmn1r* multigene family,[16] tend to be located in the epithelium's more apical layer. By contrast, a molecularly distinct group of VSNs is $G\alpha_o$-positive and expresses members of the unrelated *Vmn2r* family.[17–19] These neurons are commonly referred to as basal cells, although the concept of apical *versus* basal neurons is by no means absolute.[20] Recently, a third family of putative VNO chemoreceptors has been identified: formyl peptide receptor (FPR)-like proteins.[21,22] With the notable exception of *Fpr-rs1*, which is coexpressed with $G\alpha_o$, the remaining four vomeronasal *Fpr-rs* genes are expressed by VSNs located in the $G\alpha_{i2}$-/PDE4A-positive apical layer of the neuroepithelium (Figures 1(C) and 2).

All members of the *Vmn1r*, *Vmn2r*, and *Fpr-rs* gene families encode G protein–coupled receptors (GPCRs) that show a putative seven-transmembrane topology. With the exception of family-C *Vmn2rs*, all candidate chemoreceptor proteins are monogenically, most likely monoallelically, expressed with respect to both their own and counterpart chemoreceptor families.[21–24] All three GPCR families are almost exclusively expressed in vomeronasal tissue and, as expected, the receptor proteins are highly enriched in the VSN microvillar dendritic endings, the site of receptor–ligand interaction. Family size, however, varies considerably. Although approximately 150 and 120 potentially functional members constitute the V1R and V2R receptor families,[25–27] respectively, the FPR-rs protein family has only five members. Moreover, V1Rs and FPRs are class A (i.e., rhodopsin-like) GPCRs, whereas V2Rs are typical class C (i.e., glutamate receptor-like) GPCRs that share a large hydrophobic amino (N)-terminal domain that likely forms the extracellular ligand-binding site[28,29] (Figure 2).

Classified into 12 relatively isolated families,[25] the unusually divergent and polymorphic *Vmn1r* genes share intron-free coding regions and are clustered on most chromosomes.[26,30] Deletion of a single *Vmn1r* cluster that contained all (but one) *Vmn1ra* and *Vmn1rb* family genes has demonstrated the link between V1R expression and VSN chemoresponsivity.[4,31]

(A)   **(B)**

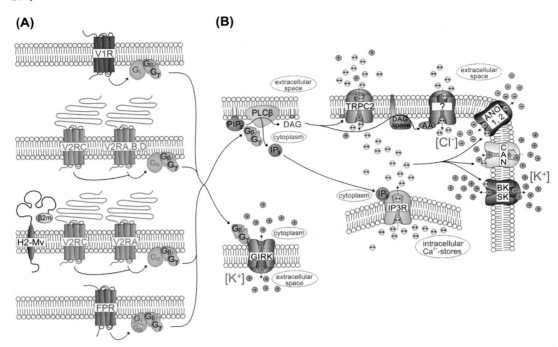

FIGURE 2 **Signal transduction mechanisms involved in vomeronasal signaling.** (A) Schematic view of putative signal transduction pathways implemented in the mouse vomeronasal organ. (B) Four general "receptor identities" are found in vomeronasal sensory neurons (VSNs): V1R- and G protein α-subunit ($G\alpha_{i2}$)-expressing neurons, VSNs coexpressing two V2Rs, and $G\alpha_o$—this neuronal population is subdivided into cells either expressing a family-A, -B, or -D receptor together with a broadly expressed member of the V2RC subfamily, or neurons coexpressing a family-A and family-C V2R in concert with a nonclassical major histocompatibility complex protein (*H2-Mv*; associated with β2-microglobulin)—as well as formyl peptide receptor (FPR)-rs-expressing VSNs that, depending on the specific receptor type, coexpress either $G\alpha_{i2}$ or $G\alpha_o$. Downstream receptor activation, current models presume G-protein-dependent activation of phospholipase C (PLCβ), cleavage of phosphatidylinositol-4,5-bisphosphate (PIP$_2$) into soluble inositol-1,4,5-trisphosphate (IP$_3$) and membrane-bound diacylglycerol (DAG), eventually resulting in gating of primary transduction channels that are formed, in part, by the transient receptor potential channel TRPC2. This channel conducts an unspecific cation current that, in addition to providing an electrical signal (depolarization), serves a biochemical role by increasing the cytosolic Ca^{2+} concentration. As a secondary event, the rise in Ca^{2+} is likely boosted by IP$_3$-dependent Ca^{2+} release from internal stores and gating of an unknown Ca^{2+} conductance by a polyunsaturated fatty acid metabolite of DAG, probably arachidonic acid (AA). As a multifaceted messenger, Ca^{2+} has been shown to act on several targets, including Ca^{2+}-activated Cl^- channels, likely ANO1 and/or ANO2, Ca^{2+}-activated cation channels (CAN), and Ca^{2+}-dependent K^+ channels (BK, SK). The electrical effects induced by Cl^- and K^+ channel gating, respectively, depend on the relative intracellular *versus* extracellular concentrations of both ions (red). This, of course, is also true for the proposed G protein-dependent opening of GIRK K^+ channels. The relative importance of each transduction protein for primary *versus* secondary signaling processes depends, in part, on its subcellular localization. Microvillar membranes are indicated by yellow lipid polar head groups (filled circles), whereas putative soma/dendritic membranes are marked by open circles.

Vmn1r and *Vmn2r* genes share no significant sequence homology. By contrast, the V2Rs are distantly related to Ca^{2+}-sensing receptors, metabotropic glutamate receptors, and T1R taste receptors (see Chapter 13).[15,28] Clusters of *Vmn2rs* are found on most chromosomes. Individual genes are grouped into four families: A, B, D, and C (commonly known as *V2r2*).[27,32,33] The seven highly homologous family-C proteins are found in most, if not all, $G\alpha_o$-positive VSNs.[34] Gene deletion studies provided direct proof of V2R chemoresponsivity. Knockout of either *Vmn2r26* (*V2r1b*) or *Vmn2r116* (*V2rp5*) results in dramatically reduced sensitivity to behaviorally relevant peptide stimuli—*Vmn2r26*-deficient mice lack major histocompatibility complex (MHC) class I peptide sensitivity,[7] and knockout of *Vmn2r116* disrupts responses to the male-specific exocrine gland-secreting peptide 1 (ESP1).[35]

Some V2R-positive basal neurons additionally express nonclassical class Ib MHC genes, referred to as *H2-Mv* or M10[36,37] (Figure 2). Early models of H2-Mv protein function suggested a chaperone role for proper V2R surface trafficking.[15,37] However, both the lack of *H2-Mv* gene expression in a substantial fraction of V2R-positive VSNs[38] and the remaining chemoresponsivity in basal VSNs after *H2-Mv* gene cluster deletion[39] questioned an essential role of H2-Mv molecules as V2R escort proteins. Nonetheless, nonrandom combinatorial coexpression of one family—A/B/D *Vmn2r* gene with a single family-C gene and either none or one of the nine *H2-Mv* genes—is likely to bestow a unique functional identity on a given basal VSN.[4]

For both V1Rs and V2Rs, deorphanization efforts have largely failed. The resulting lack of knowledge regarding VNO receptor structure—function relationships represents a, if not the, bottleneck in vomeronasal signal transduction research. However, recent reports of recombinant V2R plasma membrane targeting[40] have raised hope that direct experimental insight into vomeronasal receptor function could soon become available.

RECEPTOR-DEPENDENT TRANSDUCTION PATHWAYS AND SECONDARY SIGNALING PROCESSES IN VSNs

Following ligand detection by V1R, V2R, or FPR-rs receptors, complex biochemical transduction cascades transform the chemical energy of receptor—ligand binding into an electrical signal that, encoded as a train of action potentials, is relayed to central nuclei of the accessory olfactory pathway. The underlying transduction mechanisms are only partly understood. Accordingly, several general aspects of VNO physiology remain elusive. It is well documented, however, that upon exposure to rich natural sources of pheromonal stimuli—e.g., to urine or lacrimal gland secretions—VSNs show substantial membrane depolarization. This receptor potential is translated into an increased rate of action potential discharge and, consequently, a transient elevation of cytosolic Ca^{2+} (Figure 2).[20,31,35,41–55] Vomeronasal neurons are exquisitely sensitive to electrical stimulation. Because of an extraordinarily high input resistance of several gigaohms,[56–59] receptor currents of only a few picoamperes generate robust repetitive discharge. Somewhat surprisingly, whether VSNs are rather hyperpolarized or relatively depolarized at rest is still subject to debate. Resting membrane potentials of between −60 and −75 mV[57,59,60] have been reported. This physiologically most relevant parameter, ultimately, has to be determined *in vivo* to interpret the available electrophysiological *in vitro* findings.

Ion Channels Mediating Secondary Signaling Events

VSNs express a diverse repertoire of ion channels. Complementing "standard" Hodgkin-Huxley type voltage-activated conductances, such as those of tetrodotoxin-sensitive Na^+ currents or delayed rectifier K^+ currents,[61] VSN electrophysiology and, more specifically, their input—output relationship is shaped by expression of several somewhat unconventional ion channels. T- and L-type voltage-gated Ca^{2+} currents[57] generate low-threshold Ca^{2+} spikes that partly drive action potential discharge.[59] Interestingly, although T- and L-type Ca^{2+} currents do not appear to differ between FPR-rs3-positive neurons and control VSNs, both N- and P/Q-type currents show somewhat different properties in FPR-rs3-expressing neurons.[62] Functional coupling of voltage-gated Ca^{2+} currents to large-conductance Ca^{2+}-sensitive K^+ (BK) channels[63] has been proposed. Somatic BK channel activation was suggested to either maintain persistent action potential discharge[59] or, somewhat to the contrary, to underlie sensory adaptation via arachidonic acid—dependent BK channel recruitment.[64] Both processes, however, might occur in parallel—each in distinct subcellular compartments (i.e., soma *versus* knob/microvilli). Various other K^+ channels have also been implicated in vomeronasal signaling, either as components of primary transduction pathways or as mediators of secondary processes. The small-conductance Ca^{2+}-sensitive K^+ channel SK3 as well as the G protein—activated K^+ channel GIRK1 were proposed to be part of an independent pathway for VNO activation[65] (Figure 2). The corresponding homozygous null mice ($Kcnn3^{-/-}$ and $Kcnj3^{-/-}$) showed deficits in mating and aggressive behaviors.[65] As also noted by the authors, it is important to keep in mind that the observed behavioral deficits emerged in nonconditional knockouts of the SK3 and GIRK1 channel, respectively. Silencing of these genes in brain regions other than the VNO may thus contribute to the observed deficits in mating and aggression.[65] Another K^+ channel that controls sensory output of V2R-expressing basal VSNs, rather than being involved in primary signaling cascades, is the ether-à-go-go-related (ERG) channel.[56] Layer-specific and activity-dependent ERG K^+ channel expression serves a homeostatic function by adjusting the dynamic range of the stimulus—response function in VSNs,[56] thereby establishing a target output range in a layer-specific and use-dependent manner. In addition, members of the hyperpolarization-activated cyclic nucleotide-gated (HCN) channel family, most likely HCN2 and/or HCN4, represent yet another voltage-dependent conductance that controls VSN excitability.[66] Recently, Cichy and coworkers suggested HCN channel-dependent vomeronasal gain control of social chemosignaling.[60] The authors observed cycle stage-correlated variations in urinary pH among female mice and identified extracellular acidification as a potent activator of the vomeronasal hyperpolarization-activated current I_h,[60] thus, revealing a potential mechanistic basis for stimulus pH detection in rodent chemosensory communication.

Primary Signal Transduction Pathway(s)

Early on, speculation about a functional role of $G\alpha_{i2}$ and $G\alpha_o$ in apical and basal signaling pathways, respectively, was fueled by the strictly layer-specific coexpression of both G-protein α-subunits in the dendritic tips and microvilli of V1R- and V2R-expressing VSNs.[67–69] Yet, direct functional proof of this attractive model emerged only recently—at least for $G\alpha_o$—from conditional olfactory neuron-specific gene deletion studies. Chamero and coworkers demonstrated that functional expression of $G\alpha_o$ was critical for maintenance of basal VSN sensitivity to MHC class I antigenic peptides, major urinary proteins (MUPs) and the secreted peptide chemosignal ESP1[42] (see Chapter 1). $G\alpha_o$ was also found to be essential for VSN detection of mitochondrially encoded FPR-rs1 ligands,[42] whereas signals evoked by fMLF, a formylated peptide shown to activate at least some of the four $G\alpha_{i2}$-coupled FPRs,[22] were not significantly altered in conditional $G\alpha_o$ null mice.[42] Comparable direct evidence for a functional role of $G\alpha_{i2}$ in V1R-dependent signaling is still lacking. Unambiguously VNO-dependent phenotype(s) were not observed after constitutive deletion of $G\alpha_{i2}$,[70] at least in part because physiological VSN recordings were not performed in those $G\alpha_{i2}$ null mice. As noted by Chamero and coworkers, global deletion of an abundantly expressed and relatively promiscuous signaling component such as $G\alpha_{i2}$[70] or $G\alpha_o$[71] can be expected to induce a variety of defects.[42]

The matter of vomeronasal G protein signaling is further complicated by the apparently heterogeneous expression profiles of the β- and γ-subunits that ultimately constitute the functional heterotrimeric G protein. Among the 5 β- and 12 γ-subunits identified so far, transcripts of multiple Gβγ subunits were detected in the VNO throughout development.[72] For example, Sathyanesan and colleagues identified four different Gγ subunit transcripts—γ_2, γ_3, γ_8, and γ_{13}—in the apical $G\alpha_{i2}$-positive epithelial layer, whereas $G\gamma_8$ (*Gng8*) was the only transcript found in basal $G\alpha_o$-expressing VSNs.[72] Restricted expression of $G\gamma_8$ in basal VSNs confirmed previous data[73–75] and prompted Montani and coworkers to generate mice with a homozygous deficiency in the $G\gamma_8$-encoding gene.[76] Strikingly, both male and female $Gng8^{-/-}$ mice displayed reduced pheromone-mediated aggression, whereas other sociosexual behaviors remained essentially unchanged.[76]

β Isoforms of phospholipase C (PLCβ) appear to act as the primary target of vomeronasal G protein signaling.[15,29,77] Analogous to phosphoinositide turnover in other cell types, canonical PLCβ stimulation in VSNs is mediated predominantly by the non−GTP-binding Gβγ complex that dissociates from the activated α-subunit upon receptor−ligand interaction.[78] Dey and colleagues have recently pinned down VSN type-specific expression of PLCβ isoforms. Strikingly, PLCβ2 is a primary signal transduction element in MUP-responsive neurons, whereas other isoforms—most likely dominated by PLCβ4—are essential for detection of non-MUP stimuli.[79] Notably PLCβ2 also appears to be regulated by the estrous cycle-specific hormone progesterone, providing a direct link between endocrine state and sensory perception.[79]

Activation of PLCβ promotes cleavage of phosphatidylinositol-4,5-bisphosphate into the membrane-bound lipid diacylglycerol (DAG) and the soluble messenger inositol-1,4,5--trisphosphate (IP_3). DAG either directly targets membrane proteins or is enzymatically metabolized to polyunsaturated fatty acids, such as arachidonic acid, IP_3 signaling is usually coupled to massive Ca^{2+} release from internal stores (Figure 2). Various models for the actions of either or all products of PLC-dependent lipid turnover have been proposed.[45,49,52,64,65,80,81] They all share a common denominator: a central role of cytosolic Ca^{2+} elevations and an important, though not indispensable,[82] function of the transient receptor potential (TRP) channel isoform TRPC2[83] (Figure 2). The TRPC2 protein is highly enriched in VNO sensory microvilli, but early notions of strict VNO specificity have recently been challenged.[84] $Trpc2^{-/-}$ mice display severe defects in various sexual and social behaviors.[85–87] However, some important phenotypic differences have been noted between

$Trpc2^{-/-}$ mice and animals subject to surgical VNO ablation.[88,89] Moreover, (residual) stimulus-evoked VSN activity has been reported in $Trpc2^{-/-}$ neurons.[45,64,65,81,82]

BREAKOUT BOX

VOMERONASAL INFORMATION PROCESSING IN THE AOB

VSN axons target a specialized region at the dorso-caudal end of the olfactory bulb—the AOB. Here, the information encoded in VSN activity is processed and directly relayed to brain nuclei that control endocrine state and social behavior. In the AOB, the structural dichotomy of the VNO is retained: V1R-expressing neurons synapse on mitral cells in the rostral AOB, whereas V2R-positive cells target mitral cells in the caudal AOB.[118,119] In sharp contrast to the stereotypic map of one or two receptor-specific glomeruli in the main olfactory bulb, VSNs that share the same receptor "identity" converge onto mitral cell dendrites in 6—10 broadly distributed glomeruli.[120] This organization of VSN—mitral cell connectivity appears ideally suited to integrate sensory information.[23,24]

Many basic principles of AOB (neuro) biology remain largely unexplored. Any analogies with the main olfactory bulb are, for the most part, purely speculative.[121] Consequently, extrapolation of organizational and physiological principles from the main to the accessory bulb has frequently been premature, mostly arbitrary and, as such, has hampered an unbiased assessment of AOB biology. At the anatomical level, the relatively thick AOB glomerular layer is composed of small, clustered, and less defined confluent glomeruli[122] that are surrounded by few periglomerular cells.[108] The somata of AOB projecting neurons are rarely mitral shaped[122] and "conventional"

plexiform layers are largely missing.[2] Accordingly, AOB mitral cells lack extensive lateral dendritic trees, but elaborate branched apical dendrites that terminate in up to 12 different glomeruli.[10,123—125]

A homotypic wiring scheme[120,126] and a heterotypic connectivity model[121,127] have been independently proposed to underlie glomerular map formation in the AOB. Both at the glomerular level and along the proximal primary dendritic trees mitral cells synapse with inhibitory interneurons.[128] The majority of such interneurons are granule cells,[129] which form dendrodendritic reciprocal synapses with mitral cells. This peculiar synapse has been thoroughly investigated in the main olfactory bulb,[130,131] but has received considerably less attention in the AOB, regardless of its proposed essential role in pheromonal information processing.[132—134] Based on their "strategic" location along the basal segments of AOB mitral cell apical dendrites, the dendrodendritic synapses seem to be primarily involved in GABAergic recursive self-inhibition, rather than spreading lateral inhibition.[135] By sharpening mitral cell response selectivity,[136] various groups have proposed a relatively simple scheme of relay-like recursive self-inhibition that serves a sensory "gating" function during olfactory memory formation and one-trial learning,[133,137—139] phenomena that occur, for example, in the context of selective pregnancy failure (Bruce effect[140]).

Elevations in cytosolic Ca^{2+}, resulting from TRPC2-mediated influx[49] and/or IP_3-dependent depletion of endoplasmic reticulum stores,[45,81] will affect different cascade proteins, both primary and secondary. Increased Ca^{2+} levels exert both positive and negative feedback regulation.[4] In both hamster[90] and mouse[91] VSNs, a Ca^{2+}-activated nonselective cation current (I_{CAN}) has been identified. Other known Ca^{2+}-dependent mechanisms display striking similarities to regulatory processes implemented in canonical sensory neurons of the main olfactory epithelium. Analogous to effective inhibition of cyclic nucleotide-gated channels in olfactory sensory neurons (OSNs)[92–96] (see Chapter 7), Ca^{2+}/calmodulin mediates adaptation and gain control in VSN by inhibition of TRPC2.[91] Again, similar to the Ca^{2+}-activated chloride conductance that amplifies receptor currents in OSNs,[97–103] a substantial portion of stimulus-evoked VSN activity seems to be carried by a Ca^{2+}-activated chloride current.[45,80,81] The molecular correlate of this current has recently been identified as TMEM16A/anoctamin1,[104] a member of the recently identified anoctamin family of Ca^{2+}-activated chloride channels.[105,106] Although various groups have demonstrated expression and TRPC2 colocalization of both anoctamin1 and anoctamin2 in VSNs,[80,101,102,107] Ca^{2+}-activated currents were abolished in VSNs of TMEM16A/anoctamin1 conditional knockout mice.[104] For physiological interpretation of these results, however, a major unresolved issue is whether chloride channels contribute a depolarizing current (as in OSNs) or if TMEM16A/anoctamin1 activation leads to membrane hyperpolarization. The answer to this question depends on the *in vivo* chloride equilibrium potential at the microvillar VSN membrane, a physiological parameter presently unknown.

The interplay of the different Ca^{2+}-dependent ionic conductances reported in VSNs will, to a large extent, be determined by the spatiotemporal profile of stimulus-induced Ca^{2+} elevations. Activation of CAN channels, for example, requires much higher Ca^{2+} concentrations than typically reported for activation of a Ca^{2+}-activated chloride conductance.[108] The polarized morphology and cytoplasmic compartmentalization of VSNs pose a challenge to orchestrate discrete Ca^{2+}-sensitive responses. Their reliability, specificity, and speed depend not only on Ca^{2+} release and influx mechanisms, but also on cytoplasmic buffers that limit Ca^{2+} diffusion. In a "textbook" neuron, bulk cytosolic Ca^{2+} is maintained at resting levels of $\sim100-150$ nM by the coordinated activity of the plasma membrane Ca^{2+} ATPases, the Na^+/Ca^{2+} exchangers, and the sarco/endoplasmic reticulum Ca^{2+} pump.[109,110] In addition, mitochondria accumulate Ca^{2+} via the mitochondrial Ca^{2+} uniporter.[111–115] In OSN cilia, the resting Ca^{2+} concentration is ~40 nM,[116] but local elevations to micromolar concentrations occur during odor responses.[117] A similarly detailed description of the vomeronasal Ca^{2+} response profile will be mandatory to gain a complete picture of vomeronasal signal transduction.

References

1. Jacobson L, Trotier D, Døving KB. Anatomical description of a new organ in the nose of domesticated animals by Ludvig Jacobson (1813). *Chem Senses*. 1998;23(6):743–754. Available at: http://www.ncbi.nlm.nih.gov/pubmed/9915121.
2. Salazar I, Sanchez-Quinteiro P, Cifuentes JM, Fernandez De Troconiz P. General organization of the perinatal and adult accessory olfactory bulb in mice. *Anat Rec A Discov Mol Cell Evol Biol*. 2006;288(9):1009–1025. http://dx.doi.org/10.1002/ar.a.20366.

3. Salazar I, Quinteiro PS, Aleman N, Cifuentes JM, Fernandez De Troconiz P. Diversity of the vomeronasal system in Mammals: the singularities of the sheep model. *Microsc Res Tech*. March 2007;70:752−762. http://dx.doi.org/10.1002/jemt.

4. Chamero P, Leinders-Zufall T, Zufall F. From genes to social communication: molecular sensing by the vomeronasal organ. *Trends Neurosci*. 2012;2:1−10. http://dx.doi.org/10.1016/j.tins.2012.04.011.

5. Halpern M, Martinez-Marcos A. Structure and function of the vomeronasal system: an update. *Prog Neurobiol*. 2003;70(3):245−318. http://dx.doi.org/10.1016/S0301-0082(03)00103-5.

6. Meredith M. Sensory processing in the main and accessory olfactory systems: comparisons and contrasts. *J Steroid Biochem Mol Biol*. 1991;39(4):601−614.

7. Leinders-Zufall T, Ishii T, Mombaerts P, Zufall F, Boehm T. Structural requirements for the activation of vomeronasal sensory neurons by MHC peptides. *Nat Neurosci*. 2009;12(12):1551−1558. http://dx.doi.org/10.1038/nn.2452.

8. Brann JH, Firestein S. A lifetime of neurogenesis in the olfactory system. *Front Neurosci*. June 2014;8:1−11. http://dx.doi.org/10.3389/fnins.2014.00182.

9. Meredith M, O'Connell RJ. Efferent control of stimulus access to the hamster vomeronasal organ. *J Physiol*. 1979;286:301−316.

10. Ben-Shaul Y, Katz LC, Mooney R, Dulac C. In vivo vomeronasal stimulation reveals sensory encoding of conspecific and allospecific cues by the mouse accessory olfactory bulb. *Proc Natl Acad Sci USA*. 2010;107(11):5172−5177. http://dx.doi.org/10.1073/pnas.0915147107.

11. Luo M, Fee MS, Katz LC. Encoding pheromonal signals in the accessory olfactory bulb of behaving mice. *Science*. February 2003;299:1196−1201. http://dx.doi.org/10.1126/science.1082133.

12. Wysocki CJ, Wellington JL, Beauchamp GK. Access of urinary nonvolatiles to the mammalian vomeronasal organ. *Science*. 1980;207(4432):781−783.

13. Mucignat-Caretta C. The rodent accessory olfactory system. *J Comp Physiol A Neuroethol Sens Neural Behav Physiol*. 2010;196(10):767−777. http://dx.doi.org/10.1007/s00359-010-0555-z.

14. Halpern M. The organization and function of the vomeronasal organ. *Annu Rev Neurosci*. 1987;10:325−362.

15. Dulac C, Torello AT. Molecular detection of pheromone signals in mammals: from genes to behaviour. *Nat Rev Neurosci*. 2003;4(7):551−562. http://dx.doi.org/10.1038/nrn1140.

16. Dulac C, Axel R. A novel family of genes encoding putative pheromone receptors in mammals. *Cell*. 1995;83(2):195−206. Available at: http://www.ncbi.nlm.nih.gov/pubmed/7585937.

17. Herrada G, Dulac C. A novel family of putative pheromone receptors in mammals with a topographically organized and sexually dimorphic distribution. *Cell*. 1997;90(4):763−773. Available at: http://www.ncbi.nlm.nih.gov/pubmed/9288755.

18. Matsunami H, Buck LB. A multigene family encoding a diverse array of putative pheromone receptors in mammals. *Cell*. 1997;90(4):775−784. Available at: http://www.ncbi.nlm.nih.gov/pubmed/9288756.

19. Ryba NJP, Tirindelli R. A new multigene family of putative pheromone receptors. *Neuron*. 1997;19(2):371−379. Available at: http://www.ncbi.nlm.nih.gov/pubmed/9288756.

20. Leinders-Zufall T, Lane AP, Puche AC, et al. Ultrasensitive pheromone detection by mammalian vomeronasal neurons. *Nature*. June 15, 2000;405:792−796. http://dx.doi.org/10.1038/35015572.

21. Liberles SD, Horowitz LF, Kuang D, et al. Formyl peptide receptors are candidate chemosensory receptors in the vomeronasal organ. *Proc Natl Acad Sci USA*. 2009;106(24):9842−9847. http://dx.doi.org/10.1073/pnas.0904464106.

22. Rivière S, Challet L, Fluegge D, Spehr M, Rodriguez I. Formyl peptide receptor-like proteins are a novel family of vomeronasal chemosensors. *Nature*. 2009;459(7246):574−577. http://dx.doi.org/10.1038/nature08029.

23. Belluscio L, Koentges G, Axel R, Dulac C. A map of pheromone receptor activation in the mammalian brain. *Cell*. 1999;97(2):209−220. Available at: http://www.ncbi.nlm.nih.gov/pubmed/10219242.

24. Rodriguez I, Feinstein P, Mombaerts P. Variable patterns of axonal projections of sensory neurons in the mouse vomeronasal system. *Cell*. April 16, 1999;97:199−208. Available at: http://www.ncbi.nlm.nih.gov/pubmed/10219241.

25. Rodriguez I, Del Punta K, Rothman A, Ishii T, Mombaerts P. Multiple new and isolated families within the mouse superfamily of V1r vomeronasal receptors. *Nat Neurosci*. 2002;5(2):134−140. http://dx.doi.org/10.1038/nn795.

26. Roppolo D, Vollery S, Kan C, Lüscher C, Broillet M-C, Rodriguez I. Gene cluster lock after pheromone receptor gene choice. *EMBO J*. 2007;26(14):3423−3430. http://dx.doi.org/10.1038/sj.emboj.7601782.

27. Young JM, Trask BJ. V2R gene families degenerated in primates, dog and cow, but expanded in opossum. *Trends Genet.* 2007;23(5):209—212. http://dx.doi.org/10.1016/j.tig.2007.02.010.

28. Mombaerts P. Genes and ligands for odorant, vomeronasal and taste receptors. *Nat Rev Neurosci.* 2004;5(4):263—278. http://dx.doi.org/10.1038/nrn1365.

29. Spehr M, Munger SD. Olfactory receptors: G protein-coupled receptors and beyond. *J Neurochem.* 2009;109(6):1570—1583. http://dx.doi.org/10.1111/j.1471-4159.2009.06085.x.

30. Capello L, Roppolo D, Jungo VP, Feinstein P, Rodriguez I. A common gene exclusion mechanism used by two chemosensory systems. *Eur J Neurosci.* 2009;29(4):671—678. http://dx.doi.org/10.1111/j.1460-9568.2009.06630.x.

31. Del Punta K, Leinders-Zufall T, Rodriguez I, et al. Deficient pheromone responses in mice lacking a cluster of vomeronasal receptor genes. *Nature.* September 5, 2002;419:70—74.

32. Silvotti L, Moiani A, Gatti R, Tirindelli R. Combinatorial co-expression of pheromone receptors, V2Rs. *J Neurochem.* 2007;103(5):1753—1763. http://dx.doi.org/10.1111/j.1471-4159.2007.04877.x.

33. Silvotti L, Cavalca E, Gatti R, Percudani R, Tirindelli R. A recent class of chemosensory neurons developed in mouse and rat. *PLoS One.* 2011;6(9):e24462. http://dx.doi.org/10.1371/journal.pone.0024462.

34. Martini S, Silvotti L, Shirazi A, Ryba NJP, Tirindelli R. Co-expression of putative pheromone receptors in the sensory neurons of the vomeronasal organ. *J Neurosci.* 2001;21(3):843—848. Available at: http://www.ncbi.nlm.nih.gov/pubmed/11157070.

35. Haga S, Hattori T, Sato T, et al. The male mouse pheromone ESP1 enhances female sexual receptive behaviour through a specific vomeronasal receptor. *Nature.* 2010;466(7302):118—122. http://dx.doi.org/10.1038/nature09142.

36. Ishii T, Hirota J, Mombaerts P. Combinatorial coexpression of neural and immune multigene families in mouse vomeronasal sensory neurons. *Curr Biol.* 2003;13(5):394—400. Available at: http://www.ncbi.nlm.nih.gov/pubmed/12620187.

37. Loconto J, Papes F, Chang E, et al. Functional expression of murine V2R pheromone receptors involves selective association with the M10 and M1 families of MHC class Ib molecules. *Cell.* 2003;112(5):607—618. Available at: http://www.ncbi.nlm.nih.gov/pubmed/12628182.

38. Ishii T, Mombaerts P. Expression of nonclassical class I major histocompatibility genes defines a tripartite organization of the mouse vomeronasal system. *J Neurosci.* 2008;28(10):2332—2341. http://dx.doi.org/10.1523/JNEUROSCI.4807-07.2008.

39. Leinders-Zufall T, Ishii T, Chamero P, et al. A family of nonclassical class I MHC genes contributes to ultra-sensitive chemodetection by mouse vomeronasal sensory neurons. *J Neurosci.* 2014;34(15):5121—5133. http://dx.doi.org/10.1523/JNEUROSCI.0186-14.2014.

40. Dey S, Matsunami H. Calreticulin chaperones regulate functional expression of vomeronasal type 2 pheromone receptors. *Proc Natl Acad Sci USA.* 2011;108(40):16651—16656. http://dx.doi.org/10.1073/pnas.1018140108.

41. Chamero P, Marton TF, Logan DW, et al. Identification of protein pheromones that promote aggressive behaviour. *Nature.* 2007;450(7171):899—902. http://dx.doi.org/10.1038/nature05997.

42. Chamero P, Katsoulidou V, Hendrix P, et al. G protein G(alpha)o is essential for vomeronasal function and aggressive behavior in mice. *Proc Natl Acad Sci USA.* 2011;108(31):12898—12903. http://dx.doi.org/10.1073/pnas.1107770108.

43. Inamura K, Kashiwayanagi M, Kurihara K. lnositol-1,4,5-trisphosphate induces responses in receptor neurons in rat vomeronasal sensory slices. *Chem Senses.* 1997;22:93—103.

44. Inamura K, Kashiwayanagi M. Inward current responses to urinary substances in rat vomeronasal sensory neurons. *Eur J Neurosci.* 2000;12(10):3529—3536. Available at: http://www.ncbi.nlm.nih.gov/pubmed/11029622.

45. Kim S, Ma L, Yu CR. Requirement of calcium-activated chloride channels in the activation of mouse vomeronasal neurons. *Nat Commun.* May 2011;2:365. http://dx.doi.org/10.1038/ncomms1368.

46. Inamura K, Matsumoto Y, Kashiwayanagi M, Kurihara K. Laminar distribution of pheromone-receptive neurons in rat vomeronasal epithelium. *J Physiol.* 1999;517(Pt 3):719—731. Available at: http://www.pubmedcentral.nih.gov/articlerender.fcgi?artid=2269374&tool=pmcentrez&rendertype=abstract.

47. Kimoto H, Sato K, Nodari F, Haga S, Holy TE, Touhara K. Sex- and strain-specific expression and vomeronasal activity of mouse ESP family peptides. *Curr Biol.* 2007;17(21):1879—1884. http://dx.doi.org/10.1016/j.cub.2007.09.042.

48. Leinders-Zufall T, Brennan PA, Widmayer P, et al. MHC class I peptides as chemosensory signals in the vomeronasal organ. *Science.* 2004;306(5698):1033—1037. http://dx.doi.org/10.1126/science.1102818.

49. Lucas P, Ukhanov K, Leinders-Zufall T, Zufall F. A diacylglycerol-gated cation channel in vomeronasal neuron dendrites is impaired in TRPC2 mutant mice: mechanism of pheromone transduction. *Neuron*. 2003;40(3):551–561. Available at: http://www.ncbi.nlm.nih.gov/pubmed/14642279.
50. Nodari F, Hsu F-F, Fu X, et al. Sulfated steroids as natural ligands of mouse pheromone-sensing neurons. *J Neurosci*. 2008;28(25):6407–6418. http://dx.doi.org/10.1523/JNEUROSCI.1425-08.2008.
51. Papes F, Logan DW, Stowers L. The vomeronasal organ mediates interspecies defensive behaviors through detection of protein pheromone homologs. *Cell*. 2010;141(4):692–703. http://dx.doi.org/10.1016/j.cell.2010.03.037.
52. Spehr M, Hatt H, Wetzel CH. Arachidonic acid plays a role in rat vomeronasal signal transduction. *J Neurosci*. 2002;22(19):8429–8437. Available at: http://www.ncbi.nlm.nih.gov/pubmed/12351717.
53. Holy TE, Dulac C, Meister M. Responses of vomeronasal neurons to natural stimuli. *Science*. 2000;289(5484):1569–1572. Available at: http://www.ncbi.nlm.nih.gov/pubmed/10968796.
54. Arnson HA, Holy TE. Chemosensory burst coding by mouse vomeronasal sensory neurons. *J Neurophysiol*. 2011;106(1):409–420. http://dx.doi.org/10.1152/jn.00108.2011.
55. Turaga D, Holy TE. Organization of vomeronasal sensory coding revealed by fast volumetric calcium imaging. *J Neurosci*. 2012;32(5):1612–1621. http://dx.doi.org/10.1523/JNEUROSCI.5339-11.2012.
56. Hagendorf S, Fluegge D, Engelhardt CH, Spehr M. Homeostatic control of sensory output in basal vomeronasal neurons: activity-dependent expression of ether-à-go-go-related gene potassium channels. *J Neurosci*. 2009;29(1):206–221. http://dx.doi.org/10.1523/JNEUROSCI.3656-08.2009.
57. Liman ER, Corey DP. Electrophysiological characterization of chemosensory neurons from the mouse vomeronasal organ. *J Neurosci*. 1996;16(15):4625–4637. Available at: http://www.ncbi.nlm.nih.gov/pubmed/8764651.
58. Shimazaki R, Boccaccio A, Mazzatenta A, Pinato G, Migliore M, Menini A. Electrophysiological properties and modeling of murine vomeronasal sensory neurons in acute slice preparations. *Chem Senses*. 2006;31(5):425–435. http://dx.doi.org/10.1093/chemse/bjj047.
59. Ukhanov K, Leinders-Zufall T, Zufall F. Patch-clamp analysis of gene-targeted vomeronasal neurons expressing a defined V1r or V2r Receptor: Ionic mechanisms underlying persistent firing. *J Neurophysiol*. 2007;98(4):2357–2369. http://dx.doi.org/10.1152/jn.00642.2007.
60. Cichy A, Ackels T, Tsitoura C, et al. Extracellular pH regulates excitability of vomeronasal sensory neurons. *J Neurosci*. 2015;35(9):4025–4039. http://dx.doi.org/10.1523/JNEUROSCI.2593-14.2015.
61. Bean BP. The action potential in mammalian central neurons. *Nat Rev Neurosci*. 2007;8(6):9579–9967. http://dx.doi.org/10.1038/nrn2148.
62. Ackels T, von der Weid B, Rodriguez I, Spehr M. Physiological characterization of formyl peptide receptor expressing cells in the mouse vomeronasal organ. *Front Neuroanat*. November 2014;8:134. http://dx.doi.org/10.3389/fnana.2014.00134.
63. Fakler B, Adelman JP. Control of K(Ca) channels by calcium nano/microdomains. *Neuron*. 2008;59(6):873–881. http://dx.doi.org/10.1016/j.neuron.2008.09.001.
64. Zhang P, Yang C, Delay RJ. Urine stimulation activates BK channels in mouse vomeronasal neurons. *J Neurophysiol*. 2008;100(4):1824–1834. http://dx.doi.org/10.1152/jn.90555.2008.
65. Kim S, Ma L, Jensen KL, et al. Paradoxical contribution of SK3 and GIRK channels to the activation of mouse vomeronasal organ. *Nat Neurosci*. July 2012:1–11. http://dx.doi.org/10.1038/nn.3173.
66. Dibattista M, Mazzatenta A, Grassi F, Tirindelli R, Menini A. Hyperpolarization-activated cyclic nucleotide-gated channels in mouse vomeronasal sensory neurons. *J Neurophysiol*. 2008;100(2):576–586. http://dx.doi.org/10.1152/jn.90263.2008.
67. Berghard A, Buck LB. Sensory transduction in vomeronasal neurons: evidence for G alpha o, G alpha i2, and adenylyl cyclase II as major components of a pheromone signaling cascade. *J Neurosci*. 1996;16(3):909–918. Available at: http://www.ncbi.nlm.nih.gov/pubmed/8558259.
68. Halpern M, Shapiro LS, Jia C. Differential localization of G proteins in the opossum vomeronasal system. *Brain Res*. 1995;677(1):157–161. Available at: http://www.ncbi.nlm.nih.gov/pubmed/7606461.
69. Matsuoka M, Yoshida-Matsuoka J, Iwasaki N, Norita M, Costanzo RM, Ichikawa M. Immunocytochemical study of Gi2alpha and Goalpha on the epithelium surface of the rat vomeronasal organ. *Chem Senses*. 2001;26(2):161–166. Available at: http://www.ncbi.nlm.nih.gov/pubmed/11238246.
70. Norlin EM, Gussing F, Berghard A. Vomeronasal phenotype and behavioral alterations in Gai2 mutant mice. *Curr Biol*. July 15, 2003;13:1214–1219. http://dx.doi.org/10.1016/S.

71. Tanaka M, Treloar HB, Kalb RG, Greer CA, Strittmatter SM. G(o) protein-dependent survival of primary accessory olfactory neurons. *Proc Natl Acad Sci USA.* 1999;96(24):14106–14111. Available at: http://www.pubmedcentral.nih.gov/articlerender.fcgi?artid=24198&tool=pmcentrez&rendertype=abstract.

72. Sathyanesan A, Feijoo AA, Mehta ST, Nimarko AF, Lin W. Expression profile of G-protein βγ subunit gene transcripts in the mouse olfactory sensory epithelia. *Front Cell Neurosci.* June 2013;7:84. http://dx.doi.org/10.3389/fncel.2013.00084.

73. Tirindelli R, Ryba NJP. The G-protein gamma-subunit G gamma 8 is expressed in the developing axons of olfactory and vomeronasal neurons. *Eur J Neurosci.* 1996;8(11):2388–2398.

74. Ryba NJP, Tirindelli R. A novel GTP-binding protein γ-subunit, Gγ8, is expressed during neurogenesis in the olfactory and vomeronasal neuroepithelia. *J Biol Chem.* 1995;270(12):6757–6767. http://dx.doi.org/10.1074/jbc.270.12.6757.

75. Rünnenburger K, Breer H, Boekhoff I. Selective G protein beta gamma-subunit compositions mediate phospholipase C activation in the vomeronasal organ. *Eur J Cell Biol.* 2002;81(10):539–547.

76. Montani G, Tonelli S, Sanghez V, et al. Aggressive behaviour and physiological responses to pheromones are strongly impaired in mice deficient for the olfactory G-protein -subunit G8. *J Physiol.* 2013;591(Pt 16):3949–3962. http://dx.doi.org/10.1113/jphysiol.2012.247528.

77. Spehr M, Spehr J, Ukhanov K, Kelliher KR, Leinders-Zufall T, Zufall F. Parallel processing of social signals by the mammalian main and accessory olfactory systems. *Cell Mol Life Sci.* 2006;63(13):1476–1484. http://dx.doi.org/10.1007/s00018-006-6109-4.

78. Smrcka AV. G protein β/γ subunits: central mediators of G protein-coupled receptor signaling. *Cell Mol Life Sci.* 2008;65(14):2191–2214. http://dx.doi.org/10.1007/s00018-008-8006-5.

79. Dey S, Chamero P, Pru JK, et al. Cyclic regulation of sensory perception by a female hormone alters behavior. *Cell.* 2015;161(6):1334–1344. http://dx.doi.org/10.1016/j.cell.2015.04.052.

80. Dibattista M, Amjad A, Maurya DK, et al. Calcium-activated chloride channels in the apical region of mouse vomeronasal sensory neurons. *J Gen Physiol.* 2012;140(1):3–15. http://dx.doi.org/10.1085/jgp.201210780.

81. Yang C, Delay RJ. Calcium-activated chloride current amplifies the response to urine in mouse vomeronasal sensory neurons. *J Gen Physiol.* 2010;135(1):3–13. http://dx.doi.org/10.1085/jgp.200910265.

82. Kelliher KR, Spehr M, Li X-H, Zufall F, Leinders-Zufall T. Pheromonal recognition memory induced by TRPC2-independent vomeronasal sensing. *Eur J Neurosci.* 2006;23(12):3385–3390. http://dx.doi.org/10.1111/j.1460-9568.2006.04866.x.

83. Liman ER, Corey DP, Dulac C. TRP2: a candidate transduction channel for mammalian pheromone sensory signaling. *Proc Natl Acad Sci USA.* 1999;96(10):5791–5796. http://dx.doi.org/10.1073/pnas.96.10.5791.

84. Omura M, Mombaerts P. Trpc2-Expressing sensory neurons in the main olfactory epithelium of the mouse. *Cell Rep.* 2014:1–13. http://dx.doi.org/10.1016/j.celrep.2014.06.010.

85. Leypold BG, Yu CR, Leinders-Zufall T, Kim MM, Zufall F, Axel R. Altered sexual and social behaviors in trp2 mutant mice. *Proc Natl Acad Sci USA.* 2002;99(9):6376–6381. http://dx.doi.org/10.1073/pnas.082127599.

86. Stowers L, Holy TE, Meister M, Dulac C, Koentges G. Loss of sex discrimination and male-male aggression in mice deficient for TRP2. *Science.* 2002;295(5559):1493–1500. http://dx.doi.org/10.1126/science.1069259.

87. Kimchi T, Xu J, Dulac C. A functional circuit underlying male sexual behaviour in the female mouse brain. *Nature.* 2007;448(7157):1009–1014. http://dx.doi.org/10.1038/nature06089.

88. Pankevich DE, Baum MJ, Cherry JA. Olfactory sex discrimination persists, whereas the preference for urinary odorants from estrous females disappears in male mice after vomeronasal organ removal. *J Neurosci.* 2004;24(42):9451–9457. http://dx.doi.org/10.1523/JNEUROSCI.2376-04.2004.

89. Yu CR. TRICK or TRP? what Trpc2$^{-/-}$ mice tell us about vomeronasal organ mediated innate behaviors. *Front Neurosci.* June 2015;9:1–7. http://dx.doi.org/10.3389/fnins.2015.00221.

90. Liman ER. Regulation by voltage and adenine nucleotides of a Ca^{2+}-activated cation channel from hamster vomeronasal sensory neurons. *J Physiol.* 2003;548(Pt 3):777–787. http://dx.doi.org/10.1113/jphysiol.2002.037119.

91. Spehr J, Hagendorf S, Weiss J, Spehr M, Leinders-Zufall T, Zufall F. Ca^{2+}-calmodulin feedback mediates sensory adaptation and inhibits pheromone-sensitive ion channels in the vomeronasal organ. *J Neurosci.* 2009;29(7):2125–2135. http://dx.doi.org/10.1523/JNEUROSCI.5416-08.2009.

92. Bradley J, Reuter D, Frings S. Facilitation of calmodulin-mediated odor adaptation by cAMP-gated channel subunits. *Science.* 2001;294(5549):2176–2178. http://dx.doi.org/10.1126/science.1063415.

93. Bradley J, Bönigk W, Yau K-W, Frings S. Calmodulin permanently associates with rat olfactory CNG channels under native conditions. *Nat Neurosci.* 2004;7(7):705–710. http://dx.doi.org/10.1038/nn1266.

94. Song Y, Cygnar KD, Sagdullaev BT, et al. Olfactory CNG channel desensitization by Ca^{2+}/CaM via the B1b subunit affects response termination but not sensitivity to recurring stimulation. *Neuron.* 2008;58(3):374–386. http://dx.doi.org/10.1016/j.neuron.2008.02.029.

95. Zufall F, Leinders-Zufall T. The cellular and molecular basis of odor adaptation. *Chem Senses.* 2000;25(4):473–481. Available at: http://www.ncbi.nlm.nih.gov/pubmed/10944513.

96. Munger SD, Lane AP, Zhong H, et al. Central role of the CNGA4 channel subunit in Ca^{2+}-calmodulin-dependent odor adaptation. *Science.* 2001;294(5549):2172–2175. http://dx.doi.org/10.1126/science.1063224.

97. Stephan AB, Shum EY, Hirsh S, Cygnar KD, Reisert J, Zhao H. ANO2 is the cilial calcium-activated chloride channel that may mediate olfactory amplification. *Proc Natl Acad Sci USA.* 2009;106(28):11776–11781. http://dx.doi.org/10.1073/pnas.0903304106.

98. Ponissery-Saidu S, Stephan AB, Talaga AK, Zhao H, Reisert J, Ponissery Saidu S. Channel properties of the splicing isoforms of the olfactory calcium-activated chloride channel Anoctamin 2. *J Gen Physiol.* 2013;141(6):1–13. http://dx.doi.org/10.1085/jgp.201210937.

99. Sagheddu C, Boccaccio A, Dibattista M, Montani G, Tirindelli R, Menini A. Calcium concentration jumps reveal dynamic ion selectivity of calcium-activated chloride currents in mouse olfactory sensory neurons and TMEM16b-transfected HEK 293T cells. *J Physiol.* 2010;588(Pt 21):4189–4204. http://dx.doi.org/10.1113/jphysiol.2010.194407.

100. Henkel B, Drose DR, Ackels T, Oberland S, Spehr M, Neuhaus EM. Co-expression of anoctamins in cilia of olfactory sensory neurons. *Chem Senses.* 2014:73–87. http://dx.doi.org/10.1093/chemse/bju061.

101. Dauner K, Lißmann J, Jeridi S, Frings S, Möhrlen F. Expression patterns of anoctamin 1 and anoctamin 2 chloride channels in the mammalian nose. *Cell Tissue Res.* 2012;347(2):327–341. http://dx.doi.org/10.1007/s00441-012-1324-9.

102. Billig GM, Pál B, Fidzinski P, Jentsch TJ. Ca^{2+}-activated Cl^- currents are dispensable for olfaction. *Nat Neurosci.* 2011;14(6):763–769. http://dx.doi.org/10.1038/nn.2821.

103. Pifferi S, Dibattista M, Sagheddu C, et al. Calcium-activated chloride currents in olfactory sensory neurons from mice lacking bestrophin-2. *J Physiol.* 2009;587(17):4265–4279. http://dx.doi.org/10.1113/jphysiol.2009.176131.

104. Amjad A, Hernandez-Clavijo A, Pifferi S, et al. Conditional knockout of TMEM16A/anoctamin1 abolishes the calcium-activated chloride current in mouse vomeronasal sensory neurons. *J Gen Physiol.* 2015;1:1–17. http://dx.doi.org/10.1085/jgp.201411348.

105. Caputo A, Caci E, Ferrera L, et al. TMEM16A, a membrane protein associated with calcium-dependent chloride channel activity. *Science.* 2008;322(5901):590–594. http://dx.doi.org/10.1126/science.1163518.

106. Schroeder BC, Cheng T, Jan YN, Jan LY. Expression cloning of TMEM16A as a calcium-activated chloride channel subunit. *Cell.* 2008;134(6):1019–1029. http://dx.doi.org/10.1016/j.cell.2008.09.003.

107. Rasche S, Toetter B, Adler J, et al. Tmem16b is specifically expressed in the cilia of olfactory sensory neurons. *Chem Senses.* 2010;35(3):239–245. http://dx.doi.org/10.1093/chemse/bjq007.

108. Tirindelli R, Dibattista M, Pifferi S, Menini A. From pheromones to behavior. *Physiol Rev.* 2009;89:921–956. http://dx.doi.org/10.1152/physrev.00037.2008.

109. Berridge MJ, Bootman MD, Roderick HL. Calcium signalling: dynamics, homeostasis and remodelling. *Nat Rev Mol Cell Biol.* 2003;4(7):517–529. http://dx.doi.org/10.1038/nrm1155.

110. Clapham DE. Calcium signaling. *Cell.* 2007;131(6):1047–1058. http://dx.doi.org/10.1016/j.cell.2007.11.028.

111. Baughman JM, Perocchi F, Girgis HS, et al. Integrative genomics identifies MCU as an essential component of the mitochondrial calcium uniporter. *Nature.* 2011;476(7360):341–345. http://dx.doi.org/10.1038/nature10234.

112. Kirichok Y, Krapivinsky G, Clapham DE. The mitochondrial calcium uniporter is a highly selective ion channel. *Nature.* 2004;427(6972):360–364. http://dx.doi.org/10.1038/nature02246.

113. Mallilankaraman K, Doonan P, Cárdenas C, et al. MICU1 is an essential gatekeeper for MCU-mediated mitochondrial Ca(2+) uptake that regulates cell survival. *Cell.* 2012;151(3):630–644. http://dx.doi.org/10.1016/j.cell.2012.10.011.

114. Fluegge D, Moeller LM, Cichy A, et al. Mitochondrial Ca^{2+} mobilization is a key element in olfactory signaling. *Nat Neurosci.* February 2012:1–11. http://dx.doi.org/10.1038/nn.3074.

115. Veitinger S, Veitinger T, Cainarca S, et al. Purinergic signalling mobilizes mitochondrial Ca^{2+} in mouse Sertoli cells. *J Physiol*. 2011;589(Pt 21):5033—5055. http://dx.doi.org/10.1113/jphysiol.2011.216309.

116. Leinders-Zufall T, Greer CA, Shepherd GM, Zufall F. Imaging odor-induced calcium transients in single olfactory cilia: specificity of activation and role in transduction. *J Neurosci*. 1998;18(15):5630—5639. Available at: http://www.ncbi.nlm.nih.gov/pubmed/9671654.

117. Bradley J, Reisert J, Frings S. Regulation of cyclic nucleotide-gated channels. *Curr Opin Neurobiol*. 2005;15(3):343—349. http://dx.doi.org/10.1016/j.conb.2005.05.014.

118. Jia C, Halpern M. Segregated populations of mitral/tufted cells in the accessory olfactory bulb. *Neuroreport*. 1997;8(8):1887—1890. Available at: http://www.ncbi.nlm.nih.gov/pubmed/9223071.

119. Mori K, von Campenhausen H, Yoshihara Y. Zonal organization of the mammalian main and accessory olfactory systems. *Philos Trans R Soc Lond B Biol Sci*. 2000;355(1404):1801—1812. http://dx.doi.org/10.1098/rstb.2000.0736.

120. Del Punta K, Puche AC, Adams NC, Rodriguez I, Mombaerts P. A divergent pattern of sensory axonal projections is rendered convergent by second-order neurons in the accessory olfactory bulb. *Neuron*. 2002;35(6):1057—1066. Available at: http://www.ncbi.nlm.nih.gov/pubmed/12354396.

121. Dulac C, Wagner S. Genetic analysis of brain circuits underlying pheromone signaling. *Annu Rev Genet*. August 2006;40:449—467. http://dx.doi.org/10.1146/annurev.genet.39.073003.093937.

122. Larriva-Sahd J. The accessory olfactory bulb in the adult rat: a cytological study of its cell types, neuropil, neuronal modules, and interactions with the main olfactory system. *J Comp Neurol*. 2008;510(3):309—350. http://dx.doi.org/10.1002/cne.21790.

123. Meeks JP, Arnson HA, Holy TE. Representation and transformation of sensory information in the mouse accessory olfactory system. *Nat Neurosci*. 2010;13(6):723—730. http://dx.doi.org/10.1038/nn.2546.

124. Takami S, Graziadei PP. Light microscopic golgi study of mitral/tufted cells in the accessory olfactory bulb of the adult rat. *J Comp Neurol*. 1991;311(1):65—83. http://dx.doi.org/10.1002/cne.903110106.

125. Urban NN, Castro JB. Tuft calcium spikes in accessory olfactory bulb mitral cells. *J Neurosci*. 2005;25(20):5024—5028. http://dx.doi.org/10.1523/JNEUROSCI.0297-05.2005.

126. Hovis KR, Ramnath R, Dahlen JE, et al. Activity regulates functional connectivity from the vomeronasal organ to the accessory olfactory bulb. *J Neurosci*. 2012;32(23):7907—7916. http://dx.doi.org/10.1523/JNEUROSCI.2399-11.2012.

127. Wagner S, Gresser AL, Torello AT, Dulac C. A multireceptor genetic approach uncovers an ordered integration of VNO sensory inputs in the accessory olfactory bulb. *Neuron*. 2006;50(5):697—709. http://dx.doi.org/10.1016/j.neuron.2006.04.033.

128. Brennan PA, Kendrick KM. Mammalian social odours: attraction and individual recognition. *Philos Trans R Soc Lond B Biol Sci*. 2006;361(1476):2061—2078. http://dx.doi.org/10.1098/rstb.2006.1931.

129. Keverne EB, Brennan PA. Olfactory recognition memory. *J Physiol*. 1996;90(5—6):503—508. Available at: http://www.ncbi.nlm.nih.gov/pubmed/9089523.

130. Price JL, Powell TPS. The synaptology of the granule cells of the olfactory bulb. *J Cell Sci*. 1970;7(1):125—155. Available at: http://www.ncbi.nlm.nih.gov/pubmed/5476853.

131. Schoppa NE, Urban NN. Dendritic processing within olfactory bulb circuits. *Trends Neurosci*. 2003;26(9):501—506. http://dx.doi.org/10.1016/S0166-2236(03)00228-5.

132. Taniguchi M, Kaba H. Properties of reciprocal synapses in the mouse accessory olfactory bulb. *Neuroscience*. 2001;108(3):365—370.

133. Hayashi Y, Momiyama A, Takahashi T, et al. Role of metabotropic glutamate receptors in synaptic modulation in the accessory olfactory bulb. *Nature*. December 16, 1993;366:687—690.

134. Jia C, Chen WR, Shepherd GM. Synaptic organization and neurotransmitters in the rat accessory olfactory bulb. *J Neurophysiol*. 1999;81(1):345—355. Available at: http://www.ncbi.nlm.nih.gov/pubmed/9914294.

135. Luo M, Katz LC. Encoding pheromonal signals in the mammalian vomeronasal system. *Curr Opin Neurobiol*. 2004;14(4):428—434. http://dx.doi.org/10.1016/j.conb.2004.07.001.

136. Hendrickson RC, Krauthamer S, Essenberg JM, Holy TE. Inhibition shapes sex selectivity in the mouse accessory olfactory bulb. *J Neurosci*. 2008;28(47):12523—12534. http://dx.doi.org/10.1523/JNEUROSCI.2715-08.2008.

137. Brennan PA, Keverne EB. Neural mechanisms of mammalian olfactory learning. *Prog Neurobiol*. 1997;51(4):457—481. Available at: http://www.ncbi.nlm.nih.gov/pubmed/9106902.

138. Castro JB, Hovis KR, Urban NN. Recurrent dendrodendritic inhibition of accessory olfactory bulb mitral cells requires activation of group I metabotropic glutamate receptors. *J Neurosci*. 2007;27(21):5664–5671. http://dx.doi.org/10.1523/JNEUROSCI.0613-07.2007.
139. Kaba H, Hayashi Y, Nakanishi S. Induction of an olfactory memory by the activation of a metabotropic glutamate receptor. *Science*. July 8, 1994;265:262–264.
140. Bruce HM. A block to pregnancy in the mouse caused by proximity of strange males. *J Reprod Fertil*. 1960;1:96–103.

CHAPTER

12

Comparative Olfactory Transduction

Elizabeth A. Corey[1,3], Barry W. Ache[1,2,3]

[1]Whitney Laboratory for Marine Bioscience, St. Augustine, FL, USA; [2]Departments of Biology and Neuroscience, Gainesville, FL, USA; [3]Center for Smell and Taste, McKnight Brain Institute, University of Florida, Gainesville, FL, USA

INTRODUCTION

As far as we know, all animals—including unicellular and multicellular, invertebrate and vertebrate, aquatic and terrestrial—use some type of olfactory receptor to detect chemical cues at a distance from the source. Olfaction mediates diverse behavioral responses including foraging, food choice, mating, communication, predator detection, and escape. Although olfaction is typically considered in the context of terrestrial organisms, aquatic animals use it to detect water-soluble compounds for the same purposes and with basically the same neural machinery as those that live on land and detect airborne odors (see Chapter 5).

Olfactory perception begins when odorants interact with and activate signal transduction through the olfactory receptor. In multicellular organisms, olfactory receptors are expressed in primary receptor neurons referred to as olfactory sensory neurons (OSNs), which are located in dedicated end organs such as the olfactory sensilla on the antennae of arthropods and in the main olfactory epithelium (MOE) of vertebrates. OSNs provide the primary input to the olfactory system, such that disruption of the cellular processes leading to the activation of OSNs inevitably compromises olfactory function, an outcome that can be beneficial in the context of pest management or detrimental in the form of decreased quality of life.

OLFACTORY SYSTEMS: SIMILARITY AND DIVERSITY

Olfactory systems have evolved to extract as much relevant chemical information from the environment as possible. Relevant odorants differ among species, as would be expected given their differing environments and capacities to rely on other senses such as vision and audition. Thus, critical odors for one species may have no behavioral salience for or not even be detected by another. Given the wide range of environments and odorants, it would not be surprising to find diversity in the functional organization and signal transduction mechanisms of the olfactory systems of different types of animals. Indeed, the olfactory periphery is emerging to be much more diverse and complex than originally believed, even at the level of the nature of the olfactory receptor itself.

Despite the differences in the protein structures and downstream signal transduction pathways associated with different olfactory receptor types, the animals that use them all exist in the same olfactory world with the same challenges of identifying odorants and responding with the appropriate behavior. Given the fundamental requirements of olfaction, such as those for sensitivity and speed to allow for rapid behavioral responses, substantial similarities resulting from evolutionary convergence are also likely to occur. Although mammals and other terrestrial animals are exposed primarily to volatile, airborne odors, the odorant molecules must enter the aqueous environment surrounding the cilia of the OSNs to gain access to the olfactory receptors. The comparative study of olfactory signal transduction mechanisms in different types of animals, whether among one group such as arthropods (e.g., Refs 1,2) or between types such as vertebrates and insects (e.g., Refs 3—5), can reveal both similar and diverse mechanisms that underlie the important functional characteristics of olfactory systems in general.

FUNCTIONAL ORGANIZATION OF OLFACTORY SYSTEMS

Olfactory transduction is best understood in vertebrates and arthropods, where signal transduction occurs in specific subsystems. The mammalian sense of smell is organized into functional subsystems involving different receptor types in addition to the MOE, including the vomeronasal organ (see Chapters 10, 11), the septal organ of Masera, the Grueneberg ganglion (see Chapter 8), and the trigeminal system (see Chapter 21). Other vertebrates have their sense of smell organized into anatomically segregated subsystems as well (as reviewed in Ref. 6), but as in the case of amphibians where two have been identified, with one for general odorants and the other for species-specific odorants (pheromones), they may be less complex.[7] Olfaction in insects, and other arthropods such as crustaceans, has been found to use parallel subsystems with multiple receptor types as well.[8,9] In insects, the olfactory system is organized into different types of sensilla. These sensilla, which in the fruit fly *Drosophila melanogaster* include basiconic, trichoid, and coeloconic, encase the OSNs on the antennae and maxillary palp.[10] These sensilla types have been differentiated based on their morphology and anatomical distribution as well as on the odorants to which they respond.[10–12] In crustaceans, the systems may be based on olfactory receptor type with some OSNs expressing insect-like gustatory receptors (Grs; see Chapter 14) and others gluta-mate receptor-like ionotropic odorant receptors (Irs; see Chapter 6),[13–15] but this remains to be explored. The conservation of the use of parallel systems for processing olfactory informa-tion across multiple types of animals suggests that it is likely an important functional char-acteristic of sensory systems that may provide adaptive advantages.

FUNCTIONAL SUBSETS OF OSNs

In addition to different subsystems, there are also functionally different subsets of OSNs within them. The canonical OSNs in the nose of vertebrates are bipolar neurons that each have a single dendrite ending in a dendritic knob with cilia extending into the nasal mucus (see Chapter 7). Each has a single basal axon that projects to a glomerulus of the main olfac-tory bulb (MOB) where the first synapses occur. Different subsets of cells within the MOE can be identified, include transient amino acid receptor (TAAR)-expressing OSNs (see Chapter 4), receptor guanylyl cyclase type D–expressing OSNs (see Chapter 8), transient receptor poten-tial (TRP) channel-expressing OSNs, and type 1 vomeronasal receptor (see Chapter 10)-expressing OSNs, among others. Each of these subsets plays a different, yet sometimes over-lapping, function in mammalian chemosensation that may provide an adaptive advantage by creating dedicated sets of cells for identification of critical olfactory cues while including some redundancy in the event that a particular system fails.

Although arthropod OSNs are more compartmentalized than those of mammals, they are similarly structured, with sensory cilia (dendrites) extending into the sensilla of the olfactory antenna and basal axons projecting to olfactory glomeruli in the antennal lobe, which corre-sponds to the MOB of mammals. Although the general cell morphology is conserved, the fine details of OSNs can be adaptive in that they can vary by habitat rather than by species. OSNs in crustaceans that live in terrestrial environments as adults, such as the giant robber crab and other terrestrial hermit crabs, have a morphology that is more insect-like than it is similar to marine crustaceans.[16,17] Yet, olfactory sensilla in both types of crustaceans contain receptor lymph to provide a moist environment critical to odorant detection in terrestrial crustaceans and to protect the OSNs against changes in salinity in marine crustaceans.[17,18]

In both vertebrates and arthropods, it has been demonstrated that each glomerulus re-ceives input primarily from OSNs that express the same olfactory receptor type, with each receptor type represented by one to a small number of glomeruli.[19,20] Within the glomeruli, OSNs form their first synapse, with projection neurons in arthropods and mitral and tufted cells in vertebrates, to send olfactory information to higher centers in the central nervous system. The number of glomeruli per species is highly variable. Although it was originally speculated that there would be a close correlation between the number of olfac-tory receptors encoded by a species and the number of glomeruli, this has not held true for all animals. Although rodents appear to have a convergence ratio of 2:1, humans express approximately 350 olfactory receptors, yet have been found to have more than 5000 glomeruli, giving a ratio closer to 16:1.[21] Given that odors are generally mixtures of individual odorants that can activate more than one olfactory receptor and that OSNs can be activated by more than one odorant, the odor identity is encoded by the combination of activated glomeruli, a process known as the combinatorial code of olfaction.[22] Glomerular organization at the first synaptic relay along with combinatorial coding is likely to have arisen from convergent evolution, and their conservation suggests that they constitute an effective mechanism of coding odors.

OLFACTORY RECEPTORS

Signal transduction begins in all olfactory systems with ligand binding to the olfactory receptor, but the type of receptor and transduction mechanism can vary across species. The first olfactory receptor gene family, the so-called canonical odorant receptors (ORs), was identified in mammals as a large group of G protein–coupled receptors (GPCRs) expressed primarily in the MOE (see Chapter 3).[23] Based on phylogenetic analyses, canonical mammalian ORs are classified into two different groups, class I (fish-like ORs) and class II (tetrapod ORs), that differ primarily in the sequence of the extracellular loops.[24] Although in most cases each OSN expresses a single OR, some cells may express multiple receptors.[25] Ligand binding to canonical mammalian ORs activates the olfactory heterotrimeric G protein, consisting of subunits $G\alpha_{olf}$, $G\beta_1$, and $G\gamma_{13}$, which in turn activates adenylyl cyclase type III (ACIII) (see Chapter 7). ACIII activation results in an increase in cyclic adenosine monophosphate (cAMP), which binds to and opens a cyclic nucleotide–gated (CNG) cation channel generating an inward current that may be further amplified by a calcium-activated Cl^- channel. This transduction cascade depolarizes the OSN, which fires action potentials to transmit the odor information to the brain. Although some other subtypes of mammalian OSNs, such as those expressing TAARs (see Chapter 4), signal using this canonical pathway, others such as those expressing TRPC2 may use additional pathways.[26]

Genome and transcriptome sequencing have led to a rapid increase in the identification of the olfactory receptor gene repertoires of several species. It was originally believed that invertebrate olfactory receptors were also GPCRs. Indeed some OSNs, such as those in the nematode *Caenorhabditis elegans*,[27] have been found to express this type of olfactory receptor. However, two large gene families encoding ionotropic olfactory receptors, referred to as Ors and Irs, have been identified in insects and other arthropods (Figure 1; see Chapter 6).[28,29] The first family identified consists of seven transmembrane domain receptors that function as heteromers formed from at least one ligand-specific Or and a coreceptor subunit, Orco.[30–32] Although Orco itself is highly conserved among insects, the sequences of the ligand-specific subunits are disparate, likely because of differences in ecology and ligands. Despite their similarity to mammalian ORs in that they have similar numbers of transmembrane domains (and thus the original belief that they would act as GPCRs), they have been found to have a different orientation in the membrane[33,34] and a very different ionotropic mechanism of signaling. Insect Grs (see Chapter 14) are distantly related to Ors and are thought to have a similar topology and mechanism of signaling, but at least in insects are expressed primarily in taste organs.[35,36] It is important to note that Ors appear to be absent in some insects[8] as well

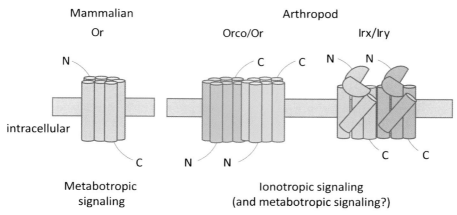

FIGURE 1 **Comparison of olfactory receptors and signal transduction mechanisms.** Canonical mammalian olfactory transduction begins with odorant binding to odorant receptors (Ors), which are G protein—coupled receptors that activate metabotropic signaling pathways. In contrast, arthropod olfactory sensory neurons express ionotropic receptors, referred to as olfactory receptors (Ors) and ionotropic receptors (Irs), and odorant binding activates ionotropic signaling (and, although it remains controversial, possibly metabotropic signaling). Arthropod Ors form heteromers of a coreceptor (Orco) and a ligand-specific subunit (Or). Irs also function as heteromers (Irx/Iry), although the coreceptor may vary with the ligand-specific subunit. In some cases, Irs may function as homomeric receptors. Orientation in the membrane is indicated by the amino (N) and carboxyl (C) termini.

as in some crustaceans,[13–15] despite the conservation of sensitive responses to olfactory cues, indicating that in these animals a second set of receptors, including Irs and possibly GRs,[14] mediates olfaction.

As with insect Ors, Irs are thought to function as heteromers formed from ligand-binding subunits and a least one conserved and broadly expressed subunit.[37] However, in some cases they may be able to function as monomers, as may be the case for Ir76b in salt detection.[38] Although using two different classes of olfactory receptors expands the ability of insects to detect different types of ligands, more recent work has demonstrated that Ors and Irs also exhibit intrinsic differences in temporal kinetics, with Ors responding faster and with higher sensitivity to short odor exposures and IRs responding better to higher concentrations with longer durations.[39] Although many insects coexpress Ors and IRs, the ratio between the two types varies, and in some cases only IRs are expressed.[8] Based on the lack of detection of expression of Or orthologues in crustaceans and some insects, their expression appears to be an evolutionarily later adaption in arthropods.[8] Given that Irs may be the only olfactory receptor type expressed in some crustaceans, much may be learned about them from the already extensively studied lobster model.

Although Ors and Orco have not been detected in the olfactory tissue of crustaceans,[13,15] Grs have been identified in the genome of the water flea *Daphnia pulex*.[14] It has been speculated that Grs may mediate olfactory-related behaviors in crustaceans, but their expression was not detected in lobster or hermit crab OSNs.[13,15] Although it is possible that Grs may act as a second chemosensory subsystem in crustaceans that involves other tissue types, this possibility remains to be explored because even the Grs originally identified in *Daphnia* have yet to be localized to a particular cell type.

NONCANONICAL ODORANT-EVOKED SIGNALING PATHWAYS

Genetic ablation of any of the main components of the canonical olfactory signaling pathways, such as ACIII-based cyclic nucleotide signaling in mammals[40,41] or Orco in insects,[42] results in severe deficits in olfactory-related behaviors, suggesting that they are the basis for primary odorant-evoked excitation cascades in the canonical OSNs of these two types of animals. Expression of Irs in insect OSNs that are odorant responsive yet lack Ors, as well as their ability to mediate transduction by their predicted ligands when ectopically expressed in other neurons, provides evidence that Irs, too, serve as a canonical excitation pathway in arthropods.[37] These findings have led to considerable controversy as to the existence and roles of signaling pathways other than a canonical excitatory pathway.

In arthropods, both the Ors and Irs are thought to act primarily as ionotropic receptors, with binding of the odorant resulting in a conformational change that opens the channel and activates signaling. However, in both cases, there is also evidence for G protein coupling and metabotropic signaling. Depletion of G protein subunits or components of phosphoinositide (PI) signaling can reduce olfactory responses in *Drosophila*.[43–45] Coexpression of insect Ors with G proteins can alter signaling in heterologous systems,[45] yet other studies have suggested that G protein signaling is not involved in this system.[46,47] Pharmacological inhibitors and activators of metabotropic signaling, such as the PI and cyclic nucleotide—mediated pathways, can modulate odorant responsiveness of lobster OSNs,[48–51] which are not thought to express Ors, indicating that metabotropic signaling may be associated with OSNs expressing Irs as well.

Based on distinct conserved amino acid motifs, mammalian ORs are thought to be part of a special group of GPCRs and thus may have unique signaling constraints. However, the ligand response profile of an Or can change depending on the type of $G\alpha$ subunit to which it couples.[52,53] Many types of GPCRs can initiate responses that are not mediated by heterotrimeric G proteins, but rather by other intracellular signaling proteins such as arrestins. Proteomic analyses of mammalian olfactory cilia reveal that elements of numerous known signaling pathways occur in olfactory cilia, which at least sets the molecular stage for noncanonical signaling.[54,55] Indeed, ligand activation of mammalian ORs has been linked to multiple signaling pathways other than that directly mediated by $G\alpha_{olf}$ and ACIII, such as β-arrestin-mediated internalization and PI-mediated signal transduction.[56–58] All of these must originate from the same receptor—suggesting a network of signaling pathways such as those seen for other GPCRs.

These potential noncanonical mechanisms of odor-evoked signaling by olfactory receptors could serve multiple roles. Odorant activation of mammalian OSNs can rescue them from apoptosis and turnover, suggesting the activation of an additional signaling pathway.[59] This function has been linked to multiple signaling pathways, such as PI and mitogen-activated protein kinase cascades,[59–62] and it remains unclear which are responsible. Additionally, β-arrestin activation has been linked to odorant-evoked mammalian OR internalization and desensitization.[56] G protein activation by insect Ors has been proposed to extend the range of sensitivity to odors, with the ionotropic pathway as the sole mediator of transduction at high odorant concentrations and the metabotropic pathway amplifying the signal at lower odorant concentrations.[39] However, an important, yet often neglected, aspect of olfaction in both vertebrates and arthropods that may involve one or more types of noncanonical signaling is that there are two opponent modes of signaling. That is, odors can either excite or inhibit OSNs.

OPPONENT SIGNALING: EXCITATION AND INHIBITION

In contrast to the depth of our understanding of how odorants excite OSNs, we still know relatively little about how odorants inhibit these cells in any animal. Odorant inhibition of OSNs was originally observed in arthropods, including both insects and crustaceans.[63,64] Although it was initially dismissed in mammalian OSNs, recent studies have identified ligands that can inhibit odorant activation of the excitatory pathway at the level of both the cell and the receptor in mammals as well (Figure 2) (as reviewed in Ref. 65). The conservation of this phenomenon across diverse animal species, and thus structurally different olfactory receptor types, suggests that opponent input is an important functional characteristic of olfactory systems in general and presumably integral to olfactory transduction.

Although odorant-specific opponent input to OSNs is often referred to as "inhibition," and we use that term here, other terminology such as suppression, hypoadditivity, and odor masking have also been used to describe odorant-evoked decreases in the output in OSNs.[66–69] It remains to be determined if this is just semantics or whether there are functionally different "inhibitory" processes with different underlying mechanisms. For example, odorants can inhibit mammalian OSNs by directly blocking elements involved in activating the cell through direct interactions not involving the OR, including voltage-gated channels[70] and the olfactory CNG channel.[71] In insect Ors, the channel-forming Orco subunit can be blocked by antagonistic compounds such as trace amines.[72] It is unclear, however, how these

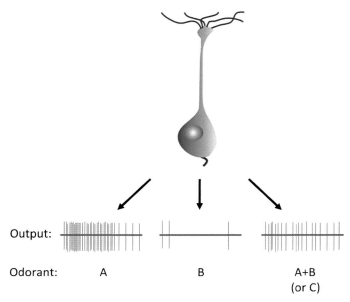

FIGURE 2   **Diagram of a hypothetical olfactory sensory neuron (OSN) that is excited by odorant A, inhibited by odorant B, and excited at a lower level when exposed to a mixture of odorants A and B (or to a partial agonist, odorant C).** The output of the OSN would reflect the integration of the relative proportion of odorants A and B in the mixture.

actions could account for reductions in output of specific OSNs without altering others because canonical OSNs are all assumed to express the same, or at least a similar, complement of signaling elements including their ion channels.

Odor-evoked inhibition in the context of opponent input has been clearly demonstrated to be a property of a particular odorant–olfactory receptor combination and not a property of the OSN. Inhibitory odorants for mammalian ORs have been identified in vitro when ORs are heterologously expressed,[52,73,74] indicating that the "antagonist" is interacting with the OR. Similarly in insects, ectopic expression of nonnative Ors, either from the same or a different species, in *Drosophila* OSNs transfers the odorant-evoked inhibition that was observed in the donor OSN,[37,75,76] indicating that it was not a feature of the OSN or Orco, but of the Or.

The most direct mechanism of inhibition is through the action of a competitive antagonist that blocks binding of the activating agonist to the orthosteric binding site on the receptor. Given that odorants typically are complex mixtures, inhibition can result from antagonists, or even just partial agonists, competing with strong agonists for the same olfactory receptor binding site to prevent full activation. For many years, olfactory antagonists were unknown, but the revelation that mammalian ORs are GPCRs stimulated the search for compounds that would interact competitively with these receptors. There has been a parallel search for odorants that can block activation of insect chemosensory receptors used for host detection. Several studies have now identified odorants that can induce the parallel, saturating rightward shift in the dose–response function in native mammalian OSNs and in heterologous cells expressing ORs that is strongly suggestive of direct competition for the orthosteric binding site.[52,73] Inhibition appears to be relatively common in insect OSNs, but greater spontaneous activity in these cells may make inhibition easier to detect. The incidence of inhibition in mammalian OSNs has been reported to be low,[77] which has led to the idea that inhibition could be relatively unimportant even if it played a role in coding. However, the incidence of inhibition in mammalian OSNs has not been rigorously tested.

NONCOMPETITIVE MECHANISMS OF INHIBITION

Noncompetitive interactions may also be important for shaping odor responses. When the experimentally determined responses of mammalian OSNs to binary odorant mixtures were modeled, the responses of almost 50% of the cells best fit a model in which noncompetitive interactions modulate signaling through the OR.[78] These results are supported by findings that single odorants can inhibit the spontaneous output of specific OSNs,[79,80] which clearly cannot result from competitive inhibition unless contaminating background odorants drive the ongoing discharge. Single odorants can also inhibit the activation of rodent OSNs by forskolin (K. Ukhanov, personal communication), which activates ACIII independently of the OR, excluding the possibility of competitive inhibition at the level of the OR. Together, these modeling and experimental results support the involvement of one or more noncompetitive mechanisms of inhibition in mammalian odor responses.

It has been proposed that noncompetitive inhibition could be caused by odorant stabilization of insect olfactory receptors in an inactive state.[11,81] In the case of ionotropic insect

Ors and Irs, this could involve binding at an allosteric site distinct from the orthosteric site and stabilization of the receptor in a conformation that can no longer interact with the primary agonist or in which the intrinsic channel is unable to open in response to agonist binding. Similarly, inhibition of signaling by a mammalian OR could result if the receptor were locked in a state that prevented activation of $G\alpha_{olf}$. A closely related idea is that the inhibitory odorant could stabilize the OR in a conformation that interacts with a specific intracellular signaling complex that suppresses the excitatory pathway. Indeed, this is consistent with our growing understanding of GPCRs. Rather than viewing ligand binding as consistently eliciting a single intracellular signal, it has become increasingly clear that the nature of the ligand and the dynamically changing intracellular environment shape the downstream signaling network for many different types of receptors. Many GPCRs can activate a hierarchy of signaling thought to be imposed in part by the conformation of the ligand—GPCR complex such that different ligands shift the balance of the network toward different pathways, the process referred to by various names such as ligand-induced selective signaling (LiSS; see Breakout Box) and biased agonism (e.g., Refs 82—84).

BREAKOUT BOX

LIGAND-INDUCED SELECTIVE SIGNALING BY G PROTEIN—COUPLED RECEPTORS

A single G protein—coupled receptor (GPCR) may couple to more than one heterotrimeric G protein isoform, signal through G protein—independent pathways, undergo modification by regulatory processes such as phosphorylation, be regulated by interactions with other proteins, including other GPCRs, and interact with scaffolding proteins. This functional versatility of GPCRs is not congruent with the classical two-state, on—off model of receptor activation. Instead, there is increasing evidence that most GPCRs are structurally dynamic proteins and that specific ligands stabilize distinct receptor conformations resulting in biased activation of a particular set of signaling pathways, processes that are often referred to using terms such as ligand-induced selective signaling or biased agonism. Thus, it is the ligand—receptor complex, not the receptor alone, which specifies the cellular response. In effect, the receptor bound to a biased ligand is a different functional entity with different signaling characteristics than the same receptor bound to a conventional orthosteric ligand.

Much of what we know about GPCR structure and signaling mechanisms is based on a small set of well-characterized receptors, such as the β2-adrenergic receptor and angiotensin II type 1 receptor (AT_1R), but increasing evidence suggests that the concepts learned from them will broadly apply. Early studies revealed that AT_1R couples primarily to $G\alpha_q$ as well as to other G proteins such as $G\alpha_i$.[96] However, it is now known that the AT_1R signaling can be biased toward a G protein—independent, β-arrestin-mediated pathway resulting in different cellular responses.[97] The search for biased ligands that allow targeting of beneficial pathways while minimizing activation of those that are detrimental is a current strategy of drug development, and biased agonists targeting receptors such as the AT_1R are currently in clinical trials.[98]

The potential involvement of LiSS in mammalian olfaction has received little attention, notwithstanding pharmacological evidence that activation of mammalian ORs can involve ACIII as well PI-dependent signaling and that the G protein–dependent activation of the two signaling pathways can be ligand selective.[57,79,80,85,86] It is well established that in vitro ORs can couple to $G\alpha_{olf}$ to activate a cyclic nucleotide pathway and to $G\alpha_{15}$ to activate PI signaling, and that the extent of activation of the two pathways can be ligand-specific.[52,53,87] There is also compelling evidence that G protein–coupled PI signaling, through phosphoinositide 3-kinase (PI3K) in particular, mediates inhibitory input to OSNs when activated by the OR in a ligand-selective manner, thus playing an important role in shaping the output of the OSN.[57,85] For some antagonistic odorant pairs, inhibition can be relieved and the agonist strength of the weaker, otherwise inhibitory, member of the pair can be increased by PI3K blockade, suggesting the inhibition was mediated through a PI3K--dependent pathway in a selective manner. Evidence suggests that the "weaker" odorant interacts with the OR, perhaps allosterically, in a specific conformation that selectively activates PI3K and opposes excitation downstream of the OR. Exogenous phosphatidylinositol (3,4,5)-trisphosphate (PIP_3), the primary product of PI3K signaling, can suppress activation of the olfactory CNG channel by cAMP,[88,89] suggesting a potential downstream mechanism through which LiSS could mediate inhibition (Figure 3). Screening a panel of odorants for PI3K-dependent inhibition found that many odorants, including many of those occurring together naturally in an odor object, have the capacity to activate the PI3K pathway in native OSNs, indicating that LiSS may be an inherent property of mammalian ORs.[80]

Thus, we would suggest that an important noncompetitive mechanism underlying opponent input in mammalian OSNs is that the inhibitory odorant can stabilize the OR in a conformation that interacts with an alternate intracellular signaling complex that, when activated, suppresses the canonical cyclic nucleotide–mediated excitatory pathway. Although there is controversial evidence for activation of nonionotropic signaling by arthropod Ors and Irs and it has been demonstrated that insect OSNs generate ligand-biased output (i.e., different ligands evoke different patterns of output in the same cell),[90,91] it remains to be determined if different modes of signaling by these receptors are activated selectively to generate the output patterning.

GAINING INSIGHT FROM THE COMPARATIVE STUDY OF OLFACTORY TRANSDUCTION

Detailed study and comparison of the signaling mechanisms used by diverse species of animals can reveal important fundamental characteristics of olfaction. There are striking similarities across species in olfaction from the overall functional organization of the olfactory system to the types of downstream signaling. These common functional characteristics span a phylogenetically broad array of animals and together serve to define olfaction in general. Such conservation also implies that there is an optimal solution to the problem of detecting and discriminating odor quality and quantity, if not for other sensory dimensions such as locating the odor source in space and time. Either these solutions evolved early and were retained despite the use of strikingly different olfactory receptor types and signaling pathways or, more likely, animals convergently evolved similar solutions to the problem of

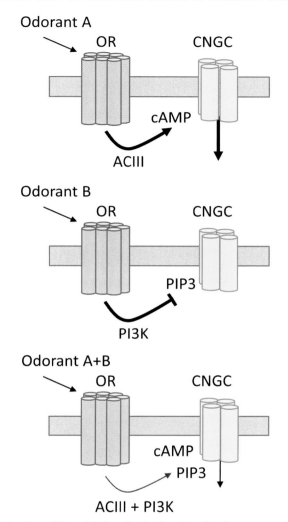

FIGURE 3   **Potential mechanism of ligand-induced selective signaling by a mammalian odorant receptor (OR).** Odorant A acts as an excitatory ligand that activates adenylyl cyclase type III (ACIII) to produce cyclic adenosine monophosphate (cAMP), with little or no activation of phosphoinositide 3-kinase (PI3K), and excites the cell through the olfactory cyclic nucleotide–gated channel (CNGC). Odorant B acts as an inhibitory ligand that primarily activates PI3K to produce phosphatidylinositol (3,4,5)-trisphosphate (PIP$_3$), with little or no activation of ACIII, inhibiting opening of the CNGC. The mixture of odorants A and B activates both ACIII and PI3K such that both cAMP and PIP$_3$ influence the opening of the CNGC, resulting in a reduced level of excitation in comparison to odorant A alone (denoted by thinner arrows), and the degree to which each pathway is activated shapes the output of the cell.

detection and localization of chemical cues, supporting the continued use of diverse animal models to investigate the sense of smell.

Insight into numerous aspects of olfactory transduction has benefited from applying what has been learned in one animal model to another. We have learned, for example, that the molecular receptive range (MRR) of OSNs needs to include odorants that inhibit the cell, not just those that excite it, as is most commonly the case. Such more-complete MRRs have been generated for insect olfactory receptors by ectopically expressing them in native OSNs and screening with large panels of potential ligands. The task is more difficult in mammals since their OSNs exhibit lower levels of spontaneous output and assays for inhibition are limited, but its necessity is beginning to be appreciated. A more complete understanding of MRRs will allow us to more fully understand combinatorial coding in olfaction.

Also, having inhibitory input that tempers or modulates excitation requires thinking of the OSN as serving an integrating role at the input level. The signal an OSN sends to the brain reflects the integrated action of the particular subset of odorants in the odor world at that moment. The meaningful input of an OSN could be zero if the inhibitory drive exceeded the excitatory one. An exciting possibility is that it may be more appropriate to view excitation and inhibition as the extremities of a continuum of odorant-specific input to OSNs rather than as two distinct inputs. Toward that end, different single odorants evoke a variety of patterns of excitatory output in the same insect OSN.[91] The signal transduction mechanisms behind this patterning are not well understood other than that they are likely determined by the receptor[81]; whether they apply to mammalian OSNs remains to be determined. Even so, one could envision that an odorant activates different patterns of excitation for different olfactory receptors based on the extent to which it also induces inhibition. Given the large convergence ratios at the first synaptic relay, even small differences in the discharge pattern of each OSN resulting from modulation of the excitatory signaling pathway could be enhanced postconvergence.

Not touched on here but relevant to using comparative study to gain insight into olfactory transduction is the recent discovery that a novel subpopulation of OSNs in the spiny lobster has the capacity to accurately encode the time of odor encounter.[92] This may allow the animal to determine the spatial dimension of the odor signal as it changes in temporal structure with distance from the odor source. Such dynamic encoding of the spatiotemporal dimension of olfaction uses a nonconventional type of "bursting" OSN that has spontaneous, rhythmic oscillations that are phase-locked or "entrained" by the odor stimulus and respond to odors with a phase-dependent burst of action potentials.[93] Similar subpopulations of bursting OSNs have been identified in other animals, including mammals and amphibians,[94,95] suggesting that dynamic encoding of the spatiotemporal dimension of odors can be considered fundamental to olfaction.

Thus, as these few examples illustrate, broad insight into the nature and mechanisms of olfactory transduction has come from studying diverse animal models, each with its own advantages and disadvantages. As the number of sequenced genomes and olfactory transcriptomes continues to increase, the potential of a wide range of animal models to contribute to our understanding of olfactory processes hopefully will be increasingly accepted and used in our search for the common fundamental features that together serve to define olfaction.

References

1. Martin JP, Beyerlein A, Dacks AM, et al. The neurobiology of insect olfaction: sensory processing in a comparative context. *Prog Neurobiol.* 2011;95(3):427–447. http://dx.doi.org/10.1016/j.pneurobio.2011.09.007.
2. Clifford MR, Riffell JA. Mixture and odorant processing in the olfactory systems of insects: a comparative perspective. *J Comp Physiol A.* 2013;199(11):911–928. http://dx.doi.org/10.1007/s00359-013-0818-6.

3. Ache BW, Young JM. Olfaction: diverse species, conserved principles. *Neuron.* 2005;48(3):417–430. http://dx.doi.org/10.1016/j.neuron.2005.10.022.

4. Ihara S, Yoshikawa K, Touhara K. Chemosensory signals and their receptors in the olfactory neural system. *Neuroscience.* 2013;254:45–60. http://dx.doi.org/10.1016/j.neuroscience.2013.08.063.

5. Kaupp UB. Olfactory signalling in vertebrates and insects: differences and commonalities. *Nat Rev Neurosci.* 2010;11(3):188–200. http://dx.doi.org/10.1038/nrn2789.

6. Munger SD, Leinders-Zufall T, Zufall F. Subsystem organization of the mammalian sense of smell. *Annu Rev Physiol.* 2009;71:115–140. http://dx.doi.org/10.1146/annurev.physiol.70.113006.100608.

7. Taniguchi K, Saito S, Taniguchi K. Phylogenic outline of the olfactory system in vertebrates. *J Vet Med Sci.* 2011;73(2):139–147. http://dx.doi.org/10.1292/jvms.10-0316.

8. Missbach C, Dweck HK, Vogel H, et al. Evolution of insect olfactory receptors. *Elife.* 2014;3:e02115. http://dx.doi.org/10.7554/eLife.02115.

9. Galizia CG, Rössler W. Parallel olfactory systems in insects: anatomy and function. *Annu Rev Entomol.* 2010;55(1):399–420. http://dx.doi.org/10.1146/annurev-ento-112408-085442.

10. Shanbhag S, Müller B, Steinbrecht R. Atlas of olfactory organs of *Drosophila melanogaster. Int J Insect Morphol Embryol.* 1999;28(4):377–397. http://dx.doi.org/10.1016/S0020-7322(99)00039-2.

11. Yao CA, Ignell R, Carlson JR. Chemosensory coding by neurons in the coeloconic sensilla of the Drosophila antenna. *J Neurosci.* 2005;25(37):8359–8367. http://dx.doi.org/10.1523/JNEUROSCI.2432-05.2005.

12. Couto A, Alenius M, Dickson BJ. Molecular, anatomical, and functional organization of the Drosophila olfactory system. *Curr Biol.* 2005;15(17):1535–1547. http://dx.doi.org/10.1016/j.cub.2005.07.034.

13. Corey EA, Bobkov Y, Ukhanov K, Ache BW. Ionotropic crustacean olfactory receptors. *PLoS One.* 2013;8(4):e60551. http://dx.doi.org/10.1371/journal.pone.0060551.

14. Peñalva-Arana DC, Lynch M, Robertson HM. The chemoreceptor genes of the waterflea *Daphnia pulex*: many Grs but no Ors. *BMC Evol Biol.* 2009;9:79. http://dx.doi.org/10.1186/1471-2148-9-79.

15. Groh KC, Vogel H, Stensmyr MC, Grosse-Wilde E, Hansson BS. The hermit crab's nose-antennal transcriptomics. *Front Neurosci.* 2013;7:266. http://dx.doi.org/10.3389/fnins.2013.00266.

16. Stensmyr MC, Erland S, Hallberg E, Wallén R, Greenaway P, Hansson BS. Insect-like olfactory adaptations in the terrestrial giant robber crab. *Curr Biol.* 2005;15(2):116–121. http://dx.doi.org/10.1016/j.cub.2004.12.069.

17. Tuchina O, Groh KC, Talarico G, et al. Morphology and histochemistry of the aesthetasc-associated epidermal glands in terrestrial hermit crabs of the genus *Coenobita* (Decapoda: Paguroidea). *PLoS One.* 2014;9(5):e96430. http://dx.doi.org/10.1371/journal.pone.0096430.

18. Gleeson RA, McDowell LM, Aldrich HC. Structure of the aesthetasc (olfactory) sensilla of the blue crab, *Callinectes sapidus*: transformations as a function of salinity. *Cell Tissue Res.* 1996;284(2):279–288. http://dx.doi.org/10.1007/s004410050588.

19. Vosshall LB, Wong AM, Axel R. An olfactory sensory map in the fly brain. *Cell.* 2000;102(2):147–159. http://www.ncbi.nlm.nih.gov/pubmed/10943836. Accessed 25.08.15.

20. Wang F, Nemes A, Mendelsohn M, Axel R. Odorant receptors govern the formation of a precise topographic map. *Cell.* 1998;93(1):47–60. http://dx.doi.org/10.1016/S0092-8674(00)81145-9.

21. Maresh A, Rodriguez Gil D, Whitman MC, Greer CA. Principles of glomerular organization in the human olfactory bulb—implications for odor processing. *PLoS One.* 2008;3(7):e2640. http://dx.doi.org/10.1371/journal.pone.0002640.

22. Malnic B, Hirono J, Sato T, Buck LB. Combinatorial receptor codes for odors. *Cell.* 1999;96(5):713–723. http://www.ncbi.nlm.nih.gov/pubmed/10089886. Accessed 01.02.15.

23. Buck L, Axel R. A novel multigene family may encode odorant receptors: a molecular basis for odor recognition. *Cell.* 1991;65(1):175–187. http://dx.doi.org/10.1016/0092-8674(91)90418-X.

24. Mezler M, Fleischer J, Breer H. Characteristic features and ligand specificity of the two olfactory receptor classes from *Xenopus laevis. J Exp Biol.* 2001;204(17):2987–2997. http://jeb.biologists.org/content/204/17/2987.long. Accessed 10.04.15.

25. Rawson NE, Eberwine J, Dotson R, Jackson J, Ulrich P, Restrepo D. Expression of mRNAs encoding for two different olfactory receptors in a subset of olfactory receptor neurons. *J Neurochem.* 2001;75(1):185–195. http://dx.doi.org/10.1046/j.1471-4159.2000.0750185.x.

26. Omura M, Mombaerts P. Trpc2-expressing sensory neurons in the mouse main olfactory epithelium of type B express the soluble guanylate cyclase Gucy1b2. *Mol Cell Neurosci.* 2015;65:114–124. http://dx.doi.org/10.1016/j.mcn.2015.02.012.

II. OLFACTORY TRANSDUCTION

27. Sengupta P, Chou JH, Bargmann CI. odr-10 encodes a seven transmembrane domain olfactory receptor required for responses to the odorant diacetyl. *Cell.* 1996;84(6):899–909. http://dx.doi.org/10.1016/S0092-8674(00)81068-5.

28. Clyne PJ, Warr CG, Freeman MR, Lessing D, Kim J, Carlson JR. A novel family of divergent seven-transmembrane proteins. *Neuron.* 1999;22(2):327–338. http://dx.doi.org/10.1016/S0896-6273(00)81093-4.

29. Benton R, Vannice KS, Gomez-Diaz C, Vosshall LB. Variant ionotropic glutamate receptors as chemosensory receptors in Drosophila. *Cell.* 2009;136(1):149–162. http://dx.doi.org/10.1016/j.cell.2008.12.001.

30. Wicher D, Schäfer R, Bauernfeind R, et al. Drosophila odorant receptors are both ligand-gated and cyclic-nucleotide-activated cation channels. *Nature.* 2008;452(7190):1007–1011. http://dx.doi.org/10.1038/nature06861.

31. Sato K, Pellegrino M, Nakagawa T, Nakagawa T, Vosshall LB, Touhara K. Insect olfactory receptors are heteromeric ligand-gated ion channels. *Nature.* 2008;452(7190):1002–1006. http://dx.doi.org/10.1038/nature06850.

32. Neuhaus EM, Gisselmann G, Zhang W, Dooley R, Störtkuhl K, Hatt H. Odorant receptor heterodimerization in the olfactory system of *Drosophila melanogaster. Nat Neurosci.* 2005;8(1):15–17. http://dx.doi.org/10.1038/nn1371.

33. Lundin C, Käll L, Kreher SA, et al. Membrane topology of the Drosophila OR83b odorant receptor. *FEBS Lett.* 2007;581(29):5601–5604. http://dx.doi.org/10.1016/j.febslet.2007.11.007.

34. Benton R, Sachse S, Michnick SW, Vosshall LB. Atypical membrane topology and heteromeric function of Drosophila odorant receptors in vivo. *PLoS Biol.* 2006;4(2):e20. http://dx.doi.org/10.1371/journal.pbio.0040020.

35. Zhang H-J, Anderson AR, Trowell SC, Luo A-R, Xiang Z-H, Xia Q-Y. Topological and functional characterization of an insect gustatory receptor. *PLoS One.* 2011;6(8):e24111. http://dx.doi.org/10.1371/journal.pone.0024111.

36. Clyne PJ. Candidate taste receptors in Drosophila. *Science.* 2000;287(5459):1830–1834. http://dx.doi.org/10.1126/science.287.5459.1830.

37. Abuin L, Bargeton B, Ulbrich MH, Isacoff EY, Kellenberger S, Benton R. Functional architecture of olfactory ionotropic glutamate receptors. *Neuron.* 2011;69(1):44–60. http://dx.doi.org/10.1016/j.neuron.2010.11.042.

38. Zhang YV, Ni J, Montell C. The molecular basis for attractive salt-taste coding in Drosophila. *Science.* 2013;340(6138):1334–1338. http://dx.doi.org/10.1126/science.1234133.

39. Getahun MN, Wicher D, Hansson BS, Olsson SB. Temporal response dynamics of Drosophila olfactory sensory neurons depends on receptor type and response polarity. *Front Cell Neurosci.* 2012;6:54. http://dx.doi.org/10.3389/fncel.2012.00054.

40. Wong ST, Trinh K, Hacker B, et al. Disruption of the type III adenylyl cyclase gene leads to peripheral and behavioral anosmia in transgenic mice. *Neuron.* 2000;27(3):487–497. http://dx.doi.org/10.1016/S0896-6273(00)00060-X.

41. Belluscio L, Gold GH, Nemes A, Axel R. Mice deficient in Golf are anosmic. *Neuron.* 1998;20(1):69–81. http://dx.doi.org/10.1016/S0896-6273(00)80435-3.

42. Larsson MC, Domingos AI, Jones WD, Chiappe ME, Amrein H, Vosshall LB. Or83b encodes a broadly expressed odorant receptor essential for Drosophila olfaction. *Neuron.* 2004;43(5):703–714. http://dx.doi.org/10.1016/j.neuron.2004.08.019.

43. Chatterjee A, Roman G, Hardin PE. Go contributes to olfactory reception in *Drosophila melanogaster. BMC Physiol.* 2009;9(1):22. http://dx.doi.org/10.1186/1472-6793-9-22.

44. Kain P, Chandrashekaran S, Rodrigues V, Hasan G. Drosophila mutants in phospholipid signaling have reduced olfactory responses as adults and larvae. *J Neurogenet.* 2009;23(3):303–312. http://dx.doi.org/10.1080/01677060802372494.

45. Deng Y, Zhang W, Farhat K, Oberland S, Gisselmann G, Neuhaus EM. The stimulatory Gα(s) protein is involved in olfactory signal transduction in Drosophila. *PLoS One.* 2011;6(4):e18605. http://dx.doi.org/10.1371/journal.pone.0018605.

46. Yao CA, Carlson JR. Role of G-proteins in odor-sensing and CO_2-sensing neurons in Drosophila. *J Neurosci.* 2010;30(13):4562–4572. http://dx.doi.org/10.1523/JNEUROSCI.6357-09.2010.

47. Smart R, Kiely A, Beale M, et al. Drosophila odorant receptors are novel seven transmembrane domain proteins that can signal independently of heterotrimeric G proteins. *Insect Biochem Mol Biol.* 2008;38(8):770–780. http://dx.doi.org/10.1016/j.ibmb.2008.05.002.

48. Doolin RE, Ache BW. Cyclic nucleotide signaling mediates an odorant-suppressible chloride conductance in lobster olfactory receptor neurons. *Chem Senses.* 2005;30(2):127–135. http://dx.doi.org/10.1093/chemse/bji008.

49. Corey EA, Bobkov Y, Pezier A, Ache BW. Phosphoinositide 3-kinase mediated signaling in lobster olfactory receptor neurons. *J Neurochem.* 2010;113(2):341–350. http://dx.doi.org/10.1111/j.1471-4159.2010.06597.x.

50. Bobkov YV, Pezier A, Corey EA, Ache BW. Phosphatidylinositol 4,5-bisphosphate-dependent regulation of the output in lobster olfactory receptor neurons. *J Exp Biol.* 2010;213(Pt 9):1417–1424. http://dx.doi.org/10.1242/jeb.037234.

51. Zhainazarov AB, Doolin R, Herlihy JD, Ache BW. Odor-stimulated phosphatidylinositol 3-kinase in lobster olfactory receptor cells. *J Neurophysiol.* 2001;85(6):2537–2544. http://www.ncbi.nlm.nih.gov/pubmed/11387399. Accessed 16.04.15.

52. Shirokova E, Schmiedeberg K, Bedner P, et al. Identification of specific ligands for orphan olfactory receptors. G protein-dependent agonism and antagonism of odorants. *J Biol Chem.* 2005;280(12):11807–11815. http://dx.doi.org/10.1074/jbc.M411508200.

53. Ukhanov K, Bobkov Y, Corey EA, Ache BW. Ligand-selective activation of heterologously-expressed mammalian olfactory receptor. *Cell Calcium.* 2014;56(4):245–256. http://dx.doi.org/10.1016/j.ceca.2014.07.012.

54. Kuhlmann K, Tschapek A, Wiese H, et al. The membrane proteome of sensory cilia to the depth of olfactory receptors. *Mol Cell Proteomics.* 2014;13(7):1828–1843. http://dx.doi.org/10.1074/mcp.M113.035378.

55. Mayer U, Ungerer N, Klimmeck D, et al. Proteomic analysis of a membrane preparation from rat olfactory sensory cilia. *Chem Senses.* 2008;33(2):145–162. http://dx.doi.org/10.1093/chemse/bjm073.

56. Mashukova A, Spehr M, Hatt H, Neuhaus EM. Beta-arrestin2-mediated internalization of mammalian odorant receptors. *J Neurosci.* 2006;26(39):9902–9912. http://dx.doi.org/10.1523/JNEUROSCI.2897-06.2006.

57. Ukhanov K, Brunert D, Corey EA, Ache BW. Phosphoinositide 3-kinase-dependent antagonism in mammalian olfactory receptor neurons. *J Neurosci.* 2011;31(1):273–280. http://dx.doi.org/10.1523/JNEUROSCI.3698-10.2011.

58. Bush CF, Jones SV, Lyle AN, Minneman KP, Ressler KJ, Hall RA. Specificity of olfactory receptor interactions with other G protein-coupled receptors. *J Biol Chem.* 2007;282(26):19042–19051. http://dx.doi.org/10.1074/jbc.M610781200.

59. Watt WC, Sakano H, Lee Z-Y, Reusch JE, Trinh K, Storm DR. Odorant stimulation enhances survival of olfactory sensory neurons via MAPK and CREB. *Neuron.* 2004;41(6):955–967. http://dx.doi.org/10.1016/S0896-6273(04)00075-3.

60. Kim SY, Yoo S-J, Ronnett GV, Kim E-K, Moon C. Odorant stimulation promotes survival of rodent olfactory receptor neurons via PI3K/Akt activation and Bcl-2 expression. *Mol Cells.* 2015;38(6):535–539. http://dx.doi.org/10.14348/molcells.2015.0038.

61. Kim SY, Mammen A, Yoo S-J, et al. Phosphoinositide and Erk signaling pathways mediate activity-driven rodent olfactory sensory neuronal survival and stress mitigation. *J Neurochem.* 2015;134(3):486–498. http://dx.doi.org/10.1111/jnc.13131.

62. Moon C, Liu BQ, Kim SY, et al. Leukemia inhibitory factor promotes olfactory sensory neuronal survival via phosphoinositide 3-kinase pathway activation and Bcl-2. *J Neurosci Res.* 2009;87(5):1098–1106. http://dx.doi.org/10.1002/jnr.21919.

63. De Bruyne M, Clyne PJ, Carlson JR. Odor coding in a model olfactory organ: the Drosophila maxillary palp. *J Neurosci.* 1999;19(11):4520–4532. http://www.jneurosci.org/content/19/11/4520.long. Accessed 19.08.15.

64. Michel WC, Ache BW. Odor-evoked inhibition in primary olfactory receptor neurons. *Chem Senses.* 1994;19(1):11–24. http://dx.doi.org/10.1093/chemse/19.1.11.

65. Ache BW. Odorant-specific modes of signaling in mammalian olfaction. *Chem Senses.* 2010;35(7):533–539. http://dx.doi.org/10.1093/chemse/bjq045.

66. Kurahashi T, Lowe G, Gold GH. Suppression of odorant responses by odorants in olfactory receptor cells. *Science.* 1994;265(5168):118–120. http://www.ncbi.nlm.nih.gov/pubmed/8016645. Accessed 13.04.15.

67. Sanhueza M, Schmachtenberg O, Bacigalupo J. Excitation, inhibition, and suppression by odors in isolated toad and rat olfactory receptor neurons. *Am J Physiol Cell Physiol.* 2000;279(1):C31–C39. http://www.ncbi.nlm.nih.gov/pubmed/10898714. Accessed 13.04.15.

68. Duchamp-Viret P, Duchamp A, Chaput MA. Single olfactory sensory neurons simultaneously integrate the components of an odour mixture. *Eur J Neurosci.* 2003;18(10):2690–2696. http://www.ncbi.nlm.nih.gov/pubmed/14656317. Accessed 13.04.15.

69. Takeuchi H, Ishida H, Hikichi S, Kurahashi T. Mechanism of olfactory masking in the sensory cilia. *J Gen Physiol.* 2009;133(6):583–601. http://dx.doi.org/10.1085/jgp.200810085.

70. Sanhueza M, Bacigalupo J. Odor suppression of voltage-gated currents contributes to the odor-induced response in olfactory neurons. *Am J Physiol.* 1999;277(6 Pt 1):C1086—C1099. http://www.ncbi.nlm.nih.gov/pubmed/10600760. Accessed 13.04.15.

71. Chen T-Y, Takeuchi H, Kurahashi T. Odorant inhibition of the olfactory cyclic nucleotide-gated channel with a native molecular assembly. *J Gen Physiol.* 2006;128(3):365—371. http://dx.doi.org/10.1085/jgp.200609577.

72. Chen S, Luetje CW. Trace amines inhibit insect odorant receptor function through antagonism of the co-receptor subunit. *F1000Res.* 2014;3:84. http://dx.doi.org/10.12688/f1000research.3825.1.

73. Oka Y, Nakamura A, Watanabe H, Touhara K. An odorant derivative as an antagonist for an olfactory receptor. *Chem Senses.* 2004;29(9):815—822. http://dx.doi.org/10.1093/chemse/bjh247.

74. Bavan S, Sherman B, Luetje CW, Abaffy T. Discovery of novel ligands for mouse olfactory receptor MOR42-3 using an in silico screening approach and in vitro validation. *PLoS One.* 2014;9(3):e92064. http://dx.doi.org/10.1371/journal.pone.0092064.

75. Pelz D, Roeske T, Syed Z, de Bruyne M, Galizia CG. The molecular receptive range of an olfactory receptor in vivo (*Drosophila melanogaster* Or22a). *J Neurobiol.* 2006;66(14):1544—1563. http://dx.doi.org/10.1002/neu.20333.

76. Dobritsa AA, van der Goes van Naters W, Warr CG, Steinbrecht RA, Carlson JR. Integrating the molecular and cellular basis of odor coding in the Drosophila antenna. *Neuron.* 2003;37(5):827—841. http://www.ncbi.nlm.nih.gov/pubmed/12628173. Accessed 13.04.15.

77. Duchamp-Viret P, Chaput MA, Duchamp A. Odor response properties of rat olfactory receptor neurons. *Science.* 1999;284(5423):2171—2174. http://www.ncbi.nlm.nih.gov/pubmed/10381881. Accessed 13.04.15.

78. Rospars J-P, Lansky P, Chaput M, Duchamp-Viret P. Competitive and noncompetitive odorant interactions in the early neural coding of odorant mixtures. *J Neurosci.* 2008;28(10):2659—2666. http://dx.doi.org/10.1523/JNEUROSCI.4670-07.2008.

79. Ukhanov K, Corey EA, Brunert D, Klasen K, Ache BW. Inhibitory odorant signaling in mammalian olfactory receptor neurons. *J Neurophysiol.* 2010;103(2):1114—1122. http://dx.doi.org/10.1152/jn.00980.2009.

80. Ukhanov K, Corey EA, Ache BW. Phosphoinositide 3-kinase dependent inhibition as a broad basis for opponent coding in mammalian olfactory receptor neurons. *PLoS One.* 2013;8(4):e61553. http://dx.doi.org/10.1371/journal.pone.0061553.

81. Hallem EA, Ho MG, Carlson JR. The molecular basis of odor coding in the Drosophila antenna. *Cell.* 2004;117(7):965—979. http://dx.doi.org/10.1016/j.cell.2004.05.012.

82. Wisler JW, Xiao K, Thomsen ARB, Lefkowitz RJ. Recent developments in biased agonism. *Curr Opin Cell Biol.* 2014;27:18—24. http://dx.doi.org/10.1016/j.ceb.2013.10.008.

83. Kenakin T. Ligand-selective receptor conformations revisited: the promise and the problem. *Trends Pharmacol Sci.* 2003;24(7):346—354. http://dx.doi.org/10.1016/S0165-6147(03)00167-6.

84. Luttrell LM. Minireview: more than just a hammer: ligand "bias" and pharmaceutical discovery. *Mol Endocrinol.* 2014;28(3):281—294. http://dx.doi.org/10.1210/me.2013-1314.

85. Brunert D, Klasen K, Corey EA, Ache BW. PI3Kgamma-dependent signaling in mouse olfactory receptor neurons. *Chem Senses.* 2010;35(4):301—308. http://dx.doi.org/10.1093/chemse/bjq020.

86. Spehr M, Wetzel CH, Hatt H, Ache BW. 3-Phosphoinositides modulate cyclic nucleotide signaling in olfactory receptor neurons. *Neuron.* 2002;33(5):731—739. http://www.ncbi.nlm.nih.gov/pubmed/11879650. Accessed 13.04.15.

87. Kato A, Katada S, Touhara K. Amino acids involved in conformational dynamics and G protein coupling of an odorant receptor: targeting gain-of-function mutation. *J Neurochem.* 2008;107(5):1261—1270. http://dx.doi.org/10.1111/j.1471-4159.2008.05693.x.

88. Zhainazarov AB, Spehr M, Wetzel CH, Hatt H, Ache BW. Modulation of the olfactory CNG channel by Ptdlns(3,4,5)P3. *J Membr Biol.* 2004;201(1):51—57. http://www.ncbi.nlm.nih.gov/pubmed/15635812. Accessed 13.04.15.

89. Brady JD, Rich ED, Martens JR, Karpen JW, Varnum MD, Brown RL. Interplay between PIP3 and calmodulin regulation of olfactory cyclic nucleotide-gated channels. *Proc Natl Acad Sci USA.* 2006;103(42):15635—15640. http://dx.doi.org/10.1073/pnas.0603344103.

90. Raman B, Joseph J, Tang J, Stopfer M. Temporally diverse firing patterns in olfactory receptor neurons underlie spatiotemporal neural codes for odors. *J Neurosci.* 2010;30(6):1994—2006. http://dx.doi.org/10.1523/JNEUROSCI.5639-09.2010.

91. Nagel KI, Wilson RI. Biophysical mechanisms underlying olfactory receptor neuron dynamics. *Nat Neurosci.* 2011;14(2):208—216. http://dx.doi.org/10.1038/nn.2725.

92. Park IM, Bobkov YV, Ache BW, Príncipe JC. Intermittency coding in the primary olfactory system: a neural substrate for olfactory scene analysis. *J Neurosci*. 2014;34(3):941–952. http://dx.doi.org/10.1523/JNEUROSCI.2204-13.2014.

93. Bobkov YV, Ache BW. Intrinsically bursting olfactory receptor neurons. *J Neurophysiol*. 2007;97(2):1052–1057. http://dx.doi.org/10.1152/jn.01111.2006.

94. Sicard G. Electrophysiological recordings from olfactory receptor cells in adult mice. *Brain Res*. 1986;397(2):405–408. http://dx.doi.org/10.1016/0006-8993(86)90648-7.

95. Reisert J, Matthews HR. Responses to prolonged odour stimulation in frog olfactory receptor cells. *J Physiol*. 2001;534(Pt 1):179–191. http://www.pubmedcentral.nih.gov/articlerender.fcgi?artid=2278694&tool=pmcentrez&rendertype=abstract. Accessed 08.04.15.

96. Crawford K, Frey E, Cote T. Angiotensin II receptor recognized by DuP753 regulates two distinct guanine nucleotide-binding protein signaling pathways. *Mol Pharmacol*. 1992;41(1):154–162. http://molpharm.aspetjournals.org/content/41/1/154.abstract. Accessed 10.04.15.

97. Wei H, Ahn S, Shenoy SK, et al. Independent beta-arrestin 2 and G protein-mediated pathways for angiotensin II activation of extracellular signal-regulated kinases 1 and 2. *Proc Natl Acad Sci USA*. 2003;100(19):10782–10787. http://dx.doi.org/10.1073/pnas.1834556100.

98. Kim K-S, Abraham D, Williams B, Violin JD, Mao L, Rockman HA. β-Arrestin-biased AT1R stimulation promotes cell survival during acute cardiac injury. *Am J Physiol Heart Circ Physiol*. 2012;303(8):H1001–H1010. http://dx.doi.org/10.1152/ajpheart.00475.2012.

GUSTATORY TRANSDUCTION

CHAPTER

13

G Protein–Coupled Taste Receptors

Maik Behrens, Wolfgang Meyerhof

Department of Molecular Genetics, German Institute of Human Nutrition
Potsdam-Rehbruecke, Nuthetal, Germany

INTRODUCTION

The sense of taste is important for the survival and well-being of vertebrates because it allows the evaluation of the chemical composition of consumed food.[1] The anatomical structures responsible for the detection of tastants, the taste buds, are located within the oral cavity and are constantly monitoring for the presence of foodborne substances.[2] The taste buds contain the chemosensory cells that are devoted to specifically detect tastants corresponding to the five basic taste qualities: salty, sour, sweet, umami, and bitter.[3] Whereas salty and sour tastes are mediated by ion channels (see Chapter 16), the latter three taste qualities rely on G protein–coupled receptors (GPCRs; see also Chapter 15). Starting with the identification of the first taste GPCRs in 1999,[4] research on taste receptor identification and characterization has been very successful and become increasingly complex. In recent years, this complexity was increased by the discovery that numerous tissues outside the oral cavity express taste GPCRs including, but not limited to, respiratory epithelia, the gastrointestinal tract, thyroid gland, and heart (for reviews of this rapidly expanding area of investigation, see previous work[5–7]). Another level of complexity has been added by the rapidly increasing number of genome projects that have been analyzed for taste receptor sequences. Some of the exciting findings on taste GPCRs across a wide range of vertebrates have been included in this chapter, including a section on the highly variable number of bitter taste receptor genes in different species. The ongoing discussion on the hypothesized existence of additional taste qualities prompted us to include a brief section on GPCRs responsive to fatty acids, the prime stimulus believed to facilitate the orosensory perception of fat components. However, the controversy over whether or not the perception of fatty acids can be considered as a distinct taste quality remains to be resolved.

TYPE 1 TASTE RECEPTORS

Tas1r genes were the first G protein–coupled taste receptors discovered.[4] The type 1 taste receptors (TAS1Rs; a.k.a., T1Rs) belong to the class C family of GPCRs, which possess large extracellular amino termini preceding the seven-transmembrane domain (7-TMD) region. The amino terminus, which schematically resembles the shape of the leaves of the venus flytrap plant and therefore is named venus flytrap domain (VFTD), is connected by an extracellular, cysteine-rich domain (CRD) to the carboxy terminal 7-TMD. In analogy to other receptors from the group C of GPCRs, such as the metabotropic glutamate receptor or γ-aminobutyric acid type B receptors—which function as homo- or heterodimers, respectively[8–12]—the TAS1Rs function as obligatory heterooligomers (Figure 1).

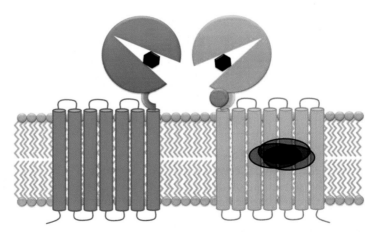

FIGURE 1 **Schematic of the sweet taste receptor heteromer.** The sweet taste receptor–specific subunit TAS1R2 (blue) and the common subunit TAS1R3 (orange) are shown embedded in the lipid bilayer of the plasma membrane (green). Binding sites for mono- and disaccharides within the extracellular venus flytrap domains of the TAS1R2 and TAS1R3 subunits are depicted (black hexagons) as well as the binding site for sweet proteins (red circle) to the cysteine-rich domain of the TAS1R3 monomer. Additional binding sites for artificial activators and the inhibitor lactisole are indicated by dark ellipses.

The mutually exclusive expression in distinct subsets of rodent taste receptor cells detected by either TAS1R1- or TAS1R2-specific in situ hybridization probes suggested previously that both genes would be engaged in the detection of separate taste modalities.[4] However, it took the efforts of several laboratories and additional years to identify the third *Tas1r* gene, *Tas1r3*, which codes for the common subunit of two functional taste receptors.[13–18] Whereas the TAS1R1 and the TAS1R3 subunits combine to form a functional receptor for the detection of amino acid stimuli,[19,20] the TAS1R2 and TAS1R3 subunits constitute the receptor for sweet-tasting substances.[17] Over the years, this beautifully simple way to detect the building blocks of two important macronutrients, proteins and carbohydrates, became more complex than previously anticipated. For example, teleostean fish possess multiple Tas1r-based L-amino acid receptors: the products of multiple *Tas1r2* genes that exist in these species combine with the *Tas1r3* gene product to form receptor proteins responsive to amino acids but not to sweet compounds (see Chapter 5).[21] Conversely, during the evolution of hummingbirds, which like other avian species lack a functional *Tas1r2* gene, the *Tas1r1* gene acquired the crucial sensitivity to sugars and therefore acts de facto as a sweet taste receptor.[22] Some rodent taste receptor cells have been reported to express TAS1R3, but neither the TAS1R1 or TAS1R2 subunits, indicating that a TAS1R3 homodimer can function as a high-affinity receptor for mono- and disaccharides.[23] Other rodent taste receptor cells have been found to coexpress all three Tas1r subunits and to consequently respond to both sweet and umami stimuli.[24] Even though these novel findings challenge the strict functional and cellular segregation of sweet and umami receptors and detection pathways, in the following sections we will describe these two receptor systems in a rather canonical fashion.

SWEET TASTE RECEPTOR

The vertebrate sweet taste receptor is a heterodimer formed by the sweet taste receptor-specific subunit TAS1R2 and the "common" TAS1R3 subunit[17,19] shared with the umami receptor. However, unlike other class C GPCRs, which are highly selective to specific high-affinity ligands, the sweet taste receptor interacts with numerous sweet compounds with comparatively low sensitivity. Indeed, the TAS1R2/TAS1R3 heteromer is activated by all compounds that elicit a sweet taste in humans. The binding sites for several sweet agonists of human and rodent receptors have been mapped to different parts of the receptor. Many of these experiments were guided by the discovery of species-specific activation patterns that enabled the coexpression of heterospecific hybrid sweet taste receptors, the subsequent expression of interspecies chimeras, and site-directed point mutagenesis to pinpoint crucial interaction sites (see, for example Refs 25–30). These and other experiments revealed that both subunits contribute agonist interaction sites and that different subdomains within the subunits engage in the binding of sweeteners. Heterologous expression of the extracellular VFTDs of mouse Tas1r2 and Tas1r3 and subsequent biophysical measurements showed that both subunits interact with natural mono- and disaccharides as well as the artificial sweetener sucralose.[31] In marked contrast, mapping of the binding sites for aspartame, neotame,[30] and monellin[27] demonstrated selective interaction of these compounds with the amino terminal domain of the TAS1R2 subunit, whereas the sweet modifying protein neoculin interacts with the TAS1R3 amino terminal domain.[32] Interestingly, another region where several effector substances were shown to interact with the sweet taste receptor is the TMD of TAS1R3, the common subunit of sweet and umami receptors. Here, the sweeteners cyclamate[26,30] and neohesperidin dihydrochalcone[28] as well as the specific inhibitor of sweetness in humans, lactisole,[25,30] interact to activate, or in case of lactisole to inhibit, the sweet taste receptor. Unsurprisingly, cyclamate and lactisole exert effects on the human umami receptor, which shares the TAS1R3 subunit (see the following section).[30] The sweet protein brazzein activates the sweet taste receptor via interactions with the CRD located between the VFTD and the 7-TMD domains.[27] Hence, even though only a single sweet taste receptor exists to mediate the responses of numerous sweet-tasting molecules, the existence of multiple activation sites within the receptor allows for the selective interaction with all of these structurally diverse compounds (Figure 1).

UMAMI TASTE RECEPTOR(S)

In contrast to sweet taste, the taste of L-amino acids is believed to be mediated by several receptors. When the taste of umami was first proposed to represent the fifth basic taste quality decades ago, a highly specific enhancement of the taste of the prototypical umami stimulus L-glutamate by 5′-ribonucleotides such as inosine monophosphate (IMP) was evident in human and rodents; this enhancement was considered to represent a hallmark of umami taste.[33,34] This qualifies the TAS1R1/TAS1R3 heteromer as the prime receptor for umami taste (Figure 2). Additional receptors for the gustatory sensation of L-glutamate have been identified and contribute to this taste quality (Figure 2). These alternative receptors will be discussed at the end of this section.

The umami-specific subunit of the TAS1R1/TAS1R3 heteromer, TAS1R1, was discovered at the same time with the sweet taste receptor-specific TAS1R2 subunit. The expression of the two genes in separate taste receptor cells indicated that they would be involved in the sensing mechanisms of different taste qualities.[4] The search and discovery of the *Tas1r3* gene at the *sac*-locus (which had previously been linked to differences in taste responsiveness to saccharin[35]) not only allowed the functional verification of the sweet taste receptor (see the previous section), but also enabled characterization of the TAS1R1/TAS1R3 heteromer.[19,20] It turned out that the human umami taste receptor

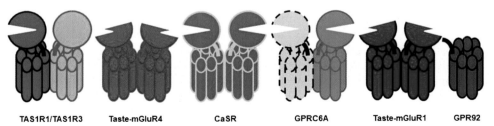

FIGURE 2   **Numerous receptors are implicated in the orosensory detection of amino acids and peptides.** The schematic structure of the umami receptor heteromer (TAS1R1/TAS1R3) as well as the structures of additional receptors responding to L-amino acids or peptides, which were detected in mammalian taste epithelia are shown. CaSR, extracellular-calcium-sensing receptor.

responded to L-glutamate and, less pronounced, to L-aspartate, whereas the rodent counterpart was more widely tuned to recognize an array of L-amino acids. Both the human and the rodent receptor, however, showed a substantial enhancement in the presence of 5'-ribonucleotides in heterologous expression assays, thus exhibiting features anticipated for a bona fide umami taste receptor. The overall architecture of the receptor is closely related to that of the sweet taste receptor: the two subunits TAS1R1 and TAS1R3 consist of amino terminal VFTDs, CRDs as well as the carboxy terminal 7-TMD. The binding of L-glutamate occurs at the umami-specific TAS1R1 subunit within the VFTD.[19,20] The 5'-ribonucleotides such as IMP also interact with the VFTD of TAS1R1.[36] It has been proposed that the agonist L-glutamate first binds to the hinge region of this domain while subsequent binding of IMP stabilizes the active conformation, thus explaining the observed enhancement of receptor responses.[36]

The recent discovery of residual L-glutamate responses in a *Tas1r1* knockout mouse model reignited the idea that, although TAS1R1/TAS1R3 might well be the prime umami taste receptor, other gustatory receptors with selectivity for L-glutamate exist.[24,37–39] Although previous evidence from two independent knockout mouse models has revealed contrasting evidence for TAS1R1/TAS1R3 functioning as the sole umami receptor,[23,37] these new data support the existence of additional routes for oral L-glutamate detection.

One of the proposed alternative receptors is a truncated, taste-specific variant of the metabotropic glutamate receptor type 4 (taste-mGluR4). This receptor represents the first identified candidate receptor for umami substances in rodents.[40] This receptor is expressed in taste receptor cells and displays a lower sensitivity to L-glutamate than does the full-length variant, placing its concentration-response function in a range that is consistent with known taste responses to L-glutamate. Another truncated mGluR, a variant of mGluR1, was identified in the gustatory epithelium of rats.[41] Again, the partial truncation of the extracellular amino terminal domain resulted in a reduced sensitivity to L-glutamate compared with the full-length receptor, shifting its activation spectrum into the taste-relevant range. These two alternative umami receptors fail to exhibit enhanced responses in the presence of 5'-ribonucleotides, distinguishing them from the TAS1R1/TAS1R3 heteromer. Indeed, *Tas1r1* knockout mice lack IMP-synergism of L-glutamate responses, whereas retaining residual but diminished responses to L-glutamate alone. Recently, mGluR4 knockout mice were generated and characterized, implicating an involvement of TAS1R1/TAS1R3, mGluR4, and mGluR1 in the umami tasting abilities of mice.[42]

Three additional GPCRs have been implicated in amino acid and peptide sensing. Two of these are group C GPCRs: the extracellular-calcium-sensing (CaSR) receptor[43] and the GPCR, class C, group 6 subtype A (GPRC6A) receptor.[44] A third, GPR92 (e.g., GPR93 or LPAR5), belongs to the rhodopsin-like GPCR class A.[45] GPR92 responds to protein-hydrolyzates and free L-amino acids[46,47] and was found in stomach enteroendocrine cells of mouse, pig, and human[48] before its presence in mouse TAS1R1-expressing taste receptor cells suggested a gustatory function as well.[49] The GPRC6A responds to several L-amino acids, with basic L-amino acids being the most potent stimuli followed by small aliphatic L-amino acids.[50–52] In addition to Ca^{2+},[53] the CaSR exhibits a broad and diverse ligand spectrum that includes L-amino acids, with aromatic L-amino acids being the most potent of this class of stimuli.[54] The expression of both of these class C GPCRs in mouse taste bud cells suggests an involvement in gustatory L-amino acid sensing.[55]

BITTER TASTE RECEPTORS

The taste receptor 2 gene (*Tas2r* or *TAS2R*) family of bitter taste receptors also belongs to the GPCR superfamily. Their low amino acid sequence homology with other GPCRs does not allow them to be reliably grouped into one of the existing clades of the phylogenetic tree. They share, however, some stretches of homology with frizzled receptors, which prompted Frederiksson and colleagues to group them together with these receptors.[56] However, TAS2Rs share common structural features of GPCRs (e.g., they possess seven TMDs connected by three extracellular and three intracellular loops), their short amino terminus is located at the extracellular side and their carboxy termini resides intracellularly (Figure 3). TAS2Rs are glycoproteins with one highly conserved site for Asn-linked glycosylation centered in the second extracellular loop; additional sites for *N*-glycosylation are present in some TAS2Rs. Glycosylation was confirmed and shown to be important for receptor function in heterologous expression assays.[57] Also analogous to many other GPCRs, human TAS2Rs form homo- and heterooligomers in vitro. However, no functional consequences of these complex formations have been detected to date.[58] Similar to vertebrate odorant receptors (ORs; see Chapter 3) and to the TAS1Rs, the TAS2Rs are difficult to functionally express in heterologous cell systems (see Breakout Box), leading to the speculation that native cells contain cofactors necessary for receptor trafficking and function. Indeed, cofactors that can improve in vitro function of some TAS2Rs have been identified.[59] They belong to the same group of receptor transporting proteins and receptor expression enhancing proteins that were initially found to enhance OR trafficking[60] and, more recently, as auxiliary factors for the sweet taste receptor.[61]

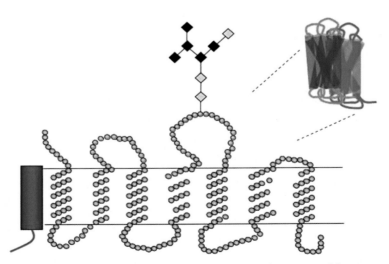

FIGURE 3 **Schematic of bitter taste receptor structure.** The amino acid sequence of the receptor is depicted as a snake plot in the center. The location of a highly conserved N-linked oligosaccharide structure is schematically indicated. Auxiliary proteins implicated in bitter taste receptor biosynthesis and trafficking are indicated as blue cylinder. The ability of type 2 taste receptors to form homo- and heterooligomers is depicted by the insert at the top right of the figure.

BREAKOUT BOX

FUNCTIONAL HETEROLOGOUS EXPRESSION OF TASTE RECEPTORS

An indispensable method for characterizing taste receptors and other chemoreceptors is the functional expression of those receptors in cell lines that allow compound screening and, if successful, the identification of agonist receptor interactions. Although several assay systems have been used successfully for taste receptor deorphanization and characterization, the dominant assay system relies on calcium imaging analyses of recombinantly-expressed receptors. In general, the receptor complementary DNAs are heterologously expressed in nongustatory cell lines such as HEK 293 cells, although immortalized human taste cells equipped with several endogenously expressed taste receptors have been established.[133] Depending on the focus of the study the receptor complementary DNA may be stably and inducibly introduced into the genome of the recipient cell line or transiently transfected (see Ref. 134 for the parallel use of both methods). This allows stringent controls for receptor-dependent calcium signals during screening and characterization procedures. Although mammalian cell lines in general possess endogenously expressed G proteins for the coupling of G protein—coupled receptors, the cell lines used for functional assays are frequently engineered to include exogenous G proteins or G protein chimeras such as $G\alpha_{16gust44}$.[135] These exogenous G proteins improve coupling to the endogenous calcium signaling pathways. After loading the cells with calcium-sensitive dyes, agonist-dependent fluorescence changes can be monitored using single cell calcium imaging setups[73,78] for detailed analyses or fluorescence plate readers for high throughput experiments (see, for example Ref. 72). Although the receptor characteristics such as threshold concentrations or half-maximal effective concentration values do not always match in vivo data from psychophysical experiments, the rank order of agonist potencies and activating concentration ranges usually correlate well with determined in vivo taste data.[78,84] Because such experiments can be combined with other sophisticated molecular biological tools (e.g., the generation of point-mutated receptors or receptor chimeras[27,30,97]) or adapted for the screening of receptor antagonists,[134,136] functional heterologous expression assays have become an indispensable tool for taste research.

The *Tas2r* Gene Repertoires of Vertebrates

TAS2R genes first emerge in euteleostomes.[62] The rapidly growing number of mature genome sequencing projects along with individual cloning efforts for additional organisms have dramatically expanded our knowledge about the sizes of *Tas2r* gene repertoires in vertebrates. The reported numbers of putatively functional *Tas2r* genes range from zero[63,64] to almost 70.[65] To date, *Tas2r* gene repertoires of several ray-finned fish species,[21,66,68] birds,[64,66,67] amphibians,[66] reptiles,[68] prototherians and marsupials,[68] aquatic mammals,[63,69] rodents,[68] and primates[68,72,73] have been assessed (Figure 4). Whether the

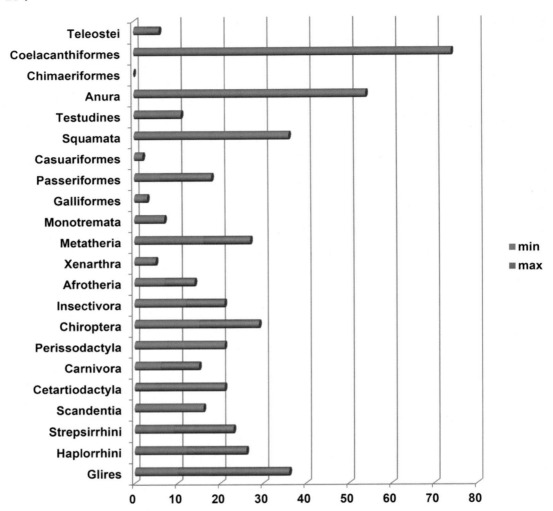

FIGURE 4 **The size of bitter taste receptor gene repertoires is highly variable in vertebrates.** The number of putatively functional bitter taste receptor genes (x-axis) for the so far analyzed taxonomic animal groups (xy-axis) is shown by columns. Blue represents the reported minimal number of *Tas2r* genes; red refers to the maximal number.

number of functional bitter taste receptors is linked to dietary habits such as carnivorous, omnivorous, or herbivorous lifestyles,[63,68] or whether it is specific to, for example, bitter toxin-rich habitats or not tightly connected to challenges by bitter substances because deviations in gene numbers can be partly compensated by adjustments of tuning breadths of TAS2Rs is currently a matter of debate.[66]

Receptive Ranges of Tas2rs

Until recently, the pharmacological characterization of bitter taste receptors was strongly biased toward human TAS2Rs. This has changed somewhat because more data on the agonist profiles of additional vertebrates have become available. Along with the deorphanized 21 human TAS2Rs, at least one bitter taste receptor of 12 vertebrate species has been associated with activating bitter substances (Table 1).

TABLE 1   Currently Deorphaned Vertebrate Bitter Taste Receptors

Species	Tas2r	Agonist	Refs	Species	Tas2r	Agonist	Refs
Human	TAS2R1	Peptides	70	Macaque	Tas2r16	Salicin	71
	TAS2R3	Chloroquine	72	Langur	Tas2r16	Salicin	71
	TAS2R4	Colchicine	72,73	Cat	Tas2r38	PTC	74
	TAS2R5	1,10-Phenanthroline	72		Tas2r43	Aloin	74
	TAS2R7	Papaverine	75	Mouse	Tas2r105	Cycloheximide	73
	TAS2R8	Chloramphenicol	72,76		Tas2r108	PROP	73
	TAS2R9	Pirenzepine	77	Rat	Tas2r105	Cycloheximide	78
	TAS2R10	Strychnine	78	Chicken	Tas2r1	Chloroquine	66
	TAS2R13	Denatonium	72		Tas2r2	Caffeine	66
	TAS2R14	Picrotoxinin	79		Tas2r7	Colchicine	66
	TAS2R16	Salicin	78	Turkey	Tas2r3	Colchicine	66
	TAS2R20	Cromolyn	72		Tas2r4	Caffeine	66
	TAS2R30	Denatonium	80	Zebra finch	Tas2r5	Chloroquine	66
	TAS2R31	Aristolochic acid	81		Tas2r6	Diphenidol	66
	TAS2R38	PROP/PTC	82,83		Tas2r7	Denatonium	66
	TAS2R39	Amarogentin	72	W. Clawed frog	Tas2r5	Coumarin	66
	TAS2R40	Humulones	84		Tas2r9a	Colchicine	66
	TAS2R41	Chloramphenicol	85		Tas2r11	Noscapine	66
	TAS2R43	Aristolochic acid	81		Tas2r20	Yohimbine	66
	TAS2R46	Absinthin	86		Tas2r29	Andrographolide	66
	TAS2R50	Andrographolide	87		Tas2r37	Quassin	66
Chimpanzee	Tas2r16	Salicin	71	Zebrafish	T2R5	Denatonium	21
	Tas2r38	PROP/PTC	88	Medaka fish	T2R1	Denatonium	21

For each deorphaned bitter taste receptor, an exemplary agonist is provided along with a reference.

The functional experiments performed with these additional vertebrate bitter taste receptors have confirmed many findings previously made for human TAS2Rs. The in vitro range of agonist potency for human TAS2Rs ranges from high nanomolar to low millimolar concentrations, with the majority of receptor-agonist interactions occurring in the micromolar range. This is also true for all other vertebrate bitter taste receptors tested so far. Hence, the notion that bitter taste receptors are rather insensitive compared to other GPCRs, such as those that recognize neurotransmitters or hormones, is true for all bitter taste receptors investigated so far. Nevertheless, it is important to note that bitter taste receptors are designed by nature to function as close-proximity sensors to alert organisms to the presence of potentially dangerous food ingredients, a job that they perform efficiently. On the contrary, the activation of TAS2Rs by harmless, minute amounts of bitter compounds such as those contained in most vegetables would limit the availability of food resources appearing safe for consumption and therefore could negatively affect survival. Hence, the concentration ranges at which bitter taste receptors are activated are well-balanced to allow species to maintain a healthy diet but avoid poisoning.

Studies of human TAS2Rs have also revealed how these receptors can recognize an enormous number of structurally diverse bitter-tasting substances. Whereas some TAS2Rs exhibit very broad tuning, allowing them to recognize about one-third of all bitter substances tested in large-scale screenings,[72] other TAS2Rs are activated only by single or very few bitter-tasting compounds. The majority of TAS2Rs exhibit intermediate tuning profiles (Figure 5). Analyses of six of the 54 putatively functional Tas2rs of the frog *Xenopus tropicalis* revealed that both narrowly and broadly tuned receptors also exist in the TAS2R repertoires of other vertebrates. In contrast to the large number of putatively functional *Tas2r* genes in frogs, two bird species, chicken and turkey, possess only three and two functional *Tas2rs*, respectively. Each of these avian TAS2Rs is broadly tuned, suggesting the small number of TAS2Rs is to some degree compensated for by the large number of agonists per receptor. Not all bird species have so few *Tas2rs* (Figure 4), and some (e.g., the white-throated sparrow, which has 18 *Tas2rs*) approach the numbers typically found in mammals. Interestingly, the three zebra finch receptors most closely related to the chicken receptor TAS2R1 exhibit narrower tuning than the chicken ortholog, suggesting that receptor specialization may accompany expanded *Tas2r* repertoires.

Structure—Function Relationships of Bitter Taste Receptors

After the discovery of the astonishing capacity of some human TAS2Rs to interact with so many structurally diverse bitter substances, considerable research efforts have been devoted to identifying the structural determinants for agonist binding of these receptors. Approaches included in silico predictions of receptor structures, functional experiments, or combinations of both methods (for a recent review, see Ref. 89). Receptor structures analyzed so far include TAS2R1,[90–92] TAS2R4,[93] TAS2R10,[94] TAS2R16,[95,96] TAS2R31 (44),[80,97] TAS2R38,[98–100] TAS2R43,[80,97] and TAS2R46.[97] Except for one study that focused on the extracellular loop regions of the two closely related receptors TAS2R31 and TAS2R43,[80] other studies mapped residues important for receptor activation by agonists mainly deeply buried within the transmembrane (TM) helical bundle of the investigated receptors. The critical residues were located in the upper half of TM helices I, II, III, V, VI, and

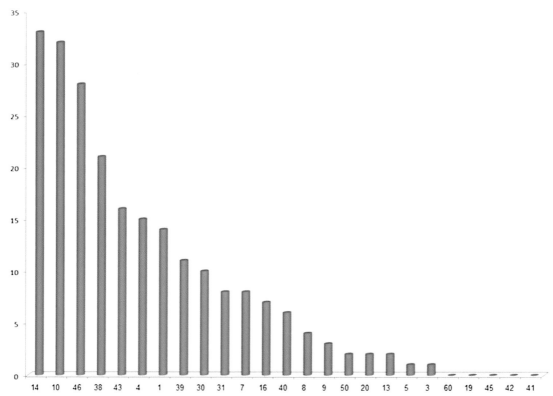

FIGURE 5   Human bitter taste receptors exhibit very different tuning breadths. The number of agonists found by screening 104 bitter substances[72] is indicated on the *y*-axis. The TAS2Rs are labeled by their numbers on the *x*-axis.

VII, although differences are seen between individual receptor isoforms (for a recent review, see Ref. 89). Rather few of the identified positions overlap between the analyzed receptors, indicating considerable flexibility of receptor-agonist contact sites. However, a clear clustering of agonist contact points is evident in TMIII, TMVI, and TMVII.[89] Functional experiments combined with site-directed mutagenesis and in silico modeling and ligand docking experiments revealed that even broadly tuned TAS2Rs possess only a single agonist-binding pocket that is able to establish overlapping but distinct contacts to its various agonists.[97] Moreover, the broadly tuned TAS2R10 is not streamlined to detect individual agonists with the highest possible sensitivity, but rather tailored to interact with many different agonists.[94] Similar experiments were performed with the human TAS2R38, a receptor that is found in two functionally distinct variants, the agonist-responsive "taster" variant and the nonresponsive "nontaster" variant. Interestingly, the amino acid positions associated with the differences in agonist responsiveness are not involved in agonist binding.[99] This suggests that the receptor activation mechanism of the nontaster TAS2R38 variant is impaired and that therefore alternative agonists for the nontaster variant are unlikely to exist.[101]

FATTY ACID RECEPTORS

Recently, numerous studies have demonstrated that both humans and rodents perceive fat constituents within the oral cavity.[102–109] Consequently, there is an ongoing discussion whether a sixth basic taste, "fat taste," may exist (for a recent review on human orosensory fat perception, see Ref. 110). Although some candidate fat taste receptors are GPCRs, others are not. Among the molecules believed to contribute to fat perception in the oral cavity are members of the delayed-rectifying potassium channels, which were the first identified in taste buds of rodents.[111] Another non-GPCR protein that is associated with the orosensory perception of fats is the fatty acid transporter CD36.[107,112] In light of the findings that free fatty acids represent the dominant stimuli for the perception of fats by rodents and humans,[105,113] the GPCRs known to respond to free fatty acids—GPR40, GPR41, GPR43, GPR84, and GPR120—have been implicated in the mechanism of fat perception as well.[102,105,114,115] Functionally, these receptors respond to fatty acids of different chain lengths: two receptors, GPR41 and GPR43, respond to short-chain fatty acids with chain lengths of up to C7.[116,117] The receptor GPR84 is activated by medium-chain fatty acids ranging between C9 and C14.[118] GPR40 is activated by fatty acids of medium to long chain lengths (≥C7),[119] whereas GPR120 exhibits responses when challenged with fatty acids with chain lengths of C8 or longer[105,120] (Figure 6). Of these previously mentioned candidate receptors, the potassium channels, CD36, GPR40, and

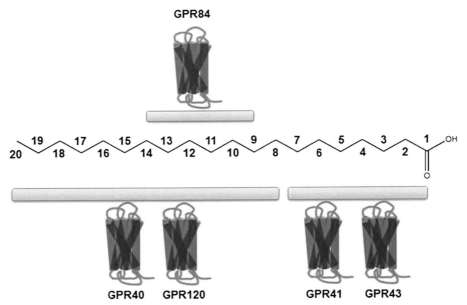

FIGURE 6　**Chain-length specificity of fatty acid responsive G protein–coupled receptors.** For illustration purposes the structure of the saturated C20 fatty acid, arachidic acid (n-eicosanoic acid; Internation Union of Pure and Applied Chemistry name, icosanoic acid), is depicted in the center of the figure. The blue bars indicate the fatty acid chain length experimentally shown to evoke responses for the receptors shown below or above the bars.

GPR120 were each detected in taste epithelia.[102,105,107,111,112,114,115] There is, however, contrasting evidence provided for the presence of GPR40 in tongue epithelium. Whereas GPR40 was detected in taste tissue of mice,[102] its absence from rat and human gustatory epithelium was reported.[105,115] To clarify the contribution of some of the molecules involved in orosensory fatty acid perception, several knockout mouse models have been analyzed. Although CD36 knockout mice repeatedly were shown to exhibit reduced preference for oral lipid stimuli,[107,121,122] contrasting evidence arose for GPR40 and GPR120 knockout models. Whereas one study demonstrated a diminished preference of GPR40 and GPR120 knockout mice for long-chain fatty acids,[102] other studies suggested that activation of the two GPCRs by oral lipid stimuli are not sufficient to drive preference for lipids.[123,124] Future research is needed to clarify the individual roles of these putative fat taste receptors and whether the findings obtained in mice apply to human oral fat perception in an analogous fashion.

OUTLOOK

Even though many of the functional properties of taste receptors have been resolved, a number of questions remain. For example, what is the identity of the taste receptor—specific modulators and of the receptor sites that serve as binding sites for them? Despite this lack of knowledge, taste receptor assays have already been used to discover novel, useful taste compounds. These include new potent sweeteners that can help to reduce calories, bitter blockers that minimize off-tastes of medicines and healthy foods, and umami substitutes to replace glutamate. Of particular interest are positive allosteric activators of sweet and umami receptors that are themselves tasteless but that can enhance agonist-dependent taste receptor activity.[125] Indeed, several promising compounds have been discovered and exploited for nutritional use.[36,126—132] Moreover, taste receptor assays have been used to identify the molecular basis for the organoleptic properties of taste-active molecules present in, or added to, food. These assays will be more frequently used in the future to explain perceptual properties of food constituents on the basis of the biochemical and pharmacological properties of taste receptors.

Acknowledgments

We are grateful for funding by the European Union's Seventh Framework Programme for research, technological development and demonstration (to WM #613879; SynSignal).

References

1. Lindemann B. Taste reception. *Physiol Rev.* 1996;76:718—766.
2. Miller Jr IJ. Anatomy of the peripheral taste system. In: Doty RL, ed. *Handbook of Olfaction and Gustation.* New York: Dekker; 1995.
3. Yarmolinsky DA, Zuker CS, Ryba NJ. Common sense about taste: from mammals to insects. *Cell.* 2009;139: 234—244.
4. Hoon MA, Adler E, Lindemeier J, Battey JF, Ryba NJ, Zuker CS. Putative mammalian taste receptors: a class of taste-specific GPCRs with distinct topographic selectivity. *Cell.* 1999;96:541—551.
5. Behrens M, Prandi S, Meyerhof W. Springer Berlin Heidelberg; 2014:1—34.

6. Li F. Taste perception: from the tongue to the testis. *Mol Hum Reprod.* 2013;19:349−360.

7. Tizzano M, Finger TE. Chemosensors in the nose: guardians of the airways. *Physiology.* 2013;28:51−60.

8. Jones KA, Borowsky B, Tamm JA, et al. GABA(B) receptors function as a heteromeric assembly of the subunits GABA(B)R1 and GABA(B)R2. *Nature.* 1998;396:674−679.

9. Kaupmann K, Malitschek B, Schuler V, et al. GABA(B)-receptor subtypes assemble into functional heteromeric complexes. *Nature.* 1998;396:683−687.

10. Kuner R, Kohr G, Grunewald S, Eisenhardt G, Bach A, Kornau HC. Role of heteromer formation in GABAB receptor function. *Science.* 1999;283:74−77.

11. Kunishima N, Shimada Y, Tsuji Y, et al. Structural basis of glutamate recognition by a dimeric metabotropic glutamate receptor. *Nature.* 2000;407:971−977.

12. White JH, Wise A, Main MJ, et al. Heterodimerization is required for the formation of a functional GABA(B) receptor. *Nature.* 1998;396:679−682.

13. Bachmanov AA, Li X, Reed DR, et al. Positional cloning of the mouse saccharin preference (Sac) locus. *Chem Senses.* 2001;26:925−933.

14. Kitagawa M, Kusakabe Y, Miura H, Ninomiya Y, Hino A. Molecular genetic identification of a candidate receptor gene for sweet taste. *Biochem Biophys Res Commun.* 2001;283:236−242.

15. Max M, Shanker YG, Huang L, et al. Tas1r3, encoding a new candidate taste receptor, is allelic to the sweet responsiveness locus Sac. *Nat Genet.* 2001;28:58−63.

16. Montmayeur JP, Liberles SD, Matsunami H, Buck LB. A candidate taste receptor gene near a sweet taste locus. *Nat Neurosci.* 2001;4:492−498.

17. Nelson G, Hoon MA, Chandrashekar J, Zhang Y, Ryba NJ, Zuker CS. Mammalian sweet taste receptors. *Cell.* 2001;106:381−390.

18. Sainz E, Korley JN, Battey JF, Sullivan SL. Identification of a novel member of the T1R family of putative taste receptors. *J Neurochem.* 2001;77:896−903.

19. Li X, Staszewski L, Xu H, Durick K, Zoller M, Adler E. Human receptors for sweet and umami taste. *Proc Natl Acad Sci USA.* 2002;99:4692−4696.

20. Nelson G, Chandrashekar J, Hoon MA, et al. An amino-acid taste receptor. *Nature.* 2002;416:199−202.

21. Oike H, Nagai T, Furuyama A, et al. Characterization of ligands for fish taste receptors. *J Neurosci.* 2007;27:5584−5592.

22. Baldwin MW, Toda Y, Nakagita T, et al. Sensory biology. Evolution of sweet taste perception in hummingbirds by transformation of the ancestral umami receptor. *Science.* 2014;345:929−933.

23. Zhao GQ, Zhang Y, Hoon MA, et al. The receptors for mammalian sweet and umami taste. *Cell.* 2003;115:255−266.

24. Kusuhara Y, Yoshida R, Ohkuri T, et al. Taste responses in mice lacking taste receptor subunit T1R1. *J Physiol.* 2013;591:1967−1985.

25. Jiang P, Cui M, Zhao B, et al. Lactisole interacts with the transmembrane domains of human T1R3 to inhibit sweet taste. *J Biol Chem.* 2005;280:15238−15246.

26. Jiang P, Cui M, Zhao B, et al. Identification of the cyclamate interaction site within the transmembrane domain of the human sweet taste receptor subunit T1R3. *J Biol Chem.* 2005;280:34296−34305.

27. Jiang P, Ji Q, Liu Z, et al. The cysteine-rich region of T1R3 determines responses to intensely sweet proteins. *J Biol Chem.* 2004;279:45068−45075.

28. Winnig M, Bufe B, Kratochwil NA, Slack JP, Meyerhof W. The binding site for neohesperidin dihydrochalcone at the human sweet taste receptor. *BMC Struct Biol.* 2007;7:66.

29. Winnig M, Bufe B, Meyerhof W. Valine 738 and lysine 735 in the fifth transmembrane domain of rTas1r3 mediate insensitivity towards lactisole of the rat sweet taste receptor. *BMC Neurosci.* 2005;6:22.

30. Xu H, Staszewski L, Tang H, Adler E, Zoller M, Li X. Different functional roles of T1R subunits in the heteromeric taste receptors. *Proc Natl Acad Sci USA.* 2004;101:14258−14263.

31. Nie Y, Vigues S, Hobbs JR, Conn GL, Munger SD. Distinct contributions of T1R2 and T1R3 taste receptor subunits to the detection of sweet stimuli. *Curr Biol.* 2005;15:1948−1952.

32. Koizumi A, Nakajima K, Asakura T, et al. Taste-modifying sweet protein, neoculin, is received at human T1R3 amino terminal domain. *Biochem Biophys Res Commun.* 2007;358:585−589.

33. Lindemann B. A taste for umami. *Nat Neurosci.* 2000;3:99−100.

34. Lindemann B, Ogiwara Y, Ninomiya Y. The discovery of umami. *Chem Senses.* 2002;27:843−844.

35. Lush IE. The genetics of tasting in mice. VI. Saccharin, acesulfame, dulcin and sucrose. *Genet Res*. 1989;53: 95–99.
36. Zhang F, Klebansky B, Fine RM, et al. Molecular mechanism for the umami taste synergism. *Proc Natl Acad Sci USA*. 2008;105:20930–20934.
37. Damak S, Rong M, Yasumatsu K, et al. Detection of sweet and umami taste in the absence of taste receptor T1r3. *Science*. 2003;301:850–853.
38. Delay ER, Hernandez NP, Bromley K, Margolskee RF. Sucrose and monosodium glutamate taste thresholds and discrimination ability of T1R3 knockout mice. *Chem Senses*. 2006;31:351–357.
39. Maruyama Y, Pereira E, Margolskee RF, Chaudhari N, Roper SD. Umami responses in mouse taste cells indicate more than one receptor. *J Neurosci*. 2006;26:2227–2234.
40. Chaudhari N, Landin AM, Roper SD. A metabotropic glutamate receptor variant functions as a taste receptor. *Nat Neurosci*. 2000;3:113–119.
41. San Gabriel A, Uneyama H, Yoshie S, Torii K. Cloning and characterization of a novel mGluR1 variant from vallate papillae that functions as a receptor for L-glutamate stimuli. *Chem Senses*. 2005;30(Suppl. 1):i25–i26.
42. Yasumatsu K, Manabe T, Yoshida R, et al. Involvement of multiple taste receptors in umami taste: analysis of gustatory nerve responses in metabotropic glutamate receptor 4 knockout mice. *J Physiol*. 2015;593: 1021–1034.
43. Brown EM, Gamba G, Riccardi D, et al. Cloning and characterization of an extracellular Ca(2+)-sensing receptor from bovine parathyroid. *Nature*. 1993;366:575–580.
44. Wellendorph P, Brauner-Osborne H. Molecular cloning, expression, and sequence analysis of GPRC6A, a novel family C G-protein-coupled receptor. *Gene*. 2004;335:37–46.
45. Lee DK, Nguyen T, Lynch KR, et al. Discovery and mapping of ten novel G protein-coupled receptor genes. *Gene*. 2001;275:83–91.
46. Choi S, Lee M, Shiu AL, Yo SJ, Aponte GW. Identification of a protein hydrolysate responsive G protein-coupled receptor in enterocytes. *Am J Physiol Gastrointest Liver Physiol*. 2007;292:G98–G112.
47. Choi S, Lee M, Shiu AL, Yo SJ, Hallden G, Aponte GW. GPR93 activation by protein hydrolysate induces CCK transcription and secretion in STC-1 cells. *Am J Physiol Gastrointest Liver Physiol*. 2007;292:G1366–G1375.
48. Haid DC, Jordan-Biegger C, Widmayer P, Breer H. Receptors responsive to protein breakdown products in g-cells and d-cells of mouse, swine and human. *Front Physiol*. 2012;3:65.
49. Haid D, Widmayer P, Voigt A, Chaudhari N, Boehm U, Breer H. Gustatory sensory cells express a receptor responsive to protein breakdown products (GPR92). *Histochem Cell Biol*. 2013;140:137–145.
50. Christiansen B, Hansen KB, Wellendorph P, Brauner-Osborne H. Pharmacological characterization of mouse GPRC6A, an L-alpha-amino-acid receptor modulated by divalent cations. *Br J Pharmacol*. 2007;150:798–807.
51. Wellendorph P, Burhenne N, Christiansen B, Walter B, Schmale H, Brauner-Osborne H. The rat GPRC6A: cloning and characterization. *Gene*. 2007;396:257–267.
52. Wellendorph P, Hansen KB, Balsgaard A, Greenwood JR, Egebjerg J, Brauner-Osborne H. Deorphanization of GPRC6A: a promiscuous L-alpha-amino acid receptor with preference for basic amino acids. *Mol Pharmacol*. 2005;67:589–597.
53. Brown EM, MacLeod RJ. Extracellular calcium sensing and extracellular calcium signaling. *Physiol Rev*. 2001;81:239–297.
54. Conigrave AD, Quinn SJ, Brown EM. L-amino acid sensing by the extracellular Ca^{2+}-sensing receptor. *Proc Natl Acad Sci USA*. 2000;97:4814–4819.
55. Bystrova MF, Romanov RA, Rogachevskaja OA, Churbanov GD, Kolesnikov SS. Functional expression of the extracellular-Ca^{2+}-sensing receptor in mouse taste cells. *J Cell Sci*. 2010;123:972–982.
56. Fredriksson R, Lagerstrom MC, Lundin LG, Schioth HB. The G-protein-coupled receptors in the human genome form five main families. Phylogenetic analysis, paralogon groups, and fingerprints. *Mol Pharmacol*. 2003;63: 1256–1272.
57. Reichling C, Meyerhof W, Behrens M. Functions of human bitter taste receptors depend on N-glycosylation. *J Neurochem*. 2008;106:1138–1148.
58. Kuhn C, Bufe B, Batram C, Meyerhof W. Oligomerization of TAS2R bitter taste receptors. *Chem Senses*. 2010;35:395–406.
59. Behrens M, Bartelt J, Reichling C, Winnig M, Kuhn C, Meyerhof W. Members of RTP and REEP gene families influence functional bitter taste receptor expression. *J Biol Chem*. 2006;281:20650–20659.

60. Saito H, Kubota M, Roberts RW, Chi Q, Matsunami H. RTP family members induce functional expression of mammalian odorant receptors. *Cell*. 2004;119:679–691.

61. Ilegems E, Iwatsuki K, Kokrashvili Z, Benard O, Ninomiya Y, Margolskee RF. REEP2 enhances sweet receptor function by recruitment to lipid rafts. *J Neurosci*. 2010;30:13774–13783.

62. Grus WE, Zhang J. Origin of the genetic components of the vomeronasal system in the common ancestor of all extant vertebrates. *Mol Biol Evol*. 2009;26:407–419.

63. Jiang P, Josue J, Li X, et al. Major taste loss in carnivorous mammals. *Proc Natl Acad Sci USA*. 2012;109: 4956–4961.

64. Zhao H, Li J, Zhang J. Molecular evidence for the loss of three basic tastes in penguins. *Curr Biol*. 2015;25: R141–R142.

65. Syed AS, Korsching SI. Positive Darwinian selection in the singularly large taste receptor gene family of an 'ancient' fish, *Latimeria chalumnae*. *BMC Genomics*. 2014;15:650.

66. Behrens M, Korsching SI, Meyerhof W. Tuning properties of avian and frog bitter taste receptors dynamically fit gene repertoire sizes. *Mol Biol Evol*. 2014;31:3216–3227.

67. Davis JK, Lowman JJ, Thomas PJ, et al. Evolution of a bitter taste receptor gene cluster in a new world sparrow. *Genome Biol Evol*. 2010;2:358–370.

68. Li D, Zhang J. Diet shapes the evolution of the vertebrate bitter taste receptor gene repertoire. *Mol Biol Evol*. 2014;31:303–309.

69. Zhu K, Zhou X, Xu S, et al. The loss of taste genes in cetaceans. *BMC Evol Biol*. 2014;14:218.

70. Maehashi K, Matano M, Wang H, Vo LA, Yamamoto Y, Huang L. Bitter peptides activate hTAS2Rs, the human bitter receptors. *Biochem Biophys Res Commun*. 2008;365:851–855.

71. Imai H, Suzuki N, Ishimaru Y, et al. Functional diversity of bitter taste receptor TAS2R16 in primates. *Biol Lett*. 2012;8:652–656.

72. Meyerhof W, Batram C, Kuhn C, et al. The molecular receptive ranges of human TAS2R bitter taste receptors. *Chem Senses*. 2010;35:157–170.

73. Chandrashekar J, Mueller KL, Hoon MA, et al. T2Rs function as bitter taste receptors. *Cell*. 2000;100:703–711.

74. Sandau MM, Goodman JR, Thomas A, Rucker JB, Rawson NE. A functional comparison of the domestic cat bitter receptors Tas2r38 and Tas2r43 with their human orthologs. *BMC Neurosci*. 2015;16:33.

75. Sainz E, Cavenagh MM, Gutierrez J, Battey JF, Northup JK, Sullivan SL. Functional characterization of human bitter taste receptors. *Biochem J*. 2007;403:537–543.

76. Pronin AN, Xu H, Tang H, Zhang L, Li Q, Li X. Specific alleles of bitter receptor genes influence human sensitivity to the bitterness of aloin and saccharin. *Curr Biol*. 2007;17:1403–1408.

77. Dotson CD, Zhang L, Xu H, et al. Bitter taste receptors influence glucose homeostasis. *PLoS One*. 2008;3:e3974.

78. Bufe B, Hofmann T, Krautwurst D, Raguse JD, Meyerhof W. The human TAS2R16 receptor mediates bitter taste in response to beta-glucopyranosides. *Nat Genet*. 2002;32:397–401.

79. Behrens M, Brockhoff A, Kuhn C, Bufe B, Winnig M, Meyerhof W. The human taste receptor hTAS2R14 responds to a variety of different bitter compounds. *Biochem Biophys Res Commun*. 2004;319:479–485.

80. Pronin AN, Tang H, Connor J, Keung W. Identification of ligands for two human bitter T2R receptors. *Chem Senses*. 2004;29:583–593.

81. Kuhn C, Bufe B, Winnig M, et al. Bitter taste receptors for saccharin and acesulfame K. *J Neurosci*. 2004;24: 10260–10265.

82. Bufe B, Breslin PA, Kuhn C, et al. The molecular basis of individual differences in phenylthiocarbamide and propylthiouracil bitterness perception. *Curr Biol*. 2005;15:322–327.

83. Kim UK, Jorgenson E, Coon H, Leppert M, Risch N, Drayna D. Positional cloning of the human quantitative trait locus underlying taste sensitivity to phenylthiocarbamide. *Science*. 2003;299:1221–1225.

84. Intelmann D, Batram C, Kuhn C, Haseleu G, Meyerhof W, Hofmann T. Three TAS2R bitter taste receptors mediate the psychophysical responses to bitter compounds of hops (*Humulus lupulus* L.) and beer. *Chem Percept*. 2009;2:118–132.

85. Thalmann S, Behrens M, Meyerhof W. Major haplotypes of the human bitter taste receptor TAS2R41 encode functional receptors for chloramphenicol. *Biochem Biophys Res Commun*. 2013;435:267–273.

86. Brockhoff A, Behrens M, Massarotti A, Appendino G, Meyerhof W. Broad tuning of the human bitter taste receptor hTAS2R46 to various sesquiterpene lactones, clerodane and labdane diterpenoids, strychnine, and denatonium. *J Agric Food Chem*. 2007;55:6236–6243.

87. Behrens M, Brockhoff A, Batram C, Kuhn C, Appendino G, Meyerhof W. The human bitter taste receptor hTAS2R50 is activated by the two natural bitter terpenoids andrographolide and amarogentin. *J Agric Food Chem.* 2009;57:9860−9866.

88. Wooding S, Bufe B, Grassi C, et al. Independent evolution of bitter-taste sensitivity in humans and chimpanzees. *Nature.* 2006;440:930−934.

89. Behrens M, Meyerhof W. Bitter taste receptor research comes of age: from characterization to modulation of TAS2Rs. *Semin Cell Dev Biol.* 2013;24:215−221.

90. Dai W, You Z, Zhou H, Zhang J, Hu Y. Structure-function relationships of the human bitter taste receptor hTAS2R1: insights from molecular modeling studies. *J Recept Signal Transduct Res.* 2011;31:229−240.

91. Singh N, Pydi P, Upadhyaya J, Chelikani P. Structural basis of activation of bitter taste receptor T2R1 and comparison with class A GPCRs. *J Biol Chem.* 2011;286:36032−36041.

92. Upadhyaya J, Pydi SP, Singh N, Aluko RE, Chelikani P. Bitter taste receptor T2R1 is activated by dipeptides and tripeptides. *Biochem Biophys Res Commun.* 2010;398:331−335.

93. Pydi SP, Sobotkiewicz T, Billakanti R, Bhullar RP, Loewen MC, Chelikani P. Amino acid derivatives as bitter taste receptor (T2R) blockers. *J Biol Chem.* 2014;289:25054−25066.

94. Born S, Levit A, Niv MY, Meyerhof W, Behrens M. The human bitter taste receptor TAS2R10 is tailored to accommodate numerous diverse ligands. *J Neurosci.* 2013;33:201−213.

95. Sakurai T, Misaka T, Ishiguro M, et al. Characterization of the beta-D-glucopyranoside binding site of the human bitter taste receptor hTAS2R16. *J Biol Chem.* 2010;285:28373−28378.

96. Sakurai T, Misaka T, Ueno Y, et al. The human bitter taste receptor, hTAS2R16, discriminates slight differences in the configuration of disaccharides. *Biochem Biophys Res Commun.* 2010;402:595−601.

97. Brockhoff A, Behrens M, Niv MY, Meyerhof W. Structural requirements of bitter taste receptor activation. *Proc Natl Acad Sci USA.* 2010;107:11110−11115.

98. Biarnes X, Marchiori A, Giorgetti A, et al. Insights into the binding of phenyltiocarbamide (PTC) agonist to its target human TAS2R38 bitter receptor. *PLoS One.* 2010;5.

99. Marchiori A, Capece L, Giorgetti A, et al. Coarse-grained/molecular mechanics of the TAS2R38 bitter taste receptor: experimentally-validated detailed structural prediction of agonist binding. *PLoS One.* 2013;8:e64675.

100. Tan J, Abrol R, Trzaskowski B, Goddard WA. The 3D structure prediction of TAS2R38 bitter receptors bound to agonists phenylthiocarbamide (PTC) and 6-n-propylthiouracil (PROP). *J Chem Inf Model.* 2012;52:1875−1885.

101. Wooding S. Phenylthiocarbamide: a 75-year adventure in genetics and natural selection. *Genetics.* 2006;172: 2015−2023.

102. Cartoni C, Yasumatsu K, Ohkuri T, et al. Taste preference for fatty acids is mediated by GPR40 and GPR120. *J Neurosci.* 2010;30:8376−8382.

103. Chale-Rush A, Burgess JR, Mattes RD. Multiple routes of chemosensitivity to free fatty acids in humans. *Am J Physiol Gastrointest Liver Physiol.* 2007;292:G1206−G1212.

104. Fukuwatari T, Shibata K, Iguchi K, et al. Role of gustation in the recognition of oleate and triolein in anosmic rats. *Physiol Behav.* 2003;78:579−583.

105. Galindo MM, Voigt N, Stein J, et al. G protein-coupled receptors in human fat taste perception. *Chem Senses.* 2012;37:123−139.

106. Hiraoka T, Fukuwatari T, Imaizumi M, Fushiki T. Effects of oral stimulation with fats on the cephalic phase of pancreatic enzyme secretion in esophagostomized rats. *Physiol Behav.* 2003;79:713−717.

107. Laugerette F, Passilly-Degrace P, Patris B, et al. CD36 involvement in orosensory detection of dietary lipids, spontaneous fat preference, and digestive secretions. *J Clin Invest.* 2005;115:3177−3184.

108. Takeda M, Sawano S, Imaizumi M, Fushiki T. Preference for corn oil in olfactory-blocked mice in the conditioned place preference test and the two-bottle choice test. *Life Sci.* 2001;69:847−854.

109. Voigt N, Stein J, Galindo MM, et al. The role of lipolysis in human orosensory fat perception. *J Lipid Res.* 2014;55:870−882.

110. Mattes RD. Accumulating evidence supports a taste component for free fatty acids in humans. *Physiol Behav.* 2011;104:624−631.

111. Gilbertson TA, Fontenot DT, Liu L, Zhang H, Monroe WT. Fatty acid modulation of K+ channels in taste receptor cells: gustatory cues for dietary fat. *Am J Physiol.* 1997;272:C1203−C1210.

112. Fukuwatari T, Kawada T, Tsuruta M, et al. Expression of the putative membrane fatty acid transporter (FAT) in taste buds of the circumvallate papillae in rats. *FEBS Lett.* 1997;414:461−464.

III. GUSTATORY TRANSDUCTION

113. Kawai T, Fushiki T. Importance of lipolysis in oral cavity for orosensory detection of fat. *Am J Physiol Regul Integr Comp Physiol*. 2003;285:R447–R454.

114. Matsumura S, Eguchi A, Mizushige T, et al. Colocalization of GPR120 with phospholipase-Cbeta2 and alpha-gustducin in the taste bud cells in mice. *Neurosci Lett*. 2009;450:186–190.

115. Matsumura S, Mizushige T, Yoneda T, et al. GPR expression in the rat taste bud relating to fatty acid sensing. *Biomed Res*. 2007;28:49–55.

116. Brown AJ, Goldsworthy SM, Barnes AA, et al. The orphan G protein-coupled receptors GPR41 and GPR43 are activated by propionate and other short chain carboxylic acids. *J Biol Chem*. 2003;278:11312–11319.

117. Le Poul E, Loison C, Struyf S, et al. Functional characterization of human receptors for short chain fatty acids and their role in polymorphonuclear cell activation. *J Biol Chem*. 2003;278:25481–25489.

118. Wang J, Wu X, Simonavicius N, Tian H, Ling L. Medium-chain fatty acids as ligands for orphan G protein-coupled receptor GPR84. *J Biol Chem*. 2006;281:34457–34464.

119. Briscoe CP, Tadayyon M, Andrews JL, et al. The orphan G protein-coupled receptor GPR40 is activated by medium and long chain fatty acids. *J Biol Chem*. 2003;278:11303–11311.

120. Hirasawa A, Tsumaya K, Awaji T, et al. Free fatty acids regulate gut incretin glucagon-like peptide-1 secretion through GPR120. *Nat Med*. 2005;11:90–94.

121. Gaillard D, Laugerette F, Darcel N, et al. The gustatory pathway is involved in CD36-mediated orosensory perception of long-chain fatty acids in the mouse. *Faseb J*. 2008;22(5):1458–1468.

122. Sclafani A, Ackroff K, Abumrad NA. CD36 gene deletion reduces fat preference and intake but not post-oral fat conditioning in mice. *Am J Physiol Regul Integr Comp Physiol*. 2007;293:R1823–R1832.

123. Godinot N, Yasumatsu K, Barcos ME, et al. Activation of tongue-expressed GPR40 and GPR120 by non caloric agonists is not sufficient to drive preference in mice. *Neuroscience*. 2013;250:20–30.

124. Sclafani A, Zukerman S, Ackroff K. GPR40 and GPR120 fatty acid sensors are critical for postoral but not oral mediation of fat preferences in the mouse. *Am J Physiol Regul Integr Comp Physiol*. 2013;305:R1490–R1497.

125. Servant G, Tachdjian C, Li X, Karanewsky DS. The sweet taste of true synergy: positive allosteric modulation of the human sweet taste receptor. *Trends Pharmacol Sci*. 2011;32:631–636.

126. Hillmann H, Mattes J, Brockhoff A, Dunkel A, Meyerhof W, Hofmann T. Sensomics analysis of taste compounds in balsamic vinegar and discovery of 5-acetoxymethyl-2-furaldehyde as a novel sweet taste modulator. *J Agric Food Chem*. 2012;60:9974–9990.

127. Ley JP, Blings M, Paetz S, Krammer GE, Bertram HJ. New bitter-masking compounds: hydroxylated benzoic acid amides of aromatic amines as structural analogues of homoeriodictyol. *J Agric Food Chem*. 2006;54: 8574–8579.

128. Ley JP, Dessoy M, Paetz S, et al. Identification of enterodiol as a masker for caffeine bitterness by using a pharmacophore model based on structural analogues of homoeriodictyol. *J Agric Food Chem*. 2012;60:6303–6311.

129. Ley JP, Krammer G, Reinders G, Gatfield IL, Bertram HJ. Evaluation of bitter masking flavanones from herba santa (*Eriodictyon californicum* (H. and A.) Torr., Hydrophyllaceae). *J Agric Food Chem*. 2005;53:6061–6066.

130. Servant G, Tachdjian C, Tang XQ, et al. Positive allosteric modulators of the human sweet taste receptor enhance sweet taste. *Proc Natl Acad Sci USA*. 2010;107:4746–4751.

131. Suess B, Brockhoff A, Degenhardt A, Billmayer S, Meyerhof W, Hofmann T. Human taste and umami receptor responses to chemosensorica generated by Maillard-type N(2)-alkyl- and N(2)-arylthiomethylation of guanosine 5′-monophosphates. *J Agric Food Chem*. 2014;62:11429–11440.

132. Zhang F, Klebansky B, Fine RM, et al. Molecular mechanism of the sweet taste enhancers. *Proc Natl Acad Sci USA*. 2010;107:4752–4757.

133. Hochheimer A, Krohn M, Rudert K, et al. Endogenous gustatory responses and gene expression profile of stably proliferating human taste cells isolated from fungiform papillae. *Chem Senses*. 2014;39:359–377.

134. Slack JP, Brockhoff A, Batram C, et al. Modulation of bitter taste perception by a small molecule hTAS2R antagonist. *Curr Biol*. 2010;20:1104–1109.

135. Ueda T, Ugawa S, Yamamura H, Imaizumi Y, Shimada S. Functional interaction between T2R taste receptors and G-protein alpha subunits expressed in taste receptor cells. *J Neurosci*. 2003;23:7376–7380.

136. Brockhoff A, Behrens M, Roudnitzky N, Appendino G, Avonto C, Meyerhof W. Receptor agonism and antagonism of dietary bitter compounds. *J Neurosci*. 2011;31:14775–14782.

Mechanism of Taste Perception in *Drosophila*

Hubert Amrein

Department of Molecular and Cellular Medicine, College of Medicine,
Texas A&M Health Science Center, Bryan, TX, USA

O U T L I N E

Chemosensory Transduction
http://dx.doi.org/10.1016/B978-0-12-801694-7.00014-7

245

INTRODUCTION

Insects first appeared about 400 million years ago, more than 100 million years after the first vertebrates evolved. Although they cannot challenge vertebrates in total biomass, insect species far outnumber vertebrates, with approximately 900,000 described species. In comparison, bony fish, easily the most speciose vertebrate class, are represented by ~28,000 different species. Speculations are abound as to why insects have been so successful in diversification over a relatively short period of time: these include their ability (1) to adapt more rapidly when facing slow but constant environmental changes and (2) to recover more expeditiously to catastrophic mass extinction events that have impacted the planet several times over the last several hundred million years (see Chapter 2).

Our knowledge of insects derives from the study of a tiny fraction of species, and how insect biology is embedded in the genetic blueprint is known for only a few of them. None of these has been better characterized than the fruit fly, *Drosophila melanogaster*. The fruit fly was first discovered as a potential tool for genetic studies more than 100 years ago by the geneticist Thomas Hunt Morgan, and over only about 15 years, he and his students established the concept of chromosomes and genes as the physical basis of inheritance.[1] This pioneering work led the foundation of today's detailed understanding of genetic networks and the specific roles that genes play in all aspects of development, growth, sex determination, sensory perception, motor function, and behavior in general, etc. Although early anatomical descriptions and neurophysiological studies were performed in many insects, it was the availability of genetic maps and countless gene mutations assembled by Morgan and the many *Drosophila* researchers following in his footsteps that made the fly a mainstay in virtually every single biological subdiscipline. Along with the advent of molecular biology in the 1970s, which was quickly applied to *Drosophila*, the technological improvements in microscopy and innovations in molecular genetics in the last decade have put the fly firmly at the forefront of molecular and cellular neurobiology research, most prominently in visual and chemosensory perception. Indeed, there are comparatively few examples, and essentially none in arthropods, where genetic studies have led to the establishment of direct functional links between genes and chemosensory physiology and behavior, and it is these connections that have catapulted the fruit fly so far ahead of any other animal model system. With recent innovations in molecular biology (e.g., CRISPR), the ability to acquire—ever faster than before—genome information and large-scale expression data, rapid progress in our understanding of sensory perception, including taste, is now conceivable in many other insect species.

It is often overlooked that arthropods, especially insects, have two extremely different life stages associated with drastically different challenges. This review will mainly focus on taste of the adult fly, which has different and more complex feeding challenges compared to the larva. However, taste in the *Drosophila* larvae will be briefly discussed towards the end of this chapter.

ADULT INSECT GUSTATORY SYSTEM

Many insects, such as flies, mosquitoes, moths and butterflies, lack strong mandibles, which are necessary to break down solid foods before ingestion. Instead, they use a tubular feeding organ, the proboscis, to suck up liquid or semiliquid foods such as overripe fruits, nectar, and honeydew. In flies, the anterior tips of the proboscis are formed by two epithelial sheets, the labial palps, which harbor an array of chemosensory cells and represent the fly's main taste organ[2,3] (Figure 1). However, flies—and insects in general—are characterized by a broadly dispersed gustatory system, with taste sensory structures located on legs and wings (Figure 1). This "loose" organization provides insects with the ability to sense chemicals without direct contact by the feeding apparatus, a topic returned to in the section Behavioral Analysis of Gustatory Function.

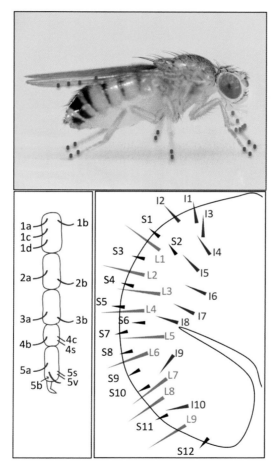

FIGURE 1 **Gustatory organs in the fruit fly, _Drosophila melanogaster._** Taste sensilla are distributed over many appendages. The main organ, the labial palps, and three clusters of internal tastes sensilla (green dots). The fly also has numerous taste sensilla on the last few segments of each leg (tarsi), as well as on the margins of the wings (purple dots). Taste hairs on the legs are thinner and more curved than the more numerous mechanosensory hairs. The labial sensilla are mostly chemosensory. The bottom drawings show all chemosensory bristles on a female foreleg and a labial palp (both lateral view). S=small, I=intermediate, L=large. Note that tarsal sensilla occur as bilaterally symmetrical pairs (lateral and medial), of which only the lateral is shown. (Image of the _Drosophila_ reprinted with permission from Solvin Zankl photography).

III. GUSTATORY TRANSDUCTION

Insect taste cells are primary sensory neurons, and most of these neurons are associated with bristles, also referred to as taste or gustatory sensilla.[2] In *D. melanogaster*, taste sensilla are stereotypically arranged on the surface of the two labial palps and on the tarsi and tibia of each leg (Figure 1). Mechano- and chemosensory sensilla are easily distinguished by shape, the latter being curved and thin, whereas the former being more straight and thick (Figure 1). In some insects such as hymenoptera, the antenna, which in flies is considered an olfactory organ, can also bear taste sensilla. As discussed in the section Gustatory Receptors Beyond Taste, the anatomical separation of olfactory and gustatory input may have to be reassessed in *Drosophila* as well.

Insects Have a Distributed Gustatory System

Each of the labial palps harbors 31 taste sensilla, which are arranged in three main lateral rows. The sensilla are classified into three types, based on length of the bristle as L (long), S (small), and I (intermediate) (Figure 1), whereas about 30, similar-sized taste sensilla are located on each leg. In addition, each wing features about 20 taste sensilla on the anterior margin; thus, a fly harbors about 280 taste sensilla. Last, a small number of taste neurons are organized in taste pegs, located between the pseudotrachae at the medial side of each palp (i.e., the opening to the pharynx), and as three smaller clusters located within the pharynx (Figure 1).

It is the distribution of taste sensilla on appendages not involved in food intake, however, that is the most distinct feature of insect taste systems, compared with that of most other animals, where taste cells are generally located exclusively within the mouth. Availability of "remote" taste sensilla provides the advantage of sampling potential food for undesirable and toxic substances without the prospect of inadvertent ingestion. The function of tarsal taste sensilla is elegantly demonstrated in behavioral assays, whereby their stimulation with food chemicals induces a stereotypic feeding reflex that is easily quantified (see section Behavioral Analysis of Gustatory Function).

Organization of Taste Sensilla

Different taste sensilla vary in composition and number of gustatory receptor neurons (GRNs). L and S sensilla harbor four gustatory neurons, whereas I sensilla contain only two neurons (Figure 2). Initially, taste in *Drosophila* was divided into four modalities, associated with one of the four neurons of S/L sensilla (only two of these modalities, sweet [sugar] and high salt are represented in I sensilla). This association was based on initial electrophysiological studies by Dethier on the blowfly[4–7] using sugar, high salt, low salt and water stimulation, and were later extended to *Drosophila*.[8–10] However, this simple view of insect taste has become more complex over the past several years as a result of more extensive electrophysiological analyses and the identification and functional characterization of *Gustatory receptor (Gr)* genes.[11,12] We now know that *Drosophila* can also sense many chemically complex compounds that induce aversive behavioral responses (i.e., taste bitter to humans)[13–18]; additional nutritious food chemicals, such as amino acids and fatty acids[19–21]; and compounds that lack nutritious value yet enhance feeding (acids).[22,23] Furthermore, the strict association of a chemical class with a specific neuron type no longer applies in the fruit

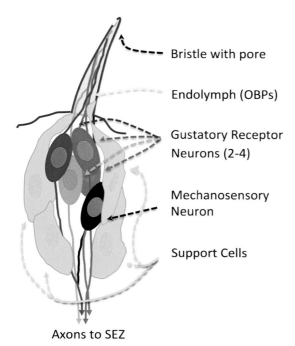

Bristle with pore

Endolymph (OBPs)

Gustatory Receptor
Neurons (2-4)

Mechanosensory
Neuron

Support Cells

Axons to SEZ

FIGURE 2 **Cellular organization of taste sensilla.** L and S sensilla of the labial palp harbor four gustatory receptor neurons (GRNs; red, purple, blue and green), each tuned to different classes of chemicals, and one mechanosensory neuron (black). Each GRN extends a dendrite into the cavity of the bristle filled with endolymph and containing odorant binding proteins (OBP). Each bipolar GRN sends an axon to the subesophageal zone (SEZ) in the brain. The OBPs are expressed in and secreted from the support cells.

fly. This association, which was extended to *Drosophila* based on studies in mammalian systems (specifically mouse, rat, and humans) is commonly known as the labeled line hypothesis. This hypothesis posits that specific taste cells are dedicated to a single taste modality (sweet, umami, sour, salty, or bitter in humans). In flies, it is now well-established that many bitter compounds are sensed by GRNs that also respond to high concentrations of table salt (NaCl).[16–18] Likewise, at least some sweet neurons appear to also mediate appetitive responses to fatty acids, a group of nutritious chemicals structurally unrelated to sugars.[20] Together, these observations suggest that some GRNs are tuned to compounds from distinct chemical groups and may express different types of gustatory receptors (see section Molecular Basis of Different Taste Modalities).

In addition to GRNs, which are the central and most critical components of the taste sensilla, several additional cells are necessary for the functional interaction of chemicals with receptor proteins. Specifically, there are several accessory support cells encapsulating the soma of GRNs (Figure 2).[2] These cells provide the sensillar lymph, a hydrophilic secretion (comparable to the mucosa surrounding the taste pore in mammalian taste buds) that maintains dendritic GRN processes in a protective extracellular environment. Although many taste chemicals are water-soluble and are easily absorbed into the lymph (sugars, most amino acids, salts), others are hydrophobic (many bitter chemicals, fatty acids) and must therefore be solubilized. This is accomplished by the taste stimuli coupling to odorant binding proteins (OBPs). OPBs, which are encoded by a large gene family, were first discovered in the olfactory system where they are expressed in and secreted from support cells present in olfactory sensilla and facilitate solubilization of hydrophobic odorant molecules.[24] Many OBPs are also expressed in support cells of taste sensilla,[25] where they serve a similar function in solubilizing hydrophobic taste chemicals, especially those activating bitter sensing GRNs.[26] Lastly, all taste sensilla harbor a single mechanosensory neuron.[2] Whether this neuron plays any role in evaluating the "texture" of potential food is currently unknown.

Subesophageal Zone

Insect sensory taste cells are primary sensory neurons that directly communicate with the taste processing center in the brain, the subesophageal zone (SEZ). Unlike the antennal lobes, the primary olfactory processing center, which is characterized by a distinct glomerular organization, the SEZ exhibits little, if any, discrete anatomical substructures.[10,27] Indeed, unlike olfactory neurons expressing a given olfactory receptor (see Chapter 6) and converging to a single glomerulus, axon termini of GRNs expressing a specific *Gr* gene show a relative broad spatial distribution in subregions of the SEZ, which are interspersed with termini from other GRNs. However, neurons expressing receptors tuned to either bitter or sweet taste show distinct, complex projection patterns in the SEZ.[16,17] The functional significance of these anatomical features were confirmed by Ca^{2+} imaging studies, showing that bitter chemicals and sugars activate receptor neurons that project their axons to their respective domains in the SEZ.[28] In addition, there is also a marked segregation of termini from GRNs sensing the same chemicals but located in different taste organs. For example, sweet-sensing GRNs located in internal mouthparts project to different regions in the SEZ than sweet GRNs located in the labial palps or legs.[28] This suggests that the fly brain differentiates between stimulations of different taste organs by the same type of ligands.

CELLULAR ANALYSIS OF GUSTATORY FUNCTION

Two basic strategies have been developed to characterize the function of insect taste neurons: electrophysiological recordings of whole-taste sensilla and Ca^{2+} imaging methods of both single cells and collections of taste neuron synapses in the brain.

Tip Recordings

The most established method to analyze taste cell properties is the "whole-sensilla tip recording" method, in which a recording pipette containing the taste solution is placed over the tip of the sensilla of choice.[29] Although this method has been of critical importance in the identification of different receptor neuron types within taste sensilla, it has certain limitations: first, activity of all (up to four) neurons associated with the capped sensillum is registered simultaneously, thus requiring the investigator to sort spikes according to shape and amplitude, a sometimes difficult if not impossible proposition[30]; and second, the stimulus solution cannot be changed during the recording, thereby providing no information about the basic firing rate and recovery dynamics of GRNs. At least the second caveat has been overcome with the tungsten microelectrode recording method, in which a recording electrode is inserted into the base of the sensillum and hence separated from the stimulus pipette.[30] Nevertheless, tip recordings from select sensilla were of immense value, identifying neurons tuned to sugars, water, and aversive ($>$400 mM) and appetitive ($<$100 mM) concentration of salt. These studies were expanded to chemicals perceived as bitter by humans, which were shown to activate the high salt responsive neuron and induce aversive feeding responses (see section Behavioral Analysis of Gustatory Function).

Ca^{2+} Imaging

A more recent development for monitoring taste neuron activity involves live Ca^{2+} imaging, which is either applied to whole flies or to dissected taste organ preparations.[31–33] The basic principle is that a genetically engineered Ca^{2+} sensor, generally green fluorescent—calmodulin—M13 fusion protein (GCaMP)[34] is expressed in a specific taste cell type, usually by means of the GAL4/UAS expression system under the control of a taste receptor gene promoter. GCaMP changes its conformation (and its fluorescence) in a Ca^{2+}dependent manner; thus, when Ca^{2+} increases in the cytosol as a consequence of receptor activation, GRN activity can be directly monitored using fluorescent microscopy. The major advantages of Ca^{2+} imaging versus electrophysiological recordings are the ability to unequivocally associate neural response to a particular cell and to continuously record activity while taste substrates are removed or exchanged. However, temporal resolution in Ca^{2+} imaging is poor, and although it is quantitative, it does not provide information about action potential firing rates, as electrophysiological recordings do.

BEHAVIORAL ANALYSIS OF GUSTATORY FUNCTION

One of the major advantages of the *Drosophila* model system is the availability of powerful behavioral assays that allow investigators to link the function of specific genes to taste behavior. These assays quantify different aspects of taste function, including (1) the appetitive and aversive quality of taste ligands (2), the short-term ingestive activity elicited by specific taste ligands or foods, and (3) the relative preference of flies between two (or more) ligands/foods.

Proboscis Extension Reflex Assay

This elegant behavioral assay was developed more than 50 years ago by the insect physiologist Dethier.[3] Dethier observed that flies, upon stimulation of their tarsi or labial palps, respond with a reflexive extension of their proboscis, a motor behavior that is necessary for food intake (Figure 3(A)). The proboscis extension reflex (PER) index is therefore a quantitative measure of the probability to extend the proboscis upon stimulation (see Figure 3(A)). The tight association of PER and feeding intention is exemplified by the fact that hunger or increasing sugar concentration increases the PER index, whereas the presence of aversive chemicals in a nutritious food reduces it.[16,22,35] Importantly, PER can as easily be applied to evaluate behavior toward aversive chemical stimuli. Behavioral responses to such chemicals are assayed by mixing them into a sugar solution and measuring the reduction of PER relative to the uncontaminated, pure sugar substrate.[16] PER has been very informative in the study of taste receptor genes, especially those encoding sugar receptors[31,36] (see section

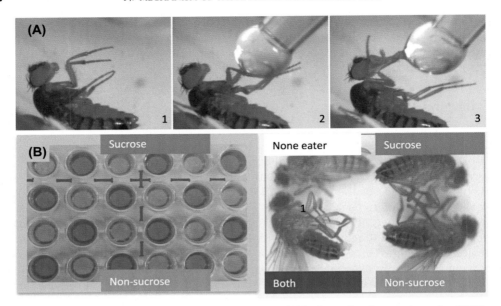

FIGURE 3   **Behavioral assays for measuring taste responses.** (A) In the proboscis extension reflex (PER) assay, a hungry fly is glued on a slide by its back (nonfeeding state) (1). When the tarsi are stimulated with a nutritious, sugar-containing solution (2), the fly responds reflexively by extending its proboscis (3), which is necessary to suck up the solution. (B) In the two-choice feeding assay, the wells of a multiwell plate are filled with agar containing two different feeding substrates (also containing two distinct food dyes) in an alternate pattern. After a 1-h feeding period, the flies' abdomens are inspected to determine food preference/intake. For example, if the blue-labeled agar contains 50 mM sucrose, whereas the pink-labeled agar contains only 2 mM sucrose, the majority (close to 90%) of hungry flies will typically feed only from the more concentrated sugar and have therefore a blue abdomen. A preference index (PI) can simply be calculated with the formula PI = N[blue] + 0.5[purple]/N[blue] + N[pink] + N[purple], N reflecting the number of flies feeding only from 50 mM sucrose (blue), 2 mM sucrose (red), or both concentrations (purple) of sugar containing wells. *Reprinted with permission from Ref. 38.*

Molecular Basis of Different Taste Modalities). An often overlooked, but distinct advantage of PER compared with other behavioral assays, especially the two-choice feeding assay, is that the taste response can be entirely uncoupled from ingestion, and hence, secondary (i.e., post-prandial) effects can be ruled out as contributing to the observed behavior.

Two-Choice Feeding Assay

This high-throughput behavioral assay allows evaluation of a relatively large number of flies. Flies are given the choice to feed on two different substrates (sugar 1 vs sugar 2, sugar vs no sugar, etc.) for a short period and their preference is quantified with the help of food dyes (red and blue) added to either of the two substrates (Figure 3(B)).[37,38] The number of flies feeding from one versus the other substrate is then determined semiquantitatively by visually scoring the intestine based on the abdomen color (Figure 3(B)), or quantitatively by using colorimetric analysis of gut extracts. Like PER, the two choice preference assay can also be used to address the aversive effects of undesirable (i.e., bitter) compounds, usually in the context of a single sugar provided at two different concentrations. For example, although wild-type flies will feed preferentially on agar containing 5 mM sucrose as opposed to 1 mM sucrose, the addition of an aversive bitter compound to the 5 mM sucrose-agar will reverse the preference toward the 1 mM sucrose-agar.[17] The two-choice assay has been applied to a range of mutant strains, allowing researchers to link specific Gr proteins to bitter compounds.[14,15,39] In contrast to PER, postingestive effects can contribute to the feeding preference in the two choice assay, especially if feeding times exceed 15 min. These effects may be mediated through nutrient sensing in the gut or brain and the resulting release of various neuropeptides that modulate feeding circuits and thereby alter feeding activity during the assay.

Capillary Feeding Assay

Although the two-choice feeding assay allows precise evaluation of food preference between two or more substrates, it does not provide a quantitative measure for how much of a given nutrient has been ingested. Such information is relevant when investigating taste preference or when studying feeding activity in the context of energy homeostasis and expenditure. The preferred assay for such cases is the capillary feeding (CAFE) assay, where single or a small number of flies are provided with a nutrient solution of defined concentration.[40] The amount of food consumed over time is recorded using graded feeding capillaries and can be compared between flies of different genotypes or flies raised under different dietary conditions. CAFE assays are performed over many hours or even days. Therefore, they primarily evaluate feeding preference, which is heavily influenced by postingestive mechanisms, rather than taste preference.[33]

Oviposition Assay

This assay is a modification of the two-choice feeding assay. Rather than measuring the food choice between two different substrates, it is designed to evaluate females' preference for suitable egg-laying substrates.[41] Because hatched larvae are confined to the food substrate on which they were deposited as eggs, this assay evaluates the food choice of females made for their offspring. Not surprisingly, females will lay more eggs on agar containing sugar than on agar alone. Interestingly however, food substrates preferred by adult flies and substrates chosen by females for their offspring can vary. For example, adult flies will feed preferentially on acetic acid-free sugar, as opposed to sugar with acetic acid, whereas females prefer to lay eggs on acid-containing sugar.[41] This suggests that acid provides a benefit to growing larvae. One possibility is that low pH prevents excessive growth of microorganisms on natural substrates (fruits),[22,41] which are rich in carboxylic acids.[42]

MOLECULAR BASIS OF DIFFERENT TASTE MODALITIES

The identification of the Gr gene family, reported by three different laboratories about 15 years ago, represented a major breakthrough in the study of insect taste behavior and physiology.[43−45] The fruit fly genome harbors 68 Gr genes,[46] a few of which are highly conserved across arthropods.[47–52] However, *Gr* genes have evolved rapidly and expanded at vastly different rates; in some species, like the honeybee, only a handful of *Gr* genes are found,[52] whereas in others, such as mosquitoes, almost 100 appear to encode functional taste receptors.[51] With the exception of *Drosophila*, however, little is known about their expression and specific functions in any arthropod. In the fly, the use of the GAL4/UAS system led to detailed expression maps for almost all *Gr* genes in both labial palp and tarsal taste sensilla.[13,36,43,44,53] Based on these studies, the *Gr* genes can be subdivided into two major groups. Expression of the majority of *Gr* genes (~35) is confined to the bitter/high-salt neurons,[13] whereas a smaller group, composed of only eight genes, is collectively expressed in the sweet neurons present in each taste sensillum.[36] Several *Gr* genes, however, fall in neither group and appear to have acquired different or additional chemosensory roles (see section Gustatory Receptors Beyond Taste).

BREAKOUT BOX

MOLECULAR DIVERSITY OF *DROSOPHILA* TASTE RECEPTORS

Appropriate food choices are critical for the acquisition of nutrients for growth, energy expenditure and reproduction. *Drosophila* has served as an important model system to dissect the molecular basis of many sensory processes, including taste perception. The main food sources of *Drosophila*—fruits and yeast—contain mixtures of macronutrients (carbohydrates, amino acids, fats), essential micronutrients (vitamins, cholesterol), salts, and nonnutritious, but mostly harmless compounds such as carboxylic acids. However, especially ripe fruit is often colonized by microorganisms, which produce a range of metabolites (alkaloids, terpenoids, and phenols), many of which are toxic and thereby pose a health hazard for animals feeding on them. Thus, a central line of investigation in taste research is concerned with the identification of receptors and sensory cells that are involved in the detection of various kinds of nutritious taste chemicals, as well as nonnutritious and possibly toxic compounds.

In the past 15 years, the focus of *Drosophila* taste research has heavily centered on the role of the Gustatory receptor (Gr) protein family, and the different taste cells that express respective genes. It is now well-established that Gr proteins fall into two main categories, receptors for sugars and receptors for bitter chemicals, and that functional sugar and bitter receptors are heteromeric complexes of two and possibly up to four different Gr proteins. However, it has become evident that numerous other receptors and channels play critical roles in the detection of food chemicals, including many members encoded by a second large gene family, the *Ionotropic receptor* (*Ir*) genes, and a select genes of the *Transient Receptor Potential* (*TRP*) and *pickpocket* (*ppt*) gene families. One of the main future avenues of taste research will include the identification of specific members of the *Ir* gene family that might mediate the taste of amino acids/proteins, carboxylic acids and fatty acids, all compounds that are abundantly present in *Drosophila* foods. Finally, important functions have been established for numerous taste receptor genes in internal nutrient sensing, temperature sensing and light sensing, broadly expanding their roles into other sensory processes.

Sugar Taste

The sweet taste of sugars is the best-studied *Drosophila* taste modality. In contrast to mammals, in which a single heterodimeric G protein–coupled receptor (GPCR) complex mediates the sweet taste of all sugars and even sweet-tasting proteins[54,55] (see Chapter 13), there are eight Gr proteins involved in sweet taste in the fly.[36] These proteins are encoded by a conserved *Gr* subfamily composed of *Gr5a, Gr61a,* and a cluster of six genes, *Gr64a–Gr64f.* The first mutation with a specific taste phenotype, severely reducing behavioral responses to the sugar trehalose, was described long before the *Gr* genes were identified.[37] It was later shown that the mutation in this gene mapped to *Gr5a*, which was found to be expressed in all sweet GRNs[56,57] as well as many additional taste neurons.[36] Subsequent genetic and electrophysiological studies implicated at least some of the other *Gr* genes in sweet taste,[35,58–60] albeit their specific roles remained unclear because of a lack of cellular expression data and availability of single gene mutations. These deficits were comprehensively addressed with a gene replacement strategy, in which individual coding sequences of six sugar *Gr* genes were substituted with *LEXA* or *GAL4* (Figure 4(A)). These studies not only provided single gene mutations, but also revealed the precise expression pattern of each of these genes.[36] Specifically, it was found that all sugar *Gr* genes are expressed in tarsi and all but *Gr64a* are expressed in the labial palps. However, with the exception of *Gr64f*, which appears to be expressed in the sweet GRN of each taste sensillum, the remaining sugar *Gr* genes are expressed in a smaller subset of sweet GRNs. Furthermore, no two *Gr* genes are completely coexpressed (Figure 4(C)). Consequently, the sweet GRNs can be subdivided into at least eight different expression subclasses (and probably more, as expression of *Gr64d* has not yet been determined), likely providing each of them with different sugar response profiles.

The GAL4/LEX alleles are also null alleles. Hence, they allowed, for the first time, a systematic assessment of both behavioral and cellular responses of flies lacking a single sugar *Gr* gene.[36] Surprisingly, the behavioral effects of single sugar *Gr* mutations are modest and generally restricted to a small fraction of sugars (Table 1).[36] The strongest phenotypes are observed in *Gr5a^LEXA* and *Gr64f^LEXA* mutant flies, in which PER responses to about half of the sugars tested were affected; however, none of these mutations led to a more than 50% reduction in PER to any sugar, arguing for redundancy between different sugar taste receptors.[36] Another genetically complex fly strain has provided additional, important insights into the molecular nature of sugar receptors. In this strain, mutations of all eight sugar *Gr* genes were combined to generate octuple mutant flies that are sugar-blind[61] (these flies retain weak tarsal fructose responses due to expression of the Gr43a nutrient sensor in a pair of sweet GRNs; see section Gustatory Receptors Beyond Taste). Importantly, expression of any of the eight sugar *Gr* genes failed to restore either cellular or behavioral responses to any sugar, whereas expression of pairwise combinations restored responses to one or a few select sugars[61] (Table 1). These observations extend previous studies and indicate that functional sugar receptors are heteromeric complexes, consisting of two (or for some sugars possibly more) Gr subunits (Figure 5). Interestingly, the phylogenetically related Odorant receptors (Ors) (see Chapter 6) also form heteromeric complexes, composed of the broadly expressed Olfactory receptor coreceptors and a ligand binding, cell type–specific Or subunit.[62] However, it appears that some Gr proteins, such as Gr43a, are capable of functioning on their own, perhaps as homomultimers[33] (see section Gustatory Receptors Beyond Taste).

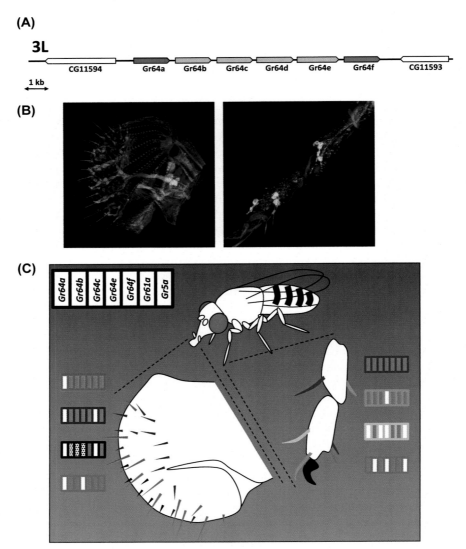

FIGURE 4   **Expression code for sugar receptors.** (A) Six of the eight classical sugar receptors are tightly clustered in the *Gr64* locus, and their expression and function were determined by precise LEXA/GAL4 replacements using homologous recombination. (B) The images show coexpression of the first and the last gene in the *Gr64* locus. Gr64a (red) is entirely absent in the labial palp taste sensilla, whereas Gr64f (green) is expressed in the sweet GRN of each sensillum. In contrast, both genes are coexpressed in all sweet GRNs of the fourth and fifth segments of the forelegs (shown by the dominantly yellow fluorescent color from merging the green and red channel). (C) The sweet GRNs in labial and tarsal taste sensilla fall into four distinct groups, based on expressed sugar *Gr* genes. The most complete expression profiles are found in the sweet neurons associated with seven of the 31 labial bristles (pink; six sugar *Gr* genes) and six tarsal bristles in the last two segments (purple, seven sugar *Gr* genes; orange, six sugar *Gr* genes). The sweet GRNs associated with the remaining bristles express between three and five sugar *Gr* genes. Gray, expression; white, lack of expression; dots, stochastic expression. Note that only the lateral bristle of each pair is shown. *Reprinted with permission from Ref. 36.*

TABLE 1   Taste Receptor Genes with Known Function, Based on Gene Knockouts Only, in Taste Perception and Other Chemosensory Processes

Gene	Function	Ligand	Partner	Comments	References
Gr5a	Sugar R	Tre	Gr64f		36,56—58
Gr8a	Bitter R	L-cav			62
Gr21a	Olfactory R	CO_2	Gr63a	Expressed in antenna	105,106
Gr32a	Pheromone R R, bitter R	(z)-7-Tricosene Caf, DEET, Pap, Str, Lob			14,65,66
Gr33a	Bitter R Pheromone R	(z)-7-Tricosene Caf, DEET, Pap, Stry, Lob		Core sub U	64
Gr43a	Sugar R	Fru, Suc	None	Brain nutrient sensor	33
Gr61a	Sugar R	Gluc			31
Gr64a	Sugar R	Mal, Fru	Gr64e		36,61
Gr64b	Sugar R	Glyc	Gr64e		36,61
Gr64c	Sugar R	Suc, Mal, Ara			36,61
Gr64e	Sugar R	Glyc	Gr64a Gr64b		36,61
Gr64f	Sugar R	Suc, Mal, Fru, Mel, Glu, tre	Gr5a		36,58,61
Gr66a	Bitter R	Caf, DEET, Pap, Stry, Lob		Core sub U	39,64
Gr93a	Bitter R	Caf			15
Ir76b	Salt C	NaCl			79
Ppk28	Water C	Water	none	Deg/ENaC	75
TrpA1	Irritant R	A-iso, Ari			69—71
Pain	Irritant R	A-iso			68
DmRX	Bitter R	L-cav	none	GPCR	86

A-iso, allyl isothiocyanate; Ari, aristolochoic acid; Caf, caffeine; L-cav, L-canavanine; Fru, fructose; Glu, glucose; Glyc, glycerol; pap, papaverine; Lob, lobeline; Mal, maltose; Mel, melezitose; Stry, strychnine; Suc, sucrose; tre, trehalose.

Bitter Taste

Similar to mammals, *Drosophila* and other insects are highly attuned to detecting potentially harmful chemicals, which taste bitter to humans and lack nutritional value. Insects are exposed to many bitter chemicals through a variety of sources. For example, many plants produce them in the form of secondary metabolites that are used as defense against herbivore insects. For *Drosophila* and other frugivores, sources of such chemicals derive often from microorganisms that colonize decaying fruit. Bitter chemicals include a large range of structurally diverse compounds such as alkaloids, terpenoids and phenols. Not surprisingly, most taste receptors ($\sim$35) are expressed in the bitter/high-salt—responsive GRNs and are thought

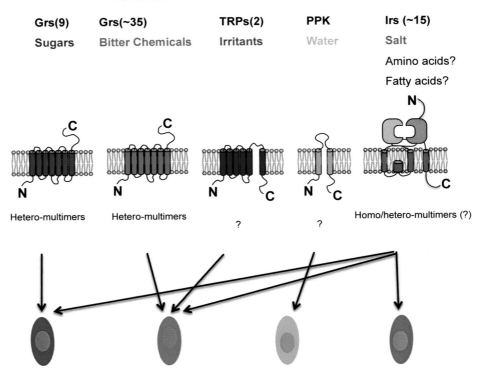

FIGURE 5   **Structure of different taste receptor types in adult flies.** At least four different types of receptor/ channels are found in taste neurons. More than 40 of the 68 *Gr* genes encode bitter and sugar receptors and are expressed almost exclusively in the bitter/high-salt and sweet GRNs, respectively. Two *TRP* genes were shown to encode receptors for chemical irritants (aristolochic acid and allyl isothiocyanate) and are also expressed in aversive bitter/high salt neurons. At least one pickpocket channel protein (PPK28) is necessary for water taste and expressed in the GRNs specifically that is activated by water/low osmolarity solutions. The role of the *Ir* genes is largely unknown, but at last one is involved in low salt taste. However, 15 *Ir* genes are expressed in taste neurons, including some in the bitter and sweet GRN. Not shown are two PPK proteins (PPK11 and PPK19) suggested to play a role in larval salt taste, and the single representative of GPCRs, DmRX.

to be activated by this wide range of bitter compounds.[13] However, only about four of the putative bitter *Gr* genes have been functionally characterized (Table 1). Null mutations in these genes revealed that each of them is necessary for normal cellular responses to various bitter chemicals.[14,15,39,63] For example, Gr93a and Gr8a are necessary for the detection of select bitter compounds, such as caffeine and L-canavanine (a plant insecticide), respectively, whereas Gr33a and Gr66a are required for sensing a broader range of chemicals, including caffeine, papaverine, and the insect pesticide DEET (Table 1).[14,15,39,63,64] Because single bitter *Gr* genes cannot convey receptor-specific responses when expressed in sweet neurons, it appears likely that bitter receptors are also composed of multiple Gr subunits, and that Gr33a and/or Gr66a are core subunits of such multimeric receptors (Figure 5). We note that some bitter *Gr* genes (*Gr32a* and *Gr33a*) are also implicated in the perception of the male inhibitory pheromone (z)-7-tricosene , which suppresses male courtship toward other males[65,14] and mediates male—male aggressive behavior.[66] Thus, it appears that activation of bitter taste

GRNs can lead to differential behavioral responses, depending on the context in which a stimulus is received.

In addition to *Gr* genes, two *transient receptor potential (TRP)* channel genes have been implicated in repulsive behavior and feeding avoidance. *TRP* channels are well known for their role in temperature sensing, mechanoreception, and chemonociception (see Chapter 21).[67] *Drosophila TRP* genes are expressed in diverse sets of neurons, and a few were found in subsets of bitter GRNs. Specifically, *TRPA1* and *painless (PAIN)* were shown to mediate repulsion to aristolochoic acid and allyl isothiocyanate (i.e., the noxious heat sensation of wasabi).[68–71] Of note is that TRP channels act as bimodal signal detectors: they can function as receptor-operated excitatory channels (activated by second messengers) or be activated by binding to specific ligands (i.e., as ionotropic receptors). In the case of TRPA1, the binding domain for allyl isothiocyanate resides in the N-terminal tail, which is located in the cytoplasm.[72] Thus, in contrast to ligands for Gr chemoreceptors, ligands for some TRP channels must enter the taste cells to activate it.

Water Taste

Detection and consumption of water is critical for all animals in maintaining osmotic homeostasis. Interestingly, some insects, including *Drosophila*, have evolved the ability to detect water through their taste sensory system, which was recognized in early electrophysiological studies.[9,73,74] These investigations revealed that one of the GRNs in most taste sensilla fires bursts of action potentials upon stimulation with pure water, or water containing low concentration of various solutes. A candidate "receptor" for water taste was identified by Cameron and colleagues,[75] who used a differential molecular screening strategy to compare transcript abundance between labial palps of wild-type (with intact number of taste neurons) and *poxN* mutant flies (lacking taste neurons, but having additional mechanosensory neurons).[76,77] In addition to many enriched *Gr* genes, they identified a number of transmembrane proteins belonging to the *pickpocket* family (related to the mammalian degenerin/epithelial sodium channel family (Deg/ENaC)).[78] Cellular expression analysis and Ca^{2+} imaging showed that one of these candidates, *pickpocket 28 (ppk28)*, is expressed in water neurons of the labial palps, and that *ppk28* mutations abolished water responses in these neurons, directly associating this channel protein with water sensing.[75]

Salt Taste

Just like mammals, *Drosophila* exhibits differential responses to NaCl; at concentrations >200 mM, flies exhibit aversive behavioral responses, whereas at a low concentration (<100 mM), NaCl elicits appetitive behavior[6,7] (see Chapter 16). These observations correlate with electrophysiological studies, which identified a high (maximum firing rate at [NaCl] > 500 mM) and a low (maximum firing rate ~50 mM) salt responsive GRN.[2,9,18,74] A candidate gene for low salt taste in adult flies was recently identified.[79] Using whole sensilla recordings, the authors of this study showed that *Ir76b* mutant flies had severely reduced firing rates when stimulated with a low (50 mM) concentration of NaCl, and lost feeding preference for this attractant stimulus.[79] Of note, many Ir proteins, which are generally believed to function as tetrameric, ligand-gated ion channels, are expressed in gustatory neurons (see below), which are generally believed to function as tetrameric, ligand-gated ion channels. Curiously, *Ir76b* is also expressed in many olfactory neurons, where its function is required, along with other *Ir* genes, for the detection of numerous odorant molecules.[80]

Amino Acids, Fatty Acids, Carboxylic Acids

Several observations indicate that the Gr proteins do not mediate all major taste modalities. Specifically, no link has been established between any specific *Gr* gene and the detection of nutritious chemicals other than sugars, including amino acids, carboxylic acids, and fatty acids. Although the cellular bases for amino and carboxylic acid sensing has yet to be established, the taste of fatty acids is mediated by sweet GRNs.[20] Even so, flies lacking any of the known sugar *Gr* genes exhibit normal responses to these chemicals at both the cellular and behavioral level.[81] Because no other *Gr* genes have been reported to be expressed in sweet GRNs, it seems likely that other receptors sense fatty acids, amino acids, and carboxylic acids.

Candidate receptors for some of these chemicals are members of the Ir protein family (Figure 5) (see Chapter 6). Ir proteins, which are related to ionotropic glutamate receptors, are encoded by a large gene family that rapidly expanded in the arthropod linage.[82] In *Drosophila*, 66 *Ir* genes were identified, one-quarter of which is expressed in olfactory neurons of trichoid sensilla.[12] Molecular genetic studies and electrophysiological recordings have linked some Ir proteins to the detection of amines, volatile carboxylic acids, and foul-smelling putrescine and cadaverine, breakdown products of dead animal matter.[80,83,84]

Intriguingly, a subset of about 15 *Ir* genes is expressed in GRNs in the labial palps and tarsi.[85] This strongly suggests that Ir-based taste receptors complement the Gr-based taste receptors in sensing food compounds and possibly aversive chemicals. Although a potential role of *Ir76b* has been proposed in salt taste,[79] recent expression studies indicate that *Ir76b* expression is found in as many as three GRNs per sensillum, including some bitter and sweet GRNs[81] (Figure 6). This observation indicates that Ir76b has a role in several different taste modalities, perhaps as a generic subunit of multimeric, Ir-based taste receptors. Regardless, detailed expression and molecular genetic analyses of *Ir* genes will be necessary to establish their specific roles in the gustatory system.

DmXR, a lone candidate taste receptor belonging to the superfamily of GPCRs and encoded by the gene *mangetout* (*mtt*), was shown to be necessary for sensing the naturally occurring plant insecticide L-canavanine.[86] DmXR is phylogenetically related to metabotropic glutamate receptors and was shown to bind L-canavanine, but not glutamate, in a heterologous expression system. A GAL4 enhancer trap line for the *mtt* gene revealed expression in bitter GRNs, where DmXR is necessary for sensing L-canavanine. Moreover, a Gα protein, encoded by the *Gαo47A* gene, was shown to be necessary for DmXR mediated avoidance of L-canavanine.[87]

TASTE SIGNAL TRANSDUCTION

The signaling mechanism of both Gr- or Ir-based taste receptors is poorly understood and the least investigated area of *Drosophila* taste physiology. Two main reasons account for this paucity of knowledge: first, it has not been possible to carry out patch clamp recordings from

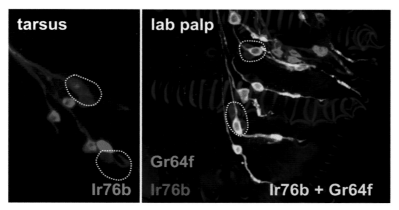

FIGURE 6 **Expression of Ir76b in labial and tarsal GRNs.** The left image shows tarsal sensilla harboring several Ir76b expressing GRNs (marked by *Ir76b-GAL4*). The neurons of one sensillum are indicated (white dotted line). In the labial palps (right), many senilla also contain multiple *Ir76b* expression neurons (red, marked bywith Ir76b-GAL4), one of which also expresses *Gr64f* (green, marked by *Gr64fLEXA*) and renders the cell yellow. However, several cells lack expression of *Gr64f.*

Drosophila taste neurons (or olfactory neurons), which limits our ability to gain insights into the neurophysiological processes underlying their activation; second, attempts to functionally express taste receptors in heterologous systems have been largely unsuccessful. One report of successful heterologous expression involved Gr43a in HEK 293 cells and in *Xenopus* oocytes.[88] In these studies, it was shown that Gr43a appears to function as a ligand-gated ion channel, because pharmacological inhibition of G protein signaling did not interfere with Ca^{2+} influx into the cells. Similar observations were reported for the related Or proteins.[89] However, a second study using also a heterologous expression system for Or proteins suggested that Or complexes may function both as GPCRs and as ligand-gated ion channels.[90] In this context, a number of different G proteins have been implicated in sugar taste.[91–94] However, the cellular requirement for these genes in relevant sweet GRNs has not yet been firmly established. Thus, whether Gr proteins can function as GPCRs will require further critical analysis.

Even less is known about the signaling mechanism of Ir-based taste receptors. Their close relationship to Ir proteins in the olfactory system and their similarity to ionotropic glutamate receptors suggest that Ir-based taste receptors are multimeric complexes. Ionotropic glutamate receptors in mammals function as ion channels that allow influx of both Na^+ and Ca^{2+} ions upon glutamate binding, leading to membrane depolarization and the generation of action potentials. However, nothing is known about the signaling mechanism that leads to the activation of olfactory neurons following odorant binding to Ir-based olfactory receptors.

Finally, *No receptor potential A*, which encodes a phospholipase C essential for visual transduction in the eye, has also been implicated in fatty acid sensing[20] and aristolochoic acid detection by TRPA1,[95] although the receptors for these stimuli are not yet known.

In conclusion, a comprehensive knowledge of Gr, TRP, and Ir-based receptor signaling is currently missing. The development of more direct electrophysiological recording methods, such as patch clamp recordings of GRNs and more broadly applicable heterologous expression systems will ultimately be required for substantive progress in our understanding of taste receptor signaling mechanisms.

GUSTATORY PERCEPTION IN LARVAE

In most insects, the gustatory system of the larvae is rather simple compared with that of the imago (the adult stage). This difference reflects the varied lifestyles of adult and larval insects: adult life is characterized by extensive locomotion activity in search of mating partners and food, and escaping predators and other unfavorable conditions, whereas during the larval growth stage the animal is confined to a small microhabitat, where its main activity is feeding. Indeed, *Drosophila* larvae stay on a single fruit, where they are deposited as eggs.

Both neuroanatomical investigations and expression studies of *Gr* genes are evidence for the simplicity of the *Drosophila* larval gustatory system, compared with that of the adult (Figure 6). The approximately 70 taste neurons of a *Drosophila* larva are located in three external sense structures, the dorsal, terminal, and ventral organs, and in three small pharyngeal clusters (Figure 7(A)).[2,96] Consistent with their predetermined living habitat, larvae do not express any of the eight sugar *Gr* genes,[96] but rely on the nutrient sensor *Gr43a*, which is expressed in a few taste neurons and several neurons in the brain (Figure 7(B) and (C)).[97] Thus, larvae probably employ taste neurons for the detection of fructose and postingestive chemosensing mechanisms for the detection of other nutritious sugars, which—upon ingestion—are converted into circulating fructose that activates Gr43a expressing brain neurons.[97] In contrast, many bitter *Gr* genes are expressed in larvae,[96] reflecting their need to avoid areas of a fruit that harbors potentially toxic by-products from colonizing microbes.

Like adult flies, larvae also exhibit similar acceptance and repulsive behaviors toward low and high concentration of salt.[98,99] Surprisingly, different candidate genes for salt sensing have been proposed for larvae than for adult flies. Members of the degenerin/epithelial Na channel (Deg/ENaC) protein family were implicated in mammalian salt taste,[100,101] prompting Liu and colleagues to investigate the potential roles for their insect orthologs, the pickpocket (PPK) proteins, in larval salt taste.[102] By expressing dominant negative PPK isoforms, these authors found that the *ppk11* and *ppk19* genes are required for avoidance behavior to high concentration of salt, an observation that was recently confirmed for *ppk19* by another group.[103] However, in adult flies, *ppk11* is not expressed in taste neurons, but found in support cells of taste sensilla,[32,79] suggesting that the two different life stages employ different salt-sensing mechanisms. Lastly, a number of *Ir* genes are also expressed in the larva,[104] which implies that this receptor gene family plays a critical role in taste perception during both major life stages.

GUSTATORY RECEPTORS BEYOND TASTE

As expression analyses came to encompass essentially all *Gr* genes and included tissues other than taste organs, some surprising observations were reported: several *Gr* genes are expressed in olfactory neurons[36,43,105,106] (see Chapter 6), the brain, proventriculus and

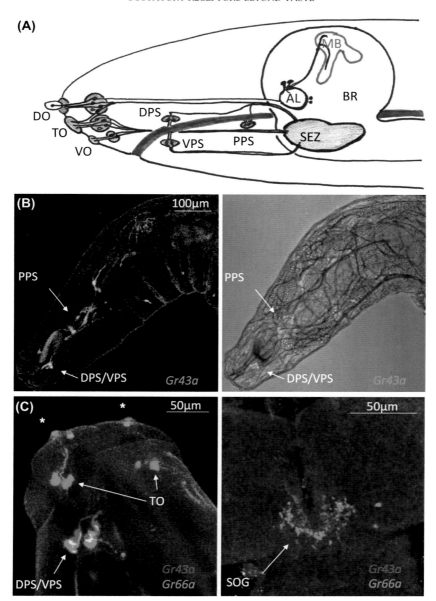

FIGURE 7 **Larval gustatory system.** (A) Diagram of the gustatory system in the larva. The dorsal, terminal, and ventral organs (DO, TO, VO) harbor peripheral taste neurons, whereas the dorsal, ventral, and posterior pharyngeal sense organs (DPS, VPS, and PPS) are internally located along the pharynx. All these neurons project to the sub-esophageal zone (SEZ) in the brain (BR). The DO also harbors olfactory neurons, which project to the antennal lobes (AL). MB, mushroom bodies. (B) Fluorescent (left) and bright field (right) image of a larva expressing the fructose sensor Gr43a. Gr43a (green) is expressed in all internal taste organs (DPS, VPS, and PPS), but not in the peripheral taste organs. (C) The bitter receptor Gr66a (green) is expressed in distinct DPS/VPS neurons than Gr43a (red), and the respective neurons project to different areas in the SEZ (left). *Reprinted with permission from Ref. 97.*

ovaries,[33] the gut,[107] and many sensory neurons not associated with chemoreception.[108] Molecular genetic analysis of mutations of respective genes established novel, nongustatory related functions in several cases.[109,110] Notably, all *Gr* genes shown or suspected to have nongustatory roles are highly conserved across species, suggesting that they serve important processes common to arthropods.

Gr Genes in the Olfactory System

Gr21a and *Gr63a* were the first two genes to show noncanonical expression.[43,105,106] They are both exclusively expressed in the ab1C olfactory neurons, which are narrowly tuned to carbon dioxide (CO_2). High CO_2 concentration is a stress signal for flies and leads to avoidance behavior.[111] *Gr21a* mutations lead to a loss of CO_2-induced action potentials in ab1C olfactory neurons. Moreover, coexpression (but not single expression) of *Gr21a* and *Gr63a* in ab3A olfactory neurons lacking endogenously expressed Or genes conveys CO_2 sensitivity, indicating that the two proteins are necessary and sufficient to form a heteromeric CO_2 receptor.[105,106] *Gr21a* and *Gr63* are highly conserved in most insects, including mosquitoes, where they are required for sensing CO_2, a generic host attractant for blood feeding insects.[98,112,113]

In addition to *Gr21a* and *Gr63a*, the three sugar *Gr* genes *Gr5a*, *Gr64b*, and *Gr64f* are also expressed in olfactory neurons.[36] Their function in olfaction, however, has yet to be determined.One intriguing possibility is that *Drosophila* OSNs in the antenna can sense sugars, which is not that far-fetched, as the antenna in hymenoptera serve also as taste organs. However, it is also conceivable that Or and Gr proteins combine to form novel chemoreceptors, or that OSNs express Gr-based chemosensors at postsynaptic terminals for the detection of signals to modulate OSN activity.[36]

Gr Proteins as Brain Nutrient Sensors

Noncanonical expression of *Gr* genes is not restricted to the olfactory system. The most striking example is *Gr43a*, which is expressed in a small number of neurons in the brain.[33] Here, Gr43a functions as an internal nutrient sensors, activated by raising levels of fructose in the hemolymph upon dietary fructose intake or intake of dietary sugars (glucose, sucrose, etc.), a fraction of which is converted into fructose upon ingestion. Activation of Gr43a neurons modulates feeding behavior of flies in a satiation-dependent manner. Specifically, Gr43a is essential for efficient feeding in hungry flies and for cessation of feeding in satiated flies.[33] Intriguingly, Gr43a brain neurons also express a second sugar receptor gene, *Gr64a*.[36] However, *Gr64a* is not required for their activation by fructose, suggesting that a single set of brain neurons in the protocerebrum "read" internal levels of two different sugars.

Other, Noncanonical Expression of Gr Genes

In addition to the brain and the olfactory system, *Gr* genes are expressed in several other tissues, including the proventriculus and uterus (*Gr43a*),[33,97] the gut (many bitter *Gr* genes),[107] and a diverse array of both neuronal and nonneuronal cell types (*Gr28a*, *Gr28b.a*—*Gr28b.e*).[108] The six highly conserved *Gr28* genes are of particular interest, given their largely nonoverlapping expression in the brain, many somatosensory neurons in larvae and adults, oenocytes and cells in the gut.[108] Members of this gene family have been implicated in larval light sensing via multidendritic neurons[110] and temperature sensing mediated by sensory neurons of the ariste.[109] Thus, identification of Gr28 ligands should be very informative in elucidating the many different roles these genes are likely to have. As for *Ir* genes, no expression beyond the olfactory and gustatory systems has been reported, albeit a detailed examination of internal tissues remains to be undertaken.

References

1. Morgan TH, et al. *The Mechanism of Mendelian Heredity*. New York: Henry Holt; 1915.
2. Stocker RF. The organization of the chemosensory system in *Drosophila melanogaster*: a review. *Cell Tissue Res.* 1994;275(1):3—26.
3. Dethier VG. *The Hungry Fly: A Physiological Study of the Behavior Associated with Feeding*. Cambridge MA: Harvard University Press; 1976.
4. Dethier VG, Goldrich-Rachman N. Anesthetic stimulation of insect water receptors. *Proc Natl Acad Sci USA.* 1976;73(9):3315—3319.
5. Omand E, Dethier VG. An electrophysiological analysis of the action of carbohydrates on the sugar receptor of the blowfly. *Proc Natl Acad Sci USA.* 1969;62(1):136—143.
6. Dethier VG, Hanson FE. Electrophysiological responses of the chemoreceptors of the blowfly to sodium salts of fatty acids. *Proc Natl Acad Sci USA.* 1968;60(4):1296—1303.
7. Dethier VG. Chemosensory input and taste discrimination in the blowfly. *Science.* 1968;161(3839):389—391.
8. Morita H. Transduction process and impulse initiation in insect contact chemoreceptors. *Zool Sci.* 1992;9:1—16.
9. Fujishiro N, Kijima H, Morita H. Impulse frequency and action potential amplitude in labellar chemosensory neurons of *Drosophila melanogaster*. *J Insect Physiol.* 1984;30(4):317—325.
10. Nayak SV, Singh RN. Sensilla on the tarsal segments and mouth parts of adult *Drosophila melaogaster* meigen. *Int J Insect Morphol Embryol.* 1983;12:273—291.
11. Liman ER, Zhang YV, Montell C. Peripheral coding of taste. *Neuron.* 2014;81(5):984—1000.
12. Vosshall LB, Stocker RF. Molecular architecture of smell and taste in *Drosophila*. *Annu Rev Neurosci.* 2007;30:505—533.
13. Weiss LA, et al. The molecular and cellular basis of bitter taste in *Drosophila*. *Neuron.* 2011;69(2):258—272.
14. Moon SJ, et al. A *Drosophila* gustatory receptor essential for aversive taste and inhibiting male-to-male courtship. *Curr Biol.* 2009;19(19):1623—1627.
15. Lee Y, Moon SJ, Montell C. Multiple gustatory receptors required for the caffeine response in *Drosophila*. *Proc Natl Acad Sci USA.* 2009;106(11):4495—4500.
16. Wang Z, et al. Taste representations in the *Drosophila* brain. *Cell.* 2004;117(7):981—991.
17. Thorne N, et al. Taste perception and coding in *Drosophila*. *Curr Biol.* 2004;14(12):1065—1079.
18. Meunier N, et al. Peripheral coding of bitter taste in *Drosophila*. *J Neurobiol.* 2003;56(2):139—152.
19. Toshima N, et al. Genetic variation in food choice behaviour of amino acid-deprived *Drosophila*. *J Insect Physiol.* 2014;69:89—94.
20. Masek P, Keene AC. *Drosophila* fatty acid taste signals through the PLC pathway in sugar-sensing neurons. *PLoS Genet.* 2013;9(9):e1003710.
21. Toshima N, Tanimura T. Taste preference for amino acids is dependent on internal nutritional state in *Drosophila melanogaster*. *J Exp Biol.* 2012;215(Pt 16):2827—2832.
22. Chen Y, Amrein H. Enhancing perception of contaminated food through acid-mediated modulation of taste neuron responses. *Curr Biol.* 2014;24(17):1969—1977.
23. Fischler W, et al. The detection of carbonation by the *Drosophila* gustatory system. *Nature.* 2007;448(7157):1054—1057.

24. Vogt RG, Prestwich GD, Lerner MR. Odorant-binding-protein subfamilies associate with distinct classes of olfactory receptor neurons in insects. *J Neurobiol*. 1991;22(1):74–84.
25. Galindo K, Smith DP. A large family of divergent *Drosophila* odorant-binding proteins expressed in gustatory and olfactory sensilla. *Genetics*. 2001;159(3):1059–1072.
26. Jeong YT, et al. An odorant-binding protein required for suppression of sweet taste by bitter chemicals. *Neuron*. 2013;79(4):725–737.
27. Stocker RF, Schorderet M. Cobalt filling of sensory projections from internal and external mouthparts in *Drosophila*. *Cell Tissue Res*. 1981;216(3):513–523.
28. Marella S, et al. Imaging taste responses in the fly brain reveals a functional map of taste category and behavior. *Neuron*. 2006;49(2):285–295.
29. Hodgson ES, Roeder KD. Physiology of primary chemoreceptor unit. *Science*. 1955;122:417–418.
30. Tanimura T, et al. Neurophysiology of gustatory receptor neurones in *Drosophila*. *Seb Exp Biol Ser*. 2009;63:59–76.
31. Miyamoto T, et al. Identification of a *Drosophila* glucose receptor using Ca(2+) imaging of single chemosensory neurons. *PLoS One*. 2013;8(2):e56304.
32. Thistle R, et al. Contact chemoreceptors mediate male-male repulsion and male-female attraction during *Drosophila* courtship. *Cell*. 2012;149(5):1140–1151.
33. Miyamoto T, et al. A fructose receptor functions as a nutrient sensor in the *Drosophila* brain. *Cell*. 2012;151(5):1113–1125.
34. Nakai J, Ohkura M, Imoto K. A high signal-to-noise Ca(2+) probe composed of a single green fluorescent protein. *Nat Biotechnol*. 2001;19(2):137–141.
35. Slone J, Daniels J, Amrein H. Sugar receptors in *Drosophila*. *Curr Biol*. 2007;17(20):1809–1816.
36. Fujii S, et al. *Drosophila* sugar receptors in sweet taste perception, olfaction, and internal nutrient sensing. *Curr Biol*. 2015;25(5):621–627.
37. Tanimura T, et al. Genetic dimorphism in the taste senitivity to trehalose in *Drosophila melanogaster*. *J Comp Physiol*. 1982;141:433–473.
38. Amrein H, Thorne N. Gustatory perception and behavior in *Drosophila melanogaster*. *Curr Biol*. 2005;15(17):R673–R684.
39. Moon SJ, et al. A taste receptor required for the caffeine response in vivo. *Curr Biol*. 2006;16(18):1812–1817.
40. Ja WW, et al. Prandiology of *Drosophila* and the CAFE assay. *Proc Natl Acad Sci USA*. 2007;104(20):8253–8256.
41. Joseph RM, et al. Oviposition preference for and positional avoidance of acetic acid provide a model for competing behavioral drives in *Drosophila*. *Proc Natl Acad Sci USA*. 2009;106(27):11352–11357.
42. Bridges M, Mattice M. Over two thousand estimations of the pH of representative foods. *Am J Dig Dis*. 1939;6(7):440–449.
43. Scott K, et al. A chemosensory gene family encoding candidate gustatory and olfactory receptors in *Drosophila*. *Cell*. 2001;104(5):661–673.
44. Dunipace L, et al. Spatially restricted expression of candidate taste receptors in the *Drosophila* gustatory system. *Curr Biol*. 2001;11(11):822–835.
45. Clyne PJ, Warr CG, Carlson JR. Candidate taste receptors in *Drosophila*. *Science*. 2000;287(5459):1830–1834.
46. Robertson HM, Warr CG, Carlson JR. Molecular evolution of the insect chemoreceptor gene superfamily in *Drosophila melanogaster*. *Proc Natl Acad Sci USA*. 2003;100(suppl 2):14537–14542.
47. Robertson HM. The insect chemoreceptor superfamily in *Drosophila pseudoobscura*: molecular evolution of ecologically-relevant genes over 25 million years. *J Insect Sci*. 2009;9:18.
48. Penalva-Arana DC, Lynch M, Robertson HM. The chemoreceptor genes of the waterflea *Daphnia pulex*: many Grs but no Ors. *BMC Evol Biol*. 2009;9:79.
49. Kent LB, Robertson HM. Evolution of the sugar receptors in insects. *BMC Evol Biol*. 2009;9:41.
50. Wanner KW, Robertson HM. The gustatory receptor family in the silkworm moth *Bombyx mori* is characterized by a large expansion of a single lineage of putative bitter receptors. *Insect Mol Biol*. 2008;17(6):621–629.
51. Kent LB, Walden KK, Robertson HM. The Gr family of candidate gustatory and olfactory receptors in the yellow-fever mosquito *Aedes aegypti*. *Chem Senses*. 2008;33(1):79–93.
52. Robertson HM, Wanner KW. The chemoreceptor superfamily in the honey bee, *Apis mellifera*: expansion of the odorant, but not gustatory, receptor family. *Genome Res*. 2006;16(11):1395–1403.

53. Ling F, et al. The molecular and cellular basis of taste coding in the legs of *Drosophila*. *J Neurosci*. 2014;34(21):7148−7164.

54. Nelson G, et al. Mammalian sweet taste receptors. *Cell*. 2001;106(3):381−390.

55. Montmayeur JP, et al. A candidate taste receptor gene near a sweet taste locus. *Nat Neurosci*. 2001; 4(5):492−498.

56. Ueno K, et al. Trehalose sensitivity in *Drosophila* correlates with mutations in and expression of the gustatory receptor gene Gr5a. *Curr Biol*. 2001;11(18):1451−1455.

57. Dahanukar A, et al. A Gr receptor is required for response to the sugar trehalose in taste neurons of *Drosophila*. *Nat Neurosci*. 2001;4(12):1182−1186.

58. Jiao Y, et al. Gr64f is required in combination with other gustatory receptors for sugar detection in *Drosophila*. *Curr Biol*. 2008;18(22):1797−1801.

59. Jiao Y, Moon SJ, Montell C. A *Drosophila* gustatory receptor required for the responses to sucrose, glucose, and maltose identified by mRNA tagging. *Proc Natl Acad Sci USA*. 2007;104(35):14110−14115.

60. Dahanukar A, et al. Two Gr genes underlie sugar reception in *Drosophila*. *Neuron*. 2007;56(3):503−516.

61. Yavuz A, et al. A genetic tool kit for cellular and behavioral analyses of insect sugar receptors. *Fly (Austin)*. 2014;8(4):189−196.

62. Benton R, et al. Atypical membrane topology and heteromeric function of *Drosophila* odorant receptors in vivo. *PLoS Biol*. 2006;4(2):e20.

63. Lee Y, et al. Gustatory receptors required for avoiding the insecticide L-canavanine. *J Neurosci*. 2012;32(4):1429−1435.

64. Lee Y, Kim SH, Montell C. Avoiding DEET through insect gustatory receptors. *Neuron*. 2010;67(4):555−561.

65. Miyamoto T, Amrein H. Courtship suppression by a *Drosophila* pheromone receptor. *Nat Neurosci*. 2008;11:874−876.

66. Wang L, et al. Hierarchical chemosensory regulation of male-male social interactions in *Drosophila*. *Nat Neurosci*. 2011;14(6):757−762.

67. Julius D. TRP channels and pain. *Annu Rev Cell Dev Biol*. 2013;29:355−384.

68. Al-Anzi B, Tracey Jr WD, Benzer S. Response of *Drosophila* to wasabi is mediated by painless, the fly homolog of mammalian TRPA1/ANKTM1. *Curr Biol*. 2006;16(10):1034−1040.

69. Kang K, et al. Analysis of *Drosophila* TRPA1 reveals an ancient origin for human chemical nociception. *Nature*. 2010;464(7288):597−600.

70. Kwon Y, et al. *Drosophila* TRPA1 channel is required to avoid the naturally occurring insect repellent citronellal. *Curr Biol*. 2010;20(18):1672−1678.

71. Kang K, et al. Modulation of TRPA1 thermal sensitivity enables sensory discrimination in *Drosophila*. *Nature*. 2012;481(7379):76−80.

72. Jordt SE, et al. Mustard oils and cannabinoids excite sensory nerve fibres through the TRP channel ANKTM1. *Nature*. 2004;427(6971):260−265.

73. Inoshita T, Tanimura T. Cellular identification of water gustatory receptor neurons and their central projection pattern in *Drosophila*. *Proc Natl Acad Sci USA*. 2006;103(4):1094−1099.

74. Hiroi M, et al. Two antagonistic gustatory receptor neurons responding to sweet-salty and bitter taste in *Drosophila*. *J Neurobiol*. 2004;61(3):333−342.

75. Cameron P, et al. The molecular basis for water taste in *Drosophila*. *Nature*. 2010;465(7294):91−95.

76. Awasaki T, Kimura K. Pox-neuro is required for development of chemosensory bristles in *Drosophila*. *J Neurobiol*. 1997;32(7):707−721.

77. Boll W, Noll M. The Drosophila Pox neuro gene: control of male courtship behavior and fertility as revealed by a complete dissection of all enhancers. *Development*. 2002;129(24):5667−5681.

78. Kellenberger S, Schild L. Epithelial sodium channel/degenerin family of ion channels: a variety of functions for a shared structure. *Physiol Rev*. 2002;82(3):735−767.

79. Zhang YV, Ni J, Montell C. The molecular basis for attractive salt-taste coding in *Drosophila*. *Science*. 2013;340(6138):1334−1338.

80. Abuin L, et al. Functional architecture of olfactory ionotropic glutamate receptors. *Neuron*. 2011;69(1):44−60.

81. Ahn JE, Chen Y, Amrein H. *Identification of an Ionotropic Receptor Necessary for the Taste of Fatty Acids*. 2015 [submitted for publication].

82. Croset V, et al. Ancient protostome origin of chemosensory ionotropic glutamate receptors and the evolution of insect taste and olfaction. *PLoS Genet*. 2010;6(8):e1001064.

83. Benton R, et al. Variant ionotropic glutamate receptors as chemosensory receptors in *Drosophila*. *Cell*. 2009;136(1):149−162.

84. Yao CA, Ignell R, Carlson JR. Chemosensory coding by neurons in the coeloconic sensilla of the *Drosophila* antenna. *J Neurosci*. 2005;25(37):8359−8367.

85. Koh TW, et al. The *Drosophila* IR20a clade of ionotropic receptors are candidate taste and pheromone receptors. *Neuron*. 2014;83(4):850−865.

86. Mitri C, et al. Plant insecticide L-canavanine repels *Drosophila* via the insect orphan GPCR DmX. *PLoS Biol*. 2009;7(6):e1000147.

87. Devambez I, et al. Galphao is required for L-canavanine detection in *Drosophila*. *PLoS One*. 2013;8(5):e63484.

88. Sato K, Tanaka K, Touhara K. Sugar-regulated cation channel formed by an insect gustatory receptor. *Proc Natl Acad Sci USA*. 2011;108(28):11680−11685.

89. Sato K, et al. Insect olfactory receptors are heteromeric ligand-gated ion channels. *Nature*. 2008;452(7190):1002−1006.

90. Wicher D, et al. *Drosophila* odorant receptors are both ligand-gated and cyclic-nucleotide-activated cation channels. *Nature*. 2008;452(7190):1007−1011.

91. Bredendiek N, et al. Go alpha is involved in sugar perception in *Drosophila*. *Chem Senses*. 2011;36(1):69−81.

92. Kain P, et al. Mutants in phospholipid signaling attenuate the behavioral response of adult *Drosophila* to trehalose. *Chem Senses*. 2010;35(8):663−673.

93. Ueno K, et al. Gsalpha is involved in sugar perception in *Drosophila melanogaster*. *J Neurosci*. 2006;26(23):6143−6152.

94. Ishimoto H, et al. G-protein gamma subunit 1 is required for sugar reception in *Drosophila*. *EMBO J*. 2005;24(18):3259−3265.

95. Kim SH, et al. *Drosophila* TRPA1 channel mediates chemical avoidance in gustatory receptor neurons. *Proc Natl Acad Sci USA*. 2010;107(18):8440−8445.

96. Kwon JY, et al. Molecular and cellular organization of the taste system in the *Drosophila* larva. *J Neurosci*. 2011;31(43):15300−15309.

97. Mishra D, et al. The molecular basis of sugar sensing in *Drosophila* larvae. *Curr Biol*. 2013;23(15):1466−1471.

98. Miyakawa Y. Behavioural evidence for the existence of sugar, salt and amino acid taste receptor cells and some of their propterties in *Drosophila* larvae. *J Insect Physiol*. 1982;28:405−410.

99. Heimbeck G, et al. Smell and taste perception in *Drosophila melanogaster* larva: toxin expression studies in chemosensory neurons. *J Neurosci*. 1999;19(15):6599−6609.

100. Lin W, et al. Epithelial Na$^+$ channel subunits in rat taste cells: localization and regulation by aldosterone. *J Comp Neurol*. 1999;405(3):406−420.

101. Kretz O, et al. Differential expression of RNA and protein of the three pore-forming subunits of the amiloride-sensitive epithelial sodium channel in taste buds of the rat. *J Histochem Cytochem*. 1999;47(1):51−64.

102. Liu L, et al. Contribution of *Drosophila* DEG/ENaC genes to salt taste. *Neuron*. 2003;39(1):133−146.

103. Alves G, et al. High-NaCl perception in *Drosophila melanogaster*. *J Neurosci*. 2014;34(33):10884−10891.

104. Stewart S, et al. Candidate ionotropic taste receptors in the *Drosophila* larva. *Proc Natl Acad Sci USA*. 2015;112(14):4195−4201.

105. Jones WD, et al. Two chemosensory receptors together mediate carbon dioxide detection in *Drosophila*. *Nature*. 2007;445(7123):86−90.

106. Kwon JY, et al. The molecular basis of CO_2 reception in *Drosophila*. *Proc Natl Acad Sci USA*. 2007;104(9):3574−3578.

107. Park JH, Kwon JY. Heterogeneous expression of *Drosophila* gustatory receptors in enteroendocrine cells. *PLoS One*. 2011;6(12):e29022.

108. Thorne N, Amrein H. Atypical expression of *Drosophila* gustatory receptor genes in sensory and central neurons. *J Comp Neurol*. 2008;506(4):548−568.

109. Ni L, et al. A gustatory receptor paralogue controls rapid warmth avoidance in *Drosophila*. *Nature*. 2013;500(7464):580−584.

110. Xiang Y, et al. Light-avoidance-mediating photoreceptors tile the *Drosophila* larval body wall. *Nature.* 2010;468(7326):921—926.

111. Suh GS, et al. A single population of olfactory sensory neurons mediates an innate avoidance behaviour in *Drosophila. Nature.* 2004;431(7010):854—859.

112. DeGennaro M, et al. Orco mutant mosquitoes lose strong preference for humans and are not repelled by volatile DEET. *Nature.* 2013;498(7455):487—491.

113. Turner SL, et al. Ultra-prolonged activation of CO_2-sensing neurons disorients mosquitoes. *Nature.* 2011;474(7349):87—91.

CHAPTER

15

G Protein—Coupled Taste Transduction

Sue C. Kinnamon

Department of Otolaryngology, University of Colorado Medical School, Aurora, CO, USA

OUTLINE

INTRODUCTION

Taste buds, the transducing elements of gustatory sensation, are housed in connective tissue papillae on the surface of the tongue and scattered throughout the epithelium of the soft palate, larynx, and epiglottis. Each taste bud contains roughly 50—100 elongate taste cells that extend from the basal lamina to the surface of the epithelium, where receptors on the apical microvilli detect and transduce sapid chemicals and ultimately evoke release of transmitters from the taste cell to activate gustatory nerve fibers. The taste system discriminates at least five distinct modalities—sweet, salty, bitter, sour, and umami. Salty (sodium chloride

Chemosensory Transduction
http://dx.doi.org/10.1016/B978-0-12-801694-7.00015-9

(NaCl) and other salts) and sour (acids) stimuli are ionic in nature and primarily rely on apically located ion channels for transduction. These mechanisms are considered in Chapter 16. Sweet (sugars and sweeteners), bitter (alkaloids and other compounds), and umami (glutamate and 5' ribonucleotides) stimuli are more structurally complex and require G protein–coupled receptors (GPCRs) for transduction. Because mouse and human genomic databases became available in the early 1990s, the molecular identities of the taste GPCRs were identified and characterized, largely by gene mining. These receptors are covered in Chapter 13, whereas the present chapter focuses on the intracellular signaling effectors (G proteins, effector enzymes, second messengers, and targets) downstream of the taste receptors and the role of adenosine triphosphate (ATP) as a transmitter to activate gustatory afferent fibers.

TASTE CELL TYPES AND INNERVATION

Although this chapter focuses on transduction, some background on taste buds and their innervation is required to place the physiological studies into the appropriate context. Taste buds in the anterior two-thirds of the tongue are housed in fungiform papillae and are innervated by the chorda tympani branch of the facial nerve (CN VII), whereas taste buds in the posterior tongue are housed in foliate and circumvallate papillae innervated predominantly by the glossopharyngeal nerve (CN IX). Taste buds in the soft palate are innervated by a different branch of the facial nerve, the greater superficial petrosal nerve, while taste buds in the larynx are innervated by a branch of the vagus nerve (CN X), the superior laryngeal nerve.

Regardless of the location of the taste buds, each taste bud contains three types of elongate cells and a population of proliferative basal cells, which gives rise to the new taste cells. Type I, or "glial-like" cells, comprise about half of the taste bud and have primarily a support function, similar to astrocytic glial cells of the nervous system. The membranes of the type I cells wrap around the other cells in the taste bud in a glial-like fashion and express enzymes for uptake and degradation of transmitters.[1] Type II, or "receptor cells," the primary focus of this chapter, contain the taste GPCRs and downstream effectors for bitter, sweet, and umami stimuli.[2,3] Type II cells comprise 30–40% of the cells of the taste bud, the remaining 10–15% being type III cells. Type III, or "presynaptic cells," are so named because they are the only cells in the taste bud to form conventional synapses with afferent nerve fibers that contain pre- and postsynaptic specializations.[4,5] These cells respond to sour,[6] carbonation,[7] and some salty tastants[8] (Figure 1). Both types II and III cells are electrically excitable, possess voltage-gated sodium ion (Na^+) and potassium ion (K^+) channels,[9–12] and regularly generate action potentials to taste stimuli.[13] It was once believed that, because type III cells had the only conventional synaptic connections with the nerve fibers in the taste bud, the type III cells would be required for transmission of all taste qualities. However, when type III cells were ablated with diphtheria toxin, only sour and a portion of salty responses were abolished,[8,14] confirming that type II cells also signal directly to afferent nerve fibers, but likely using an unconventional synaptic mechanism.

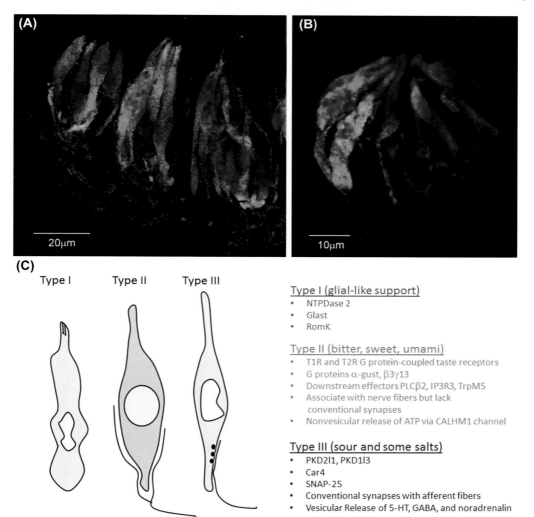

FIGURE 1 Confocal z-stack images of taste buds from a foliate papilla, (A) and the soft palate, (B) of a transgenic mouse expressing green fluorescent protein (GFP) from the *Trpm5* promoter. Type II cells (green) are labeled with GFP, whereas afferent nerve fibers (red) are labeled with an antibody against the purinergic receptor P2X3. *(Figure courtesy of Dr Thomas Finger, Rocky Mountain Taste and Smell Center.)* (C) Diagrammatic representation of types I, II, and III taste cells, showing the predominant markers for each cell type. The glial-like type I cells express enzymes for the degradation and uptake of transmitters as well as ROMK, an apically located K$^+$ channel that regulates K$^+$ levels surrounding taste buds. Type II cells express the receptors and downstream effectors for bitter, sweet, and umami taste transduction and release adenosine triphosphate via the ion channel CALHM1. Type III cells express putative receptors and effectors for sour, carbonation (mediated by carbonic anhydrase 4 (Car4)), and possibly salty taste, although the transduction mechanisms are not yet clear.

G PROTEIN–COUPLED RECEPTORS

GPCRs serve as receptors, not only for taste, but also for visual and olfactory stimuli as well as for many neuromodulators and hormones. All GPCRs have seven transmembrane domains, with the extracellular N-terminal region involved in ligand binding and the intracellular C-terminal associated with a heterotrimeric G protein, which consists of an alpha (α) subunit and a beta-gamma ($\beta\gamma$) heterodimer (see Figure 2). The Gα subunit has a nucleotide binding domain, and in the inactive (unbound) state of the receptor the Gα subunit is bound to guanosine diphosphate (GDP). Upon ligand binding, the GDP is replaced by guanosine triphosphate (GTP) and the α and $\beta\gamma$ subunits dissociate from the receptor and each other. Both α and $\beta\gamma$ subunits can target effector enzymes, which produce second messengers that in turn target ion channels and other intracellular proteins such as kinases. GPCR signaling is terminated by the endogenous guanosine triphosphatase (GTPase) activity of the α subunit, which can be regulated by G protein modulators. A common feature of all GPCR signaling is signal amplification, in which a single ligand molecule can effectively produce hundreds of second messenger molecules, each of which can modulate ion channel function.

Two families of taste GPCRs have been cloned: the type 2 taste receptors (TAS2Rs) for bitter taste and the type 1 taste receptors (TAS1Rs) for sweet and umami taste (see Chapter 13). In

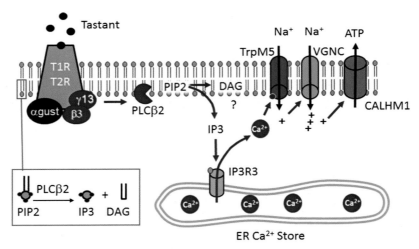

FIGURE 2　G protein beta gamma (G$\beta\gamma$) pathway mediates bitter, sweet, and umami taste transduction. Upon receptor binding, G$\beta\gamma$ dissociates from the taste receptor (either a type 2 taste receptor or type 1 taste receptor) to activate phospholipase C β2 (PLCβ2), which cleaves the membrane lipid phosphatidylinositol 4,5-bisphosphate (PIP$_2$) into diacylglycerol (DAG), which remains in the membrane, and inositol trisphosphate (IP$_3$), which is cytosolic and binds to the type III IP$_3$ receptor (IP$_3$R3) located on the membrane of the endoplasmic reticulum (ER). Upon binding to IP$_3$R3, Ca^{2+} is released from the Ca^{2+} store, causing an increase in intracellular Ca^{2+}, which subsequently directly activates the monovalent cation channel TRPM5. Na$^+$ influx via TRPM5 activates voltage-gated Na$^+$ channels (VGNC), which further depolarize the membrane by generating action potentials. The strong depolarization activates the voltage-gated adenosine triphosphate (ATP) release channel, CALHM1, which releases ATP to activate purinergic receptors P2X2 and P2X3 on afferent nerve fibers (not shown). The enlarged inset at the bottom illustrates the enzymatic activity of PLCβ2 on PIP$_2$, producing IP$_3$ (showing the three phosphates) and DAG.

general, compared with other GPCRs, taste receptors are low-affinity receptors with ligand--binding affinities in the high micromolar to millimolar range, but within the range of most nutrients in foods. Because of this low affinity, taste receptors could not be identified easily by biochemical means and were identified molecularly by mapping the mouse and human genomes, using taste polymorphisms and quantitative trait loci linkages.[15,16] The taste receptors are covered in detail in Chapter 13, but an overview is presented below.

TAS2Rs

The TAS2R bitter receptors are encoded by a family of about 30 *Tas2r* genes in mouse, which were identified independently by the Zuker[16,17] and Buck[18] laboratories. Some TAS2Rs are narrowly tuned to structurally similar bitter compounds, whereas others are broadly tuned, responding to many bitter compounds.[19] It was originally believed that each bitter-sensitive type II taste cell expressed every TAS2R isoform,[17] but more recent studies suggest that TAS2Rs can be expressed differentially, allowing for the possible discrimination of bitter compounds.[20,21] Recent data suggest that TAS2R receptors could be involved in detecting high (aversive) concentrations of salts, including NaCl,[8] however, how GPCRs might be activated by these high concentrations of salt is not yet clear.

TAS1Rs

Three TAS1R receptors, products of the *Tas1r* genes, have been identified molecularly: TAS1R1, TAS1R2, and TAS1R3.[22–26] These receptors function as heterodimers, with TAS1R3 an obligate partner of both the umami (TAS1R1/TAS1R3) and sweet (TAS1R2/TAS1R3) receptors. These receptors are classical "C" type GPCRs, with large N-terminal domains that exhibit a venus flytrap binding module similar to other C type receptors, such as the metabotropic glutamate receptors. Heterologous expression of the sweet receptor TAS1R2/TAS1R3 shows that it responds to sugars, synthetic sweeteners, D-amino acids, and some sweet-tasting proteins. Binding involves primarily the TAS1R2 subunit, but some large sweet proteins, sweeteners, and sugars also bind to regions of TAS1R3. Heterologous expression studies of rodent TAS1R1/TAS1R3 show that the primary ligand-binding subunit is TAS1R1 and that it binds glutamate as well as most L-amino acids[24]; however, the human receptor is highly selective for glutamate over other L-amino acids.[27] Both the human and rodent T1R1 subunits also allosterically bind 5′-ribonucleotides such as guanosine monophosphate and inosine monophosphate, which strongly potentiate the umami taste response.

DOWNSTREAM SIGNALING EFFECTORS

Although bitter, sweet, and umami taste receptors are expressed in largely nonoverlapping subsets of type II taste cells, they all couple to the same downstream signaling effectors.[28] The canonical pathway (see Figure 2) involves receptor binding, followed by activation of a heterotrimeric G protein, which consists of Gα-gustducin ($G\alpha_{gust}$)[29] and its partner, Gβ1γ13.[30] The βγ leg of the pathway predominates in driving the receptor response and will be considered first.

Gβγ Signaling

Taste receptor binding results in Gβγ activation of phospholipase C β2 (PLCβ2),[31] which cleaves the membrane lipid phosphatidylinositol 4,5-bisphosphate (PIP$_2$) into the second messengers inositol 1,4,5-trisphosphate (IP$_3$) and diacylglycerol (DAG). Although the function of DAG in taste cells is unclear, IP$_3$ is cytosolic and binds to the type 3 IP$_3$ receptor (IP$_3$R3),[32,33] located on the endoplasmic reticulum, causing a release of Ca^{2+} from intracellular stores and subsequent Ca^{2+}-dependent activation of the monovalent-selective cation channel, transient receptor potential channel M5 (TRPM5).[28,34–36] Influx of Na$^+$ through TRPM5 depolarizes the membrane, leading to activation of voltage-gated Na$^+$ channels,[37] generation of action potentials,[38] and release of ATP through a voltage-gated ATP release channel, likely calcium homeostasis modulator 1 (CALHM1)[39,40] (Figure 2).

Evidence for this pathway comes from molecular and immunocytochemical studies showing that the downstream signaling effectors PLCβ2, TRPM5, and IP$_3$R3 are coexpressed in the same subsets of type II cells.[2,3,32] In addition, labeled type II taste cells from transgenic mice expressing green fluorescent protein (GFP) from either the *Trpm5*, *Plcb2*, or *Itrp3* (the latter of which encodes IP$_3$R3) promoters respond to bitter, sweet, or umami taste stimuli with increases in intracellular Ca^{2+} that do not require extracellular Ca^{2+}.[41–43] These Ca^{2+} responses are inhibited by U73122, a phospholipase C antagonist, and thapsigargin, an inhibitor of the Ca^{2+} ATPase responsible for the refilling of Ca^{2+} stores.[41,44] Further evidence for the importance of this GPCR cascade in taste responses is that knockout of either *Trpm5*, *Plcb2*,[28,45] or *Itrp3*[46] results in greatly reduced or abolished responses to bitter, sweet, and umami taste stimuli. Likewise, genetic deletion of the transcription factor Skn-1A, which is required for development of type II cells,[47] abolishes responses to these same modalities due to the absence of type II cells in these mice.

The role of TRPM5 in this process is to couple the increase in intracellular Ca^{2+} due to IP$_3$-mediated release from intracellular stores, to a depolarization sufficient to activate voltage-gated Na$^+$ channels and evoke action potentials, likely required for opening the voltage-gated ATP release channels.[13] Because the TRPM5 channels show desensitization in response to prolonged exposure to intracellular Ca^{2+},[48] Ca^{2+} levels must be tightly regulated in the taste cells to ensure adequate transduction in the face of changing ionic conditions. Intracellular Ca^{2+} regulation involves both a constitutive Ca^{2+} influx as well as several mechanisms to extrude Ca^{2+} from taste cells.[49,50] Although the constitutive influx pathway has not been identified, extrusion mechanisms involve mitochondrial buffering,[51] Na$^+$/Ca^{2+} exchangers,[52] and the sarco-/endoplasmic reticulum Ca^{2+}-ATPases (SERCA), responsible for refilling Ca^{2+} stores following IP$_3$-mediated release of Ca^{2+}. Two isoforms have been identified in taste cells. SERCA3 is expressed in TAS2R-expressing (bitter) taste cells, whereas SERCA2 is predominately expressed in TAS1R (sweet and umami)-expressing taste cells.[53] Whether these different isoforms have physiological significance is not known.

Gα Signaling

Although Gβ1γ13 signaling seems to be required for bitter, sweet, and umami taste receptors in all taste fields, several Gα subunits expressed in taste buds are differentially expressed among the different taste fields[54–56] (Table 1). Gα–gustducin (Gα$_{gust}$), which was the first

TABLE 1   G Protein-α Subunits Expressed in Taste Cells

Gα subunit	Effector enzyme	Second messenger	Fields expressed	References
α-gustducin	PDE[a]	↓cAMP	T1R, anterior tongue T2R, all fields	29
α-transducin	PDE	↓cAMP	All fields	54
α-i2	AC[b]	↓cAMP	All fields	54,56
α-i3	AC	↓cAMP	All fields	54
α-14	PLC[c]	↑Ca^{2+}	T1R2/posterior tongue	54,68
α-15	PLC	↑Ca^{2+}	All fields	55
α-q	PLC	↑Ca^{2+}	All fields	55
α-s	AC	↑cAMP	All fields	54,56

cAMP, cyclic adenosine monophosphate.
[a]*Phosphodiesterase.*
[b]*Adenylyl cyclase.*
[c]*Phospholipase C.*

taste transduction element to be cloned,[29] has been the best studied. Gα-gust is structurally homologous to Gα-transducins, the Gα subunits that mediate phototransduction in rods and cones. $G\alpha_{gust}$ is coexpressed with TAS2Rs throughout the tongue and palate, but with TAS1Rs only in the anterior tongue and palate.[16,57,58] By analogy with Gα-transducin, which is also expressed in taste cells,[59] $G\alpha_{gust}$ should activate one or more phosphodiesterases (PDEs) to decrease intracellular cyclic adenosine monophosphate (cAMP) levels. Both bitter[60] and umami stimuli[61] do decrease intracellular cAMP levels in taste tissue, and the bitter activation of PDE is inhibited by antibodies against $G\alpha_{gust}$.[60] These data clearly implicate a role of $G\alpha_{gust}$ in bitter and possibly umami transduction.

How does a PDE-mediated decrease in intracellular cAMP affect taste signal transduction? cAMP could directly gate ion channels, as is the case for olfactory receptor cells (see Chapter 7). Although a cyclic nucleotide–gated ion channel is expressed in taste cells (CNG-gust),[62] there is no physiological evidence that it is functional or involved in taste transduction. Another potential target of cAMP is cAMP-dependent protein kinase (PKA), which phosphorylates both PLCβ2 and IP3R3 in other cell types.[63] To determine if PKA phosphorylation is involved in regulating Ca^{2+} signaling in taste cells, biochemical assays were performed on isolated lingual tissues from $G\alpha_{gust}$ (*Gnat3*) knockout and control mice. Surprisingly, the knockout mice had highly elevated cAMP levels compared to their wild-type littermates.[64] These data were obtained in the absence of taste stimulation, indicating that either the taste receptors or $G\alpha_{gust}$ is tonically active in taste cells. The elevated cAMP was likely activating PKA, because H-89, a specific PKA inhibitor, rescued Ca^{2+} responses to a bitter stimulus in isolated taste cells of $G\alpha_{gust}$ knockout mice.[64] These data suggest that $G\alpha_{gust}$ tonically activates PDE to keep cAMP levels low to prevent phosphorylation of the Ca^{2+} signaling effectors and subsequent adaptation to bitter stimuli. Whether $G\alpha_{gust}$ plays a similar role in umami and sweet taste transduction remains to be determined (Figure 3).

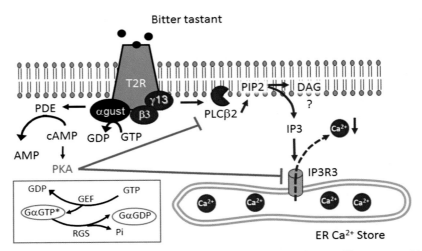

FIGURE 3   Proposed role of Gα-gustducin (Gα$_{gust}$) in bitter taste transduction. Upon tastant binding, Gα$_{gust}$ exchanges guanosine triphosphate (GTP) for guanosine diphosphate (GDP) and separates from the taste receptor and G protein beta gamma heterodimer. Gα$_{gust}$ activates phosphodiesterase (PDE) to decrease intracellular levels of cyclic adenosine monophosphate (cAMP). One target of cAMP is protein kinase A (PKA), which is known to phosphorylate both type 3 IP$_3$ receptor (IP$_3$R3) and phospholipase C β2 (PLCβ2), decreasing Ca^{2+} release from intracellular stores and therefore the Ca^{2+}-dependent activation of TRPM5. Thus by keeping cAMP levels low during bitter taste signaling, PKA will not be activated and taste cells will produce more robust Ca^{2+} signals and subsequently more depolarization and ATP release. Recent data from Gα$_{gust}$ (Gnat3) knockout mice suggest that Gα$_{gust}$ is tonically active even in the absence of bitter stimuli, keeping cAMP levels low and taste cells in a ready state to respond maximally to taste stimulation. Whether Gα$_{gust}$ plays a similar role in sweet and umami transduction is not known. The enlarged inset at the bottom left shows effects of G protein modulators. Guanine nucleotide exchange factors (GEFs), such as Ric-8, accelerate exchange of GTP for GDP, thus speeding up the G protein–coupled receptor signaling. RGS proteins activate GTPase activity, thus shortening the lifetime of the activated Gα subunit.

In contrast to the situation for bitter and umami, the role of Gα-gust in sweet transduction is less clear. Gα$_{gust}$ knockout mice are compromised to bitter, umami, and sweet taste stimuli,[65] but the effect for bitter and umami is more pronounced than for sweet.[66,67] Part of the reason for the lesser effect for sweet in the knockout is that sweet receptors in circumvallate taste buds tend to be coexpressed with Gα$_{14}$, rather than Gα$_{gust}$.[68] Whether Gα$_{14}$ plays a role in sweet taste remains to be determined.

Further evidence for the complicated role of Gα$_{gust}$ in sweet transduction is that sweet stimuli tend to increase, rather than decrease, intracellular cAMP levels,[69–71] and exogenously applied cAMP mimics sweet taste activity in the anterior tongue of hamsters.[72,73] The increase in cAMP may result from activation of Gα$_s$, which activates adenylyl cyclase to produce cAMP. However, whether Gα$_s$ is specifically associated with sweet receptors has not been examined. Alternatively, cAMP could be produced as a secondary consequence of Gβγ activation, which elevates intracellular Ca^{2+} levels because of release from intracellular stores. The Ca^{2+}-activated adenylyl cyclase AC8 is expressed in taste tissue,[71] but sweet stimuli still activated cAMP in taste tissue when intracellular Ca^{2+} signaling was blocked with a PLC inhibitor,[71] suggesting the increase in cAMP is not a downstream consequence of Ca^{2+} signaling. Further studies will be required to fully understand the role of cAMP in sweet taste transduction.

Regulation of G Protein Signaling

Accessory proteins can act as regulators of G protein signaling by interacting directly with the Gα subunits to either accelerate or inhibit GPCR signaling (see inset of Figure 3). Regulators of G protein signaling (RGS) proteins inhibit signaling by activating GTPases that associate with Gα subunits, thus terminating the active GTP-bound Gα subunit and terminating the response. RGS21 was recently cloned from taste cells and shown to interact with Gα$_{gust}$ expressing taste cells, where it specifically opposes bitter signaling.[74,75] By contrast, guanine nucleotide exchange factors (GEFs) act as signal amplifiers by accelerating the exchange of GDP for GTP. Two GEFs were recently cloned from taste cells, Ric-8A and Ric-8B. Ric-8A interacts with Gα$_{gust}$ and Gα$_{i2}$ to amplify the signal transduction of TAS2R16, a human bitter receptor.[76] Whether accessory G protein regulatory proteins are involved with sweet and umami transduction remains an open question.

BREAKOUT BOX

TASTE RECEPTOR SIGNALING IN SOLITARY CHEMOSENSORY CELLS OF THE AIRWAY

The cloning of taste signaling effectors and the production of several lines of transgenic mice expressing green fluorescent protein from their promoters has revealed that taste signaling effectors are not limited to expression in taste buds but are also extensively expressed in other cell types,[87–89] including solitary chemosensory cells (SCCs) of the airway epithelium. These cells, which morphologically resemble taste receptor cells, are scattered singly in the epithelium rather than clustered into end organs. SCCs are abundant in the upper airway epithelium[90–92] at the entrance of the vomeronasal duct[93] and in the tracheal epithelium.[89] The SCCs of the upper airway are innervated by the trigeminal nerve, where unlike type II taste cells, they form conventional synapses with afferent fibers and release acetylcholine as a transmitter.[94,95] SCCs express all of the bitter taste signaling effectors, including the type 2 taste receptor bitter receptors and downstream effectors G protein Gα-gustducin (Gα$_{gust}$), phospholipase C β2, and M5-type transient receptor potential channel (TRPM5). Some SCCs also express type 1 taste receptors (TAS1Rs).[96,97] SCCs respond to "bitter" stimuli, which activate the trigeminal nerve thus provoking a sensation of irritation rather than taste.[98] In addition, SCCs respond to homoserine lactones (HSLs) produced by gram-negative pathogenic bacteria as quorum signaling molecules.[99] When the concentration of these molecules reaches a critical level, the bacteria become pathogenic and form a biofilm, attacking the epithelium. SCCs detect the HSLs at appropriate concentrations and activate the trigeminal nerve to produce a profound neurogenic inflammation that acts to repel the invading organisms.[95] These responses are absent in both TRPM5 and Gα$_{gust}$ knockout mice, suggesting that SCCs are required for detection of the HSL molecules.[99] Important questions that remain are the nature and role of the stimuli that activate the TAS1Rs and the physiological role of the SCCs in the lower airways. Because many of these SCCs are not innervated, activation of these cells likely results in secretory functions or mucociliary clearance mechanisms, rather than protective nerve reflexes.

ATP RELEASE AND ACTIVATION OF GUSTATORY AFFERENTS

How are action potentials in type II taste cells translated into neurotransmitter release and activation of gustatory afferent fibers? Interestingly, type II cells lack both voltage-gated Ca^{2+} channels and presynaptic SNARE proteins[3,9,10]—both normally required for conventional, vesicular-mediated release of transmitter. Instead, contacts between type II cell membranes and afferent nerve fibers are characterized by large atypical mitochondria and occasional subsurface cisternae at the presynaptic membrane.[2,77] These ultrastructural data suggested that the taste transmitter and mechanism of release were likely unconventional in type II taste cells. In 2005, Finger and colleagues provided strong evidence that ATP was a crucial transmitter responsible for communicating taste information from type II cells to afferent nerve fibers. Evidence for the role of ATP includes (1) the purinergic receptors P2X2 and P2X3 are abundantly expressed on afferent nerve fibers[78]; (2) double knockout of P2X2 and P2X3,[79] or pharmacological inhibition of P2X3,[80] blocks responses to most taste stimuli; (3) bitter taste stimuli evoke release of ATP from taste tissue[79]; and (4) the released ATP is hydrolyzed by a specific ectoATPase, NTPDase2, located on the membranes of the glial-like type I taste cells.[1] Thus, ATP meets all the requirements for a taste transmitter. Following these initial studies, several laboratories showed that ATP was specifically released from type II taste cells, and that the release mechanism was likely to be nonvesicular, because blockers of ATP release channels inhibited both taste- and voltage-evoked release.[13,81,82] Several voltage-gated ATP release channels have been identified in taste cells, including Pannexin-1,[81] Connexins 30 and 43,[82] and CALHM1. Previously, Pannexin-1 was believed to be the ATP release channel, primarily because ATP release in taste cells was inhibited by low concentrations of carbenoxolone, an inhibitor of Pannexin-1.[13,81] However, recent data now point toward CALHM1 as the primary channel responsible for ATP release. CALHM1 is the only ATP release channel specifically expressed in type II cells, and *Calhm1* knockout mice have significantly reduced responses to bitter, sweet, and umami stimuli[39] as well as blunted responses to high concentrations of salty stimuli.[83] However, small residual responses remain, suggesting other mechanisms may contribute to ATP release. The residual responses are not likely to be mediated by Pannexin-1 because *Panx1* knockout mice release ATP[84] and have normal behavioral[85] and electrophysiological[86] responses to taste stimuli. Whether connexins are involved remains to be determined.

Genetic knockout[79] and pharmacological inhibition of P2X2 and P2X3 receptors[80] block responses to all taste qualities, not just those mediated by type II taste cells. Thus, ATP is required for sour and salty stimuli, as well as bitter, sweet, and umami stimuli. However, ATP release has not yet been detected from type III taste cells, either to sour stimuli or membrane depolarization,[81] so the source of the ATP for these qualities remains unclear.

SUMMARY AND FUTURE DIRECTIONS

The canonical pathway for bitter, sweet, and umami transduction involves different subsets of type II taste cells signaling through a common downstream signaling pathway to evoke release of ATP and activate gustatory afferent fibers. More specifically, the TAS1R sweet and umami receptors and the TAS2R bitter receptors couple to a heterotrimeric G protein that consists of $G\alpha_{gust}$ and its partners, $\beta1\gamma13$. Upon receptor activation, $G\beta\gamma$ activates PLCβ2, producing IP$_3$, which causes release of Ca^{2+} from intracellular stores and Ca^{2+}-dependent activation of the transduction channel, TRPM5. The resulting depolarization elicits action potentials, which evoke release of ATP via the voltage-dependent ATP release channel, CALHM1. The released ATP binds and activates the purinergic receptors P2X2 and P2X3 on the afferent fibers to transmit taste information to the brain. The role of $G\alpha_{gust}$ in the process, at least for bitter transduction, is to tonically activate PDE to keep cAMP levels low and prevent PKA-mediated phosphorylation and subsequent desensitization of the Ca^{2+} signaling effectors. Thus, taste cells are kept in an active state so they can respond to taste stimuli with robust Ca^{2+} signals. However, whether $G\alpha_{gust}$ plays a similar role in sweet and umami transduction is not yet clear. Sweet stimuli tend to increase rather than decrease cAMP levels in biochemical studies and sweet and umami taste receptors are generally not coexpressed with $G\alpha_{gust}$ in circumvallate and foliate taste buds. Thus determining the $G\alpha$ subunits that regulate sweet and umami transduction in these taste buds should be an important target of future investigations.

Acknowledgments

I would like to thank Ms Nicole Shultz and Dr Thomas Finger of the Rocky Mountain Taste and Smell Center for the images in Figure 1. Dr Finger also provided helpful comments on the manuscript. The transgenic mice used for the image in Figure 1 were originally provided by Dr Robert Margolskee, currently at the Monell Chemical Senses Center. Support for this book chapter was provided in part by R01DC012555 (to SCK) and P30DC004657 (to D. Restrepo).

References

1. Bartel DL, Sullivan SL, Lavoie EG, Sevigny J, Finger TE. Nucleoside triphosphate diphosphohydrolase-2 is the ecto-ATPase of type I cells in taste buds. *J Comp Neurol.* July 1, 2006;497(1):1−12.
2. Clapp TR, Yang R, Stoick CL, Kinnamon SC, Kinnamon JC. Morphologic characterization of rat taste receptor cells that express components of the phospholipase C signaling pathway. *J Comp Neurol.* January 12, 2004;468(3):311−321.
3. DeFazio RA, Dvoryanchikov G, Maruyama Y, et al. Separate populations of receptor cells and presynaptic cells in mouse taste buds. *J Neurosci.* April 12, 2006;26(15):3971−3980.
4. Murray RG, Murray A. Relations and possible significance of taste bud cells. *Contrib Sens Physiol.* 1971;5:47−95.
5. Royer SM, Kinnamon JC. HVEM serial-section analysis of rabbit foliate taste buds: I. Type III cells and their synapses. *J Comp Neurol.* April 1, 1991;306(1):49−72.
6. Huang YA, Maruyama Y, Stimac R, Roper SD. Presynaptic (Type III) cells in mouse taste buds sense sour (acid) taste. *J Physiol.* June 15, 2008;586(Pt 12):2903−2912.
7. Chandrashekar J, Yarmolinsky D, von Buchholtz L, et al. The taste of carbonation. *Science.* October 16, 2009;326(5951):443−445.
8. Oka Y, Butnaru M, von Buchholtz L, Ryba NJ, Zuker CS. High salt recruits aversive taste pathways. *Nature.* February 28, 2013;494(7438):472−475.
9. Clapp TR, Medler KF, Damak S, Margolskee RF, Kinnamon SC. Mouse taste cells with G protein-coupled taste receptors lack voltage-gated calcium channels and SNAP-25. *BMC Biol.* 2006;4:7.
10. Medler KF, Margolskee RF, Kinnamon SC. Electrophysiological characterization of voltage-gated currents in defined taste cell types of mice. *J Neurosci.* April 1, 2003;23(7):2608−2617.
11. Romanov RA, Kolesnikov SS. Electrophysiologically identified subpopulations of taste bud cells. *Neurosci Lett.* March 13, 2006;395(3):249−254.

12. Vandenbeuch A, Zorec R, Kinnamon SC. Capacitance measurements of regulated exocytosis in mouse taste cells. *J Neurosci*. November 3, 2010;30(44):14695–14701.

13. Murata Y, Yasuo T, Yoshida R, et al. Action potential-enhanced ATP release from taste cells through hemi-channels. *J Neurophysiol*. August 2010;104(2):896–901.

14. Huang AL, Chen X, Hoon MA, et al. The cells and logic for mammalian sour taste detection. *Nature*. August 24, 2006;442(7105):934–938.

15. Reed DR. Gene mapping for taste related phenotypes in humans and mice. *Appetite*. October 2000;35(2): 189–190.

16. Adler E, Hoon MA, Mueller KL, Chandrashekar J, Ryba NJ, Zuker CS. A novel family of mammalian taste receptors. *Cell*. March 17, 2000;100(6):693–702.

17. Chandrashekar J, Mueller KL, Hoon MA, et al. T2Rs function as bitter taste receptors. *Cell*. March 17, 2000;100(6):703–711.

18. Matsunami H, Montmayeur JP, Buck LB. A family of candidate taste receptors in human and mouse. *Nature*. April 6, 2000;404(6778):601–604.

19. Meyerhof W, Batram C, Kuhn C, et al. The molecular receptive ranges of human TAS2R bitter taste receptors. *Chem Senses*. February 2010;35(2):157–170.

20. Caicedo A, Roper SD. Taste receptor cells that discriminate between bitter stimuli. *Science*. February 23, 2001;291(5508):1557–1560.

21. Behrens M, Foerster S, Staehler F, Raguse JD, Meyerhof W. Gustatory expression pattern of the human TAS2R bitter receptor gene family reveals a heterogenous population of bitter responsive taste receptor cells. *J Neurosci*. November 14, 2007;27(46):12630–12640.

22. Max M, Shanker YG, Huang L, et al. Tas1r3, encoding a new candidate taste receptor, is allelic to the sweet responsiveness locus Sac. *Nat Genet*. May 2001;28(1):58–63.

23. Montmayeur JP, Liberles SD, Matsunami H, Buck LB. A candidate taste receptor gene near a sweet taste locus. *Nat Neurosci*. May 2001;4(5):492–498.

24. Nelson G, Chandrashekar J, Hoon MA, et al. An amino-acid taste receptor. *Nature*. March 14, 2002;416(6877): 199–202.

25. Nelson G, Hoon MA, Chandrashekar J, Zhang Y, Ryba NJ, Zuker CS. Mammalian sweet taste receptors. *Cell*. August 10, 2001;106(3):381–390.

26. Sainz E, Korley JN, Battey JF, Sullivan SL. Identification of a novel member of the T1R family of putative taste receptors. *J Neurochem*. May 2001;77(3):896–903.

27. Li X, Staszewski L, Xu H, Durick K, Zoller M, Adler E. Human receptors for sweet and umami taste. *Proc Natl Acad Sci USA*. April 2, 2002;99(7):4692–4696.

28. Zhang Y, Hoon MA, Chandrashekar J, et al. Coding of sweet, bitter, and umami tastes: different receptor cells sharing similar signaling pathways. *Cell*. February 7, 2003;112(3):293–301.

29. McLaughlin SK, McKinnon PJ, Margolskee RF. Gustducin is a taste-cell-specific G protein closely related to the transducins. *Nature*. June 18, 1992;357(6379):563–569.

30. Huang L, Shanker YG, Dubauskaite J, et al. Ggamma13 colocalizes with gustducin in taste receptor cells and mediates IP3 responses to bitter denatonium. *Nat Neurosci*. December 1999;2(12):1055–1062.

31. Rossler P, Kroner C, Freitag J, Noe J, Breer H. Identification of a phospholipase C beta subtype in rat taste cells. *Eur J Cell Biol*. November 1998;77(3):253–261.

32. Clapp TR, Stone LM, Margolskee RF, Kinnamon SC. Immunocytochemical evidence for co-expression of Type III IP3 receptor with signaling components of bitter taste transduction. *BMC Neurosci*. 2001;2:6.

33. Miyoshi MA, Abe K, Emori Y. IP(3) receptor type 3 and PLCbeta2 are co-expressed with taste receptors T1R and T2R in rat taste bud cells. *Chem Senses*. March 2001;26(3):259–265.

34. Hofmann T, Chubanov V, Gudermann T, Montell C. TRPM5 is a voltage-modulated and Ca(2+)-activated monovalent selective cation channel. *Curr Biol*. July 1, 2003;13(13):1153–1158.

35. Perez CA, Huang L, Rong M, et al. A transient receptor potential channel expressed in taste receptor cells. *Nat Neurosci*. November 2002;5(11):1169–1176.

36. Prawitt D, Monteilh-Zoller MK, Brixel L, et al. TRPM5 is a transient Ca^{2+}-activated cation channel responding to rapid changes in [Ca2+]i. *Proc Natl Acad Sci USA*. December 9, 2003;100(25):15166–15171.

37. Gao N, Lu M, Echeverri F, et al. Voltage-gated sodium channels in taste bud cells. *BMC Neurosci*. March 12, 2009;10(1):20.

38. Yoshida R, Sanematsu K, Shigemura N, Yasumatsu K, Ninomiya Y. Taste receptor cells responding with action potentials to taste stimuli and their molecular expression of taste related genes. *Chem Senses*. January 2005;30(suppl 1):i19—20.

39. Taruno A, Vingtdeux V, Ohmoto M, et al. CALHM1 ion channel mediates purinergic neurotransmission of sweet, bitter and umami tastes. *Nature*. March 14, 2013;495(7440):223—226.

40. Hellekant G, Schmolling J, Marambaud P, Rose-Hellekant TA. CALHM1 deletion in mice affects glossopharyngeal taste responses, food intake, body weight, and life span. *Chem Senses*. April 8, 2015;40(6): 373—379.

41. Ogura T, Margolskee RF, Kinnamon SC. Taste receptor cell responses to the bitter stimulus denatonium involve Ca^{2+} influx via store-operated channels. *J Neurophysiol*. June 2002;87(6):3152—3155.

42. Kim JW, Roberts C, Maruyama Y, Berg S, Roper S, Chaudhari N. Faithful expression of GFP from the PLCbeta2 promoter in a functional class of taste receptor cells. *Chem Senses*. March 2006;31(3):213—219.

43. Starostik MR, Rebello MR, Cotter KA, Kulik A, Medler KF. Expression of GABAergic receptors in mouse taste receptor cells. *PLoS One*. 2010;5(10):e13639.

44. Hacker K, Laskowski A, Feng L, Restrepo D, Medler K. Evidence for two populations of bitter responsive taste cells in mice. *J Neurophysiol*. March 2008;99(3):1503—1514.

45. Damak S, Rong M, Yasumatsu K, et al. Trpm5 null mice respond to bitter, sweet, and umami compounds. *Chem Senses*. March 2006;31(3):253—264.

46. Hisatsune C, Yasumatsu K, Takahashi-Iwanaga H, et al. Abnormal taste perception in mice lacking the type 3 inositol 1,4,5-trisphosphate receptor. *J Biol Chem*. December 21, 2007;282(51):37225—37231.

47. Matsumoto I, Ohmoto M, Narukawa M, Yoshihara Y, Abe K. Skn-1a (Pou2f3) specifies taste receptor cell lineage. *Nat Neurosci*. June 2011;14(6):685—687.

48. Zhang Z, Zhao Z, Margolskee RF, Liman ER. The transduction channel TRPM5 is gated by intracellular calcium in taste cells. *J Neurosci*. 2007;27:5777—5786.

49. Medler KF. Calcium signaling in taste cells: regulation required. *Chem Senses*. November 2010;35(9):753—765.

50. Medler KF. Calcium signaling in taste cells. *Biochim Biophys Acta*. November 16, 2014;1853(9):2025—2032.

51. Hacker K, Medler KF. Mitochondrial calcium buffering contributes to the maintenance of basal calcium levels in mouse taste cells. *J Neurophysiol*. October 2008;100(4):2177—2191.

52. Szebenyi SA, Laskowski AI, Medler KF. Sodium/calcium exchangers selectively regulate calcium signaling in mouse taste receptor cells. *J Neurophysiol*. July 2010;104(1):529—538.

53. Iguchi N, Ohkuri T, Slack JP, Zhong P, Huang L. Sarco/endoplasmic reticulum Ca^{2+}-ATPases (SERCA) contribute to GPCR-mediated taste perception. *PLoS One*. 2011;6(8):e23165.

54. McLaughlin SK, McKinnon PJ, Spickofsky N, Danho W, Margolskee RF. Molecular cloning of G proteins and phosphodiesterases from rat taste cells. *Physiol Behav*. December 1994;56(6):1157—1164.

55. Kusakabe Y, Yamaguchi E, Tanemura K, et al. Identification of two alpha-subunit species of GTP-binding proteins, Galpha15 and Galphaq, expressed in rat taste buds. *Biochim Biophys Acta*. July 24, 1998;1403(3): 265—272.

56. Kusakabe Y, Yasuoka A, Asano-Miyoshi M, et al. Comprehensive study on G protein alpha-subunits in taste bud cells, with special reference to the occurrence of Galphai2 as a major Galpha species. *Chem Senses*. October 2000;25(5):525—531.

57. Kim MR, Kusakabe Y, Miura H, Shindo Y, Ninomiya Y, Hino A. Regional expression patterns of taste receptors and gustducin in the mouse tongue. *Biochem Biophys Res Commun*. December 12, 2003;312(2):500—506.

58. Stone LM, Barrows J, Finger TE, Kinnamon SC. Expression of T1Rs and gustducin in palatal taste buds of mice. *Chem Senses*. March 2007;32(3):255—262.

59. McLaughlin SK, McKinnon PJ, Robichon A, Spickofsky N, Margolskee RF. Gustducin and transducin: a tale of two G proteins. *Ciba Found Symp*. 1993;179:186—196. discussion 196—200.

60. Yan W, Sunavala G, Rosenzweig S, Dasso M, Brand JG, Spielman AI. Bitter taste transduced by PLC-beta(2)-dependent rise in IP(3) and alpha-gustducin-dependent fall in cyclic nucleotides. *Am J Physiol Cell Physiol*. April 2001;280(4):C742—C751.

61. Abaffy T, Trubey KR, Chaudhari N. Adenylyl cyclase expression and modulation of cAMP in rat taste cells. *Am J Physiol Cell Physiol*. June 2003;284(6):C1420—C1428.

62. Misaka T, Kusakabe Y, Emori Y, Gonoi T, Arai S, Abe K. Taste buds have a cyclic nucleotide-activated channel, CNGgust. *J Biol Chem*. September 5, 1997;272(36):22623—22629.

63. Giovannucci DR, Groblewski GE, Sneyd J, Yule DI. Targeted phosphorylation of inositol 1,4,5-trisphosphate receptors selectively inhibits localized Ca^{2+} release and shapes oscillatory Ca^{2+} signals. *J Biol Chem*. October 27, 2000;275(43):33704–33711.

64. Clapp TR, Trubey KR, Vandenbeuch A, et al. Tonic activity of Galpha-gustducin regulates taste cell responsivity. *FEBS Lett*. November 12, 2008;582(27):3783–3787.

65. Wong GT, Gannon KS, Margolskee RF. Transduction of bitter and sweet taste by gustducin. *Nature*. June 27, 1996;381(6585):796–800.

66. Danilova V, Damak S, Margolskee RF, Hellekant G. Taste responses to sweet stimuli in alpha-gustducin knockout and wild-type mice. *Chem Senses*. July 2006;31(6):573–580.

67. Glendinning JI, Bloom LD, Onishi M, et al. Contribution of alpha-gustducin to taste-guided licking responses of mice. *Chem Senses*. May 2005;30(4):299–316.

68. Tizzano M, Dvoryanchikov G, Barrows JK, Kim S, Chaudhari N, Finger TE. Expression of Galpha14 in sweet-transducing taste cells of the posterior tongue. *BMC Neurosci*. 2008;9:110.

69. Striem B, Naim M, Lindemann B. Generation of cyclic AMP in taste buds of the rat circumvallate papilla in response to sucrose. *Cell Physiol Biochem*. 1991;1:46–54.

70. Bernhardt SJ, Naim M, Zehavi U, Lindemann B. Changes in IP3 and cytosolic Ca^{2+} in response to sugars and non-sugar sweeteners in transduction of sweet taste in the rat. *J Physiol*. January 15, 1996;490(Pt 2):325–336.

71. Trubey KR, Culpepper S, Maruyama Y, Kinnamon SC, Chaudhari N. Tastants evoke cAMP signal in taste buds that is independent of calcium signaling. *Am J physiology Cell physiology*. August 2006;291(2):C237–C244.

72. Cummings TA, Daniels C, Kinnamon SC. Sweet taste transduction in hamster: sweeteners and cyclic nucleotides depolarize taste cells by reducing a K^+ current. *J Neurophysiol*. March 1996;75(3):1256–1263.

73. Cummings TA, Powell J, Kinnamon SC. Sweet taste transduction in hamster taste cells: evidence for the role of cyclic nucleotides. *J Neurophysiol*. December 1993;70(6):2326–2336.

74. von Buchholtz L, Elischer A, Tareilus E, et al. RGS21 is a novel regulator of G protein signalling selectively expressed in subpopulations of taste bud cells. *Eur J Neurosci*. March 2004;19(6):1535–1544.

75. Cohen SP, Buckley BK, Kosloff M, et al. Regulator of G-protein signaling-21 (RGS21) is an inhibitor of bitter gustatory signaling found in lingual and airway epithelia. *J Biol Chem*. December 7, 2012;287(50):41706–41719.

76. Fenech C, Patrikainen L, Kerr DS, et al. Ric-8A, a Galpha protein guanine nucleotide exchange factor potentiates taste receptor signaling. *Front Cell Neurosci*. 2009;3:11.

77. Royer SM, Kinnamon JC. Ultrastructure of mouse foliate taste buds: synaptic and nonsynaptic interactions between taste cells and nerve fibers. *J Comp Neurol*. April 1, 1988;270(1):11–24, 58-9.

78. Bo X, Alavi A, Xiang Z, Oglesby I, Ford A, Burnstock G. Localization of ATP-gated P2X2 and P2X3 receptor immunoreactive nerves in rat taste buds. *Neuroreport*. April 6, 1999;10(5):1107–1111.

79. Finger TE, Danilova V, Barrows J, et al. ATP signaling is crucial for communication from taste buds to gustatory nerves. *Science*. December 2, 2005;310(5753):1495–1499.

80. Vandenbeuch A, Larson ED, Anderson CB, et al. Postsynaptic P2X3-containing receptors in gustatory nerve fibres mediate responses to all taste qualities in mice. *J Physiol*. March 1, 2015;593(5):1113–1125.

81. Huang YJ, Maruyama Y, Dvoryanchikov G, Pereira E, Chaudhari N, Roper SD. The role of pannexin 1 hemichannels in ATP release and cell-cell communication in mouse taste buds. *Proc Natl Acad Sci USA*. April 10, 2007;104(15):6436–6441.

82. Romanov RA, Rogachevskaja OA, Bystrova MF, Jiang P, Margolskee RF, Kolesnikov SS. Afferent neurotransmission mediated by hemichannels in mammalian taste cells. *EMBO J*. February 7, 2007;26(3):657–667.

83. Tordoff MG, Ellis HT, Aleman TR, et al. Salty taste deficits in CALHM1 knockout mice. *Chem Senses*. July 2014;39(6):515–528.

84. Romanov RA, Bystrova MF, Rogachevskaya OA, Sadovnikov VB, Shestopalov VI, Kolesnikov SS. The ATP permeability of pannexin 1 channels in a heterologous system and in mammalian taste cells is dispensable. *J Cell Sci*. November 15, 2012;125(Pt 22):5514–5523.

85. Tordoff MG, Aleman TR, Ellis HT, et al. Normal taste acceptance and preference of PANX1 knockout mice. *Chem Senses*. May 18, 2015;40(7).

86. Vandenbeuch A, Anderson CB, Kinnamon SC. Mice lacking Pannexin 1 release ATP and respond normally to all taste qualities. *Chem Senses*. 2015;40(7):461–467.

87. Krasteva G, Hartmann P, Papadakis T, et al. Cholinergic chemosensory cells in the auditory tube. *Histochem Cell Biol*. April 2012;137(4):483–497.

88. Krasteva G, Canning BJ, Hartmann P, et al. Cholinergic chemosensory cells in the trachea regulate breathing. *Proc Natl Acad Sci USA.* June 7, 2011;108(23):9478—9483.
89. Kaske S, Krasteva G, Konig P, et al. TRPM5, a taste-signaling transient receptor potential ion-channel, is a ubiquitous signaling component in chemosensory cells. *BMC Neurosci.* 2007;8:49.
90. Finger TE, Bottger B, Hansen A, Anderson KT, Alimohammadi H, Silver WL. Solitary chemoreceptor cells in the nasal cavity serve as sentinels of respiration. *Proc Natl Acad Sci USA.* July 22, 2003;100(15):8981—8986.
91. Sbarbati A, Merigo F, Benati D, et al. Identification and characterization of a specific sensory epithelium in the rat larynx. *J Comp Neurol.* July 19, 2004;475(2):188—201.
92. Lin W, Ogura T, Margolskee RF, Finger TE, Restrepo D. TRPM5-expressing solitary chemosensory cells respond to odorous irritants. *J Neurophysiol.* March 2008;99(3):1451—1460.
93. Ogura T, Krosnowski K, Zhang L, Bekkerman M, Lin W. Chemoreception regulates chemical access to mouse vomeronasal organ: role of solitary chemosensory cells. *PLoS One.* 2010;5(7):e11924.
94. Krasteva G, Canning BJ, Veres T, et al. Tracheal brush cells are neuronally connected cholinergic sensory cells (abstract). *Paper Presented at: Society for Neuroscience.* 2010.
95. Saunders CJ, Christensen M, Finger TE, Tizzano M. Cholinergic neurotransmission links solitary chemosensory cells to nasal inflammation. *Proc Natl Acad Sci USA.* April 22, 2014;111(16):6075—6080.
96. Tizzano M, Cristofoletti M, Sbarbati A, Finger TE. Expression of taste receptors in solitary chemosensory cells of rodent airways. *BMC Pulm Med.* 2011;11:3.
97. Tizzano M, Finger TE. Chemosensors in the nose: guardians of the airways. *Physiology.* January 2013;28(1):51—60.
98. Gulbransen BD, Clapp TR, Finger TE, Kinnamon SC. Nasal solitary chemoreceptor cell responses to bitter and trigeminal stimulants in vitro. *J Neurophysiol.* June 2008;99(6):2929—2937.
99. Tizzano M, Gulbransen BD, Vandenbeuch A, et al. Nasal chemosensory cells use bitter taste signaling to detect irritants and bacterial signals. *Proc Natl Acad Sci USA.* February 16, 2010;107(7):3210—3215.

The Mechanisms of Salty and Sour Taste

Steven D. Munger

Center for Smell and Taste, University of Florida, Gainesville, FL, USA;
Department of Pharmacology and Therapeutics, University of Florida, Gainesville, FL, USA;
Department of Medicine, Division of Endocrinology, Diabetes and Metabolism,
University of Florida, Gainesville, FL, USA

OUTLINE

INTRODUCTION

Taste stimuli can be as large as a protein or as small as a single proton. In the case of salts and acids—which elicit salty and sour taste, respectively—the stimuli are ions. As such, the gustatory receptors for salty and sour-tasting stimuli are predicted to be ion channels. This contrasts quite starkly with the transduction of sweet-, bitter-, or umami-tasting stimuli, which rely on G protein-coupled receptors (GPCRs) and an associated enzymatic cascade (see Chapters 13 and 15). Although salty and sour taste stimuli share a similar transduction strategy, the specific molecular and cellular mechanisms employed in the gustatory detection of salts and acids are quite distinct. This chapter will explore the ion channels and other proteins that have been implicated in vertebrate salt and acid taste. For a review of these processes in invertebrates (particularly the fly), please consult this review[1] or many other excellent sources.

SALT TASTE

The gustatory system helps to maintain electrolyte homeostasis by regulating salt intake. At lower concentrations (<150 mM), table salt (NaCl) is appetitive to mammals and is preferentially consumed, whereas higher concentrations of NaCl become aversive and are not consumed under normal circumstances. However, sodium depletion increases the preference for sodium, but not potassium, salts and for salty foods, and decreases the threshold for salt taste detection.[2,3] These and other observations, some of which will be discussed in this chapter, suggest that mammals have multiple mechanisms for salt detection: one that is specific for Na^+ and mediates appetitive salt taste, and one that responds to multiple cations and mediates salt aversion.

Amiloride-Sensitive Na^+ Taste

We have a dichotomous relationship with the sodium ion. Na^+ is an important nutrient that is critical for establishing and altering cellular membrane potentials, absorbing nutrients and maintaining blood pressure and volume. Sodium deficiencies (hyponatremia) can lead to devastating effects, including seizures, coma, or death.[4] However, elevated Na^+ levels (hypernatremia), especially levels that exceed the ability of the kidneys to sufficiently excrete Na^+, can be similarly harmful.[4] Of particular note is the strong association between hypertension and excessive Na^+ intake.[5] Therefore, maintaining Na^+ homeostasis is critical for health and survival.

Based on the temporal and physiological characteristics of NaCl-evoked responses in electrophysiological recordings from rat taste nerves, Lloyd Beidler proposed that the transduction of Na^+ ions was most likely mediated by an ion channel or transporter, as opposed to an enzymatic cascade.[6] This idea was supported by studies from John Desimone and colleagues on ion transport in the lingual epithelium of the dog.[7] But it was the observation that Na^+ taste responses are at least partially sensitive to the diuretic amiloride, which inhibits transepithelial Na^+ transport in a variety of tissues, that provided the strongest clue as to the molecular identity of the Na^+ taste mechanism. Human psychophysical studies by Schiffman and colleagues found that the intensity of Na^+ and Li^+ salts (but also sweeteners) was reduced in individuals who had their tongues exposed to amiloride.[8] A similar effect was found in rats by Desimone and colleagues, where application of amiloride to the tongue reduced chorda tympani responses in those animals to NaCl but not KCl.[9] These findings were confirmed in many species, including rat,[10,11] mouse,[12] hamster,[13] macaque,[14] and chimpanzee.[15] Oral amiloride decreased consumption of normally appetitive 0.1 M NaCl by hamsters,[13] whereas sodium appetite was nearly eliminated in rats that were pretreated with amiloride.[16] Amiloride also decreases behavioral taste sensitivity of mice to sodium salts.[17] Patch clamp recordings from isolated frog fungiform taste receptor cells (TRCs) or from rat fungiform TRCs in isolated taste buds showed that amiloride inhibits Na^+ responses at the level of the TRC.[18,19] Similar conclusions were reached in studies using a Ca^{2+} imaging approach in an *ex vivo* preparation of fungiform taste buds.[20] These and other studies also clarified that there is an amiloride-insensitive component of Na^+ taste, which will be discussed later. Even so, these findings pointed to members of the amiloride-sensitive degenerin (DEG)/epithelial Na^+ channel (ENaC) family as intriguing candidates to be the major Na^+ detector in vertebrate taste.

The DEG/ENaC channel family is found in diverse animal species including mammals, molluscs, nematodes, and fruitflys.[21] These channels have two transmembrane domains joined by a large extracellular loop, function in multimeric complexes, are voltage-insensitive, and are selective for Na^+ over other cations.[21,22] The native mammalian ENaCs is heteromeric (likely a trimer) and composed of α, β, and γ subunits (Figure 1), although a δ subunit may substitute for ENaCα in some tissues.[21,23,24] ENaCs play a major role in regulating Na^+ flux across epithelia in a variety of tissues, and are particularly important in the regulation of salt and water homeostasis through their functions in the kidney.[25]

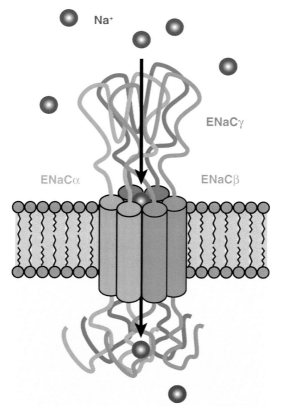

FIGURE 1 **The epithelial Na^+ channel (ENaC) mediates the appetitive taste of sodium salts.** ENaC is composed of three subunits, α (orange), β (blue), and γ (green). This heteromeric channel is highly selective for Na^+ (purple).

Several lines of evidence in addition to the effects of amiloride support a role for ENaCs in Na^+ taste. For example, taste receptor cells of the anterior tongue (those innervated by the chorda tympani branch of the facial nerve, one of three cranial nerves that carry taste information from the oral cavity to the brainstem nucleus of the solitary tract) express ENaCα, ENaCβ, and ENaCγ, as assessed at both the messenger RNA and protein level.[26–28] Vallate taste buds in the posterior tongue (innervated by the glossopharyngeal nerve) also express the three channel subunits,[26] but there seems to be little overlap between ENaCα and ENaCβ.[20] Because all three subunits are required to form a functional channel, there is unlikely to be a functional ENaC in the posterior taste field; this is consistent with observations that Na^+ responses in the rat glossopharyngeal nerve are amiloride-insensitive.[29]

Major support for the idea that ENaC functions as the primary Na^+ taste sensor *in vivo* came as the result of a conditional knockout strategy in which *Scnn1a* (the gene encoding ENaCα) was deleted in mouse TRCs, rendering ENaC nonfunctional in the peripheral taste system.[20] Chorda tympani nerve recordings from these animals showed no taste responses to low concentrations of NaCl as well as an absence of any effect of amiloride on responses to higher NaCl concentrations or other salts. Furthermore, the knockouts lacked a behavioral preference for NaCl, although they continued to avoid higher NaCl concentrations.

However, major questions remain. For example, the TRC subtype that mediates Na^+ taste has not been identified. By contrast, stimuli that elicit other taste qualities have been shown to activate distinct TRC subpopulations (i.e., "sweet," "bitter," "umami," and "sour" TRCs) that can be differentiated based on stimulus selectivity and the expression of unique molecular markers. One study found that amiloride-sensitive Na^+ channels were found in fungiform TRCs that lack voltage-dependent inward currents,[30] but the molecular and functional identity of these cells has not been further refined. It is also unclear to what extent ENaCs are important for human Na^+ taste. Amiloride appears less effective in humans than in rodents, with one study finding that amiloride suppresses only ~20% of the human taste response to NaCl[31] (this number is closer to 70% in rodents[3]). By contrast, Na^+ taste in nonhuman primates shows a significant sensitivity to amiloride.[14,15] One possible explanation for this discrepancy is the effect of diet on ENaC expression.[3] Salt is widely available in the modern human diet, but less plentiful in the diet of other mammals. The effects of amiloride on NaCl-evoked lingual surface potential (a metric of Na^+ transepithelial conductance) also varies greatly across individuals, with some subjects reporting hardly any effect and others seeing nearly a 50% decrement.[32] The observations that amiloride-sensitive Na^+ currents are found in vallate taste cells after injections of aldosterone (a steroid hormone that acts to conserve sodium)[27] suggests a mechanism by which systemic Na^+ levels could impact the expression of ENaCs in the taste bud.

Amiloride-Insensitive Salt Taste

As mentioned previously, a significant component of Na^+ taste is amiloride-insensitive. Electrophysiological recordings from taste nerves of rodents and nonhuman primates indicate that they contain Na^+-responsive fibers that are amiloride-insensitive and respond to other salts such as KCl.[15,33–36] Similar results were seen in TRCs with either patch clamp recording or Ca^{2+} imaging: amiloride-sensitive TRCs were Na^+-specific, whereas amiloride-insensitive cells responded to both higher concentrations of NaCl (tested at 300–500 mM) and to

KCl.[19,20] Mice generalize the tastes of NaCl and KCl in the presence of amiloride,[37] indicating that amiloride-insensitive salt taste is relatively nonselective for cations. As might be expected from these results, conditional knockout mice lacking ENaCα in TRCs retained taste responses to (and avoidance of) nonsodium salts and higher concentrations of NaCl.[20]

So, what is the molecular identity of the amiloride-insensitive salt taste transducer? It remains unknown. One candidate is a TRC-specific variant of the transient receptor potential V1 (TRPV1, a.k.a., vanilloid receptor-1 or VR-1) channel referred to as TRPV1t.[38] Although the TRPV1 complementary DNA can be amplified from rat fungiform TRCs,[38] immunohisto-chemistry suggests that the channel may more likely be associated with nerve fibers inner-vating the taste bud.[39,40] TRPV1 agonists inhibit chorda tympani nerve responses to salts,[38] and *Trpv1* knockout mice show no tonic responses in the chorda tympani upon treatment with benzamil (an amiloride analog).[41] However, in behavioral tests these same knockout mice show no obvious deficits in NaCl or KCl detection thresholds and maintain the ability to generalize between these two salts in conditioned taste aversion tests, arguing that TRPV1t is not required for amiloride-insensitive salt taste.[41,42]

Another model suggests that nonsodium salts and high concentrations of sodium salts conscript transduction mechanisms in "sour" and "bitter"-responsive taste cells.[43] This model has some appeal as high concentrations of NaCl, nonsodium salts, acids, and bitter stimuli are all normally aversive. In this way, the hedonic valence of amiloride-insensitive salt taste is also an important aspect of its encoding in the gustatory periphery. Even so, this model remains to be fully tested to understand how aversive salts are differentiated from sour or bitter tastants. Furthermore, potential molecular mediators of salt detection in bitter and sour TRCs remain to be identified.

SOUR TASTE

Animals need to avoid toxic, contaminated, or spoiled foods. Aversive taste qualities such as bitterness or sourness are often associated with noxious compounds[44] (although aversive tastes can become preferred in the complex context of foods[45]). This relationship is particu-larly clear for bitter-tasting compounds, many of which are plant-derived toxins. The reasons to eschew acid ingestion is less clear, but may be associated with a need to avoid unripe or spoiled foods as well as the importance of maintaining acid-base homeostasis. Indeed, lower pH in the oral cavity can promote the erosion of tooth enamel and impact dental caries, while excess acid ingestion is associated with metabolic acidosis and toxicity.[3,46,47]

BREAKOUT BOX

TURNING SOUR INTO SWEET

Sweet-tasting substances can effectively mask the aversive taste of acids, which is why sugar-sweetened lemonade is much more palatable than pure lemon juice. But miraculin and other so-called "taste-modifying proteins" were long thought to go one step further to make sour substances taste sweet. How could this be? In fact, taste-modifying proteins *do not* modify taste. Rather, they represent a novel group of ligands for the human sweet taste receptor TAS1R2:TAS1R3 (see Chapter 13) that function as antagonists (or perhaps inverse agonists[65]) at neutral pH but become agonists under acidic conditions.[66,67]

Several small tropical fruits produce proteins, ranging in size from 6.5 to nearly 100 kDa, that can act as ligands for the mammalian sweet taste receptor. Some of these proteins—the so-called "sweet proteins" that include thaumatin (from *Thaumatococcus danielli*), monellin (from *Dioscoreophyllum cumminsii*) and brazzein (from *Pentadiplandra brazzeana*)—are highly potent stimuli that are thousands of times sweeter than sucrose on a molar basis.[68] However, miraculin (from the "miracle fruit" of *Richadella dulcifica*) and neoculin (the heterodimeric form of curculin, from *Curculingo latifolia*), which are structurally unrelated to the "sweet proteins" or to each other, have little or no sweetness at neutral pH but exhibit pronounced sweetness when exposed to acids. Mutagenesis studies suggests that acids exert this effect by protonating key histidine residues on miraculin and neoculin, causing conformational changes in these ligands that result in activation of the sweet taste receptor to which they are already bound.[66,67]

We rarely experience taste stimuli in isolation, as most foods are a complex mixture of compounds that can elicit different taste perceptual qualities. Although taste-modifying proteins have been suggested as potential low-calorie sweeteners,[68] they remain at this point a novelty in the food world. But they also offer a unique tool to understand one way that stimuli of different perceptual qualities—in this case, sweet and sour—can interact at the level of the sensory cell to modify the way we perceive taste.

It is often assumed that a stronger acid has a more sour taste, but this is not the case.[48,49] Rather, weak organic acids such as acetic acid or citric acid have a more intense sour taste than do strong mineral acids such as HCl at the same pH. Weak acids are only partially dissociated in water, and thus their titratable acidity is higher than that of a fully dissociated strong acid. Indeed, it seems that titratable acidity is a better predictor of sour taste intensity than is the concentration of dissociated protons in solution.[45]

Even so, the proton (H^+, or more precisely, the hydronium ion HCO_3^+) is still believed to be the proximate stimulus for acid taste, although it may act at different sites depending on the source (Figure 2). Weak organic acids likely diffuse across the plasma membrane as neutral molecules.[50] Dissociation of the acids within the cytoplasm results in intracellular acidification. How this impacts cellular physiology is unclear. Possibly, intracellular acidification results in cellular depolarization by inhibiting membrane K^+ channels (Figure 2) or perhaps by activating other cation channels,[51,52] either through direct modification of the

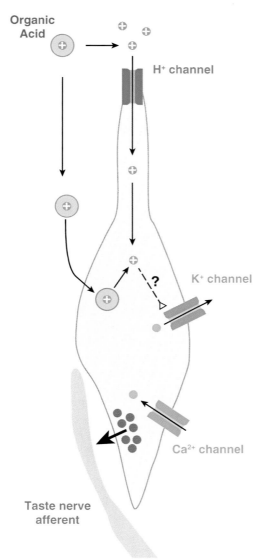

FIGURE 2   **Model for the activation of sour taste receptor cells by strong and weak acids.** Strong acids, which are fully dissociated in solution, depolarize sour TRCs via an inward proton conductance (green). This could lead to additional depolarization through activation of voltage-gated Na^+ channels (not shown) or inhibition (open triangle) of K^+ channels (blue) via intracellular acidification. Weak acids (red/green), which are only partially dissociated in solution, are thought to diffuse across the plasma membrane before dissociating intracellularly (extracellular protons would still pass through proton channels). Cytoplasmic protons could then inhibit K^+ channels. Either mechanism would result in membrane depolarization and subsequent activation of voltage-gated Ca^{2+} channels (orange), promoting the release of neurotransmitter (gray circles) onto nearby afferent nerves (purple). *Modified from Ref. 1 with permission.*

channel proteins or perhaps via cytoskeletal changes that could mechanically gate the channels. [50,53] By contrast, H^+ from strong acids, which are fully dissociated in the extracellular solution, likely enter the TRC through plasma membrane channels (see the following section). This H^+ flux can both provide an initial depolarization of the TRC as well as contribute to the intracellular acidification of other proteins, including ion channels.

Several channel and transporter types have been proposed as sour taste receptors. These include: ACCN1 (amiloride-sensitive cation channel 1); HCN1 (hyperpolarization-activated cyclic nucleotide-gated channel 1) and HCN4; the two-pore potassium channel KCNK3 (potassium channel, subfamily K, member 3; a.k.a., TASK-1); and NHE-1 (Na^+-H^+-exchanger isoform 1), but convincing proof is lacking for any of them[54]. An extracellular carbonic anhydrase isoform, CAR4, has been implicated in the transduction of an unusual sour stimulus, CO_2,[55] but CAR4 likely serves a limited role in catalyzing this sour-tasting gas to release protons that can then acidify the TRC. The most promising candidates in recent years had been members of the transient receptor potential type P channel family, PKD1L3 and PKD2L1. Both are expressed in a subset of TRCs (although PKD1L3 is restricted to vallate and foliate taste buds).[56-59] Mice in which PKD2L1-expressing TRCs were ablated using the targeted expression of the diphtheria toxin A subunit displayed normal taste responses to sweet, bitter, umami, and salty stimuli but lacked responses to both weak and strong acids, as assessed by electrophysiological recordings from the chorda tympani nerve.[58] PKD2L1 and PKD1L3 respond to acids when coexpressed in heterologous cells,[57] albeit to the removal of the stimulus rather than to its application.[60] However, both *Pkd1l3* or *Pkd2l1* knockout mice retain largely normal acid responses,[61,62] although *Pkd2l1* knockouts do exhibit a small decrease in responsiveness to acid stimuli in TRCs and in chorda tympani (but not glossopharyngeal) nerve recordings.[62] Therefore, although PKD2L1 and PKD1L3 are clearly expressed in acid-sensitive TRCs, it appears that they are not required for normal acid taste responses. However, the finding that PKD2L1 is a marker for acid-responsive TRCs has enabled researchers to better target their investigation of the transduction mechanisms underlying acid taste. For example, a proton conductance (and candidate acid taste receptor) that is specific to PKD2L1-expressing TRCs has been identified.[63,64] This conductance is evoked by acids, leads to intracellular acidification and promotes the generation of action potentials in the PKD2L1-expressing TRCs.[63,64] Although it is still a mystery what channel protein mediates this H^+ conductance, the tools may now be in hand to raise the curtain on its identity.

CONCLUSIONS

Our understanding of the molecular mechanisms and cellular mediators of salt and acid taste remains relatively poor as compared with other taste qualities. Although the principal transducer of Na^+ taste has been identified as ENaC, the broadly tuned salt receptor(s) remain unidentified as do the proton channel and other targets of intracellular acidification needed to transduce acids. Acid taste requires PKD2L1-espressing TRCs, but which TRC subset expresses the functional ENaC is not known. Amiloride-insensitive salt taste appears to co-opt bitter and sour-sensitive TRCs. The advent of new molecular biological and cellular imaging approaches offers hope for rapid progress for defining these important aspects of salt and acid taste. Such knowledge would be invaluable for positively impacting the palatability of foods, reducing Na^+ consumption, and better understanding the strategies by which the nervous system encodes and processes sensory information.

Acknowledgments

Research in the Munger laboratory is supported by grants from the National Institute on Deafness and Other Communication Disorders.

References

1. Liman ER, Zhang YV, Montell C. Peripheral coding of taste. *Neuron*. 2014;81(5):984−1000.
2. Beauchamp GK, Bertino M, Burke D, Engelman K. Experimental sodium depletion and salt taste in normal human volunteers. *Am J Clin Nutr*. 1990;51(5):881−889.
3. DeSimone JA, Lyall V. Taste receptors in the gastrointestinal tract III. Salty and sour taste: sensing of sodium and protons by the tongue. *Am J Physiol Gastrointest Liver Physiol*. 2006;291(6):G1005−G1010.
4. Sterns RH. Disorders of plasma sodium−causes, consequences, and correction. *N Engl J Med*. 2015;372(1):55−65.
5. Strazzullo P, D'Elia L, Kandala NB, Cappuccio FP. Salt intake, stroke, and cardiovascular disease: meta-analysis of prospective studies. *BMJ*. 2009;339:b4567.
6. Beidler LM. A theory of taste stimulation. *J Gen Physiol*. 1954;38(2):133−139.
7. Desimone JA, Heck GL, Mierson S, Desimone SK. The active ion transport properties of canine lingual epithelia in vitro. Implications for gustatory transduction. *J Gen Physiol*. 1984;83(5):633−656.
8. Schiffman SS, Lockhead E, Maes FW. Amiloride reduces the taste intensity of Na^+ and Li^+ salts and sweeteners. *Proc Natl Acad Sci USA*. 1983;80(19):6136−6140.
9. Heck GL, Mierson S, DeSimone JA. Salt taste transduction occurs through an amiloride-sensitive sodium transport pathway. *Science*. 1984;223(4634):403−405.
10. DeSimone JA, Ferrell F. Analysis of amiloride inhibition of chorda tympani taste response of rat to NaCl. *Am J Physiol*. 1985;249(1 Pt 2):R52−R61.
11. Brand JG, Teeter JH, Silver WL. Inhibition by amiloride of chorda tympani responses evoked by monovalent salts. *Brain Res*. 1985;334(2):207−214.
12. Ninomiya Y, Sako N, Funakoshi M. Strain differences in amiloride inhibition of NaCl responses in mice, *Mus musculus*. *J Comp Physiol A*. 1989;166(1):1−5.
13. Hettinger TP, Frank ME. Specificity of amiloride inhibition of hamster taste responses. *Brain Res*. 1990; 513(1):24−34.
14. Hellekant G, Dubois GE, Roberts TW, Vanderwel H. On the gustatory effect of amiloride in the monkey (*Macaca-Mulatta*). *Chem Senses*. 1988;13(1):89−93.
15. Hellekant G, Ninomiya Y, Danilova V. Taste in chimpanzees II: single chorda tympani fibers. *Physiol Behav*. 1997;61(6):829−841.
16. Bernstein IL, Hennessy CJ. Amiloride-sensitive sodium channels and expression of sodium appetite in rats. *Am J Physiol*. 1987;253(2 Pt 2):R371−R374.
17. Eylam S, Spector AC. Oral amiloride treatment decreases taste sensitivity to sodium salts in C57BL/6J and DBA/2J mice. *Chem Senses*. 2003;28(5):447−458.
18. Avenet P, Lindemann B. Amiloride-blockable sodium currents in isolated taste receptor cells. *J Membr Biol*. 1988;105(3):245−255.
19. Yoshida R, Horio N, Murata Y, Yasumatsu K, Shigemura N, Ninomiya Y. NaCl responsive taste cells in the mouse fungiform taste buds. *Neuroscience*. 2009;159(2):795−803.
20. Chandrashekar J, Kuhn C, Oka Y, et al. The cells and peripheral representation of sodium taste in mice. *Nature*. 2010;464(7286):297−301.
21. Ben-Shahar Y. Sensory functions for degenerin/epithelial sodium channels (DEG/ENaC). *Adv Genet*. 2011;76:1−26.
22. Snyder PM, McDonald FJ, Stokes JB, Welsh MJ. Membrane topology of the amiloride-sensitive epithelial sodium channel. *J Biol Chem*. 1994;269(39):24379−24383.

23. Canessa CM, Schild L, Buell G, et al. Amiloride-sensitive epithelial Na$^+$ channel is made of three homologous subunits. *Nature.* 1994;367(6462):463–467.

24. Kashlan OB, Kleyman TR. ENaC structure and function in the wake of a resolved structure of a family member. *Am J Physiol Ren Physiol.* 2011;301(4):F684–F696.

25. Rossier BC. Epithelial sodium channel (ENaC) and the control of blood pressure. *Curr Opin Pharmacol.* 2014;15:33–46.

26. Kretz O, Barbry P, Bock R, Lindemann B. Differential expression of RNA and protein of the three pore-forming subunits of the amiloride-sensitive epithelial sodium channel in taste buds of the rat. *J Histochem Cytochem.* 1999;47(1):51–64.

27. Lin W, Finger TE, Rossier BC, Kinnamon SC. Epithelial Na$^+$ channel subunits in rat taste cells: localization and regulation by aldosterone. *J Comp Neurol.* 1999;405(3):406–420.

28. Shigemura N, Islam AA, Sadamitsu C, Yoshida R, Yasumatsu K, Ninomiya Y. Expression of amiloride-sensitive epithelial sodium channels in mouse taste cells after chorda tympani nerve crush. *Chem Senses.* 2005; 30(6):531–538.

29. Stewart RE, DeSimone JA, Hill DL. New perspectives in a gustatory physiology: transduction, development, and plasticity. *Am J Physiol.* 1997;272(1 Pt 1):C1–C26.

30. Vandenbeuch A, Clapp TR, Kinnamon SC. Amiloride-sensitive channels in type I fungiform taste cells in mouse. *BMC Neurosci.* 2008;9:1.

31. Smith DV, Ossebaard CA. Amiloride suppression of the taste intensity of sodium chloride: evidence from direct magnitude scaling. *Physiol Behav.* 1995;57(4):773–777.

32. Feldman GM, Mogyorosi A, Heck GL, et al. Salt-evoked lingual surface potential in humans. *J Neurophysiol.* 2003;90(3):2060–2064.

33. Ninomiya Y. Reinnervation of cross-regenerated gustatory nerve fibers into amiloride-sensitive and amiloride-insensitive taste receptor cells. *Proc Natl Acad Sci USA.* 1998;95(9):5347–5350.

34. Ninomiya Y, Tanimukai T, Yoshida S, Funakoshi M. Gustatory neural responses in preweanling mice. *Physiol Behav.* 1991;49(5):913–918.

35. Formaker BK, Hill DL. Lack of amiloride sensitivity in SHR and WKY glossopharyngeal taste responses to NaCl. *Physiol Behav.* 1991;50(4):765–769.

36. Hellekant G, Danilova V, Ninomiya Y. Primate sense of taste: behavioral and single chorda tympani and glossopharyngeal nerve fiber recordings in the rhesus monkey, *Macaca mulatta. J Neurophysiol.* 1997;77(2):978–993.

37. Eylam S, Spector AC. Taste discrimination between NaCl and KCl is disrupted by amiloride in inbred mice with amiloride-insensitive chorda tympani nerves. *Am J Physiol Regul Integr Comp Physiol.* 2005;288(5):R1361–R1368.

38. Lyall V, Heck GL, Vinnikova AK, et al. The mammalian amiloride-insensitive non-specific salt taste receptor is a vanilloid receptor-1 variant. *J Physiol.* 2004;558(Pt 1):147–159.

39. Ishida Y, Ugawa S, Ueda T, Murakami S, Shimada S. Vanilloid receptor subtype-1 (VR1) is specifically localized to taste papillae. *Brain Res Mol Brain Res.* 2002;107(1):17–22.

40. Kido MA, Muroya H, Yamaza T, Terada Y, Tanaka T. Vanilloid receptor expression in the rat tongue and palate. *J Dent Res.* 2003;82(5):393–397.

41. Treesukosol Y, Lyall V, Heck GL, DeSimone JA, Spector AC. A psychophysical and electrophysiological analysis of salt taste in Trpv1 null mice. *Am J Physiol Regul Integr Comp Physiol.* 2007;292(5):R1799–R1809.

42. Smith KR, Treesukosol Y, Paedae AB, Contreras RJ, Spector AC. Contribution of the TRPV1 channel to salt taste quality in mice as assessed by conditioned taste aversion generalization and chorda tympani nerve responses. *Am J Physiol Regul Integr Comp Physiol.* 2012;303(11):R1195–R1205.

43. Oka Y, Butnaru M, von Buchholtz L, Ryba NJ, Zuker CS. High salt recruits aversive taste pathways. *Nature.* 2013;494(7438):472–475.

44. Lunceford BE, Kubanek J. Reception of aversive taste. *Integr Comp Biol.* 2015;55(3):507–517.

45. Neta ERD, Johanningsmeier SD, McFeeters RF. The chemistry and physiology of sour taste - a review. *J Food Sci.* 2007;72(2):R33–R38.

46. DeMars CS, Hollister K, Tomassoni A, Himmelfarb J, Halperin ML. Citric acid ingestion: a life-threatening cause of metabolic acidosis. *Ann Emerg Med.* 2001;38(5):588–591.

47. Tahmassebi JF, Duggal MS, Malik-Kotru G, Curzon ME. Soft drinks and dental health: a review of the current literature. *J Dent.* 2006;34(1):2–11.

48. Richards TW. The relation of the taste of acids to their degree of dissociation, II. *J Phys Chem.* 1900;4(3):207–211.

49. Richards TW. The relation of the taste of acids to their degree of dissociation. *Am J Chem*. 1898;20:121−126.
50. Lyall V, Alam RI, Phan DQ, et al. Decrease in rat taste receptor cell intracellular pH is the proximate stimulus in sour taste transduction. *Am J Physiol Cell Physiol*. 2001;281(3):C1005−C1013.
51. Richter TA, Caicedo A, Roper SD. Sour taste stimuli evoke Ca^{2+} and pH responses in mouse taste cells. *J Physiol*. 2003;547(Pt 2):475−483.
52. Lin W, Burks CA, Hansen DR, Kinnamon SC, Gilbertson TA. Taste receptor cells express pH-sensitive leak K^{+} channels. *J Neurophysiol*. 2004;92(5):2909−2919.
53. Lyall V, Pasley H, Phan TH, et al. Intracellular pH modulates taste receptor cell volume and the phasic part of the chorda tympani response to acids. *J Gen Physiol*. 2006;127(1):15−34.
54. Bachmanov AA, Beauchamp GK. Taste receptor genes. *Annu Rev Nutr*. 2007;27:389−414.
55. Chandrashekar J, Yarmolinsky D, von Buchholtz L, et al. The taste of carbonation. *Science*. 2009; 326(5951):443−445.
56. LopezJimenez ND, Cavenagh MM, Sainz E, Cruz-Ithier MA, Battey JF, Sullivan SL. Two members of the TRPP family of ion channels, Pkd1l3 and Pkd2l1, are co-expressed in a subset of taste receptor cells. *J Neurochem*. 2006;98(1):68−77.
57. Ishimaru Y, Inada H, Kubota M, Zhuang H, Tominaga M, Matsunami H. Transient receptor potential family members PKD1L3 and PKD2L1 form a candidate sour taste receptor. *Proc Natl Acad Sci USA*. 2006;103(33):12569−12574.
58. Huang AL, Chen X, Hoon MA, et al. The cells and logic for mammalian sour taste detection. *Nature*. 2006;442(7105):934−938.
59. Kataoka S, Yang R, Ishimaru Y, et al. The candidate sour taste receptor, PKD2L1, is expressed by type III taste cells in the mouse. *Chem Senses*. 2008;33(3):243−254.
60. Inada H, Kawabata F, Ishimaru Y, Fushiki T, Matsunami H, Tominaga M. Off-response property of an acid-activated cation channel complex PKD1L3-PKD2L1. *EMBO Rep*. 2008;9(7):690−697.
61. Nelson TM, Lopezjimenez ND, Tessarollo L, Inoue M, Bachmanov AA, Sullivan SL. Taste function in mice with a targeted mutation of the pkd1l3 gene. *Chem Senses*. 2010;35(7):565−577.
62. Horio N, Yoshida R, Yasumatsu K, et al. Sour taste responses in mice lacking PKD channels. *PloS One*. 2011;6(5):e20007.
63. Chang RB, Waters H, Liman ER. A proton current drives action potentials in genetically identified sour taste cells. *Proc Natl Acad Sci USA*. 2010;107(51):22320−22325.
64. Bushman JD, Ye W, Liman ER. A proton current associated with sour taste: distribution and functional properties. *FASEB J*. 2015;29(7):3014−3026.
65. Galindo-Cuspinera V, Winnig M, Bufe B, Meyerhof W, Breslin PA. A TAS1R receptor-based explanation of sweet 'water-taste'. *Nature*. 2006;441(7091):354−357.
66. Nakajima K, Morita Y, Koizumi A, et al. Acid-induced sweetness of neoculin is ascribed to its pH-dependent agonistic-antagonistic interaction with human sweet taste receptor. *FASEB J*. 2008;22(7):2323−2330.
67. Koizumi A, Tsuchiya A, Nakajima K, et al. Human sweet taste receptor mediates acid-induced sweetness of miraculin. *Proc Natl Acad Sci USA*. 2011;108(40):16819−16824.
68. Kant R. Sweet proteins−potential replacement for artificial low calorie sweeteners. *Nutr J*. 2005;4:5.

Peptide Signaling in Taste Transduction

Shingo Takai[1], Ryusuke Yoshida[1], Noriatsu Shigemura[1], Yuzo Ninomiya[1,2]

[1]Section of Oral Neuroscience, Graduate School of Dental Sciences, Kyushu University, Fukuoka, Japan; [2]Division of Sensory Physiology, Research and Development Center for Taste and Odor Sensing, Kyushu University, Fukuoka, Japan

O U T L I N E

Chemosensory Transduction
http://dx.doi.org/10.1016/B978-0-12-801694-7.00017-2

INTRODUCTION

Taste is the sensory system primarily devoted to the selection of what animals eat and drink. The taste of food is vitally important signal for the detection of nutrients, minerals, or toxins. Sweet, bitter, salty, sour, and umami (amino acid) tastes are recognized to represent the five basic taste qualities experienced by humans and appear to be conserved in rodents and many other mammals. Generally, sweet and umami tastes allow the detection of nutritious substances, bitter and sour tastes prevent the ingestion of toxic chemicals, and salty taste provides a signal for the detection of certain electrolytes. These sensory inputs from taste cells are critical for regulating nutrient absorption and energy homeostasis.

There is growing evidence that taste function is modulated in the context of animal's nutritional state, and various hormones have been suggested to underlie this modulation. Several hormones in blood circulation, such as leptin, insulin, angiotensin II (AngII), and oxytocin, could affect peripheral taste functions.[1-3] For example, leptin, which is produced in adipose tissues, has been found to selectively suppress sweet taste responses through the putative involvement of the leptin receptor Ob-Rb in sweet sensitive cells.[1] Other bioactive peptides, such as glucagon, glucagon-like peptide-1 (GLP-1), cholecystokinin (CCK), vasoactive intestinal peptide (VIP), neuropeptide Y (NPY), peptide YY (PYY), ghrelin, and galanin, are expressed in taste bud cells themselves. In some cases, their receptors are also expressed in subsets of taste cells with unique expression patterns that are correlated with specific taste cell markers, implying that these peptides may contribute to the modulation of the taste information in the taste bud through autocrine and/or paracrine (cell–cell communication) effects.[3] Gustatory nerve fibers express some peptide receptors, including the GLP-1 receptor (GLP-1R) or NPY receptors, suggesting that the peptides may function as neurotransmitters or neuromodulators.[3,4] In this chapter, we review the expressions and functions of bioactive peptides in peripheral gustatory system.

TASTE BUD CELLS

Each of the five taste qualities is thought to be detected initially by dedicated taste bud cells, which are classified into types I, II, III, or IV according to their cytological, ultrastructural, and functional characteristics.[5-7] Basic features of each type of taste bud cells are summarized in Table 1. Sweet, umami, and bitter receptors (see Chapter 13) and transduction components (see Chapter 17) such as taste receptor type 1 member 2 and 3 (T1R2 + T1R3, sweet), T1R1 + T1R3 (umami), taste receptor type 2 (T2Rs, bitter), α-gustducin, phospholipase Cβ2, and transient receptor potential cation channel subfamily M member 5 (TRPM5) are expressed in type II cells. Candidate sour taste receptor polycystic kidney disease 1-like 3 (PKD1L3) and 2-like (PKD2L1)[8,9] and synapse-related molecules such as serotonin (5-HT), synaptosomal-associated protein (SNAP25), neural cell adhesion molecule (NCAM), and glutamate decarboxylase (GAD67) are expressed in type III cells (see Chapter 15). Degrading enzymes such as glutamate aspartate transporter and nucleoside triphosphate diphosphohydrolase 2 are expressed in type I cells. Type IV cells are thought to contain taste bud progenitor cells.[10,11]

TABLE 1   Classification of Taste Bud Cells

	Type I cells	Type II cells	Type III cells	Type IV cells
	Glial-like cells	Taste-receptor cells	Presynaptic cells	Basal, nonpolarized, presumably undifferentiated cells
Cell markers	GLAST, NTPDase II, ENaC	T1Rs, T2Rs, α-gustducin, PLCβ2, IP3R3, TRPM5, CALHM1	5-HT, GAD67, SNAP25, PKD1L3/2L1, NCAM, chromogranin	Shh
Dedicated to	Salty taste?	Sweet taste, umami taste, bitter taste, lipid sensing?	Sour taste?	?

CALHM1, calcium homeostasis modulator 1; ENaC, epithelial sodium channel; GAD67, glutamate decarboxylase 67; GLAST, glutamate aspartate transporter; 5-HT, serotonin; IP3R3, inositol 1, 4,5-trisphosphate receptor type 3; NCAM, neural cell adhesion molecule; NTPDase II, nucleoside triphosphate diphosphohydrolase II; PKD1L3/2L1, polycystic kidney disease 1-like 3/polycystic kidney disease 2-like 1; PLCβ2, phospholipase Cβ2; Shh, sonic hedgehog; SNAP25, synaptosomal-associated protein 25; TIRs, taste receptors; T2R, taste receptor type 2; TRPM5, transient receptor potential cation channel subfamily M member 5.

LEPTIN

Leptin is the product of the obese gene (*ob*). It is primarily secreted from adipose cells and acts as an anorexigenic mediator that regulates food intake, energy expenditure, and body weight mainly via activation of the hypothalamic leptin receptor Ob-Rb (a.k.a. LepRb).[12–14] The *db/db* mice have defects in the leptin receptor and are hyperphagic, massively obese, and diabetic.[13,14] We found that they also show greater gustatory nerve and behavioral responses to various sweet substances than do lean control mice; this effect is not seen in response to salty, bitter, and sour substances.[15,16] By contrast, mice in which a diabetic state is induced by streptozotocin, a glucosamine-nitrosourea compound that is highly toxic to pancreatic β cells,[17] did not exhibit greater sugar responses,[15] indicating that the greater responsiveness to sweet substances in *db/db* mice is not a result of diabetic state itself. When leptin was administered into lean control mice, chorda tympani (CT) nerve responses and behavioral lick responses to sweet substances were selectively suppressed, and those to sour, salty, and bitter substances were unaffected.[18,19] In comparison, such selective suppression of sweet responses by leptin was not observed in *db/db* mice,[18,19] indicating that the leptin's effect on sweet taste sensitivity is mediated by the leptin receptor Ob-Rb. In taste tissue, Ob-Rb receptors are expressed in type II taste bud cells.[18–22] In isolated mouse taste cells, leptin could potentiate outward potassium currents leading to reduced cell excitability.[18] Indeed, about half of the sweet-sensitive taste cells in lean mice showed a significant reduction of responses to sweet stimuli during leptin application (10–20 ng/ml); the adenosine triphosphate (ATP)-sensitive K^+ (K_{ATP}) channel inhibitor glibenclamide eliminated this effect [1,23]. In CT nerve recordings, lean mice exhibited significant increases in responses to sweet compounds after intraperitoneal administration of leptin receptor antagonist (LA) mutant L39A/D40A/F41A.[24] Thus leptin reduces taste sensitivity of sweet sensitive taste cells in lean mice via activation of Ob-Rb and K_{ATP} channel expressed in these cells. In addition, the inhibitory effect of leptin on sweet taste is observed in lean mice but not in diet-induced obese (DIO) mice. The effect of LA on CT nerve responses to sweet compounds was gradually decreased as obesity progressed,[24] and leptin's effect on sweet sensitive taste cells was abolished in DIO mice.[23].

A recent study using an isolated mouse taste bud preparation demonstrated that sweet-induced calcium responses and ATP release from taste cells were significantly decreased by bath application of leptin.[25] Additionally, leptin enhanced sweet-induced 5-HT release from taste cells. These findings support the idea that leptin affects responses of taste cells to sweet tastants. ATP plays a crucial role in transmission of taste information from the taste cells to the taste nerves[26]; therefore, the reduction of sweet-induced ATP release from taste cells by leptin would lead to decrease in gustatory nerve responses to sweeteners.

However, a recent report claimed that intraperitoneally administered leptin (100 ng/g body weight [bw]) increased CT nerve responses to sucrose in both free-fed and fasted mice rather than decreased.[27] The other recent study reported that leptin had no effect on gustatory nerve responses to sweet stimuli, although plasma leptin levels were increased greater than 10-fold by intraperitoneal injection of leptin (100 ng/g bw).[22] Such disparate observations in nerve recordings may be caused by differences in experimental conditions among these studies. For example, one study used tastants dissolved in distilled water at 24 °C,[18] whereas the other studies used tastants dissolved in artificial saliva at 35 °C.[22,27] In addition, the state of the taste nerve during recordings was varied; transected and desheathed[18,27] or intact.[22] All of these conditions and techniques of surgery are possible to impact on the results. Moreover, different results in leptin's effect on behavioral responses were demonstrated. One reported that intraperitoneal injection of leptin (100 ng/g bw; 500 ng/g bw for *db/db*) reduced behavioral lick responses to sweet—bitter mixture solutions (sucrose and saccharin + 3 mM quinine-hydrochloride) in *ob/ob* and normal littermates but not in *db/db* mice in 10-s lick tests,[19] whereas the other reported that leptin treatment (100 ng/g bw) could not alter animal's licking behavior to sucrose in various concentrations (0.01—1.0 M) in brief-access testing (5 s).[22] Such differences may also be due to differences in experimental designs and conditions among these studies. Altogether, the functions of leptin in the peripheral taste system seem to be more complex than originally expected. Further studies are needed to understand the relationship between leptin and sweet taste sensitivity comprehensively.

Circulating leptin levels exhibit diurnal variation,[28,29] with plasma leptin rising before noon and peaking at midnight (23:00—01:00 h) before declining until morning in humans.[30] When diurnal patterns of plasma leptin and taste recognition threshold were measured in healthy adults (body mass index <25, male and female, aged 21—31 years), the recognition threshold for sweet compounds—but not for salty, bitter, sour, or umami compounds— exhibited a diurnal variation that paralleled the variation in plasma leptin.[31] The diurnal variation of leptin correlated with meal-related shifts. For example, when meals were shifted by 6.5 h without changing the light or sleep cycles in humans, plasma leptin levels were similarly shifted by 5—7 h.[32] The nocturnal rise of leptin does not occur if the subjects are fasted.[33] When leptin levels were phase-shifted following imposition of one or two meals per day, the diurnal variation in recognition threshold for sweet taste shifted in parallel.[31] These relationships between plasma leptin concentration and recognition threshold for sweet taste imply a regulatory role for leptin on human sweet sensitivity.

GLUCAGON

The 29-amino acid peptide hormone glucagon is produced by the α-cells of pancreatic islets during fasting conditions.[34] Glucagon is cleaved from proglucagon by the site-selective protease prohormone convertase 2 (PC2). Glucagon stimulates hepatic glycogenolysis and gluconeogenesis to increase blood glucose level by the activation of the G protein-coupled glucagon receptor (GlucR). GlucRs are also expressed in the pancreatic islets and may contribute to the regulation of glucose-stimulated insulin secretion from β cells[35] and glucagon secretion from α cells.[36,37]

In mouse taste buds, both glucagon and GlucR are coexpressed in a subset of taste cells. Most of these cells also express T1R3, PLCβ2, PC2, and its obligatory chaperone 7B2, indicating that glucagon is biosynthesized and expressed with its own receptor in sweet and/or umami taste cells.[38] 7B2 is encoded by Scg5 gene and is essential for the activation of PC2 to process glucagon from its precursor.[39] The mice lacking 7B2 (Scg5$^{-/-}$ mice) showed no glucagon staining in their taste bud cells and were less sensitive to sucrose but not salty, bitter, and sour compounds in the brief access test than their littermate controls.[38] Furthermore, acute and local disruption of glucagon signaling by the highly specific, membrane-permeable GlucR antagonist L-168,049 in wild type (WT) mice caused similar reduced responses to sucrose. Taken together, glucagon may function predominantly as an autocrine or paracrine signal in taste buds and may enhance or maintain sweet taste responsiveness.

GLUCAGON-LIKE PEPTIDE-1

GLP-1, a 30-amino acid hormone, is one of several peptide hormones proteolytically cleaved from the precursor protein, proglucagon by prohormone convertase 1/3 (PC1/3).[40] GLP-1 is originally identified as a glucoincretin hormone that is expressed in enteroendocrine L cells located mainly in lower intestine and colon.[41] The main functions of GLP-1 are to amplify glucose-dependent insulin secretion from pancreatic β cells and to inhibit glucagon secretion from α cells, thereby regulating postprandial glucose excursions. Interestingly, enteroendocrine L cells express multiple taste transduction molecules, including the sweet taste receptor T1R2 + T1R3 and the taste G-protein α-gustducin, and release GLP-1 in response to sweet compounds.[42] GLP-1 is highly susceptible to degradation by the enzyme dipeptidyl peptidase-4; therefore, only 10−15% of GLP-1 secreted from gut enters the systemic circulation.[40] Portion of intact GLP-1 in portal vein facilitates the hepatic vagal afferent activity via G protein−coupled GLP-1R, and evokes powerful neuroincretin effect.[43]

In taste tissues, GLP-1 is expressed in a subset of T1R3 or/and α-gustducin expressing type II taste cells and also in a subset of 5-HT or GAD67 expressing type III taste cells.[4,44] The other report demonstrated that 91−93% of GLP-1−expressing cells expressed TRPM5 in circumvallate papillae (CV), whereas only 23% of GLP-1−expressing cells expressed TRPM5 in fungiform papillae (FP).[45] GLP-1 is coexpressed with PC1/3 in a subset of taste cells,[44] suggesting that GLP-1 is produced and accumulated within these taste cells. In FP, a subset of sweet responsive taste cells released GLP-1 in response to sweet taste stimuli such as glucose, sucrose, and saccharin.[4] Dietary lipids also trigger GLP-1 release from mouse taste buds under putative involvement of GPR120.[46] GLP-1R is expressed in intragemmal nerve fibers in CV[44] and a subset of neurons in the geniculate ganglion,[4] indicating that gustatory nerve fibers express GLP-1R. GLP-1R$^{-/-}$ mice exhibited reduced behavioral and gustatory nerve responses to sweet compounds (sucrose and sucralose),[4,47] reduced behavioral responses to low concentrations of oil (α-linolenic acid and oleic acid)[46] and enhanced behavioral responses to umami compounds.[46] Furthermore, intravenous injection of GLP-1 produced a transient increase in spike activities in a subset of sweet-sensitive gustatory nerve fibers and these responses were significantly suppressed by preadministration of the GLP-1 receptor antagonist Exendin-4(3−39).[4] Altogether, GLP-1 in taste buds is likely to function as a sweet specific neurotransmitter to adjacent gustatory nerve fibers. It may also be involved in umami and lipid taste transduction.

The recent study suggested that GLP-1 released from taste cells underlies a portion of the cephalic-phase rise in blood GLP-1 levels.[45] It might lead to cephalic-phase insulin release observed within 1−1.5 min after oral stimulation with saccharin or glucose in rat study.[48] Thus, the tongue could serve as not only a sensory organ, but also a secretive organ and the GLP-1 from the tongue might be involved in the regulation of hormonal releases and energy homeostasis.

INSULIN

Insulin reduces blood glucose levels and plays a crucial role in the regulation of carbohydrate and fat metabolism. A recent study demonstrated that insulin from pancreatic β cells into the blood circulation may affect the peripheral salt sensitivity of mice.[49] In mice, the epithelial sodium channel (ENaC) is a candidate receptor for sodium chloride (NaCl) taste.[50,51] Insulin potentiates the open probability of ENaC[52] and transport of ENaC proteins to the membrane.[53] In some isolated FP taste cells, insulin significantly increased Na$^+$ current.[49] This increased Na$^+$ current was sensitive to amiloride and/or benzamil, both of which are inhibitors of ENaC, suggesting that insulin increases ENaC current in taste bud cells. Enhancement of ENaC function by insulin was mediated by both the phosphoinositide 3- and phosphoinositide 4 kinases' signaling pathways: insulin's effect was suppressed by pharmacological blockers for these kinases such as LY294002 and wortmannin. In taste preference tests, insulin-treated mice significantly avoided NaCl solutions at lower concentrations than the control mice; this effect of insulin was abolished by addition of amiloride to NaCl solutions.[49] These data indicate that insulin may enhance salt taste sensitivity of mice via amiloride-sensitive (ENaC-expressing), salt-responsive taste cells.

ANGIOTENSIN II

AngII, an octapeptide hormone, plays an important role in the regulation of vascular tone, cardiac function, and sodium reabsorption.[54] AngII directly evokes the constriction of blood vessels and the secretion of vasopressin and aldosterone from the posterior pituitary gland and the adrenal cortex, respectively. In the kidneys, vasopressin increases reabsorption of water and aldosterone increases salt retention.[55] Intracerebroventricular and intravenous infusion of AngII in rat produces dose-dependent salt appetite and stimulates sodium intake over a range of concentrations normally rejected,[56,57] suggesting that AngII serves as a potent stimulus for sodium appetite and preference.

The recent study revealed that one of the AngII receptor subtype the angiotensin II type I receptor (AT1) is expressed in taste cells.[2] AT1 is coexpressed with α-ENaC or T1R3 in mouse taste bud cells, suggesting the possible effect of AngII on salt, sweet, and umami taste sensitivities. Indeed, AngII suppressed amiloride-sensitive taste responses to NaCl and enhanced the responses to sweeteners (glucose, sucrose, saccharin, and SC45647) without affecting the responses to KCl, sour, bitter, and umami tastants. AngII's effect on sweet taste responses was not observed in cannabinoid receptor 1 (CB_1) knockout mice. CB_1 is required for the enhancement of sweet taste responses by endocannabinoids.[58] In Chinese hamster ovary cells, the activation of AT1 with AngII elevated the endocannabinoid (2-arachidonoyl glycerol [2-AG]) level, leading to the transactivation of CB_1 receptors via auto/paracrine mechanism.[59] Thus the effect of AngII on sweet responses might be mediated by the interaction of AT1 and CB_1 within a taste bud. The intravenous administration of AT1 antagonist CV11974 significantly blocked these effects of AngII on taste nerve responses. In behavioral tests, 23 h of water deprivation elevated plasma AngII levels in mice; CV11974 treatment reduced the stimulated high licking rate to NaCl and sweeteners in these water-restricted mice.[2] Altogether it appears that AngII signaling modulates animal's salt and sweet sensitivities via the interaction of AT1 and CB_1 receptors and contributes to regulation of sodium and calorie intakes.

CHOLECYSTOKININ

CCK is a gastrointestinal hormone secreted from enteroendocrine I-cells. CCK stimulates gastric emptying, release of pancreatic enzymes and the contraction of the gallbladder.[60] In addition, CCK is abundantly expressed in the brain and functions as a neurotransmitter.[61] There are two types of CCK receptors: CCK-Ar and CCK-Br, both of which are G protein—coupled receptors derived from different genes.[62] These receptors are expressed in various central and peripheral tissues including brain, stomach, pancreas, kidney, and vagal afferent fibers, although expression patterns and function of these receptors may be different among tissues and species.[62,63] Activation of bitter taste receptors in the intestine leads to secretion of CCK, which is proposed to help limit the absorption of dietary-derived bitter-tasting toxins.[64,65] Released CCK from the intestine may activate CCK receptors on the vagus nerve, leading to suppression of food intake and gastric emptying.

CCK and its receptor CCK-Ar are expressed in a subset of taste cells.[66—68] In CV taste buds of rat, 50—60% of CCK-positive cells expressed α-gustducin and 15% of them expressed T1R2.[67] In the posterior tongue, T2Rs are coexpressed with α-gustducin, whereas T1Rs are mostly segregated from α-gustducin.[69] Therefore, the majority of CCK-positive cells may be bitter-sensitive taste cells. Expression of CCK-Ar in taste buds completely overlapped with that of CCK.[68] Exogenous application of CCK inhibited the outward K^+ current of taste cells and induced Ca^{2+} responses, which were suppressed by lorglumide (a specific inhibitor for CCK-Ar over CCK-Br) but not by L-365-260 (CCK-Br inhibitor).[66] Many CCK-sensitive cells showed Ca^{2+} responses to bitter taste stimuli, such as quinine and caffeine.[70] Collectively, CCK may regulate bitter sensitivity of taste cells in an auto-crine manner via CCK-Ar. In our preliminary experiments, the mice genetically lacking CCK receptor (CCK-Ar$^{-/-}$, CCK-Br$^{-/-}$, and CCK-Ar$^{-/-}$Br$^{-/-}$ mice) exhibited reduced sensitivity to bitter, but not to other (sweet, umami, salty, and sour) taste compounds in short-term lick tests and CT nerve recordings.[71]. In one rat study, Otsuka Long-Evans Tokushima fatty rats, which lack functional CCK-Ar, exhibited enhanced responses to sweet and umami taste solutions in behavioral assay.[72] Thus, CCK is likely to be expressed in bitter responsive taste cells and may mainly contribute to the bitter taste sensitivity and may have some effects on sweet and umami taste sensitivities.

VASOACTIVE INTESTINAL PEPTIDE

VIP, a peptide hormone comprising 29 amino acid residues, is identified in the central and peripheral nervous system[73] and has been recognized as a widely distributed neuropeptide, acting as a neurotransmitter or neuromodulator in many organs and tissues including heart, lung, thyroid gland, kidney, immune system, urinary tract, and genital organs.[74]

VIP is expressed in the taste buds.[68] Many (about 60%) VIP-positive taste cells coex-pressed with gustducin and few (about 20%) coexpressed with T1R2 in rat CV. Both CCK and VIP were expressed in almost identical subsets of taste bud cells, suggesting that VIP-expressing cells are also bitter sensitive. VIP$^{-/-}$ mice showed elevated plasma glucose, insulin, and leptin levels without morphological changes in pancreatic islets.[21] VIP and its receptors VPAC1 and VPAC2 were expressed in a subset of PLCβ2-expressing taste cells, indicating that VIP might act as an autocrine/paracrine agent within a taste bud. In taste buds of the VIP$^{-/-}$ mice, the number of leptin receptor—expressing taste cells was significantly decreased, and the number of GLP-1—expressing cells was increased compared with WT mice. The VIP$^{-/-}$ mice exhibited enhanced taste sensitivity to sweet and bitter tastants (sucrose and denatonium benzoate) and reduced sensitivity to sour tastant (citric acid) in brief access tests. The low expression level of leptin receptor and high expression level of GLP-1 in taste buds may be involved in the alteration of sweet taste sensitivity of the VIP$^{-/-}$ mice, and that alteration may represent a part of the responsive amelioration of the VIP$^{-/-}$ mice diabetic-like pathophysiology.[21] However, how VIP signals interact with sour or bitter taste sensitivities still has not been elucidated.

NEUROPEPTIDE Y

NPY, 36-amino acid peptide, is known as the most potent stimulant inducing feeding behavior, and its wide distribution in the central nervous system accounts for a variety of behavioral effects including modulation of anxiety-related behaviors and mood disorders such as depression and anxiety.[75-77] It is also suggested that NPY might affect the animal's learning memory, circadian rhythm, bone formation, cardiovascular function, and blood pressure control.[78,79] These wide ranges of effects are mediated by NPY receptors NPY1, Y2, Y4, Y5, and y6 receptors (y6 is expressed in mice but not in humans).

In taste buds of mice, NPY immunopositive taste cells virtually always coexpressed with CCK and VIP,[80,81] suggesting that NPY expressing cells may be bitter sensitive taste cells. NPY receptor isoforms Y1, Y2, Y4, and Y5 were expressed in taste buds of mice.[82] Among them, Y1 receptors were localized with T1R3 but not NPY,[80] whereas Y4 receptors were expressed in intragemmal fibers.[82] When NPY was administrated to isolated taste bud cells of rats, inward rectifying K^+ currents in about one-third of isolated taste cells tested were enhanced,[81] indicating that a subset of taste bud cells may become less responsive to sensory stimulation by NPY. Thus, NPY may modulate the electrical excitability of sweet-sensitive taste cells in a paracrine manner. Considering the colocalization of NPY on T2R cells and Y1R on T1R3 cells, it is possible that NPY secreted from the bitter cells might act as a lateral inhibitor for the sweet cells during bitter stimulation.[80]

PEPTIDE TYROSINE–TYROSINE

PYY, a 36-amino acid polypeptide hormone, is mainly released from enteroendocrine cells in the ileum or colon and the enterochromaffin cells in the pancreas in response to feeding. PYY inhibits food intake and reduces the rate of weight gain in rodents and rhesus monkeys.[83–86] PYY also reduces gastric motility and increases water and electrolyte absorption in the colon.[87,88] The PYY-containing L cells possess GLP-1, bitter and sweet taste receptors (T2Rs and T1Rs), as well as α-gustducin, and these cells are thought to act as important chemosensors in the gastrointestinal tract.[89] The peripheral administration of the PYY$_{3-36}$, which has highly selective affinity to Y2 and Y5 receptors,[90] reduced food intake in mice, rats, and humans[83,84,91] and evoked conditioned taste aversion to PYY-paired flavors in mice.[92]

Several studies mentioned the potential involvement of PYY signaling in bitter and lipid sensitivity. PYY$^{-/-}$ mice exhibited a significant reduction in sensitivity to denatonium, quinine, intralipid emulsion, and corn oil emulsion.[93] PYY is expressed in murine taste cells and can also be found in murine and human saliva.[94,93] In taste bud cells of PYY-GFP mice, GFP fluorescence was colocalized mainly with PLCβ2 and not so frequently with NCAM.[93] In addition, the Y1R subtype of NPY receptor was expressed in PLCβ2 immunopositive type II cells. Thus, PYY may be accumulated in taste cells before being released to activate PYY receptors (Y1R) on a subset of type II cells in an autocrine/paracrine manner. This NPY signaling might be involved in the modulation of bitter taste or lipid sensing.[93] However, another study reported that Y1 receptors were colocalized with T1R3,[80] suggesting a role in sweet or umami taste. By contrast, Y2Rs are expressed at the apical pore of taste buds.[93] Salivary (endocrine-derived) PYY might activate a subset of taste cells containing these apically expressed Y2 receptors.

OXYTOCIN

The nonapeptide hormone oxytocin is produced in the paraventricular and supraoptic nuclei of the hypothalamus and stimulates parturition, lactation, and prosocial behaviors.[96,97] It also has key roles in the regulation of complex social behavior and cognition, such as attachment, social exploration, recognition, and aggression.[97] In addition, oxytocin has potent central effects on feeding behavior.[98] Oxytocin administered either intracerebroventricularly or intraperitoneally reduced food intake and feeding behavior in a dose-dependent manner in both fasted and nonfasted rats.[99–101]

Using reverse transcriptase polymerase chain reaction, oxytocin receptor (OXTR) messenger RNA was detected in FP, CV, foliate, and palate taste buds.[102] In OXTR-yellow fluorescent protein (YFP) knockin mice, which express the Venus variant of YFP in OXTR-expressing cells, YFP fluorescence was observed in NTPDase2-immunopositive cells (type I) but not in PLCβ2-expressing cells (type II) and chromogranin A-expressing cells (type III). Taste bud cells and their associated nerves did not contain oxytocin peptide itself, suggesting that OXTRs in taste bud cells may be activated by oxytocin present in blood or saliva. The application of oxytocin to isolated taste bud cells induced Ca^{2+} responses in a subset of taste bud cells and these responses were inhibited by the OXTR antagonist, L-371,257, suggesting that oxytocin actually affects a subset of taste cells via OXTR.[102] Additionally, intraperitoneally injected oxytocin (0.1 mg/kg bw) reduced sweet sensitivity of mice in brief access tests.[103] Thus, oxytocin signaling in taste buds might contribute to sweet taste sensitivity, although it seems not through direct actions on sweet-sensitive type II cells. Oxytocin$^{-/-}$ mice overconsume solutions of saccharin and carbohydrates (Polycose and starch)[104] and show enhanced consumption of NaCl solution after fluid deprivation.[105] Such behavioral responses of oxytocin$^{-/-}$ mice may be due, at least in part, to the lack of oxytocin's effect on taste sensitivities.

GHRELIN

The ghrelin precursor, prepro-ghrelin contains 117 amino acids, whereas the mature ghrelin peptide consists of 28 amino acids with a fatty acid chain modification (octanoyl group) on the third amino acid.[106] Ghrelin is primarily produced in the enteroendocrine cells of the stomach, although its expression has also been reported in other peripheral tissues such as the intestine, pancreas, ovary, adrenal cortex, and taste bud cells.[106–110] Ghrelin only becomes active when caprylic (octanoic) acid is linked posttranslationally to serine at the 3-position by ghrelin-O-acyltransferase (GOAT). Ghrelin binds to the G protein–coupled growth hormone secretagogue receptor. The functions of ghrelin signaling are well established as appetite-stimulatory signal and growth hormone–releasing signal. Preprandial and postprandial ghrelin release from the stomach is increased and decreased, respectively.[111–113]

In taste tissue, a subset of taste bud cells expressed ghrelin, its precursor peptide prepro-ghrelin, PC1/3 and GOAT.[114] Ghrelin was coexpressed with all types of taste cell markers (type I: NTPDase2; type II: PLCβ2, α-gust; type III: NCAM; type IV: Shh), and some of them are coexpressed with prepro-ghrelin, PC1/3 and GOAT, all of which are essential for producing active ghrelin. The ghrelin receptor (GHSR) is also widely distributed among the taste cells; almost all GHSR signals are colocalized with ghrelin signals. A subset of ghrelin- and GHSR-expressing taste cells also possesses ENaC. GHSR null mice showed significantly reduced taste responsivity to sour and salty taste solutions in the brief access test.[115] Another study reported that ghrelin null mice demonstrated reduced lick responses to salty solutions, whereas GOAT$^{-/-}$ mice had enhanced sensitivity to salty taste in brief access tests.[115] Both ghrelin$^{-/-}$ and GOAT$^{-/-}$ mice exhibited reduced lipid taste responsivity compared with WT controls in this study. Consistently, the expression of the potential lipid taste receptors CD36 and GPR120 were reduced in the taste buds of ghrelin$^{-/-}$ and GOAT$^{-/-}$ mice, whereas γENaC was increased in GOAT$^{-/-}$ mice compared with WT mice. These results suggest that ghrelin is biosynthesized within a taste bud and exerts multiple effects on various kinds of taste cells.

GALANIN

Galanin was first isolated from porcine intestine.[116] Galanin is expressed not only in intestine (duodenum), but also in the central nervous system (hypothalamus) and primary sensory neurons.[117–119] Galanin signaling occurs through three G protein–coupled receptors (GALR1, GALR2, and GALR3), each with different distributions in the central nervous system or in pancreas.[119–122] Galanin's function remains largely unclear. However, diverse regulatory functions have been suggested in many physiological processes, including sleep and arousal, nociception, memory, learning, feeding, water intake, metabolism, and reproduction.[123,124]

A subset of taste cells in rat CV taste buds were shown to express galanin and its receptor GALR2.[125] Almost all of PLCβ2-immunopositive and α-gustducin-immunopositive taste cells—as well as a majority of NCAM-immunopositive taste cells—expressed galanin. GALR2 was expressed in taste buds, but its coexpression with taste cell markers was not investigated. Previous studies suggested that GALR2 mediates the neurotrophic effect of galanin and affects outgrowth of sensory neurons.[126,127] Taken together, galanin may act as a paracrine modulator within a taste bud and/or serve as a neurotrophic factor to adjacent nerve fibers. However, the physiological function of galanin in taste buds remains unknown.

CONCLUSION

We discussed the various effects of peptides on taste functions (an overview of peptide signaling is shown in Figure 1, Table 2). Recently, it has been suggested that there is a bidirectional, neurohumoral communication system between the gut and the brain to regulate energy homeostasis and that the taste system may largely contribute to this system.[128,129] Peptide signaling, which in part reflects the animal's nutritional state, might link the oral–brain–gut axis and efficiently regulate feeding behavior.

However, there are many unclear points about peptide signaling in a taste bud. For example, some peptide hormones could affect the electroexcitability of taste cells, whereas the molecular basis and physiological significance of these effects are still largely unclear. How peptide hormones expressed in taste cells are released in response to taste stimuli or some other signals is also unknown. In enteroendocrine cells, gut peptides are released by exocytosis of secretory granules,[130] whereas in taste cells (especially in type II cells), there

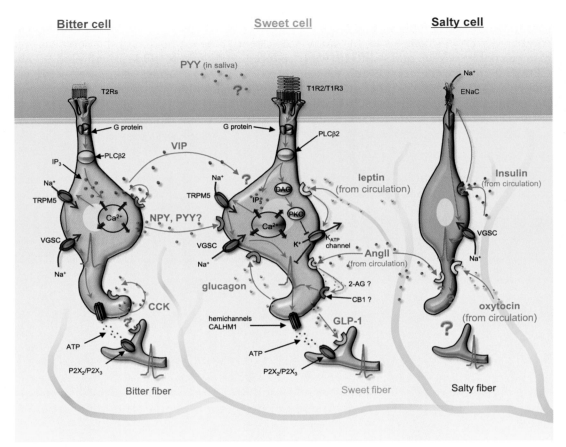

FIGURE 1 **Peptide signaling in taste transduction.** 1-AG, 1-arachidonoyl glycerol; 2-AG, 2-arachidonoyl glycerol; AngII, angiotensin II; ATP, adenosine triphosphate; CALHM1, calcium homeostasis modulator 1; CB1, cannabinoid receptor 1; CCK, cholecystokinin; ENaC, epithelial sodium channel; IP3, inositol triphosphate; K_{ATP}, ATP-gated potassium cation; NPY, neuropeptide Y; P2X2/X3, purinergic receptor 2/3; PLCB2, phospholipase Cβ2; PYY, peptide YY; T1R2, taste receptor type 1 member 2; T1R3, taste receptor type 1 member 3; TRPM5, transient receptor potential cation channel subfamily M member 5; VGSC, voltage gated sodium channel; VIP, vasoactive intestinal peptide.

is no histochemical evidence for vesicles containing such peptide hormones. These points should be answered if we are to comprehensively understand the function of peptide signaling in taste transduction.

Many studies have demonstrated an impact of peptide signaling in peripheral taste responses. However, these peptides may have other undiscovered functions in the taste system. In addition, peptides other than those discussed here might contribute to peripheral taste functions. Furthermore, various peptide signaling cascades might interact with each other, thus creating a complex regulatory system for peripheral taste. The investigation of this system has just begun.

TABLE 2 Summary of Peptide Signaling in Taste Transduction

	Localization: coexpressed molecules	Putative receptors in taste tissue	Receptor localization: coexpressed molecules	Possible functions
Leptin	Blood circulation	Ob-Rb	Taste buds	Suppress sweet taste sensitivity
Glucagon	Type II cells: T1R3, α-gustducin	GlucR	Type II cells: PLCβ2, α-gustducin, glucagon, 7B2	Maintain sweet taste sensitivity
GLP-1	Type II cells: T1R3, α-gustducin Type III cells: 5-HT, GAD67	GLP-1R	Taste nerve fibers: PGP9.5, P2X2	Maintain sweet/ umami/lipid taste sensitivity and act as a sweet-specific neurotransmitter
Insulin	Blood circulation	IR (insulin receptor)?	Type I cells?: ENaC	Enhance salty taste
AngII	Blood circulation	AT1	Type I cells: ENaC Type II cells: T1R3	Suppress salty taste and enhance sweet taste sensitivity
CCK	Type II cells: T1R2, α-gustducin, VIP, NPY	CCK-AR CCK-BR	Type II cells: CCK	Maintain bitter taste sensitivity
VIP	Type II cells: T1R2, α-gustducin, CCK, NPY	VPAC1 VPAC2	Type II cells: PLCβ2	Suppress sweet and bitter taste and enhance sour taste sensitivity
NPY	Type II cells: CCK, VIP	NPY1R NPY2R NPY4R NPY5R	Type II cells: T1R3 (NPY1R) Type III cells, taste nerve fibers: NCAM (NPY4R) Lingual epithelium	Suppress sweet taste sensitivity during bitter perception
PYY	Type II cells: CCK, VIP			Maintain bitter and lipid taste sensitivity
Oxytocin	Blood circulation	OXTR	Type I cells: NTPDase	Suppress sweet and starch consumption maintain salty taste sensitivity
Ghrelin	Type I cells: NTPDase Type II cells: PLCβ2, α-gustducin Type III cells: NCAM Type IV cells: Shh	GHSR	Type I cells: NTPDase Type II cells: PLCβ2, α-gustducin Type III cells: NCAM Type IV cells: Shh	Maintain salty and sour taste sensitivity
Galanin	Type II cells: PLCβ2, α-gustducin Type III cells: NCAM	GALR2	Taste buds	?

AngII, angiotensin II; CCK, cholecystokinin; ENaC, epithelial sodium channel; GAD67, glutamate decarboxylase 67; GALR2, Galanin receptor 2; GHSR, ghrelin receptor; GLP-1, glucagon-like peptide-1; GLP-1R, glucagon-like peptide-1 receptor; GlucR, G protein-coupled glucagon receptor; 5-HT, serotonin; NCAM, neural cell adhesion molecule ; NPY, neuropeptide Y; NPY1(2, 4 or 5)R, neuropeptide Y receptor type 1 (2, 4 or 5); NTPDase II, nucleoside triphosphate diphosphohydrolase II; Ob-Rb, leptin receptor; OXTR, oxytocin receptor; P2X2/X3, purinergic receptor 2/3; PGP9.5, ubiquitin carboxyl-terminal hydrolase L1; PLCβ2, phospholipase Cβ2; PKD1L3/2L1, polycystic kidney disease 1-like 3/2-like 1; PYY, peptide YY; Shh, sonic hedgehog; T1R2, taste receptor type 1 member 2; T1R3, taste receptor type 1 member 3; TRPM5, transient receptor potential cation channel subfamily M member 5; VIP, vasoactive intestinal peptide.

III. GUSTATORY TRANSDUCTION

BREAKOUT BOX

RECIPROCAL MODULATION OF SWEET TASTE BY OREXIGENIC AND ANOREXIGENIC MEDIATORS

Leptin, an anorexigenic hormone, suppresses sweet taste sensitivity, which would reduce palatability of sweetened foods. In contrast, orexigenic mediators endocannabinoids have the opposite action on (i.e., the enhancement of) sweet taste sensitivity. Endocannabinoids, such as anandamide [N-arachidonoylethanolamine (AEA)] and 2-arachidonoyl glycerol (2-AG), are arachidonate-based lipids but not peptide hormones that exert their effect by acting on cannabinoid receptors (CB_1) in various brain regions.[131–133] Circulating endocannabinoid levels are inversely correlated with plasma levels of leptin.[134] Defective leptin signaling is associated with elevated hypothalamic levels of endocannabinoids in obese *db/db* and *ob/ob* mice and in Zucker rats.[135] These pieces of evidence raise the possibility that endocannabinoids may oppose the effects of leptin on sweet taste sensitivity. Indeed, endocannabinoids selectively enhanced sweet taste sensitivity.[58] In taste buds, approximately 60% of taste cells expressing T1R3, a sweet receptor component, coexpressed CB_1. Intraperitoneal injection of AEA or 2-AG into wild-type (WT) mice increased sweet taste responses without affecting salty, sour, bitter, and umami taste responses in gustatory nerve recordings and short-term lick tests. This effect was observed in WT mice, but not in $CB_1^{-/-}$ mice. At the peripheral taste cell level, basolateral administration of AEA and 2-AG increased responses of sweet sensitive taste cells, and this effect was inhibited by AM251, a CB_1 receptor antagonist, but not by AM630, a CB_2 receptor antagonist. Thus, endocannabinoids enhance sweet taste sensitivity by acting on CB_1 receptors in sweet sensitive taste cells. A recent study demonstrated that hyperresponsivity to sweet compounds in *db/db* mice is due not only to the lack of leptin effect, but also elevated endocannabinoid levels in taste buds.[24] In *db/db* mice, AM251 reduced gustatory nerve responses to sweeteners to those shown by WT mice but not to salty, bitter, sour, and umami compounds.[24] Therefore, enhanced sweet taste responses in *db/db* mice may occur through tonic activation of CB_1 receptor by endocannabinoids. Such reciprocal modulation of sweet taste sensitivity by orexigenic and anorexigenic mediators may be involved in regulation of feeding behaviors and energy homeostasis.

Acknowledgments

This work was supported by KAKENHI 18077004, 18109013, 23249081, 26670810, 15H02571 (Y.N.), 20689034, 15K11044 (N.S.), 23689076, 26462815 (R.Y.) and15H06485 (S.T.) for Scientific Research from the Japan Society for the Promotion of Science.

References

1. Yoshida R, Niki M, Jyotaki M, Sanematsu K, Shigemura N, Ninomiya Y. Modulation of sweet responses of taste receptor cells. *Semin Cell Dev Biol*. 2013;24:226–231.
2. Shigemura N, Iwata S, Yasumatsu K, et al. Angiotensin II modulates salty and sweet taste sensitivities. *J Neurosci*. 2013;33:6267–6277.

3. Dotson CD, Geraedts MC, Munger SD. Peptide regulators of peripheral taste function. *Semin Cell Dev Biol.* 2013;24:232—239.
4. Takai S, Yasumatsu K, Inoue M, et al. Glucagon-like peptide-1 is specifically involved in sweet taste transmission. *FASEB J.* 2015;29(6):2268—2280.
5. Chaudhari N, Roper SD. The cell biology of taste. *J Cell Biol.* 2010;190:285—296.
6. Yoshida R, Ninomiya Y. New insights into the signal transmission from taste cells to gustatory nerve fibers. *Int Rev Cell Mol Biol.* 2010;279:101—134.
7. Chandrashekar J, Hoon MA, Ryba NJ, Zuker CS. The receptors and cells for mammalian taste. *Nature.* 2006;444:288—294.
8. Ishimaru Y, Inada H, Kubota M, Zhuang H, Tominaga M, Matsunami H. Transient receptor potential family members PKD1L3 and PKD2L1 form a candidate sour taste receptor. *Proc Natl Acad Sci USA.* 2006;103:12569—12574.
9. Huang AL, Chen X, Hoon MA, et al. The cells and logic for mammalian sour taste detection. *Nature.* 2006;442:934—938.
10. Ichimori Y, Ueda K, Okada H, Honma S, Wakisaka S. Histochemical changes and apoptosis in degenerating taste buds of the rat circumvallate papilla. *Arch Histol Cytol.* 2009;72:91—100.
11. Miura H, Scott JK, Harada S, Barlow LA. Sonic hedgehog-expressing basal cells are general post-mitotic precursors of functional taste receptor cells. *Dev Dyn.* 2014;243:1286—1297.
12. Zhang Y, Proenca R, Maffei M, Barone M, Leopold L, Friedman JM. Positional cloning of the mouse obese gene and its human homologue. *Nature.* 1994;372:425—432.
13. Lee GH, Proenca R, Montez JM, et al. Abnormal splicing of the leptin receptor in diabetic mice. *Nature.* 1996;379:632—635.
14. Chen H, Charlat O, Tartaglia LA, et al. Evidence that the diabetes gene encodes the leptin receptor: identification of a mutation in the leptin receptor gene in db/db mice. *Cell.* 1996;84:491—495.
15. Ninomiya Y, Sako N, Imai Y. Enhanced gustatory neural responses to sugars in the diabetic *db/db* mouse. *Am J Physiol.* 1995;269:R930—R937.
16. Ninomiya Y, Imoto T, Yatabe A, Kawamura S, Nakashima K, Katsukawa H. Enhanced responses of the chorda tympani nerve to nonsugar sweeteners in the diabetic *db/db* mouse. *Am J Physiol.* 1998;274:R1324—R1330.
17. Wang Z, Gleichmann H. GLUT2 in pancreatic islets: crucial target molecule in diabetes induced with multiple low doses of streptozotocin in mice. *Diabetes.* 1998;47:50—56.
18. Kawai K, Sugimoto K, Nakashima K, Miura H, Ninomiya Y. Leptin as a modulator of sweet taste sensitivities in mice. *Proc Natl Acad Sci USA.* 2000;97:11044—11049.
19. Shigemura N, Ohta R, Kusakabe Y, et al. Leptin modulates behavioral responses to sweet substances by influencing peripheral taste structures. *Endocrinology.* 2004;145:839—847.
20. Shigemura N, Miura H, Kusakabe Y, Hino A, Ninomiya Y. Expression of leptin receptor (Ob-R) isoforms and signal transducers and activators of transcription (STATs) mRNAs in the mouse taste buds. *Arch Histol Cytol.* 2003;66:253—260.
21. Martin B, Shin YK, White CM, et al. Vasoactive intestinal peptide-null mice demonstrate enhanced sweet taste preference, dysglycemia, and reduced taste bud leptin receptor expression. *Diabetes.* 2010;59:1143—1152.
22. Glendinning JI, Elson AE, Kalik S, et al. Taste responsiveness to sweeteners is resistant to elevations in plasma leptin. *Chem Senses.* 2015;40(4):223—231.
23. Yoshida R, Noguchi K, Shigemura N, et al. Leptin Suppresses Mouse Taste Cell Responses to Sweet Compounds. *Diabetes.* 2015;64:3751—3762.
24. Niki M, Jyotaki M, Yoshida R, et al. Modulation of sweet taste sensitivities by endogenous leptin and endocannabinoids in mice. *J Physiol.* 2015;593(11):2527—2545.
25. Meredith TL, Corcoran A, Roper SD. Leptin's effect on taste bud calcium responses and transmitter secretion. *Chem Senses.* 2015;40:217—222.
26. Finger TE, Danilova V, Barrows J, et al. ATP signaling is crucial for communication from taste buds to gustatory nerves. *Science.* 2005;310:1495—1499.
27. Lu B, Breza JM, Nikonov AA, Paedae AB, Contreras RJ. Leptin increases temperature-dependent chorda tympani nerve responses to sucrose in mice. *Physiol Behav.* 2012;107:533—539.
28. Saladin R, De Vos P, Guerre-Millo M, et al. Transient increase in obese gene expression after food intake or insulin administration. *Nature.* 1995;377:527—529.
29. Ahima RS, Prabakaran D, Mantzoros C, et al. Role of leptin in the neuroendocrine response to fasting. *Nature.* 1996;382:250—252.

30. Sinha MK, Sturis J, Ohannesian J, et al. Ultradian oscillations of leptin secretion in humans. *Biochem Biophys Res Commun.* 1996;228:733—738.
31. Nakamura Y, Sanematsu K, Ohta R, et al. Diurnal variation of human sweet taste recognition thresholds is correlated with plasma leptin levels. *Diabetes.* 2008;57:2661—2665.
32. Schoeller DA, Cella LK, Sinha MK, Caro JF. Entrainment of the diurnal rhythm of plasma leptin to meal timing. *J Clin Invest.* 1997;100:1882—1887.
33. Boden G, Chen X, Mozzoli M, Ryan I. Effect of fasting on serum leptin in normal human subjects. *J Clin Endocrinol Metab.* 1996;8:3419—3423.
34. Unger RH, Cherrington AD. Glucagonocentric restructuring of diabetes: a pathophysiologic and therapeutic makeover. *J Clin Invest.* 2012;122:4—12.
35. Huypens P, Ling Z, Pipeleers D, Schuit F. Glucagon receptors on human islet cells contribute to glucose competence of insulin release. *Diabetologia.* 2000;43:1012—1019.
36. Kieffer TJ, Heller RS, Unson CG, Weir GC, Habener JF. Distribution of glucagon receptors on hormone-specific endocrine cells of rat pancreatic islets. *Endocrinology.* 1996;137:5119—5125.
37. Ma X, Zhang Y, Gromada J, et al. Glucagon stimulates exocytosis in mouse and rat pancreatic alpha-cells by binding to glucagon receptors. *Mol Endocrinol.* 2005;19:198—212.
38. Elson AE, Dotson CD, Egan JM, Munger SD. Glucagon signalling modulates sweet taste responsiveness. *FASEB J.* 2010;24:3960—3969.
39. Lee SN, Lindberg I. 7B2 prevents unfolding and aggregation of prohormone convertase 2. *Endocrinology.* 2008;149:4116—4127.
40. Holst JJ. The physiology of glucagon-like peptide 1. *Physiol Rev.* 2007;87:1409—1439.
41. Bell GI, Santerre RF, Mullenbach GT. Hamster preproglucagon contains the sequence of glucagon and two related peptides. *Nature.* 1983;302:716—718.
42. Jang HJ, Kokrashvili Z, Theodorakis MJ, et al. Gut-expressed gustducin and taste receptors regulate secretion of glucagon-like peptide-1. *Proc Natl Acad Sci USA.* 2007;104:15069—15074.
43. Nishizawa M, Nakabayashi H, Uehara K, Nakagawa A, Uchida K, Koya D. Intraportal GLP-1 stimulates insulin secretion predominantly through the hepatoportal-pancreatic vagal reflex pathways. *Am J Physiol Endocrinol Metab.* 2013;305:E376—E387.
44. Shin YK, Martin B, Golden E, et al. Modulation of taste sensitivity by GLP-1 signaling. *J Neurochem.* 2008;106:455—463.
45. Kokrashvili Z, Yee KK, Ilegems E, et al. Endocrine taste cells. *Br J Nutr.* 2014;111:S23—S29.
46. Martin C, Passilly-Degrace P, Chevrot M, et al. Lipid-mediated release of GLP-1 by mouse taste buds from circumvallate papillae: putative involvement of GPR120 and impact on taste sensitivity. *J Lipid Res.* 2012;53:2256—2265.
47. Martin B, Dotson CD, Shin YK, et al. Modulation of taste sensitivity by GLP-1 signaling in taste buds. *Ann N Y Acad Sci.* 2009;1170:98—101.
48. Ionescu E, Rohner-Jeanrenaud F, Proietto J, Rivest RW, Jeanrenaud B. Taste-induced changes in plasma insulin and glucose turnover in lean and genetically obese rats. *Diabetes.* 1998;37:773—779.
49. Baquero AF, Gilbertson TA. Insulin activates epithelial sodium channel (ENaC) via phosphoinositide 3-kinase in mammalian taste receptor cells. *Am J Physiol Cell Physiol.* 2011;300:C860—C871.
50. Heck GL, Mierson S, DeSimone JA. Salt taste transduction occurs through an amiloride-sensitive sodium transport pathway. *Science.* 1984;223:403—405.
51. Chandrashekar J, Kuhn C, Oka Y, et al. The cells and peripheral representation of sodium taste in mice. *Nature.* 2010;464:297—301.
52. Blazer-Yost BL, Liu X, Helman SI. Hormonal regulation of ENaCs: insulin and aldosterone. *Am J Physiol.* 1998;274:C1373—C1379.
53. Wang J, Barbry P, Maiyar AC, et al. SGK integrates insulin and mineralocorticoid regulation of epithelial sodium transport. *Am J Physiol Ren Physiol.* 2001;280:F303—F313.
54. Mehta PK, Griendling KK. Angiotensin II cell signaling: physiological and pathological effects in the cardio-vascular system. *Am J Physiol Cell Physiol.* 2007;292:C82—C97.
55. Taubman MB. Angiotensin II: a vasoactive hormone with ever-increasing biological roles. *Circ Res.* 2003;92:9—11.
56. Avrith DB, Fitzsimons JT. Increased sodium appetite in the rat induced by intracranial administration of components of the renin-angiotensin system. *J Physiol.* 1980;301:349—364.

57. Fitts DA, Thunhorst RL. Rapid elicitation of salt appetite by an intravenous infusion of angiotensin II in rats. *Am J Physiol*. 1996;270:R1092—R1098.

58. Yoshida R, Ohkuri T, Jyotaki M, et al. Endocannabinoids selectively enhance sweet taste. *Proc Natl Acad Sci USA*. 2010;107:935—939.

59. Turu G, Simon A, Gyombolai P, et al. The role of diacylglycerol lipase in constitutive and angiotensin AT1 receptor-stimulated cannabinoid CB_1 receptor activity. *J Biol Chem*. 2007;282:7753—7757.

60. Moran TH, Kinzig KP. Gastrointestinal satiety signals II. Cholecystokinin. *Am J Physiol Gastrointest Liver Physiol*. 2004;286:G183—G188.

61. Innis RB, Snyder SH. Distinct cholecystokinin receptors in brain and pancreas. *Proc Natl Acad Sci USA*. 1980;77:6917—6921.

62. Wank SA. Cholecystokinin receptors. *Am J Physiol*. 1995;269:G628—G646.

63. Dufresne M, Seva C, Fourmy D. Cholecystokinin and gastrin receptors. *Physiol Rev*. 2006;86:805—847.

64. Jeon TI, Zhu B, Larson JL, Osborne TF. SREBP-2 regulates gut peptide secretion through intestinal bitter taste receptor signaling in mice. *J Clin Invest*. 2008;118:3693—3700.

65. Jeon TI, Seo YK, Osborne TF. Gut bitter taste receptor signalling induces ABCB1 through a mechanism involving CCK. *Biochem J*. 2011;438:33—37.

66. Herness S, Zhao FL, Lu SG, Kaya N, Shen T. Expression and physiological actions of cholecystokinin in rat taste receptor cells. *J Neurosci*. 2002;22:10018—10029.

67. Herness S, Zhao FL, Kaya N, Shen T, Lu SG. Communication routes within the taste bud by neuro-transmitters and neuropeptides. *Chem Senses*. 2005;30:i37—i38.

68. Shen T, Kaya N, Zhao FL, Lu SG, Cao Y, Herness S. Co-expression patterns of the neuropeptides vasoactive intestinal peptide and cholecystokinin with the transduction molecules alpha-gustducin and T1R2 in rat taste receptor cells. *Neuroscience*. 2005;130:229—238.

69. Kim MR, Kusakabe Y, Miura H, Shindo Y, Ninomiya Y, Hino A. Regional expression patterns of taste receptors and gustducin in the mouse tongue. *Biochem Biophys Res Commun*. 2003;312:500—506.

70. Lu SG, Zhao FL, Herness S. Physiological phenotyping of cholecystokinin-responsive rat taste receptor cells. *Neurosci Lett*. 2003;351:157—160.

71. Shin M, Yasumatsu K, Takai S, et al. Involvement of CCK in normal gustatory responses to bitter compounds Abstract for European Chemoreception Research Organization XXIVth Congress, 2014, P64.

72. Hajnal A, Covasa M, Bello NT. Altered taste sensitivity in obese, prediabetic OLETF rats lacking CCK-1 re-ceptors. *Am J Physiol Regul Integr Comp Physiol*. 2005;289:R1675—R1686.

73. Said SI, Rosenberg RN. Vasoactive intestinal polypeptide: abundant immunoreactivity in neural cell lines and normal nervous tissue. *Science*. 1976;192:907—908.

74. Henning RJ, Sawmiller DR. Vasoactive intestinal peptide: cardiovascular effects. *Cardiovasc Res*. 2001;49:27—37.

75. Leibowitz SF. Brain neuropeptide Y: an integrator of endocrine, metabolic and behavioral processes. *Brain Res Bull*. 1991;27:333—337.

76. Carvajal C, Dumont Y, Herzog H, Quirion R. Emotional behavior in aged neuropeptide Y (NPY) Y2 knockout mice. *J Mol Neurosci*. 2006;28:239—245.

77. Thorsell A. Neuropeptide Y (NPY) in alcohol intake and dependence. *Peptides*. 2007;28:480—483.

78. Herzog H. Neuropeptide Y and energy homeostasis: insights from Y receptor knockout models. *Eur J Pharmacol*. 2003;480:21—29.

79. Lin S, Boey D, Herzog H. NPY and Y receptors: lessons from transgenic and knockout models. *Neuropeptides*. 2004;38:189—200.

80. Herness S, Zhao FL. The neuropeptides CCK and NPY and the changing view of cell-to-cell communication in the taste bud. *Physiol Behav*. 2009;97:581—591.

81. Zhao FL, Shen T, Kaya N, Lu SG, Cao Y, Herness S. Expression, physiological action, and coexpression patterns of neuropeptide Y in rat taste-bud cells. *Proc Natl Acad Sci USA*. 2005;102:11100—11105.

82. Hurtado MD, Acosta A, Riveros PP, et al. Distribution of Y-receptors in murine lingual epithelia. *PLoS One*. 2012;7:e46358.

83. Batterham RL, Cowley MA, Small CJ, et al. Gut hormone PYY(3-36) physiologically inhibits food intake. *Nature*. 2002;418:650—654.

84. Batterham RL, Bloom SR. The gut hormone peptide YY regulates appetite. *Ann N Y Acad Sci*. 2003;994:162—168.

85. Moran TH, Smedh U, Kinzig KP, Scott KA, Knipp S, Ladenheim EE. Peptide YY(3-36) inhibits gastric emptying and produces acute reductions in food intake in rhesus monkeys. *Am J Physiol Regul Integr Comp Physiol.* 2005;288:R384–R388.

86. Ballantyne GH. Peptide YY(1-36) and peptide YY(3-36): Part I. Distribution, release and actions. *Obes Surg.* 2006;16:651–658.

87. Liu CD, Aloia T, Adrian TE, et al. Peptide YY: a potential proabsorptive hormone for the treatment of malabsorptive disorders. *Am Surg.* 1996;62:232–236.

88. Liu CD, Newton TR, Zinner MJ, Ashley SW, McFadden DW. Intraluminal peptide YY induces colonic absorption in vivo. *Dis Colon Rectum.* 1997;40:478–482.

89. Rozengurt N, Wu SV, Chen MC, Huang C, Sternini C, Rozengurt E. Colocalization of the alpha-subunit of gustducin with PYY and GLP-1 in L cells of human colon. *Am J Physiol Gastrointest Liver Physiol.* 2006;291:G792–G802.

90. Keire DA, Bowers CW, Solomon TE, Reeve Jr JR. Structure and receptor binding of PYY analogs. *Peptides.* 2002;23:305–321.

91. Chelikani PK, Haver AC, Reidelberger RD. Intravenous infusion of peptide YY(3-36) potently inhibits food intake in rats. *Endocrinology.* 2005;146:879–888.

92. Halatchev IG, Cone RD. Peripheral administration of PYY(3-36) produces conditioned taste aversion in mice. *Cell Metab.* 2005;1:159–168.

93. La Sala MS, Hurtado MD, Brown AR, et al. Modulation of taste responsiveness by the satiation hormone peptide YY. *FASEB J.* 2013;27:5022–5033.

94. Nguyen AD, Herzog H, Sainsbury A. Neuropeptide Y and peptide YY: important regulators of energy metabolism. *Curr Opin Endocrinol Diabetes Obes.* 2011;18:56–60.

95. Acosta A, Hurtado MD, Gorbatyuk O, et al. Salivary PYY: a putative bypass to satiety. *PLoS One.* 2011;6:e26137.

96. Russell JA, Leng G, Douglas AJ. The magnocellular oxytocin system, the fount of maternity: adaptations in pregnancy. *Front Neuroendocrinol.* 2003;24:27–61.

97. Meyer-Lindenberg A, Domes G, Kirsch P, Heinrichs M. Oxytocin and vasopressin in the human brain: social neuropeptides for translational medicine. *Nat Rev Neurosci.* 2011;12:524–538.

98. Leng G, Onaka T, Caquineau C, Sabatier N, Tobin VA, Takayanagi Y. Oxytocin and appetite. *Prog Brain Res.* 2008;170:137–151.

99. Arletti R, Benelli A, Bertolini A. Influence of oxytocin on feeding behavior in the rat. *Peptides.* 1989;10:89–93.

100. Arletti R, Benelli A, Bertolini A. Oxytocin inhibits food and fluid intake in rats. *Physiol Behav.* 1990;48:825–830.

101. Olson BR, Drutarosky MD, Chow MS, Hruby VJ, Stricker EM, Verbalis JG. Oxytocin and an oxytocin agonist administered centrally decrease food intake in rats. *Peptides.* 1991;12:113–118.

102. Sinclair MS, Perea-Martinez I, Dvoryanchikov G, et al. Oxytocin signaling in mouse taste buds. *PLoS One.* 2010;5:e11980.

103. Sinclair MS, Perea-Martinez I, Abouyared M, St John SJ, Chaudhari N. Oxytocin decreases sweet taste sensitivity in mice. *Physiol Behav.* 2015;141:103–110.

104. Sclafani A, Rinaman L, Vollmer RR, Amico JA. Oxytocin knockout mice demonstrate enhanced intake of sweet and nonsweet carbohydrate solutions. *Am J Physiol Regul Integr Comp Physiol.* 2007;292:R1828–R1833.

105. Puryear R, Rigatto KV, Amico JA, Morris M. Enhanced salt intake in oxytocin deficient mice. *Exp Neurol.* 2001;171:323–328.

106. Kojima M, Hosoda H, Date Y, Nakazato M, Matsuo H, Kangawa K. Ghrelin is a growth-hormone-releasing acylated peptide from stomach. *Nature.* 1999;402:656–660.

107. Date Y, Kojima M, Hosoda H, et al. Ghrelin, a novel growth hormone-releasing acylated peptide, is synthesized in a distinct endocrine cell type in the gastrointestinal tracts of rats and humans. *Endocrinology.* 2000;141:4255–4261.

108. Date Y, Nakazato M, Hashiguchi S, et al. Ghrelin is present in pancreatic alpha-cells of humans and rats and stimulates insulin secretion. *Diabetes.* 2002;51:124–129.

109. Gaytan F, Barreiro ML, Chopin LK, et al. Immunolocalization of ghrelin and its functional receptor, the type 1a growth hormone secretagogue receptor, in the cyclic human ovary. *J Clin Endocrinol Metab.* 2003;88:879–887.

110. Tortorella C, Macchi C, Spinazzi R, Malendowicz LK, Trejter M, Nussdorfer GG. Ghrelin, an endogenous ligand for the growth hormone-secretagogue receptor, is expressed in the human adrenal cortex. *Int J Mol Med.* 2003;12:213–217.

111. Cummings DE, Purnell JQ, Frayo RS, Schmidova K, Wisse BE, Weigle DS. A preprandial rise in plasma ghrelin levels suggests a role in meal initiation in humans. *Diabetes*. 2001;50:1714—1719.
112. Ariyasu H, Takaya K, Tagami T, et al. Stomach is a major source of circulating ghrelin, and feeding state determines plasma ghrelin-like immunoreactivity levels in humans. *J Clin Endocrinol Metab*. 2001;86:4753—4758.
113. Tschöp M, Wawarta R, Riepl RL, et al. Post-prandial decrease of circulating human ghrelin levels. *J Endocrinol Invest*. 2001;24:RC19—21.
114. Shin YK, Martin B, Kim W, et al. Ghrelin is produced in taste cells and ghrelin receptor null mice show reduced taste responsivity to salty (NaCl) and sour (citric acid) tastants. *PLoS One*. 2010;5:e12729.
115. Cai H, Cong WN, Daimon CM, et al. Altered lipid and salt taste responsivity in ghrelin and GOAT null mice. *PLoS One*. 2013;8:e76553.
116. Tatemoto K, Rökaeus A, Jörnvall H, McDonald TJ, Mutt V. Galanin - a novel biologically active peptide from porcine intestine. *FEBS Lett*. 1983;164:124—128.
117. Kaplan LM, Spindel ER, Isselbacher KJ, Chin WW. Tissue-specific expression of the rat galanin gene. *Proc Natl Acad Sci USA*. 1988;85:1065—1069.
118. Branchek TA, Smith KE, Gerald C, Walker MW. Galanin receptor subtypes. *Trends Pharmacol Sci*. 2000;21:109—117.
119. Waters SM, Krause JE. Distribution of galanin-1, -2 and -3 receptor messenger RNAs in central and peripheral rat tissues. *Neuroscience*. 2000;95:265—271.
120. Branchek T, Smith KE, Walker MW. Molecular biology and pharmacology of galanin receptors. *Ann N Y Acad Sci*. 1998;863:94—107.
121. Smith KE, Walker MW, Artymyshyn R, et al. Cloned human and rat galanin GALR3 receptors. Pharmacology and activation of G-protein inwardly rectifying K^+ channels. *J Biol Chem*. 1998;273:23321—23326.
122. Iismaa TP, Shine J. Galanin and galanin receptors. *Results Probl Cell Differ*. 1999;26:257—291.
123. Lang R, Gundlach AL, Kofler B. The galanin peptide family: receptor pharmacology, pleiotropic biological actions, and implications in health and disease. *Pharmacol Ther*. 2007;115:177—207.
124. Lang R, Gundlach AL, Holmes FE, et al. Physiology, signaling, and pharmacology of galanin peptides and receptors: three decades of emerging diversity. *Pharmacol Rev*. 2015;67:118—175.
125. Seta Y, Kataoka S, Toyono T, Toyoshima K. Expression of galanin and the galanin receptor in rat taste buds. *Arch Histol Cytol*. 2006;69:273—280.
126. Mahoney SA, Hosking R, Farrant S, et al. The second galanin receptor GalR2 plays a key role in neurite outgrowth from adult sensory neurons. *J Neurosci*. 2003;23:416—421.
127. Shi TJ, Hua XY, Lu X, et al. Sensory neuronal phenotype in galanin receptor 2 knockout mice: focus on dorsal root ganglion neurone development and pain behaviour. *Eur J Neurosci*. 2006;23:627—636.
128. Heijboer AC, Pijl H, Van den Hoek AM, Havekes LM, Romijn JA, Corssmit EP. Gut-brain axis: regulation of glucose metabolism. *J Neuroendocrinol*. 2006;18:883—894.
129. Burcelin R, Serino M, Cabou C. A role for the gut-to-brain GLP-1-dependent axis in the control of metabolism. *Curr Opin Pharmacol*. 2009;9:744—752.
130. Gunawardene AR, Corfe BM, Staton CA. Classification and functions of enteroendocrine cells of the lower gastrointestinal tract. *Int J Exp Pathol*. 2011;92:219—231.
131. Jamshidi N, Taylor DA. Anandamide administration into the ventromedial hypothalamus stimulates appetite in rats. *Br J Pharmacol*. 2001;134:1151—1154.
132. Cota D, Marsicano G, Lutz B, et al. Endogenous cannabinoid system as a modulator of food intake. *Int J Obes Relat Metab Disord*. 2003;27:289—301.
133. Kirkham TC, Williams CM, Fezza F, Di Marzo V. Endocannabinoid levels in rat limbic forebrain and hypothalamus in relation to fasting, feeding and satiation: stimulation of eating by 2-arachidonoyl glycerol. *Br J Pharmacol*. 2002;136:550—557.
134. Monteleone P, Matias I, Martiadis V, De Petrocellis L, Maj M, Di Marzo V. Blood levels of the endocannabinoid anandamide are increased in anorexia nervosa and in binge-eating disorder, but not in bulimia nervosa. *Neuropsychopharmacology*. 2005;30:1216—1221.
135. Di Marzo V, Goparaju SK, Wang L, et al. Leptin-regulated endocannabinoids are involved in maintaining food intake. *Nature*. 2001;410:822—825.

STIMULUS TRANSDUCTION IN OTHER CHEMODETECTION SYSTEMS

18

O_2 and CO_2 Detection by the Carotid and Aortic Bodies

Nanduri R. Prabhakar

Institute for Integrative Physiology and Center for Systems Biology of O_2 Sensing, Biological Sciences Division, University of Chicago, Chicago, IL, USA

O U T L I N E

INTRODUCTION

The concept of sensory organs for detecting the chemical composition of the arterial blood has originated from seminal studies by Fernando de Castro from the Cajal Institute in Madrid, Spain, and Jean-Francois Heymans and Corneille Heymans from Ghent, Belgium, in the 1920s. Fernando de Castro observed a ganglion-like structure at the bifurcation of the common carotid artery, which was named as "glomus caroticum or the carotid body." Based on the structure's morphology, de Castro proposed that carotid body functions as a sensory organ for monitoring the chemical composition of the arterial blood.[1] Independent studies by Jean-Francois Heymans and Corneille Heymans demonstrated that cardiorespiratory responses to alterations in blood gas composition are reflex in nature and originate from the carotid body.[2] A later study by von Euler et al.[3] firmly established the sensory nature of the carotid body by directly recording the sensory nerve responses to arterial hypoxemia (i.e., decrease in arterial blood oxygen [O_2] levels) and hypercapnia (i.e., increase in arterial blood carbon dioxide [CO_2] levels).

The existence of sensory organs similar to the carotid body in the aortic region was suggested by Heymans.[2] Julius Comroe[4] established a role for aortic bodies in the reflex regulation of cardiorespiratory functions by hypoxia. Thus, the carotid and aortic bodies are regarded as the major arterial chemoreceptors. Corneille Heymans was awarded the 1938 Nobel Prize for Physiology and Medicine "for the discovery of the role played by the sinus and aortic mechanisms in the regulation of respiration."

Much progress has been made in the past two decades in understanding the cellular mechanisms of sensory transduction at the carotid body as evidenced by a number of reviews on this topic.[5–8] However, relatively less intense efforts were made with regard to aortic chemoreceptors. The goal of this chapter is first to highlight the recent advances in the O_2 and CO_2 sensory transduction at the carotid body wherein much information is available, and then discuss what is currently known on the chemosensory transduction process at the aortic body.

CAROTID BODY

Anatomical Location and Morphology

The carotid bodies are located bilaterally in the neck at the bifurcation of common carotid arteries into internal and external carotid arteries. The wet weight of the carotid body ranges from $\sim 60\ \mu g$ in the rat to $600\ \mu g$ in the cat[9] and up to $\sim 13\ mg$ in the adult human.[10]

The chemoreceptor tissue comprises type I and type II cells. Type I (also called glomus) cells are the most abundant cell type in the carotid body; there are $\sim 12,000$ cells in the rat and $\sim 60,000$ cells in the cat carotid bodies.[9] Type I cells are of neural crest origin and they express a variety of neural cell markers including enolase, tyrosine hydroxylase, and trophic factors such as glial-derived nerve growth factor.[11–13] Type I cells exhibit clear as well as dense-core vesicles,[14] suggesting the storage of multiple neurotransmitters.

Type II cells represent ~20% of the carotid body cells.[15] A single type II cell is associated with three to five type I cells.[9,15,16] Unlike type I cells, type II cells lack dense-core vesicles. Type II (also called sustentacular) cells morphologically resemble glial cells of the nervous system.[9]

Innervation

The carotid sinus nerve, which is a branch of the glossopharyngeal nerve (cranial nerve IX), provides the sensory innervation to the carotid body. The cell bodies of the afferent neurons reside in petrosal ganglion. Both myelinated (A) and unmyelinated fibers (C) are found in the carotid sinus nerve.[17] Electron microscopy revealed synaptic contact between afferent nerve terminals and type I cells. The carotid sinus nerve terminals are classified into two types: one expresses tyrosine hydroxylase and the other contains neuropeptides including substance P and calcitonin gene−related peptide,[18,19] neuronal nitric oxide synthase,[20,21] and choline acetyltransferase.[20] Gap junction proteins are described in the carotid body, which enable bidirectional electrical communication between type I cells as well as between type I and type II cells.[22,23] Efferent innervation of the carotid bodies comes from ganglioglomerular nerve, which originates from the superior cervical sympathetic ganglion.[24]

Blood Flow and O_2 Consumption

Arterial blood supply to the carotid body is derived from branches of internal and external carotid, the occipital, and the pharyngeal arteries. The carotid body has high blood flow for its volume, which ranges between 1000 and 2000 ml/min/100 g of tissue weight, which[25−28] in terms of weight of the tissue is ~10 times more than the brain. Values of carotid body resting O_2 consumption range between 1.0 and 1.5 ml/min/100 g of tissue weight.[29,30] It was proposed that <3% of the delivered O_2 is consumed per minute.[28] Measurements of resting O_2 consumption in individual type I cells by reduced nicotinamide adenine dinucleotide (NAD) phosphate/nicotinamide NAD phosphate autofluorescence ratios suggest that type I cell operates near its maximal O_2 consumption.[31] O_2 consumption increases significantly during hypoxia.[29,32]

Tissue PO_2

Tissue partial pressure of O_2 (PO_2) of the carotid body is much lower than the PO_2 of venous blood draining the organ.[25,30] These findings suggest that much of the carotid blood flow, in normal conditions, bypass the chemoreceptor tissue, thus affecting the relation between total blood flow, O_2 consumption, and the chemo afferent nerve discharge.

SENSING HYPOXIA

Reflex Response to Hypoxia

Bilateral resection of carotid bodies[33] or anesthetic block of glossopharyngeal nerves[34] completely blocks ventilatory response to acute hypoxia in humans. Similar absence of ventilatory stimulation by acute hypoxia was also noted in several other mammalian species including rats,[35] cats,[36] dogs,[37] goats,[38] and pigs,[39] suggesting that carotid bodies are the primary sensors for detecting O_2 levels in arterial blood. Consequently, much attention has been focused on understanding the mechanism of O_2 sensing by the carotid bodies.

Sensory Complex and the Site of Sensory Transduction

Type I cells along with the afferent nerve ending constitute the sensory complex. The sensory complex is enveloped by the type II cell. It was proposed that the afferent nerve ending, whose cell body lies in the petrosal ganglion, is the primary site of sensory transduction.[40] However, petrosal neurons by themselves were found to be insensitive to hypoxia.[41] A substantial body of evidence suggests that type I cell is the initial site of sensory transduction. First, sensory discharge no longer increases in response to hypoxia after cryodestruction of type I cells.[42] Second, sinus nerve neuromas do not respond to hypoxia in the absence of the carotid body.[43] Third, isolated type I cells respond to hypoxia with increased exocytosis.[44] Fourth, when neurons from the petrosal ganglion are cocultured with type I cells, only those neurons that developed functional contact with type I cells respond to hypoxia with increased nerve activity.[45] Eyzaguirre and colleagues[46] transplanted cat carotid bodies into the tenuissimus muscle in the thigh of the same animal and allowed them to be reinnervated by the muscle nerve. They found that after 3—6 months, ~40% of the regenerating fibers responded to hypoxia and electron microscopy revealed synaptic contact of muscle nerve fibers with type I cells. These findings further support the notion that type I cells are the initial site of sensory transduction.

Although type I cells are critical for the sensory transduction, it is uncertain whether all type I cells respond to hypoxia and whether they are a homogeneous population of cells. For instance, only half of type I cells receive sensory innervation.[47] For a given hypoxic stimulus, some type I cells depolarize and others hyperpolarize.[31] Likewise, in response to a given level of hypoxia, some but not all type I cells respond with increase in cytosolic calcium concentration $[Ca^{2+}]_i$ and the response patterns vary amongst the responding type I cells.[48] It is likely that the transduction process is initiated at certain population of type I cells, perhaps the group of cells that receive afferent innervation or vice versa, and other population, by virtue of their intercellular gap junction connections regulate the activity of transducing type I cells.

Sensing Variable: O_2 Tension or O_2 Content

The sensory discharge of the carotid sinus is low under normoxia (arterial PO_2 ~ 100 mmHg), which increases dramatically even with a modest drop in arterial PO_2 (e.g., 80 to 60 mmHg).[49–52] The response of the carotid body is fast in vivo and occurs within 0.2—0.3 s after the onset of hypoxia.[53,54] The sensory nerve response of the carotid body is nonadapting and the stimulus response to graded hypoxia is hyperbolic, which is a mirror image of the O_2-hemoglobin dissociation curve. Decreasing O_2 content either by anemia[55] or by carbon monoxide (CO) poisoning of blood hemoglobin or mild hypotension[56,57] had no effect on carotid body sensory nerve activity. These findings suggest that the carotid body senses the fall in the PO_2 but not the reduction in O_2 content.

Transduction Mechanism

Emerging evidence suggests that complex interplay between three gases—O_2, CO, and hydrogen sulfide (H_2S)—is necessary for stimulation of the carotid body sensory nerve activity by hypoxia.[58] The following section summarizes how gas messenger signaling contributes to the sensory nerve excitation by hypoxia.

H_2S Mediates Sensory Excitation by Hypoxia

Recent studies suggest that mammalian cells endogenously generate H_2S.[59] Cystathionine-Υ-lyase (CSE) and cystathionine β-synthase (CBS) are the two major enzymes that catalyze the formation of endogenous H_2S. CBS is more abundant in the central nervous system, whereas CSE is predominantly expressed in peripheral tissues.[59] Type I cells express CSE[60,61] and possibly CBS.[62,63]

CSE-deficient mice exhibit remarkable absence of carotid body sensory nerve excitation by hypoxia, and markedly attenuated stimulation of breathing by low O_2, a hallmark of carotid body chemoreflex.[60] Similar reduction or absence of the carotid body sensory nerve responses to hypoxia was also seen in rats following pharmacological inhibition of H_2S synthesis.[60,62] Furthermore, H_2S donors, like hypoxia, stimulate the carotid body sensory nerve activity of several mammalian species including mice,[60,62] rats, rabbits, and cats.[64] Hypoxia increases H_2S levels in the carotid body in a stimulus-dependent manner and this response is nearly absent in CSE knockout mice and after pharmacological inhibition of CSE in rats and in wild-type mice.[60,65]

A recent study by Peng et al.[65] showed that stimulation of the carotid body sensory nerve by hypoxia was markedly attenuated in Brown-Norway (BN) rats, whereas it was exaggerated in spontaneously hypertensive (SH) rats even before the development of hypertension as compared with age-matched Sprague—Dawley (SD) rats. The enhanced hypoxic sensitivity in SH rats was associated with higher H_2S levels, whereas in BN rats, the reduced hypoxic sensitivity was correlated to lower H_2S levels, relative to that of SD rats. Correcting the H_2S levels restored the carotid body hypoxic sensitivity in BN and SH rats comparable to SD rats. These studies further support the notion that H_2S mediates the carotid body sensory excitation by hypoxia.

Heme-Oxygenase-2—Derived CO Regulates CSE-Dependent H_2S Generation by Hypoxia

A recent study by Yuan et al.[66] examined the mechanisms underlying increased H_2S generation by hypoxia. They found that CSE-derived H_2S generation is insensitive to changes in O_2 in a heterologous expression system, suggesting that a hypoxia-evoked increase in H_2S in the carotid body requires additional signaling molecules.

Type I cells express heme oxygenase-2 (HO-2), an enzyme that catalyzes the formation of CO,[67] and CO is a physiological inhibitor of the carotid body excitation by hypoxia.[65,67,68] Yuan et al.[66] showed that, unlike CSE, CO generation from HO-2 was remarkably sensitive to changes in O_2 availability, such that CO generation is high under normoxia and low during hypoxia. This unique O_2 sensitivity requires Cys^{265} and Cys^{282}, which lower the affinity of

HO-2 for O_2 and thereby enabling the enzyme to transduce changes in O_2 into changes in CO. They further demonstrated that CO inhibits H_2S generation by CSE through protein kinase G (PKG)-dependent phosphorylation of Ser^{377} of CSE. Hypoxia decreased inhibition of CSE by reducing CO generation resulting in increased H_2S, which stimulated carotid body neural activity. These findings suggest that low carotid body sensory nerve activity during normoxia is due to high levels of HO-2—derived CO, which inhibits H_2S generation by CSE, whereas the increased sensory nerve activity during hypoxia is due to reduced CO generation by HO-2 leading to activation of CSE resulting in increased H_2S generation (Figure 1). In carotid bodies from mice lacking HO-2, compensatory increase of neuronal nitric oxide synthase in type I cells mediate O_2 sensing through PKG-dependent regulation of H_2S by nitric oxide, thereby providing an important fail-safe redundancy for vital homeostatic process.[66]

BREAKOUT BOX

TRANSDUCTION MACHINERIES FOR DETECTING DIVERSE SENSORY MODALITIES

Vertebrates are endowed with a variety of sensory receptors for detecting diverse external stimuli including touch, thermal, visual, olfactory, and taste sensations. Two categories of sensory transduction machineries are identified to trigger the transduction cascade at these receptors. One category includes ion channels and the other G protein—coupled receptors (GPCRs). Although ion channels initiate the sensory transduction of touch and thermal sensations, GPCRs trigger the transduction cascade of visual, olfactory, and taste sensations.[125] In contrast to these sensory receptors, emerging evidence suggests that biochemical mechanisms comprising O_2-dependent enzymatic generation of the gaseous messengers CO and H_2S are necessary initial steps in triggering the transduction cascade of O_2 sensing by the carotid body. Thus, the O_2 sensory transduction machinery in the carotid body appears distinctly different from other known sensory receptors.

Cellular Basis of Gas Messenger—Mediated Carotid Body Hypoxic Sensing

The current consensus of the cellular basis of carotid body O_2 sensing suggests that hypoxia depolarizes type I cells, leading to Ca^{2+}-dependent release of excitatory neurotransmitter(s), which stimulates the afferent nerve ending and increases the sensory nerve activity.[6–8,69,70] Two theories have been proposed to account how hypoxia facilitates Ca^{2+}-dependent exocytosis from type I cells. According to the *membrane hypothesis*, K^+ channel(s) proteins are the O_2 sensors and low O_2 depolarizes type I cells by inhibiting O_2-sensitive K^+ conductance, causing an influx of Ca^{2+} through voltage-gated Ca^{2+} channels, leading to neurotransmitter release. The *metabolic or mitochondrial hypothesis* proposes a specialized mitochondrial cytochrome with low affinity for O_2 in type I cells functions as

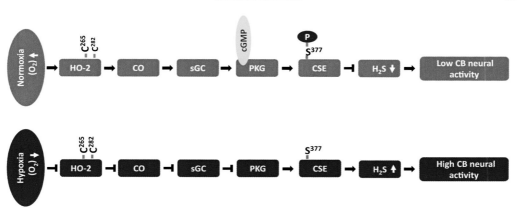

FIGURE 1 Schematic presentation of oxygen (O_2) sensing signaling pathways involving the interplay among three gases, O_2, carbon monoxide (CO), and hydrogen sulfide (H_2S), in a type I cell of the carotid body (CB) and their impact on CB neural activity. Cys^{265} and Cys^{282} are located in the heme regulatory motif of heme oxygenase (HO)-2 and Ser^{377} is the target of protein kinase G (PKG) phosphorylation in cystathionine-Υ-lyase (CSE) regulating the H_2S generation. cGMP, cyclic guanosine monophosphate; sGC, soluble guanylyl cyclase.

an O_2 sensor and hypoxia by reducing the cytochrome leads to mitochondrial depolarization and initiates the sensory transduction.[8]

Membrane Hypothesis

Supporting the membrane hypothesis, patch-clamp studies demonstrated that hypoxia inhibits a wide variety of K^+ channels in type I cells including an outward K^+ conductance,[71] a Ca^{2+}-activated K^+ conductance,[72] and a TASK (TWIK-related acid-sensitive K^+)-like conductance[73] in various mammalian species. However, attempts at recreating inherent O_2 sensitivity using recombinant voltage-gated K^+ channels in heterologous cell expression system proved inconclusive.[8] Consequently, it was suggested that the O_2 sensitivity of any particular type I cell K^+ channel is not intrinsic but instead is conferred by its coupling to an O_2-dependent, extrinsic factor(s).[74]

The following lines of evidence suggest that CO and H_2S profoundly impact ion channel conductance in type I cells. First, like O_2, CO activates voltage-dependent K^+ channel activity in type I cells,[75] whereas H_2S, like hypoxia, inhibits Ca^{2+}-activated K^+ channel activity[62,76] as well as a TASK-like K^+ conductance and depolarizes type I cells.[77] Second, an H_2S donor (NaHS) elevates $[Ca^{2+}]_i$ in type I cells; this effect was lacking in the absence of extracellular Ca^{2+}[77,78] as well as by preventing the depolarization by voltage clamping the cell at the resting membrane potential.[77] Third, nifedipine, a blocker of L-type Ca^{2+} channel, as well as T-type Cav3.2 Ca^{2+} channel blockers, prevent H_2S- and the hypoxia-evoked $[Ca^{2+}]_i$ elevation in type I cells.[78,79] Fourth, perhaps the more important is the finding that CSE-null type I cells exhibit a near absence of hypoxia-induced $[Ca^{2+}]_i$ elevation and exocytosis.[78] These findings suggest that CO-regulated H_2S transduces the hypoxic stimulus to changes in ion channel function and neurotransmitter release from type I cells to elicit sensory nerve excitation. Thus, K^+ channel proteins might function as an effector system to hypoxia accounting for the rapidity of the chemoreceptor response to low O_2 rather than as O_2 sensors.[70]

Metabolic or Mitochondrial Hypothesis

Mills and Jobsis[80] reported that carotid bodies express a putative cytochrome a3 with two O_2 affinities, one with high affinity that remains oxidized at a PO_2 levels below 7 mmHg and the other with low affinity that is half reduced at PO_2 near 90 mmHg. Based on spectral analysis, later studies[81,82] suggested that carotid bodies express cytochrome a5, which is half-reduced at PO_2 levels of 60–80 mmHg; this cytochrome is not expressed in the superior cervical and nodose ganglion. A role for mitochondria is further supported by measurements of mitochondrial membrane potential and NADH fluorescence. Duchen and Biscoe[31,83] reported an increased NADH/NAD ratio and decreased mitochondrial membrane potential in type I cells by hypoxia; these effects were not seen in other nonchemoreceptor tissues such as dorsal root ganglion. Based on these observations, it was proposed that hypoxia, by affecting the mitochondrial membrane potential, leads to inactivation of K^+ channel in type I cells.[31,83] Supporting this possibility, inhibitors of mitochondrial function, like hypoxia, cause cell depolarization,[84,85] increase Ca^{2+} influx, elevate $[Ca^{2+}]_i$,[31,83–85] and induce exocytosis.[86] Although there appears a causal link between mitochondrial depolarization and carotid body function in hypoxia besides spectral analysis, the molecular identity of the specialized cytochrome with low affinity for O_2 remains elusive, and the signaling mechanisms coupling the redox state of the mitochondria to the increased sensory nerve activity by hypoxia remain uncertain.

It is possible that the spectral changes in the carotid body previously attributed to low affinity mitochondrial cytochromes might in fact be arising from the heme–HO-2 complex because it exhibits similar redox-dependent spectral changes.[87] A recent study by Buckler[77] reported that H_2S increases NADH autofluorescence in type I cells, suggesting that H_2S might mediate its actions in part from its effects on the mitochondrial electron transport chain. Thus, it is plausible that the gas messenger signaling contributes to carotid body sensory nerve response to hypoxia via effects on ion channel activities and/or mitochondrial oxidative phosphorylation in type I cells (Figure 2).

Identity of Excitatory Neurotransmitter: A Role for Adenosine Triphosphate

Much attention has been focused in identifying the excitatory neurotransmitter released from type I cells, which contributes to the sensory excitation by hypoxia. Recent studies suggest that adenosine triphosphate (ATP) is one of the major contributors to hypoxia-evoked carotid body sensory excitation. Evidence includes: (1) hypoxia releases ATP from the carotid body and this response requires type I cell depolarization and voltage-gated Ca^{2+} entry,[88] (2) exogenous application of ATP evokes a dose-dependent increase in carotid body sensory nerve activity and reflex hyperventilation,[89] (3) ATP induces a fast-depolarizing, suramin-sensitive and slowly desensitizing inward current in rat petrosal neurons[90] and a dose-dependent increase in petrosal neuronal activity,[91] and (4) the effects of ATP on petrosal neurons is mediated by $P2X_2/P2X_3$ heteromultimeric purinoceptors.[92] These findings suggest that the stimulatory effects of ATP are due to its direct effects on afferent

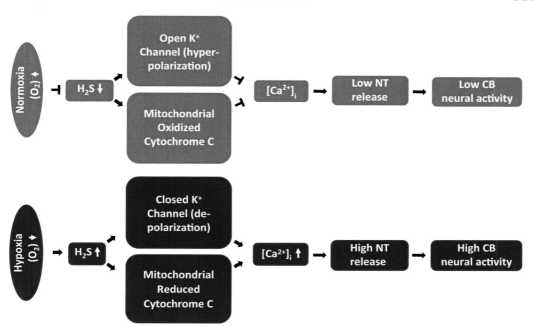

FIGURE 2 Schematic presentation of cellular targets of hydrogen sulfide (H$_2$S) in type I cell and their impact carotid body (CB) neural activity under normoxia and hypoxia. NT, neurotransmitter; [Ca^{2+}]$_i$, cytosolic calcium concentration; O$_2$, oxygen.

sinus nerve terminals and not by an indirect modulatory effect on type I cell activity. Further evidence for a role of ATP in the carotid body has come from studies on mice deficient in purinergic receptor subtypes. Mice deficient in P2X$_2$, or both P2X$_2$ and P2X$_3$, subunits exhibit impaired carotid body sensory nerve and ventilatory response to hypoxia.[93] It was proposed that ATP along with acetylcholine (ACh) might function as cotransmitters for evoking stimulation of the carotid body sensory activity by hypoxia.[90]

Interactions between Neurotransmitters

Type I cells, also express a variety of neurotransmitters/modulators besides ATP including dopamine (DA), ACh, substance P, and enkephalins (ENK).[94] For a given hypoxic stimulus, the carotid body releases multiple neurotransmitters in a Ca^{2+}-dependent manner.[94] Some of the neurotransmitters released by hypoxia such as DA are known to be potent inhibitors of the carotid body sensory nerve activity. What might be the significance of inhibitory transmitters in carotid body sensory nerve response to hypoxia? The carotid body is a slowly adapting type of sensory receptor in that the sensory nerve activity remains elevated during the entire period of hypoxia. Thus, neurotransmitters should function not only to initiate the sensory nerve excitation, but more importantly should also play a role in maintaining the increased sensory nerve discharge during the entire duration of hypoxia. It is likely that excitatory neurotransmitters such as ATP initiate the sensory nerve excitation by hypoxia, whereas the inhibitory transmitters such as DA maintain the increased sensory nerve activity by preventing the overexcitation caused by the excitatory transmitter(s). Thus, the excitatory and inhibitory neurotransmitters may work in concert with each other as a push—pull regulatory system to sustain the increased sensory nerve activity during the entire duration of hypoxia.[94]

SENSING CO_2

Role of Carotid Body in Ventilatory Response to CO_2

The carotid body contributes to 30—50% of the ventilatory stimulation by systemic hypercapnia.[95—97] A recent study suggests that ventilatory response to CO_2 is in part due to sensitization of central chemoreceptors by the sensory input from peripheral chemoreceptors, especially from the carotid body.[98] In addition, ventilatory stimulation by CO_2 is considerably delayed in carotid body—denervated animals.[99,100] These studies suggest that, although central chemoreceptors are the primary drivers of the stimulation of breathing by hypercapnia, peripheral chemoreceptors, especially the carotid body, also contribute significantly to the overall ventilatory response to CO_2.

Carotid Body Sensory Response to CO_2

It is well-established that CO_2 stimulates the carotid body sensory nerve activity. The response characteristics of CO_2 differ from the sensory response to hypoxia. An increase in arterial blood PCO_2 leads to a rapid increase in carotid body sensory activity followed by adaptation to a new steady-state level.[53] The chemosensory response to PCO_2 is linear until a PCO_2 of ~65 mmHg; thereafter, the response reaches a plateau.[49,50,101] In contrast to hypoxia, there is threshold for CO_2-induced sensory excitation. Thus, under normoxia, PCO_2 below 18—25 mmHg silences the sensory nerve activity.[50,102] The prevailing arterial blood O_2 levels affect the threshold for CO_2. Thus, the threshold is reduced by hypoxia and increased under hyperoxia.[103] Although CO_2 stimulates the carotid body, it is a less potent stimulant than hypoxia.[103—105] This reduced potency of CO_2 is ascribed to the direct inhibitory effect of hypercapnic acidosis on voltage-gated Ca^{2+} channels and exocytosis from type I cells.[106]

CO_2 Sensory Transduction

The transduction of the CO_2 stimulus by the carotid body received relatively less attention. Because the carotid body also responds to changes in arterial blood pH,[107] it is debated whether the sensory response to CO_2 is due to its molecular form and/or alteration in arterial blood pH. However, when arterial blood pH is held constant, the carotid body still responds to an increase in $PaCO_2$ with an augmented sensory discharge.[49] At a fixed level of $PaCO_2$, acidosis decreases the sensory nerve activity.[108] In an in vitro carotid body preparation, isohydric hypercapnia still augments the carotid body activity.[109] The enzyme carbonic anhydrase catalyzes the conversion of CO_2 to carbonic acid. Inhibitors of carbonic anhydrase reduce the carotid body sensory nerve activity.[110] These findings support the notion that the effects of CO_2 are due to changes in intracellular pH.

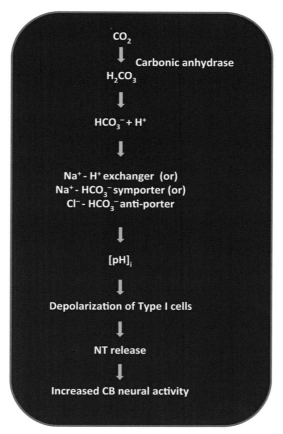

FIGURE 3 Schematic presentation of CO_2 sensing and signaling pathways associated with stimulation of the carotid body (CB) neural activity. NT, neurotransmitter.

Currently, it is thought that carotid body sensitivity to CO_2/pH is due to tight coupling of intracellular pH to extracellular pH. A number of molecules including the Na^+-H^+ exchanger,[111−113] the Na^+-HCO_3^- symporter,[111,113] as well as the DIDS (4,4′-diisothiocyana-tostilbene-2,2′-disulfonic acid)-sensitive, Cl^-/HCO_3^- antiporter,[113] and an anion-permeable channel have all been proposed to contribute to regulation of intracellular pH in type I cells (Figure 3). Nonetheless, these findings do not exclude the potential role of molecular CO_2 as a signaling molecule independent of its effects on $[pH]_i$.

AORTIC BODIES

Anatomical Location and Morphology

Aortic chemoreceptor tissue is distributed along the aorta, pulmonary arterial trunk, and subclavian arteries.[114] Of these various tissues, only the subclavian chemoreceptor tissue receives systemic arterial blood supply; these are commonly called "aortic" bodies.

The sensory function of aortic chemoreceptors has been established in cats,[56,57] whereas chemoreceptor cells are sparsely expressed in the aortic region of rats[115] and are morphologically and functionally absent in rabbit and mouse.[116] The aortic body receives sensory innervation from the aortic nerve, which is a branch of the vagus nerve. In the cat, sympathetic fibers originating from stellate ganglion[117] provide efferent innervation to the aortic body.

Type I cells are the predominant cell type in the aortic body, similar to the carotid body. The fine structure of aortic body type I cells and their innervation are indistinguishable from that seen in the carotid body.[118] Type I cells of the aortic body express catecholamines, ACh, serotonin, and ATP.[119,120] Aortic body blood flow is considered to be approximately one-sixth of the carotid body.[56,57] Because the aortic chemoreceptor tissue is diffusely arranged in small groupings of cells, little information is available on the O_2 consumption and tissue PO_2 profiles.

Reflex Regulation by the Aortic Body

The functional contribution of aortic body chemoreceptors to cardiorespiratory reflexes appears minimal in subjects with an intact functional carotid body. In carotid body–resected subjects, respiratory responses to hypoxia return, albeit much reduced, with time. The partial recovery of chemoreflex function has been attributed to augmented aortic body responses to hypoxia.[33] Aortic chemoreceptors have also been implicated in partial restoration of hypoxic sensitivity in carotid body resected nonhuman mammals, including rats,[35] cats,[36] goats,[38] and pigs.[39]

Hypoxic Sensing by the Aortic Body

Systemic hypoxia increases aortic chemoreceptor activity in cats in a stimulus-dependent manner much similar to that seen with the carotid body.[56,57,121] However, there are some notable differences between the carotid and aortic body responses to decreased O_2 content. Stimuli that reduce arterial blood O_2 content such as anemia, hypotension, and carboxyhemoglobinemia markedly stimulate the aortic body nerve activity, whereas they have little effect on the carotid body activity.[55–57] Because blood flow to aortic bodies is considered to be much less than to the carotid body, it was proposed that the reduced blood flow renders the aortic body sensitive to reduction in O_2 content of the arterial blood.[56,57] Thus, the aortic chemoreceptors, in addition to detecting changes in partial pressure of O_2, also sense reduction in arterial blood O_2 content.

Unlike the carotid body, little is known on the cellular basis of hypoxic sensing by the aortic chemoreceptors. Recent studies using an in vitro coculture preparation of aortic body type I cells and vagal neurons provided important insights into the hypoxic sensing by the aortic chemoreceptors.[122,123] These studies demonstrated that $[Ca^{2+}]_i$ responses of aortic body type I cells to hypoxia are essentially the same as that of carotid body cells.[122] It was proposed that, like the carotid body, aortic body type I cells release ACh and ATP, which act on the afferent nerve ending and stimulate the sensory nerve activity.[122,123]

How does the aortic body sense the O_2 content? Lahiri et al.[56] postulated that oxygenated hemoglobin contributes to sensing the O_2 content by the aortic body. A recent study by Piskuric et al.[123] proposed that, in response to a decrease in O_2 content, red blood cells release ATP near the vicinity of local neurons close to type I cells of the aortic body, which by acting on $P2X_2/P2X_3$ receptors, excite the neuronal activity. This increased neuronal activity is spread to other neurons via electrical coupling. Simultaneous release of ACh and ATP from type I cells further aids in neuronal excitation.

CO_2 Sensing by the Aortic Body

Much of the information on CO_2 response of the aortic chemoreceptors has come from the studies on cats. Like the carotid body, hypercapnia increases the aortic body chemoreceptor activity, albeit of lesser magnitude, and the effects of CO_2 are augmented by hypoxia.[56,57] Whereas metabolic acidosis stimulates the aortic chemoreceptor response, metabolic alkalosis reduces the response to hypoxia and CO_2.[108] Lahiri et al.[56] suggested that both aortic and carotid body share a common CO_2 sensing mechanism. However, Anand and Paintal[124] concluded that aortic chemoreceptor excitation by CO_2 is secondary to stimulation of the sympathetic nervous system. Recently, Nurse and coworkers[123] studied the effects of CO_2 on aortic body type I cells in culture and compared the responses with carotid body type I cells. They found that CO_2 increased $[Ca^{2+}]_i$ in aortic body type I cells similar to carotid body type I cells, and nickel blocks CO_2-induced Ca^{2+} influx, suggesting the involvement of voltage-gated Ca^{2+} channels. The findings by Nurse and coworkers[123] on in vitro type I cells do not support the notion that aortic chemoreceptor response to CO_2 is secondary to hypotension as proposed by Anand and Paintal.[124]

Acknowledgments

I am grateful to my colleague Professor G.K. Kumar for critical review of the manuscript. My sincere thanks to Ms. Michelle Smith-Williams for her help in formatting the references. The research in the author's laboratory is supported by grants from the National Institutes of Health, Heart, Lung and Blood Institute P01 HL-90554 and UH2-HL-123610.

References

1. De Castro F. Sur la structure et l'innervation de la glande intercarotidienne (Glomus caroticum) de l'homme et des mammiferes et sur un nouveau systeme de l'innervation autonome du nerf glossopharyngien. *Trav Lab Rech Biol.* 1926;24:365−432.

2. Heymans J, Heymans C. Sur les modifications directes et sur la regulationreflexede l'activite du centre respiratoire de la tete isolee du chien. *Arch Int Pharmacodyn Ther.* 1927;33:273−372.

3. Von Euler US, Liljestrand G, Zotterman Y. The excitation mechanism of the chemoreceptors of the carotid body. *Scand Arch Physiol.* 1939;83:132−152.

4. Comroe Jr JH. The location and function of the chemoreceptors of the aorta. *Am J Physiol.* 1939;127:176−191.

5. Fidone S, Gonzalez C. Initiation and control of chemoreceptor activity in the carotid body. In: Fishman AP, Cherniack NS, Widdicombe JG, Geiger SR, eds. *Handbook of Physiology-the Respiratory System-Control of Breathing.* Vol. II. Maryland, USA: American Physiological Society; 1986:247−312.

6. Gonzalez C, Almaraz L, Obeso A, Rigual R. Carotid body chemoreceptors: from natural stimuli to sensory discharges. *Physiol Rev.* 1994;74:829−898.

7. Prabhakar NR. Oxygen sensing by the carotid body chemoreceptors. *J Appl Physiol*. June 2000;88(6):2287–2295.

8. Kumar P, Prabhakar NR. Peripheral chemoreceptors: function and plasticity of the carotid body. *Compr Physiol*. January 2012;2(1):141–219.

9. McDonald DM. Peripheral chemoreceptors: structure-function relations of the carotid body. In: Hornbein TF, ed. *Lung Biology in Health and Disease. The Regulation of Breathing*. Vol. 17. New York: Dekker; 1981:105–319.

10. Heath D, Edwards C, Harris P. Post-mortem size and structure of the human carotid body. *Thorax*. 1970;25:129–140.

11. Kondo H, Iwanaga T, Nakajima T. Immunocytochemical study on the localization of neuron-specific enolase and S-100 protein in the carotid body of rats. *Cell Tissue Res*. 1982;227:291–295.

12. Kameda Y. Mash1 is required for glomus cell formation in the mouse carotid body. *Dev Biol*. 2005;283:128–139.

13. Izal-Azcarate A, Belzunegui S, San Sebastian W, et al. Immunohistochemical characterization of the rat carotid body. *Respir Physiol Neurobiol*. 2008;161:95–99.

14. McDonald DM, Mitchell RA. The innervation of glomus cells, ganglion cells and blood vessels in the rat carotid body: a quantitative ultrastructural analysis. *J Neurocytol*. 1975;4:177–230.

15. De Kock LL, Dunn AE. An electron microscopic study of the carotid body. *Acta Anat*. 1966;64:163–173.

16. Verna A. Ultrastructure of the carotid body in the mammals. *Int Rev Cytol*. 1979;60:271–330.

17. Eyzaguirre C, Uchizono K. Observations on the fibre content of nerves reaching the carotid body of the cat. *J Physiol*. 1961;159:268–281.

18. Finley JC, Polak J, Katz DM. Transmitter diversity in carotid body afferent neurons: dopaminergic and peptidergic phenotypes. *Neuroscience*. 1992;51:973–987.

19. Kummer W. Retrograde neuronal labelling and double-staining immunohistochemistry of tachykinin- and calcitonin gene-related peptide-immunoreactive pathways in the carotid sinus nerve of the guinea pig. *J Auton Nerv Syst*. 1988;23:131–141.

20. Wang ZZ, Bredt DS, Fidone SJ, Stensaas LJ. Neurons synthesizing nitric oxide innervate the mammalian carotid body. *J Comp Neurol*. 1993;336:419–432.

21. Prabhakar NR, Kumar GK, Chang CH, Agani FH, Haxhiu MA. Nitric oxide in the sensory function of the carotid body. *Brain Res*. 1993;625:16–22.

22. Baron M, Eyzaguirre C. Effects of temperature on some membrane characteristics of carotid body cells. *Am J Physiol*. 1977;233:C35–C46.

23. Eyzaguirre C, Abudara V. Possible role of coupling between glomus cells in carotid body chemoreception. *Biol Signals*. 1995;4:263–270.

24. Gerard MW, Billingsley PR. The innervation of the carotid body. *Anat Rec*. 1923;26:391–400.

25. Acker H, O'Regan RG. The effects of stimulation of autonomic nerves on carotid body blood flow in the cat. *J Physiol*. 1981;315:99–110.

26. Barnett S, Mulligan E, Wagerle LC, Lahiri S. Measurement of carotid body blood flow in cats by use of radioactive microspheres. *J Appl Physiol*. 1988;65:2484–2489.

27. Clarke JA, De Burgh Daly M, Ead HW. Dimensions and volume of the carotid body in the adult cat, and their relation to the specific blood flow through the organ. A histological and morphometric study. *Acta Anat (Basel)*. 1986;126:84–86.

28. De Burgh Daly M, Lambertsen CJ, Schweitzer A. Observations on the volume of blood flow and oxygen utilization of the carotid body in the cat. *J Physiol*. 1954;125:67–89.

29. Obeso A, Gonzalez C, Rigual R, Dinger B, Fidone S. Effect of low O_2 on glucose uptake in rabbit carotid body. *J Appl Physiol*. 1993;74:2387–2393.

30. Whalen WJ, Nair P. Oxidative metabolism and tissue PO_2 of the carotid body. In: Acker H, O'Regan RG, eds. *Physiology of the Peripheral Arterial Chemoreceptors*. Amsterdam: Elsevier; 1983:117–132.

31. Duchen MR, Biscoe TJ. Mitochondrial function in type I cells isolated from rabbit arterial chemoreceptors. *J Physiol*. 1992;450:13–31.

32. Rumsey WL, Iturriaga R, Spergel D, Lahiri S, Wilson DF. Optical measurements of the dependence of chemoreception on oxygen pressure in the cat carotid body. *Am J Physiol*. 1991;261:C614–C622.

33. Honda Y. Respiratory and circulatory activities in carotid body-resected humans. *J Appl Physiol (1985)*. July 1992;73(1):1–8.

34. Guz A, Noble MI, Widdicombe JG, Trenchard D, Mushin WW. Peripheral chemoreceptor block in man. *Respir Physiol*. 1966;1:38–40.

35. Martin-Body RL, Robson GJ, Sinclair JD. Restoration of hypoxic respiratory responses in the awake rat after carotid body denervation by sinus nerve section. *J Physiol*. 1986;380:61−73.
36. Smith PG, Mills E. Restoration of reflex ventilatory response to hypoxia after removal of carotid bodies in the cat. *Neuroscience*. 1980;5:573−580.
37. Rodman JR, Curran AK, Henderson KS, Dempsey JA, Smith CA. Carotid body denervation in dogs: eupnea and the ventilatory response to hyperoxic hypercapnia. *J Appl Physiol*. 2001;91:328−335.
38. Pan LG, Forster HV, Martino P, et al. Important role of carotid afferents in control of breathing. *J Appl Physiol*. 1998;85:1299−1306.
39. Lowry TF, Forster HV, Pan LG, et al. Effects on breathing of carotid body denervation in neonatal piglets. *J Appl Physiol*. 1999;l87:2128−2135.
40. Mitchell RA, Sinha AK, Mcdonald DM. Chemoreceptive properties of regenerated endings of the carotid sinus nerve. *Brain Res*. 1972;43:681−685.
41. Alcayaga J, Varas R, Arroyo J, Iturriaga R, Zapata P. Responses to hypoxia of petrosal ganglia in vitro. *Brain Res*. 1999;845:28−34.
42. Verna A, Roumy M, Leitner LM. Loss of chemoreceptive properties of the rabbit carotid body after destruction of the glomus cells. *Brain Res*. 1975;100:13−23.
43. Ponte J, Sadler CL. Studies on the regenerated carotid sinus nerve of the rabbit. *J Physiol*. 1989;410:411−424.
44. Montoro RJ, Urena J, Fernandez-Chacon R, Alvarez de Toledo G, Lopez-Barneo J. Oxygen sensing by ion channels and chemotransduction in single glomus cells. *J Gen Physiol*. 1996;107:133−143.
45. Zhong H, Zhang M, Nurse CA. Synapse formation and hypoxic signalling in co-cultures of rat petrosal neurons and carotid body type I cells. *J Physiol*. 1997;503:599−612.
46. Monti-Bloch L, Stensaas LJ, Eyzaguirre C. Carotid body grafts induce chemosensitivity in muscle nerve fibers of the cat. *Brain Res*. 1983;270:77−92.
47. McDonald DM, Mitchell RA. A quantitative analysis of synaptic connections in the rat carotid body. In: Purves MJ, ed. *The Peripheral Arterial Chemoreceptors*. London: Cambridge University Press; 1975:101−131.
48. Bright GR, Agani FH, Haque U, Overholt JL, Prabhakar NR. Heterogeneity in cytosolic calcium responses to hypoxia in carotid body cells. *Brain Res*. 1996;706:297−302.
49. Biscoe TJ, Purves MJ, Sampson SR. The frequency of nerve impulses in single carotid body chemoreceptor afferent fibres recorded in vivo with intact circulation. *J Physiol*. 1970;208:121−131.
50. Eyzaguirre C, Lewin J. Effect of different oxygen tensions on the carotid body in vitro. *J Physiol*. 1961;159:238−250.
51. Hornbein TF, Griffo ZJ, Roos A. Quantitation of chemoreceptor activity: interrelation of hypoxia and hypercapnia. *J Neurophysiol*. 1961;24:561−568.
52. Vidruk EH, Olson Jr EB, Ling L, Mitchell GS. Responses of single-unit carotid body chemoreceptors in adult rats. *J Physiol*. 2001;531:165−170.
53. Black AM, McCloskey DI, Torrance RW. The responses of carotid body chemoreceptors in the cat to sudden changes of hypercapnic and hypoxic stimuli. *Respir Physiol*. 1971;13:36−49.
54. Ponte J, Purves MJ. Frequency response of carotid body chemore-receptors in the cat to changes of $PaCO_2$, PaO_2, and pHa. *J Appl Physiol*. 1974;37:635−647.
55. Hatcher JD, Chiu LK, Jennings DB. Anemia as a stimulus to aortic and carotid chemoreceptors in cats. *J Appl Physiol*. 1978;44:696−702.
56. Lahiri S, Mulligan E, Nishino T, Mokashi A, Davies RO. Relative responses of aortic body and carotid body chemoreceptors to carboxyhemoglobinemia. *J Appl Physiol Respir Environ Exerc Physiol*. March 1981;50(3):580−586.
57. Lahiri S, Nishino T, Mokashi A, Mulligan E. Relative responses of aortic body and carotid body chemoreceptors to hypotension. *J Appl Physiol Respir Environ Exerc Physiol*. May 1980;48(5):781−788.
58. Prabhakar NR. Sensing hypoxia: physiology, genetics and epigenetics. *J Physiol*. May 1, 2013;591(Pt 9):2245−2257.
59. Gadalla MM, Snyder SH. Hydrogen sulfide as a gasotransmitter. *J Neurochem*. April 2010;113(1):14−26.
60. Peng YJ, Nanduri J, Raghuraman G, et al. H_2S mediates O_2 sensing in the carotid body. *Proc Natl Acad Sci USA*. June 8, 2010;107(23):10719−10724.
61. Mkrtchian S, Kåhlin J, Ebberyd A, et al. The human carotid body transcriptome with focus on oxygen sensing and inflammation—a comparative analysis. *J Physiol*. August 15, 2012;590(Pt 16):3807−3819.

62. Li Q, Sun B, Wang X, et al. A crucial role for hydrogen sulfide in oxygen sensing via modulating large conductance calcium-activated potassium channels. *Antioxid Redox Signal*. 2010;12:1179−1189.

63. Fitzgerald RS, Shirahata M, Chang I, Kostuk E, Kiihl S. The impact of hydrogen sulfide (H$_2$S) on neurotransmitter release from the cat carotid body. *Respir Physiol Neurobiol*. 2011;176:80−89.

64. Jiao Y, Li Q, Sun B, Zhang G, Rong W. Hydrogen sulfide activates the carotid body chemoreceptors in cat, rabbit and rat ex vivo preparations. *Respir Physiol Neurobiol*. March 2015;208:15−20.

65. Peng YJ, Makarenko V, Nanduri J, et al. Inherent variations in CO-H$_2$S-mediated carotid body O$_2$ sensing mediate hypertension and pulmonary edema. *Proc Natl Acad Sci USA*. January 21, 2014;111(3):1174−1179.

66. Yuan G, Vasavda C, Peng YJ, et al. Protein kinase G-regulated production of H$_2$S governs oxygen sensing. *Sci Sig*. 2015;8.

67. Prabhakar NR, Dinerman JL, Agani FH, Snyder SH. Carbon monoxide: a role in carotid body chemoreception. *Proc Natl Acad Sci USA*. 1995;92:1994−1997.

68. Williams SE, Wootton P, Mason HS, et al. Hemoxygenase-2 is an oxygen sensor for a calcium-sensitive potassium channel. *Science*. December 17, 2004;306(5704):2093−2097.

69. Nurse CA. Neurotransmitter and neuromodulatory mechanisms at peripheral arterial chemoreceptors. *Exp Physiol*. 2010;95:657−667.

70. Prabhakar NR, Peers C. Gasotransmitter regulation of ion channels: a key step in O$_2$ sensing by the carotid body. *Physiology*. January 2014;29(1):49−57.

71. Lopez-Barneo J, Lopez-Lopez JR, Urena J, Gonzalez C. Chemotransduction in the carotid body: K$^+$ current modulated by PO$_2$ in type I chemoreceptor cells. *Science*. 1988;241:580−582.

72. Peers C. Hypoxic suppression of K$^+$ currents in type I carotid body cells: selective effect on the Ca^{2+}-activated K$^+$ current. *Neurosci Lett*. 1990;119:253−256.

73. Buckler KJ. A novel oxygen-sensitive potassium current in rat carotid body type I cells. *J Physiol*. 1997;498(Pt 3):649−662.

74. Wyatt CN, Wright C, Bee D, Peers C. O$_2$-sensitive K$^+$ currents in carotid body chemoreceptor cells from normoxic and chronically hypoxic rats and their roles in hypoxic chemotransduction. *Proc Natl Acad Sci USA*. 1995;92:295−299.

75. Lopez-Lopez JR, Gonzalez C. Time course of K$^+$ current inhibition by low oxygen in chemoreceptor cells of adult rabbit carotid body. Effects of carbon monoxide. *FEBS Lett*. 1992;299:251−254.

76. Telezhkin V, Brazier SP, Cayzac SH, Wilkinson WJ, Riccardi D, Kemp PJ. Mechanism of inhibition by hydrogen sulfide of native and recombinant BK(Ca) channels. *Respir Physiol Neurobiol*. 2010;172: 169−178.

77. Buckler KJ. Effects of exogenous hydrogen sulphide on calcium signalling, background (TASK) K channel activity and mitochondrial function in chemoreceptor cells. *Pflügers Arch*. 2012;463:743−754.

78. Makarenko VV, Nanduri J, Raghuraman G, et al. Endogenous H$_2$S is required for hypoxic sensing by carotid body glomus cells. *Am J Physiol Cell Physiol*. November 1, 2012;303(9):C916−C923.

79. Makarenko VV, Peng YJ, Yuan G, et al. CaV3.2 T-type Ca^{2+} channels in H$_2$S-mediated hypoxic response of the carotid body. *Am J Physiol Cell Physiol*. January 15, 2015;308(2):C146−C154.

80. Mills E, Jobsis FF. Simultaneous measurement of cytochrome a3 reduction and chemoreceptor afferent activity in the carotid body. *Nature*. 1970;225:1147−1149.

81. Lahiri S, Ehleben W, Acker H. Chemoreceptor discharges and cytochrome redox changes of the rat carotid body: role of heme ligands. *Proc Natl Acad Sci USA*. August 3, 1999;96(16):9427−9432.

82. Streller T, Huckstorf C, Pfeiffer C, Acker H. Unusual cytochrome a592 with low PO$_2$ affinity correlates as putative oxygen sensor with rat carotid body chemoreceptor discharge. *FASEB J*. August 2002; 16(10):1277−1279.

83. Duchen MR, Biscoe TJ. Relative mitochondrial membrane potential and [Ca^{2+}]$_i$ in type I cells isolated from the rabbit carotid body. *J Physiol*. May 1992;450:33−61.

84. Buckler KJ, Vaughan-Jones RD. Effects of mitochondrial uncouplers on intracellular calcium, pH and membrane potential in rat carotid body type I cells. *J Physiol*. 1998;513(Pt 3):819−833.

85. Wyatt CN, Buckler KJ. The effect of mitochondrial inhibitors on membrane currents in isolated neonatal rat carotid body type I cells. *J Physiol*. 2004;556:175−191.

86. Rocher A, Obeso A, Gonzalez C, Herreros B. Ionic mechanisms for the transduction of acidic stimuli in rabbit carotid body glomus cells. *J Physiol*. 1991;433:533−548.

87. McCoubrey Jr WK, Huang TJ, Maines MD. Heme oxygenase-2 is a hemoprotein and binds heme through heme regulatory motifs that are not involved in heme catalysis. *J Biol Chem*. May 9, 1997;272(19):12568−12574.

88. Buttigieg J, Nurse CA. Detection of hypoxia-evoked ATP release from chemoreceptor cells of the rat carotid body. *Biochem Biophys Res Commun*. 2004;322:82−87.

89. McQueen DS, Bond SM, Moores C, Chessell I, Humphrey PP, Dowd E. Activation of P2X receptors for adenosine triphosphate evokes cardiorespiratory reflexes in anaesthetized rats. *J Physiol*. 1998;507 (Pt 3):843−855.

90. Zhang M, Zhong H, Vollmer C, Nurse CA. Co-release of ATP and ACh mediates hypoxic signalling at rat carotid body chemoreceptors. *J Physiol*. 2000;525(Pt 1):143−158.

91. Alcayaga J, Cerpa V, Retamal M, Arroyo J, Iturriaga R, Zapata P. Adenosine triphosphate-induced peripheral nerve discharges generated from the cat petrosal ganglion in vitro. *Neurosci Lett*. 2000;282:185−188.

92. Lewis C, Neidhart S, Holy C, North RA, Buell G, Surprenant A. Coexpression of $P2X_2$ and $P2X_3$ receptor subunits can account for ATP-gated currents in sensory neurons. *Nature*. 1995;377:432−435.

93. Rong W, Gourine AV, Cockayne DA, et al. Pivotal role of nucleotide $P2X_2$ receptor subunit of the ATP-gated ion channel mediating ventilatory responses to hypoxia. *J Neurosci*. 2003;23:11315−11321.

94. Prabhakar NR. O_2 sensing at the mammalian carotid body: why multiple O_2 sensors and multiple transmitters? *Exp Physiol*. 2006;91:17−23.

95. Bruce EN, Cherniack NS. Central chemoreceptors. *J Appl Physiol*. 1987;62:389−402.

96. Herringa J, Berkenbosch A, de Goede J, Olievier CN. Relative contribution of central and peripheral chemoreceptors to the ventilator response to CO_2 during hyperoxi. *Respir Physiol*. 1979;37(353):365−379.

97. Nattie E. CO_2, brainstem chemoreceptors and breathing. *Prog Neurobiol*. 1999;59:299−331.

98. Blain GM, Smith CA, Henderson KS, Dempsey JA. Peripheral chemoreceptors determine the respiratory sensitivity of central chemoreceptors to CO(2). *J Physiol*. July 1, 2010;588(Pt 13):2455−2471.

99. Ahmad HR, Loeschcke HH. Transient and steady state responses of pulmonary ventilation to the medullary extracellular pH after approximately rectangular changes in alveolar PCO_2. *Pflugers Arch*. December 1982;395(4):285−292.

100. Smith CA, Forster HV, Blain GM, Dempsey JA. An interdependent model of central/peripheral chemoreception: evidence and implications for ventilatory control. *Respir Physiol Neurobiol*. October 31, 2010;173(3):288−297.

101. Fitzgerald RS, Dehghani GA. Neural responses of the cat carotid and aortic bodies to hypercapnia and hypoxia. *J Appl Physiol*. 1982;52:596−601.

102. Bartels H, Witzleb E. Effect of arterial carbon dioxide pressure on chemoreceptor action potentials in the carotid sinus nerves. *Pflugers Arch*. 1956;262:466−472.

103. Lahiri S, DeLaney RG. Stimulus interaction in the responses of carotid body chemoreceptor single afferent fibers. *Respir Physiol*. 1975;24:249−266.

104. Fitzgerald RS, Parks DC. Effect of hypoxia on carotid chemoreceptor response to carbon dioxide in cats. *Respir Physiol*. 1971;12:218−229.

105. Pepper DR, Landauer RC, Kumar P. Postnatal development of CO_2−O_2 interaction in the rat carotid body in vitro. *J Physiol*. 1995;485(Pt 2):531−541.

106. Rocher A, Caceres AI, Almaraz L, Gonzalez C. EPAC signalling pathways are involved in low PO_2 chemoreception in carotid body chemoreceptor cells. *J Physiol*. 2009;587:4015−4027.

107. Eyzaguirre C, Koyano H. Effects of hypoxia, hypercapnia, and pH on the chemoreceptor activity of the carotid body in vitro. *J Physiol*. 1965;178:385−409.

108. Pokorski M, Lahiri S. Relative peripheral and central chemosensory responses to metabolic alkalosis. *Am J Physiol*. 1983;245:R873−R880.

109. Rigual R, Gonzalez E, Fidone S, Gonzalez C. Effects of low pH on synthesis and release of catecholamines in the cat carotid body in vitro. *Brain Res*. 1984;309:178−181.

110. Hanson MA, Nye PC, Rao PS, Torrance RW. Effects of acetazolamide and benzolamide on the response of the carotid chemoreceptors to CQ2 [proceedings]. *J Physiol*. 1978;284:165P−166P.

111. Buckler KJ, Vaughan-Jones RD. Application of a new pH-sensitive fluoroprobe (carboxy-SNARF-1) for intracellular pH measurement in small, isolated cells. *Pflugers Arch*. 1990;417:234−239.

112. Wilding TJ, Cheng B, Roos A. pH regulation in adult rat carotid body glomus cells. Importance of extracellular pH, sodium, and potassium. *J Gen Physiol*. 1992;100:593−608.

IV. STIMULUS TRANSDUCTION IN OTHER CHEMODETECTION SYSTEMS

113. Buckler KJ, Vaughan-Jones RD, Peers C, Nye PC. Intracellular pH and its regulation in isolated type I carotid body cells of the neonatal rat. *J Physiol.* 1991;436:107−129.

114. Coleridge H, Coleridge J, Howe A. Search for pulmonary artery chemoreceptors in the cat with a comparison of the blood supply of the aortic bodies in the newborn and adult animal. *J Physiol.* 1967;191:353−374.

115. Easton J, Howe A. The distribution of thoracic glomus tissue (aortic bodies) in the rat. *Cell Tissue Res.* 1983;232:349−356.

116. Chalmers JP, Korner PI, White SW. The relative roles of the aortic and carotid sinus nerves in the rabbit in the control of respiration and circulation during arterial hypoxia and hypercapnia. *J Physiol.* 1967;188:435−450.

117. Neil E, O'Regan RG. Efferent and afferent impulse activity recorded from few-fibre preparations of otherwise intact sinus and aortic nerves. *J Physiol.* 1971;215:33−47.

118. McDonald DM, Blewett RW. Location and size of carotid body-like organs (paraganglia) revealed in rats by the permeability of blood vessels to Evans blue dye. *J Neurocytol.* August 1981;10(4):607−643.

119. Dvorakova MC, Kummer W. Immunohistochemical evidence for species-specific coexistence of catecholamines, serotonin, acetylcholine and nitric oxide in glomus cells of rat and guinea pig aortic bodies. *Ann Anat.* September 2005;187(4):323−331.

120. Piskuric NA, Vollmer C, Nurse CA. Confocal immunofluorescence study of rat aortic body chemoreceptors and associated neurons in situ and in vitro. *J Comp Neurol.* April 1, 2011;519(5):856−873.

121. Paintal AS. Mechanism of stimulation of aortic chemoreceptors by natural stimuli and chemical substances. *J Physiol.* 1967;189:63−84.

122. Piskuric NA, Nurse CA. Effects of chemostimuli on [Ca^{2+}]$_i$ responses of rat aortic body type I cells and endogenous local neurons: comparison with carotid body cells. *J Physiol.* May 1, 2012;590(Pt 9):2121−2135.

123. Piskuric NA, Zhang M, Vollmer C, Nurse CA. Potential roles of ATP and local neurons in the monitoring of blood O$_2$ content by rat aortic bodies. *Exp Physiol.* January 2014;99(1):248−261.

124. Anand A, Paintal AS. The influence of the sympathetic outflow on aortic chemoreceptors of the cat during hypoxia and hypercapnia. *J Physiol.* January 1988;395:215−231.

125. Julius D, Nathans J. Signaling by sensory receptors. *Cold Spring Harb Perspect Biol.* 2012;4:a005991.

Chemosensation in the Ventricles of the Central Nervous System

Shuping Wen, Mari Aoki, Ulrich Boehm

Department of Pharmacology and Toxicology, University of Saarland School of Medicine, Homburg, Germany

INTRODUCTION

Within the core of the forebrain and brainstem of mammalian brains are four interconnected cavities filled with cerebrospinal fluid (CSF) known as the ventricular system.[1] The lateral ventricles are the two largest cavities and are located in the cerebellum. In the middle of the diencephalon of the forebrain lies the third ventricle, and the fourth ventricle is located between the cerebellum and pons. The third and the fourth ventricles are connected by the aqueduct.[2] Vascularized structures known as choroid plexuses (CPs) float in the lateral ventricles, the roof of the third ventricle and the fourth ventricle, where they are constantly replenishing and refreshing the CSF and also function as a blood-CSF barrier.[3] The CSF carries chemical information from peripheral organs and different parts of the brain and flows from the lateral ventricles into the third ventricle via the foramen of Monro, along the aqueduct reaching the fourth ventricle, from where it flows down the spinal cord or enters the subarachnoid space, where CSF is finally reabsorbed by arachnoid villi into the blood[4] (Figure 1).

The ventricular system not only maintains homeostasis of the central nervous system by providing an optimal environment rich in nutrients and devoid of toxic metabolites via the CSF circulation, but also acts as a medium through which peripheral systems can communicate with different brain regions.[5] Via this route, chemical information from various sources (e.g., peripheral organs, CPs, or brain parenchyma) is relayed to distant brain regions to regulate brain functions. This chemical information can be in the form of proteins, hormones, neurotransmitters, glucose, cholesterol, lipids, and other molecules and can influence a wide variety of biological processes, including brain development and behavior.[6,7]

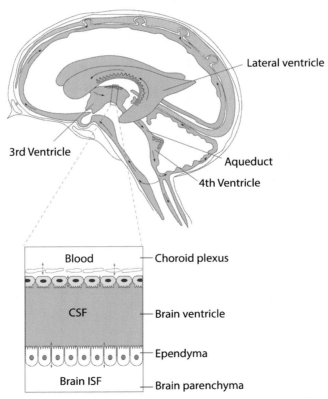

FIGURE 1   **Schematic diagram of structure of the ventricular system and the circulatory pathway of the CSF in the human brain.** Arrows indicate CSF flow directions. CSF, cerebrospinal fluid; ISF, interstitial fluid.

The composition of the CSF is highly regulated through development with distinct chemical CSF profiles observed in embryonic and adult animals.[8] Furthermore, the CSF composition is altered in certain neurological disorders.[9] Taken together, these data suggest that the composition of the CSF closely reflects the physiological and pathophysiological status of the organism.

In this chapter, we will first describe the structure and functions of the CP-CSF in the ventricles, and then review experimental evidence for a specialized chemosensory system located in the wall of the third ventricle.

CHOROID PLEXUS-CEREBROSPINAL FLUID SYSTEM

CSF: A Source of Chemosensory Cues

The CSF is a continually refreshed liquid. It not only alleviates mechanical or thermal stress of the brain, but also provides the brain with nutrients, removes waste products, and modulates brain functions via neuropeptides and other neuroactive molecules.[10] About 70% of the CSF is produced by the CP,[8] whereas the rest is generated by extrachoroidal resources such as the capillary endothelium in the parenchyma and ependymal cells located in the ventricular walls.[5] The CSF composition has been studied by proteomic analyses both in adulthood and in embryonic stages[9,11] as well as in patients with neurological diseases to look for potential biomarkers and improve diagnosis.[12,13] The pH of the CSF is neutral, it consists of 99% water and contains multiple ions (Na^+, K^+, Mg^{2+}, Ca^{2+}, Cl^-, HCO_3^-), amino acids, glucose, lipids, proteins, and many other molecules and metabolites. Various growth factors including fibroblast growth factors, insulin-like growth factors (IGFs), and retinoic acid have been identified in embryonic CSF.[7] The signaling molecules in the CSF act both on brain parenchymal cells to influence a wide range of central nervous system functions, as well as on the CP to regulate CSF secretion. Some important signaling molecules that have been found in the CSF are listed in Table 1.

Choroid Plexus

There are four CPs floating inside the ventricular cavities of the brain: one in each of the two lateral ventricles, one in the third, and one in the fourth ventricle.[1] They are extensions of the ependymal lining of the ventricular walls and form the major parts of the blood-CSF barrier, along with the arachnoid membrane and the circumventricular organs. The CP consists of a fenestrated vasculature core surrounded by a single layer of polarized cuboidal epithelium with an interstitial stromal layer in between. The epithelial cells of the CP function as both a physical and a biochemical barrier between the peripheral blood and the CSF in the brain.[14] The blood flow in the CP is 10-fold higher than that in the brain parenchyma.[15]

The main known function of the choroid plexus epithelium is to produce CSF. CSF formation in the CP is via passive filtration of peripheral blood across the choroidal capillary endothelium in the vasculature core followed by regulated active secretion across the single-layered epithelium.[16] The production of CSF is regulated by various mechanisms and at least partially controlled by autonomic noradrenergic, cholinergic, and peptidergic

TABLE 1   Selected Examples of Signaling Molecules in the CSF Related to Diseases

Signal molecule	Neural disease	Selected references
Acetylcholine	Alzheimer's disease	127,128
Adrenocorticotropin	West syndrome, infantile spasms, multiple sclerosis	129–131
Corticotropin-releasing hormone	Depression, mood disorders, fibromyalgia	132–134
Dopamine	Depression, HIV, Parkinson's disease, schizophrenia	135–138
Epinephrine	Alzheimer's disease, Parkinson's disease	139,140
Endothelin 1	Aneurysmal subarachnoid hemorrhage	141,142
β-Endorphin	Depression	143
Estrogen	Alzheimer's disease	144
Gamma-aminobutyric acid	Schizophrenia, seizures, Huntington's disease, epilepsy, etc.	145–149
Glutamate	HIV dementia, seizures, Rett syndrome, schizophrenia, etc.	148,150–152
Histamine	Multiple sclerosis, Alzheimer's disease, narcolepsy, sleep disorders, etc.	153–155
Leptin	Idiopathic intracranial hypertension, depression, obesity	156,157
Noradrenaline	Depression, schizophrenia, Parkinson's disease, etc.	158–160

innervations.[17] In addition, several hormone receptors are expressed in the CP suggesting endocrine modulation.[17] Within the ventricles, the CSF flows unidirectionally from the lateral ventricles into the third ventricle, passing through the aqueduct and finally reaching the fourth ventricle.[18] The CSF also flows—in different directions—in the subarachnoid spaces surrounding the brain. The majority of the CSF is reabsorbed by arachnoid villi from the subarachnoid space into the venous circulation and then into the systemic veins; however, some of the CSF is also reabsorbed by cerebral capillaries.[19]

Various transport proteins are expressed in the CP to regulate the access of chemical substances from the blood to the CSF. Transport of water and ions and electrolyte homeostasis in the CSF are regulated by different ion channels and cotransporters expressed in the CP.[3] The calcium-permeable transient receptor potential vanilloid 4 channel is highly expressed in CP epithelial cells, and its interaction with anoctamin 1 has been proposed to modulate water transport in the CP.[20] Specific channel-transporter signaling complexes in the CP epithelium may facilitate solute transport and electrochemical cross-talk.[21] Aldosterone (derived from the adrenal glands and the hypothalamus), angiotensin II, and arginine vasopressin produced locally in the CP also play important roles in the regulation of CSF water and monovalent electrolyte balance.[3] The CP contains transport systems to control the entry of signaling molecules originating from peripheral blood into the brain. For example, the hormone leptin from adipose tissue circulates in peripheral blood and is transported by the leptin receptor (OB-Ra) expressed in CP into the CSF to reach its final targets within the hypothalamus.[22] Receptors for prolactin,[23] insulin, and IGF-1[24] have also been detected in the CP indicating possible receptor-mediated transport of these molecules. In addition, glucose transporters are expressed in the CP in the human and rodent brain.[25,26] Moreover, expression of the taste receptor genes *Tas1r1*, *Tas1r2*, and *Tas1r3* (see Chapter 13) and the taste G protein subunit α-gustducin (see Chapter 15) has been reported in the intraventricular epithelial cells of the CP.[27] Whether other components of the taste signal transduction machinery are functionally operating in these cells is not yet known.

The CSF is not simply an ultrafiltrate of blood plasma, but is actively synthesized by the CP. Studies have shown that the CP also synthesizes a variety of bioactive factors, such as neurotrophic factors, growth factors, and hormones.[28] Most of these molecules have their own receptors expressed in the CP, indicating that their secretion is regulated in an autocrine or paracrine manner.[28] In addition to CSF production, the CP also exerts neuroprotective and detoxifying functions. Efflux transporters and metabolizing/antioxidant enzymes expressed in the CP eliminate toxic metabolites and xenobiotics from the CSF and also limit their entry into the CSF.[29]

Function of the CP-CSF System

In addition to carrying nutrients and removing metabolic waste, the CSF transports peptidergic and other signal molecules to specific target areas in the brain and plays important roles in the regulation of numerous biological functions related to neurogenesis, aging, and behavior.[8] Furthermore, changes in the composition of the CSF have been associated with neurological diseases.[30]

The CP-CSF System in Neurogenesis

The composition of the embryonic CSF is different from that of adult CSF.[31] Proteomic analysis of embryonic CSF revealed large groups of protease inhibitors, extracellular matrix proteins, and intracellular proteins involved in various brain functions affecting neuronal activity, signal transduction, cell proliferation, and differentiation.[11] It was further shown that signaling molecules secreted from specific parts of the ventricular system are essential for the development of the cortex.[32] In addition, certain molecules within the CSF regulate adult neurogenesis in the subventricular zone of the lateral ventricles, such as EGF-2 and retinoic acid, which are synthesized in the CP and released into the CSF.[4]

The CP-CSF System in Aging and Neurological Diseases

The CP-CSF system has also been implicated in aging and in multiple neurological diseases. CP senescence associated with aging is thought to disturb ion homeostasis and to increase ion-mediated toxicity, inflammation, and oxidative stress in aged individuals and in patients with Alzheimer's disease.[33] Consistent with this, dysregulation of a number of CSF molecules has been associated with various neurological diseases (some examples are listed in Table 1). Recently, a study showed that the CP itself could be a site where aging-induced type I interferon responses takes place to negatively affect brain function.[34] CSF-derived signals induce interferon-I signaling-dependent gene expression in aged mice. When this signaling is blocked within the aged brain, previously deteriorated cognitive functions and hippocampal neurogenesis are restored.

The CP-CSF System and Behavior

Stereotactic infusion of neuropeptides into specific regions of the brain such as the prefrontal cortex, periaqueductal gray, or the hypothalamus can induce various behavioral responses.[35–37] Interestingly, some of these areas border the ventricular system or subarachnoid space and thus are in close proximity to the CSF. This is consistent with the hypothesis that neuroactive substances can be sent to distant downstream brain areas via the CSF circulation and directly or indirectly activate respective brain regions to induce emotional, motivational, and behavior changes. Indeed, a number of molecules identified in the CSF have been demonstrated to regulate a wide range of behaviors including sleep, food intake, and sexual receptivity. Studies on sleep-deprived animals have, for example, shown that oleamide, a brain fatty acid present in the CSF, can induce sleep.[38] Orexin-A, an arousal-promoting factor released by neurons in the lateral hypothalamus, has also been detected in the CSF and been shown to promote wakefulness.[39] Strikingly, the strength of the response to orexin-A depends on the infusion site within the ventricles, suggesting functional heterogeneity in different areas of the ventricular walls (see below). Molecules involved in the regulation of circadian rhythms and locomotor activity such as transforming growth factor (TGF)-α[40] and cardiotrophin-like cytokine[41] have also been identified in the CSF. Melatonin, which has been shown to regulate seasonal control of luteinizing hormone release via activation of specific target cells in the hypothalamus[42] has also been detected in the CSF.[43] Interestingly, the concentration of the liver-derived hibernation-specific protein complex was found to change depending on the hibernation status.[44] Other molecules in the CSF such as insulin[45] and leptin[46] also affect food intake and influence appetite. Brain-derived gonadotropin-releasing hormone (GnRH) is also present in the CSF.[47] Although no effects on luteinizing hormone release have been reported,[47] behavioral effects of CSF GnRH on sexual receptivity have been documented.[48] The presence of corticotropin-releasing factor (CRF) and vasopressin in the CSF has prompted speculations, that these peptide hormones may be involved in fear and power-dominance motivation, respectively.[49]

How are signaling molecules in the CSF detected and how is information carried by these cues subsequently relayed to specific brain areas? Recent studies have provided experimental evidence that tanycytes, a population of specialized chemosensory cells located in the wall of the third ventricle, play a prominent role in these processes, in particular in the regulation of metabolism and reproduction as well as in adult neurogenesis.

TANYCYTES

Morphology

Tanycytes were first described by Horstmann in 1954.[50] The word tanycyte is derived from the Greek word *tanus*, which means "extended." As the name implies, tanycytes have long processes or end-feet that project into the brain parenchyma and can extend up to several 100 microns. Tanycytes are ependymoglial cells that line the lateral walls of the infundibular recess, and the floor of the third ventricle in the brain.[51] During development, the first tanycytes appear at embryonic day (E)18 in rat[52] and at E12 in mouse.[53] One unique feature of

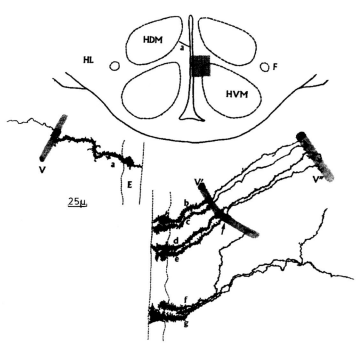

FIGURE 2   **Rapid Golgi section, adult rat. Tanycytes at the dorsal level of the ventromedial hypothalamic nucleus (HVM).** The necks of cells a and b—e have typical relations with vessels (V and V′), usually forming a cuff around the vessel wall (arrow). The tails of b—e continue into hypothalamus and terminate on vessel V″. Although the necks of cells f and g have no apparent connection with a vessel, their tail portions do (cell f). E, ependymal layer; F, fornix; HDM, dorsomedial hypothalamus; HL, lateral hypothalamus. *Reprinted, with permission, from Ref. 55.*

tanycytes is that they are in direct contact with the CSF, with the blood and with neurons (Figure 2). They have microvilli on their cell bodies facing the CSF and their processes end on capillaries.[54,55] Tanycytes are connected by tight junctions that limit intercellular transport.[56] Electron microscopy studies suggest that tanycytes have between 100 and 200 synaptoid contacts per cell and contain clear and/or dense core vesicles[57] (but see also previous work[51]), although electrophysiological approaches this far failed to demonstrate a functional significance of the synaptic contacts between hypothalamic neurons and the postsynaptic tanycytes.[58] Early studies showed that horseradish peroxidase or GnRH injected into the CSF were absorbed by hypothalamic ependymal cells and were also found around the capillaries of portal vessels.[59–61] These findings strongly suggest a transport function of tanycytes.

Tanycyte Classification

Tanycytes can be classified into four groups based on gene expression and the position of their cell bodies along the wall of the third ventricle[62] (Figure 3). α1 and α2 tanycytes are found along the ependymal surface of the ventromedial hypothalamic nucleus and the

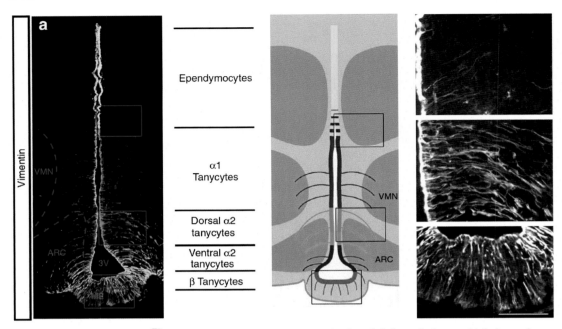

FIGURE 3   **GLAST:CreER^{T2} marks α-tanycyte subpopulations in the adult hypothalamus.** (a) Left panel: coronal section through third ventricle (3V), immunolabeled with vimentin. The vimentin-positive process distinguishes tanycytes (false-colored on the right side of image) from ependymocytes. Central panel cartoon shows position and process projection of tanycyte subtypes: purple, α1; green, dα2; blue, vα2; red, β. Note β-tanycytes divide into medial β2 and lateral β1 subsets. Right panel shows high-power magnifications of boxed regions. Ventrally, all ventricular cells appear to have a vimentin-positive process; medially, many ventricular cells have a vimentin-positive process; dorsally, the ependymocyte/α1 boundary is indistinct, with intermingling of tanycytes and ependymocytes. ARC, arcuate nucleus; ME, median eminence; VMN, ventromedial nucleus. *Reprinted, with permission, from Ref. 69.*

arcuate nucleus, respectively (Figure 3). They project to adjacent hypothalamic regions and contact blood vessels and neurons. β1 tanycytes reside in the lateral extensions of the infundibular recess whereas β2 tanycytes line the floor of the ventricle.[63,64] These ventrally located β cells contact capillaries of the median eminence.[55] Most tanycytes express vimentin, nestin, SOXB1, and a dopamine- and cyclic adenosine-3':5'-monophosphate-regulated phosphoprotein with an apparent molecular weight of 32,000 kDa (DARPP-32), whereas α1 and dorsal α2 tanycytes can be specifically distinguished from other tanycytes by immunoreactivity for glial fibrillary acidic protein.[65−69]

Functional Roles

The remarkable feature that tanycytes have access to both the CSF and to the blood puts them in prime position to monitor and integrate signals from multiple body compartments. Consistent with this, tanycytes have been shown to detect numerous chemicals and been implicated in regulating various physiological states.

Glucose Sensitivity for the Regulation of Energy Expenditure

Glucose is the primary energy source for the brain. Facilitative glucose transporter (GLUT) proteins permit glucose entry into cells.[70] In pancreatic β-cells, which act as the major glucose sensor, glucose is taken up by GLUT and subsequently followed by ATP-sensitive potassium channel closure, calcium entry and insulin secretion. GLUT isoforms were not only found in pancreatic β-cells, blood vessels, neurons, and astrocytes but also in tanycytes.[70,71] GLUT1 and GLUT2 are expressed in both α and β tanycytes.[72–75] GLUT1 is mainly localized in the processes, whereas GLUT2 is localized in the proximal region.[73] Kir6.1 and Kir6.2, pore-forming subunits of the ATP-sensing potassium channel, were also found in tanycytes and pancreas, respectively.[73,76,77] These data shed light on a potential pathway active in tanycytes that is reminiscent of the glucose-sensing pathway active in pancreatic β-cells. Direct evidence for the glucose sensitivity of tanycytes has recently been provided by calcium imaging experiments.[78] Tanycytes strongly respond to puffs of glucose and release ATP in response. Interestingly, neurons in the ventromedial hypothalamic nucleus that regulate food intake express the ATP receptor P2Y1.[79] These results suggest a possible role for glucose-sensitive tanycytes in the regulation of energy balance.

Orexigenic Control by Hormones

A lack of balance between energy intake and expenditure may cause obesity. Leptin is produced and secreted by white adipose tissue[80] and circulating leptin levels are correlated with the adipose tissue mass.[81] Leptin reaches the central nervous system by crossing the blood-brain barrier and activates neurons in hypothalamic nuclei.[82] A recent study has shown that leptin activates its receptor in median eminence tanycytes resulting in the activation of neurons in the mediobasal hypothalamus.[83] Specifically, leptin is taken up by the tanycytic end-feet in a leptin receptor-dependent manner and is subsequently transported to the cell body. Extracellular signal-regulated kinase signaling in tanycytes then triggers the activation of mediobasal hypothalamus neurons and thus plays a functional role in regulating energy expenditure in obese mice. These findings imply tanycytes as a checkpoint that allows leptin to enter the brain from the peripheral blood.

Although leptin works as a negative regulator for appetite, ghrelin has an orexigenic effect. Ghrelin is a peptide hormone produced in the stomach and released into the bloodstream.[84] It was first identified as a growth hormone-releasing peptide that acts on the pituitary.[85] Interestingly, ghrelin also regulates energy balance. Peripheral ghrelin administration increases both food intake and body weight.[86] Ghrelin activates neurons in the arcuate and ventromedial nuclei of the hypothalamus via its receptor.[87] Intravenously injected ghrelin was detected in tanycytes, suggesting that tanycytes transport ghrelin from the periphery to the brain in a similar manner as leptin.[88]

Checkpoint for Glutamate

Glutamate is a primary neurotransmitter in the hypothalamus.[89] Because excessive glutamate receptor activation is toxic to neurons, the level of extracellular glutamate concentration must be tightly regulated.[90] Glutamate transporters play an important role in controlling glutamate levels; glutamate is taken up from the synaptic cleft by glutamate transporters expressed in astrocytes and is subsequently converted to glutamine. Then neurons take up glutamine and hydrolyze it to glutamate.[91] The glutamate transporters GLT-1 and GLAST have been found in tanycytes.[92] GLT-1 is primarily expressed by tanycytes of the dorsal third ventricular wall, whereas GLAST is expressed more ventrally in the wall and the floor of the ventricle. Glutamate receptors are expressed in hypothalamic nuclei that have important roles in reproduction.[93] It is tempting to speculate that tanycytes play a role in the glutamate-glutamine cycle and that GLAST-positive tanycytes take up glutamate from the CSF or from hypothalamic nuclei and modulate neuronal activity.

In situ hybridization and immunohistochemistry analyses revealed the presence of glutamate receptor subunit GluR7 and metabotropic glutamate receptor-subunit 1α (mGluR1α) in tanycytes.[94,95] The immunoreactive GluR7 and mGluR1α signals were not only found in the cell bodies but also detected in the end-feet of tanycytes, raising the possibility that glutamate from the CSF can directly act on tanycytes and modulate their activation.

Iron Transporter

Ferrous iron (Fe^{2+}) reacts with hydrogen peroxide to generate ferric iron (Fe^{3+}), OH^-, and a highly reactive hydroxyl radical. Although iron is required for hemoglobin synthesis and cell survival, the radical damages lipid membranes, proteins, and nucleic acids; therefore, iron levels are strictly controlled. Extracellular iron binds to transferrin, resulting in an inactive form, whereas intracellular iron is stored in ferritin. Although tanycytes are not immunoreactive for the transferrin receptor,[96] they are strongly positive for iron and ferritin.[97] Tanycytes also express transferrin at low levels. Ferritin-positive tanycytes lining the wall of ventricle extend their processes to the hypothalamic region and have connections with blood vessels, consistent with a possible role for tanycytes as iron transporters.

Reproduction

Several electron microscopy studies have demonstrated that GnRH neuron axon terminals have a close structural relationship to tanycyte end-feet.[98–100] Remarkably, the association between tanycytes and GnRH neurons is estrus cycle-dependent (Figure 4). GnRH terminals are completely surrounded by tanycyte end-feet in diestrus. However, in proestrus, the tanycyte end-feet are retracted, thus allowing GnRH neurons to directly contact the pituitary portal blood, possibly resulting in the secretion of GnRH.[101,102] This dynamic change of tanycyte end-feet is regulated by TGF signaling. The level of TGF-β messenger RNA increases during proestrus in the hypothalamus.[103] Hypothalamic astrocytes express TGF receptors erbB-1, erbB-2, and erbB-4 and tanycytes express erbB-1, erbB-2, and erbB-3.[104–106] Although TGF-α treatment promotes tanycyte end-foot outgrowth, TGF-β treatment induces the retraction of processes.[107] Interestingly, long-term TGF-α treatment causes tanycyte end-foot retraction, possibly because TGF-α stimulates TGF-β release from tanycytes. Furthermore, TGF-α-induced erbB-1 activation initiates prostaglandin E_2 secretion from both astrocytes and tanycytes resulting in GnRH release from the median eminence.[107,108]

Nitric oxide (NO) is also involved in regulating tanycyte end-foot plasticity. NO, which is generated during the oxidation of L-arginine to L-citrulline by NO synthase (NOS), is a free radical gas acting as a neurotransmitter.[109] NOS is found in endothelial cells and neurons in the brain. Unlike canonical neurotransmitters, NO diffuses into adjacent cells and binds to target proteins including soluble guanylyl cyclase, methionine synthase, and hemoglobin.[110,111] Numerous studies show that NO can modulate GnRH release. The application of an NO donor to hypothalamic explants and an immortalized GnRH neuronal cell line

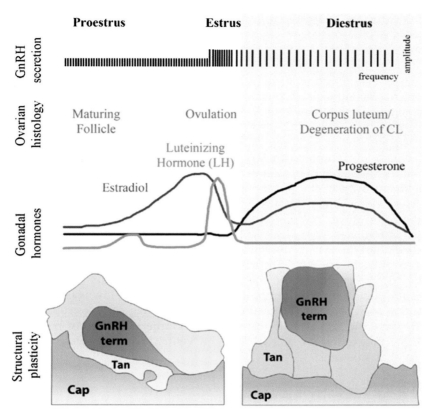

FIGURE 4   **Structural plasticity of the gonadotropin-releasing hormone (GnRH) nerve terminals and tanycytic end-feet in the median eminence of the hypothalamus during diestrus and proestrus.** The schematic highlights the causal relationships during the different phases of the ovulatory cycle with changes respectively in GnRH secretion, ovarian histology, and circulating sex hormones levels. Changes in circulating gonadal steroids are responsible for the neuro/glial structural plasticity of the median eminence. During proestrus, GnRH nerve endings (blue) sprout toward the basal lamina delineating the pericapillary space (pink, Cap), with which they eventually make direct contact, while tanycytes retract. In diestrus, under conditions of low gonadotropin output, GnRH-secreting axon terminals are distant from the pericapillary space and tanycytes enwrap GnRH nerve endings, thus impairing access of the neurohormone to the pituitary portal circulation. CL, corpus luteum. *Reprinted, with permission, from Ref. 126.*

elicits GnRH secretion. This GnRH release is inhibited by the application of the cGMP-dependent protein kinase blocker, indicating guanylyl cyclase-dependent GnRH release.[112,113] In the presence of an NOS inhibitor, reduced pulsatile GnRH secretion was observed both in vivo and in vitro.[114,115] Although these studies demonstrated a direct effect of NO on GnRH neurons, other data have shown that NO also acts on tanycytes and promotes end-foot plasticity.[116] Tanycytes cocultured with purified endothelial cells or the endothelial cell line EA.hy926 showed elevated actin cytoskeleton remodeling; this reorganization of the actin cytoskeleton was prevented by L-NAME, an NOS inhibitor.[116] L-Arginine administration induced a rapid structural change of tanycyte end-feet, allowing GnRH fibers to be in close proximity to the pericapillary space.[116] Rats showed a disrupted estrus cycle after L-NAME infusion, indicating a prominent role of tanycyte-mediated NO signaling in reproduction.[116]

Recent publications have implicated additional factors in the regulation of tanycyte end-foot plasticity. Semaphorins constitute a large gene family first identified as axon guidance molecules. They play important roles in neural circuit formation, including the GnRH neuronal system. Semaphorin7A (Sema7A) is necessary for GnRH neuron migration and also for tanycyte end-foot remodeling.[117,118] The expression level of Sema7A in tanycytes was increased on the day of diestrus, possibly regulated by progesterone.[118] Sema7A application triggers the retraction of GnRH terminals from capillaries via PlexinC1 and β1-integrin, which are receptors for Sema7A expressed in GnRH neurons.[118] Sema7A also acts in an autocrine/paracrine manner on tanycytes via β1-integrin signaling. In vitro Sema7A treatment induces the phosphorylation of β1-integrin and thus activation of focal adhesion kinase, extracellular signal–regulated kinase 1/2, and AKT, resulting in cytoskeletal remodeling and the extension of tanycyte end-feet.[118]

IGF-1 also modulates reproductive physiology. IGF-1 levels in the arcuate nucleus increase with puberty and early estrus stage.[119] Furthermore, IGF-1 administration results in accelerated puberty onset.[120] Surprisingly, tanycytes display IGF-1 immunoreactivity although they do not express IGF-1 messenger RNA.[119] Accumulation of IGF-1 was found in tanycytes after injection into the lateral cerebral ventricle at low levels in the afternoon of proestrus and at high levels in the afternoon of estrus, suggesting estrous-dependent IGF-1 uptake.[121] Because the IGF-1 receptor was found in the microvilli of tanycytes, tanycytes may detect the concentration of IGF-1 in the ventricle and regulate the availability of IGF-1 to hypothalamic cells.[122,123]

Niche for Adult Neurogenesis

Adult neurogenesis has been considered to be confined mainly to the hippocampus and to the subventricular zone. Intriguingly, some recent studies revealed that tanycytes also proliferate and are neurogenic.[124,125] Using intraperitoneal injection of bromodeoxyuridine, Lee et al. demonstrated that proliferating cells within the hypothalamus were β2 tanycytes.[124] A genetic fate mapping approach with Nestin:CreER^{T2} driver mice and ROSA26stopYFP reporter mice showed that 1 month after the Cre activation, some YFP-positive cells express the neuronal marker Hu, suggesting that β2 tanycytes generate neurons.[124] Interestingly, administering a high-fat diet (HFD) to the animals increased the rate of neurogenesis in the hypothalamus and local inhibition of neurogenesis by irradiation resulted in the attenuation of weight gain in HFD-fed mice.[124] A similar genetic approach also showed that fibroblast growth factor 10-positive tanycytes give rise to neurons in the arcuate and ventromedial nuclei of the hypothalamus that respond to fasting or leptin treatment.[125]

CONCLUDING REMARKS

Although it has long been known that the composition of the CSF closely reflects the physiological and pathophysiological status of the organism, information about individual subpopulations of chemosensory cells in CP or in the ventricular walls, and about the molecular components of the chemosensory signal transduction machinery within these cells, is just beginning to emerge. With genetic strategies to label and characterize individual populations of cells readily at hand, it seems that we will be able to tackle these questions in great detail in the future.

Acknowledgments

We thank Michael Candlish for critical comments on the manuscript. Work in the authors' laboratory is supported by the Deutsche Forschungsgemeinschaft through grants SPP1392, SFB894, SFB/TRR152, BO1743/6, and BO1743/7. U.B. is a member of the European network of investigators studying GnRH biology and reproduction (COST Action BM1105, "GnRH deficiency: Elucidation of the neuroendocrine control of human reproduction").

References

1. Mortazavi MM, Adeeb N, Griessenauer CJ, et al. The ventricular system of the brain: a comprehensive review of its history, anatomy, histology, embryology, and surgical considerations. *Childs Nerv Syst.* 2014;30(1):19−35.
2. Nicholson C. Signals that go with the flow. *Trends Neurosci.* 1999;22(4):143−145.
3. Damkier HH, Brown PD, Praetorius J. Cerebrospinal fluid secretion by the choroid plexus. *Physiol Rev.* 2013;93(4):1847−1892.
4. Lehtinen MK, Bjornsson CS, Dymecki SM, Gilbertson RJ, Holtzman DM, Monuki ES. The choroid plexus and cerebrospinal fluid: emerging roles in development, disease, and therapy. *J Neurosci.* 2013;33(45):17553−17559.
5. Skipor J, Thiery JC. The choroid plexus−cerebrospinal fluid system: undervaluated pathway of neuroendocrine signaling into the brain. *Acta Neurobiol Exp.* 2008;68(3):414−428.
6. Johanson CE, Duncan 3rd JA, Klinge PM, Brinker T, Stopa EG, Silverberg GD. Multiplicity of cerebrospinal fluid functions: new challenges in health and disease. *Cerebrospinal Fluid Res.* 2008;5:10.
7. Zappaterra MW, Lehtinen MK. The cerebrospinal fluid: regulator of neurogenesis, behavior, and beyond. *Cell Mol Life Sci.* 2012;69(17):2863−2878.
8. Redzic ZB, Preston JE, Duncan JA, Chodobski A, Szmydynger-Chodobska J. The choroid plexus-cerebrospinal fluid system: from development to aging. *Curr Top Dev Biol.* 2005;71:1−52.
9. Maurer MH. Proteomics of brain extracellular fluid (ECF) and cerebrospinal fluid (CSF). *Mass Spectrom Rev.* 2010;29(1):17−28.
10. Veening JG, Barendregt HP. The regulation of brain states by neuroactive substances distributed via the cerebrospinal fluid; a review. *Cerebrospinal Fluid Res.* 2010;7:1.
11. Zappaterra MD, Lisgo SN, Lindsay S, Gygi SP, Walsh CA, Ballif BA. A comparative proteomic analysis of human and rat embryonic cerebrospinal fluid. *J Proteome Res.* 2007;6(9):3537−3548.
12. Kroksveen AC, Opsahl JA, Guldbrandsen A, et al. Cerebrospinal fluid proteomics in multiple sclerosis. *Biochim Biophys Acta.* 2014;1854(7):746−756.
13. Oeckl P, Steinacker P, Feneberg E, Otto M. Cerebrospinal fluid proteomics and protein biomarkers in fronto-temporal lobar degeneration: current status and future perspectives. *Biochim Biophys Acta.* 2014;1854(7):757−768.
14. Wolburg H, Paulus W. Choroid plexus: biology and pathology. *Acta Neuropathol.* 2010;119(1):75−88.
15. Keep RF, Jones HC. A morphometric study on the development of the lateral ventricle choroid-plexus, choroid-plexus capillaries and ventricular ependyma in the rat. *Brain Res Dev Brain Res.* 1990;56(1):47−53.
16. Speake T, Whitwell C, Kajita H, Majid A, Brown PD. Mechanisms of CSF secretion by the choroid plexus. *Microsc Res Tech.* 2001;52(1):49−59.
17. Nilsson C, Lindvall-Axelsson M, Owman C. Neuroendocrine regulatory mechanisms in the choroid plexus-cerebrospinal fluid system. *Brain Res Brain Res Rev.* 1992;17(2):109−138.
18. Reiber H. Proteins in cerebrospinal fluid and blood: barriers, CSF flow rate and source-related dynamics. *Restor Neurol Neurosci.* 2003;21(3−4):79−96.
19. Greitz D, Hannerz J. A proposed model of cerebrospinal fluid circulation: observations with radionuclide cisternography. *Am J Neuroradiol.* 1996;17(3):431−438.
20. Takayama Y, Shibasaki K, Suzuki Y, Yamanaka A, Tominaga M. Modulation of water efflux through functional interaction between TRPV4 and TMEM16A/anoctamin 1. *FASEB J.* 2014;28(5):2238−2248.

21. Abbott GW, Tai KK, Neverisky DL, et al. KCNQ1, KCNE2, and Na$^+$-coupled solute transporters form recip-rocally regulating complexes that affect neuronal excitability. *Sci Signal*. 2014;7(315):ra22.

22. Zlokovic BV, Jovanovic S, Miao W, Samara S, Verma S, Farrell CL. Differential regulation of leptin transport by the choroid plexus and blood—brain barrier and high affinity transport systems for entry into hypothalamus and across the blood-cerebrospinal fluid barrier. *Endocrinology*. 2000;141(4):1434—1441.

23. Bakowska JC, Morrell JI. Atlas of the neurons that express mRNA for the long form of the prolactin receptor in the forebrain of the female rat. *J Comp Neurol*. 1997;386(2):161—177.

24. Bondy C, Werner H, Roberts CT, Leroith D. Cellular-pattern of type-I insulin-like growth-factor receptor gene-expression during maturation of the rat-brain - comparison with insulin-like growth factor-I and factor-II. *Neuroscience*. 1992;46(4):909—923.

25. Vannucci SJ, Clark RR, Koehler-Stec E, et al. Glucose transporter expression in brain: relationship to cerebral glucose utilization. *Dev Neurosci*. 1998;20(4—5):369—379.

26. Ueno M, Nishi N, Nakagawa T, et al. Immunoreactivity of glucose transporter 5 is located in epithelial cells of the choroid plexus and ependymal cells. *Neuroscience*. 2014;260:149—157.

27. Ren X, Zhou L, Terwilliger R, Newton SS, de Araujo IE. Sweet taste signaling functions as a hypothalamic glucose sensor. *Front Integr Neurosci*. 2009;3:12.

28. Chodobski A, Szmydynger-Chodobska J. Choroid plexus: target for polypeptides and site of their synthesis. *Microsc Res Tech*. 2001;52(1):65—82.

29. Kratzer I, Liddelow SA, Saunders NR, Dziegielewska KM, Strazielle N, Ghersi-Egea JF. Developmental changes in the transcriptome of the rat choroid plexus in relation to neuroprotection. *Fluids Barriers CNS*. 2013;10(1):25.

30. Gonzalez-Cuyar LF, Sonnen JA, Montine KS, Keene CD, Montine TJ. Role of cerebrospinal fluid and plasma biomarkers in the diagnosis of neurodegenerative disorders and mild cognitive impairment. *Curr Neurol Neurosci Rep*. 2011;11(5):455—463.

31. Gato A, Alonso MI, Martin C, et al. Embryonic cerebrospinal fluid in brain development: neural progenitor control. *Croat Med J*. 2014;55(4):299—305.

32. Miyan JA, Nabiyouni M, Zendah M. Development of the brain: a vital role for cerebrospinal fluid. *Can J Physiol Pharm*. 2003;81(4):317—328.

33. Mesquita SD, Ferreira AC, Sousa JC, et al. Modulation of iron metabolism in aging and in Alzheimer's disease: relevance of the choroid plexus. *Front Cell Neurosci*. 2012;6.

34. Baruch K, Deczkowska A, David E, et al. Aging. Aging-induced type I interferon response at the choroid plexus negatively affects brain function. *Science*. 2014;346(6205):89—93.

35. Toriya M, Maekawa F, Maejima Y, et al. Long-term infusion of brain-derived neurotrophic factor reduces food intake and body weight via a corticotrophin-releasing hormone pathway in the paraventricular nucleus of the hypothalamus. *J Neuroendocrinol*. 2010;22(9):987—995.

36. Miguel TT, Nunes-de-Souza RL. Anxiogenic and antinociceptive effects induced by corticotropin-releasing factor (CRF) injections into the periaqueductal gray are modulated by CRF1 receptor in mice. *Hormones Behav*. 2011;60(3):292—300.

37. Mena JD, Selleck RA, Baldo BA. Mu-opioid stimulation in rat prefrontal cortex engages hypothalamic orexin/hypocretin-containing neurons, and reveals dissociable roles of nucleus accumbens and hypothalamus in cortically driven feeding. *J Neurosci*. 2013;33(47):18540—18552.

38. Lerner RA, Siuzdak G, Prospero-Garcia O, Henriksen SJ, Boger DL, Cravatt BF. Cerebrodiene: a brain lipid isolated from sleep-deprived cats. *Proc Natl Acad Sci USA*. 1994;91(20):9505—9508.

39. Espana RA, Baldo BA, Kelley AE, Berridge CW. Wake-promoting and sleep-suppressing actions of hypocretin (orexin): basal forebrain sites of action. *Neuroscience*. 2001;106(4):699—715.

40. Kramer A, Yang FC, Snodgrass P, et al. Regulation of daily locomotor activity and sleep by hypothalamic EGF receptor signaling. *Science*. 2001;294(5551):2511—2515.

41. Kraves S, Weitz CJ. A role for cardiotrophin-like cytokine in the circadian control of mammalian locomotor activity. *Nat Neurosci*. 2006;9(2):212—219.

42. Malpaux B, Daveau A, Maurice-Mandon F, Duarte G, Chemineau P. Evidence that melatonin acts in the premammillary hypothalamic area to control reproduction in the ewe: presence of binding sites and stimu-lation of luteinizing hormone secretion by in situ microimplant delivery. *Endocrinology*. 1998;139(4):1508—1516.

43. Tricoire H, Moller M, Chemineau P, Malpaux B. Origin of cerebrospinal fluid melatonin and possible function in the integration of photoperiod. *Reproduction*. 2003;61:311—321.

44. Kondo N, Sekijima T, Kondo J, Takamatsu N, Tohya K, Ohtsu T. Circannual control of hibernation by HP complex in the brain. *Cell*. 2006;125(1):161—172.

45. Woods SC, Lotter EC, McKay LD, Porte Jr D. Chronic intracerebroventricular infusion of insulin reduces food intake and body weight of baboons. *Nature*. 1979;282(5738):503—505.

46. Schwartz MW, Peskind E, Raskind M, Boyko EJ, Porte Jr D. Cerebrospinal fluid leptin levels: relationship to plasma levels and to adiposity in humans. *Nat Med*. 1996;2(5):589—593.

47. Skinner DC, Caraty A, Evans NP. Does gonadotropin-releasing hormone in the cerebrospinal fluid modulate luteinizing hormone release? *Neuroendocrinology*. 1998;67(1):37—44.

48. Caraty A, Delaleu B, Chesneau D, Fabre-Nys C. Sequential role of e2 and GnRH for the expression of estrous behavior in ewes. *Endocrinology*. 2002;143(1):139—145.

49. Sewards TV, Sewards MA. Fear and power-dominance motivation: proposed contributions of peptide hormones present in cerebrospinal fluid and plasma. *Neurosci Biobehav Rev*. 2003;27(3):247—267.

50. Horstmann E. Die Faserglia des Selachiergehirns. *Z Zellforsch Mik Ana*. 1954;39(6):588—617.

51. Flament-Durand J, Brion JP. Tanycytes: morphology and functions: a review. *Int Rev Cytol*. 1985;96:121—155.

52. Rutzel H, Schiebler TH. Prenatal and early postnatal development of the glial cells in the median eminence of the rat. *Cell Tissue Res*. 1980;211(1):117—137.

53. Rakic P, Sidman RL. Subcommissural organ and adjacent ependyma: autoradiographic study of their origin in the mouse brain. *Am J Anat*. 1968;122(2):317—335.

54. Scott DE, Knigge KM. Ultrastructural changes in the median eminence of the rat following deafferentation of the basal hypothalamus. *Z Zellforsch Mikrosk Anat*. 1970;105(1):1—32.

55. Millhouse OE. A golgi study of third ventricle tanycytes in the adult rodent brain. *Z Zellforsch Mikrosk Anat*. 1971;121(1):1—13.

56. Knigge KM, Scott DE. Structure and function of the median eminence. *Am J Anat*. 1970;129(2):223—243.

57. Guldner FH, Wolff JR. Neurono-glial synaptoid contacts in the median eminence of the rat: ultrastructure, staining properties and distribution on tanycytes. *Brain Res*. 1973;61:217—234.

58. Jarvis CR, Andrew RD. Correlated electrophysiology and morphology of the ependyma in rat hypothalamus. *J Neurosci*. 1988;8(10):3691—3702.

59. Ugrumov MV, Mitskevich MS. The adsorptive and transport capacity of tanycytes during the perinatal period of the rat. *Cell Tissue Res*. 1980;211(3):493—501.

60. Wagner HJ, Pilgrim C. Extracellular and transcellular transport of horseradish peroxidase (HRP) through the hypothalamic tanycyte ependyma. *Cell Tissue Res*. 1974;152(4):477—491.

61. Uemura H, Asai T, Nozaki M, Kobayashi H. Ependymal absorption of luteinizing hormone-releasing hormone injected into the third ventricle of the rat. *Cell Tissue Res*. 1975;160(4):443—452.

62. Akmayev IG, Fidelina OV, Kabolova ZA, Popov AP, Schitkova TA. Morphological aspects of the hypothalamic-hypophyseal system. IV. Medial basal hypothalamus. An experimental morphological study. *Z Zellforsch Mikrosk Anat*. 1973;137(4):493—512.

63. Akmayev IG, Fidelina OV. Morphological aspects of the hypothalamic-hypophyseal system. VI. The tanycytes: their relation to the sexual differentiation of the hypothalamus. An enzyme-histochemical study. *Cell Tissue Res*. 1976;173(3):407—416.

64. Rodriguez EM, Gonzalez CB, Delannoy L. Cellular organization of the lateral and postinfundibular regions of the median eminence in the rat. *Cell Tissue Res*. 1979;201(3):377—408.

65. Basco E, Woodhams PL, Hajos F, Balazs R. Immunocytochemical demonstration of glial fibrillary acidic protein in mouse tanycytes. *Anat Embryol*. 1981;162(2):217—222.

66. de Vitry F, Picart R, Jacque C, Tixier-Vidal A. Glial fibrillary acidic protein. A cellular marker of tanycytes in the mouse hypothalamus. *Dev Neurosci*. 1981;4(6):457—460.

67. Chouaf L, Didier-Bazes M, Aguera M, et al. Comparative marker analysis of the ependymocytes of the subcommissural organ in four different mammalian species. *Cell Tissue Res*. 1989;257(2):255—262.

68. Meister B, Hokfelt T, Tsuruo Y, et al. DARPP-32, a dopamine- and cyclic AMP-regulated phosphoprotein in tanycytes of the mediobasal hypothalamus: distribution and relation to dopamine and luteinizing hormone-releasing hormone neurons and other glial elements. *Neuroscience*. 1988;27(2):607—622.

69. Robins SC, Stewart I, McNay DE, et al. Alpha-tanycytes of the adult hypothalamic third ventricle include distinct populations of FGF-responsive neural progenitors. *Nat Commun*. 2013;4:2049.

70. McEwen BS, Reagan LP. Glucose transporter expression in the central nervous system: relationship to synaptic function. *Eur J Pharmacol*. 2004;490(1—3):13—24.

71. Harik SI, Kalaria RN, Andersson L, Lundahl P, Perry G. Immunocytochemical localization of the erythroid glucose transporter: abundance in tissues with barrier functions. *J Neurosci.* 1990;10(12):3862–3872.

72. Arluison M, Quignon M, Nguyen P, Thorens B, Leloup C, Penicaud L. Distribution and anatomical localization of the glucose transporter 2 (GLUT2) in the adult rat brain—an immunohistochemical study. *J Chem Neuroanat.* 2004;28(3):117–136.

73. Garcia M, Millan C, Balmaceda-Aguilera C, et al. Hypothalamic ependymal-glial cells express the glucose transporter GLUT2, a protein involved in glucose sensing. *J Neurochem.* 2003;86(3):709–724.

74. Garcia MA, Carrasco M, Godoy A, et al. Elevated expression of glucose transporter-1 in hypothalamic ependymal cells not involved in the formation of the brain-cerebrospinal fluid barrier. *J Cell Biochem.* 2001;80(4):491–503.

75. Silva-Alvarez C, Carrasco M, Balmaceda-Aguilera C, et al. Ependymal cell differentiation and GLUT1 expression is a synchronous process in the ventricular wall. *Neurochem Res.* 2005;30(10):1227–1236.

76. Schuit FC, Huypens P, Heimberg H, Pipeleers DG. Glucose sensing in pancreatic beta-cells: a model for the study of other glucose-regulated cells in gut, pancreas, and hypothalamus. *Diabetes.* 2001;50(1):1–11.

77. Thomzig A, Wenzel M, Karschin C, et al. Kir6.1 is the principal pore-forming subunit of astrocyte but not neuronal plasma membrane K-ATP channels. *Mol Cell Neurosci.* 2001;18(6):671–690.

78. Frayling C, Britton R, Dale N. ATP-mediated glucosensing by hypothalamic tanycytes. *J Physiol.* 2011;589(Pt 9):2275–2286.

79. Kittner H, Franke H, Harsch JI, et al. Enhanced food intake after stimulation of hypothalamic P2Y1 receptors in rats: modulation of feeding behaviour by extracellular nucleotides. *Eur J Neurosci.* 2006;24(7):2049–2056.

80. Zhang Y, Proenca R, Maffei M, Barone M, Leopold L, Friedman JM. Positional cloning of the mouse obese gene and its human homologue. *Nature.* 1994;372(6505):425–432.

81. Friedman JM, Halaas JL. Leptin and the regulation of body weight in mammals. *Nature.* 1998;395(6704):763–770.

82. Park HK, Ahima RS. Physiology of leptin: energy homeostasis, neuroendocrine function and metabolism. *Metabolism.* 2015;64(1):24–34.

83. Balland E, Dam J, Langlet F, et al. Hypothalamic tanycytes are an ERK-gated conduit for leptin into the brain. *Cell Metab.* 2014;19(2):293–301.

84. Kojima M, Kangawa K. Ghrelin: structure and function. *Physiol Rev.* 2005;85(2):495–522.

85. Kojima M, Hosoda H, Date Y, Nakazato M, Matsuo H, Kangawa K. Ghrelin is a growth-hormone-releasing acylated peptide from stomach. *Nature.* 1999;402(6762):656–660.

86. Tschop M, Smiley DL, Heiman ML. Ghrelin induces adiposity in rodents. *Nature.* 2000;407(6806):908–913.

87. Nakazato M, Murakami N, Date Y, et al. A role for ghrelin in the central regulation of feeding. *Nature.* 2001;409(6817):194–198.

88. Collden G, Balland E, Parkash J, et al. Neonatal overnutrition causes early alterations in the central response to peripheral ghrelin. *Mol Metab.* 2015;4(1):15–24.

89. van den Pol AN, Wuarin JP, Dudek FE. Glutamate, the dominant excitatory transmitter in neuroendocrine regulation. *Science.* 1990;250(4985):1276–1278.

90. Vandenberg RJ, Ryan RM. Mechanisms of glutamate transport. *Physiol Rev.* 2013;93(4):1621–1657.

91. Broer S, Brookes N. Transfer of glutamine between astrocytes and neurons. *J Neurochem.* 2001;77(3):705–719.

92. Berger UV, Hediger MA. Differential distribution of the glutamate transporters GLT-1 and GLAST in tanycytes of the third ventricle. *J Comp Neurol.* 2001;433(1):101–114.

93. Mahesh VB, Brann DW. Regulatory role of excitatory amino acids in reproduction. *Endocrine.* 2005;28(3):271–280.

94. Eyigor O, Jennes L. Identification of kainate-preferring glutamate receptor subunit GluR7 mRNA and protein in the rat median eminence. *Brain Res.* 1998;814(1–2):231–235.

95. Tang FR, Sim MK. Metabotropic glutamate receptor subtype-1 alpha (mGluR1 alpha) immunoreactivity in ependymal cells of the rat caudal medulla oblongata and spinal cord. *Neurosci Lett.* 1997;225(3):177–180.

96. Dickinson TK, Connor JR. Immunohistochemical analysis of transferrin receptor: regional and cellular distribution in the hypotransferrinemic (hpx) mouse brain. *Brain Res.* 1998;801(1–2):171–181.

97. Benkovic SA, Connor JR. Ferritin, transferrin, and iron in selected regions of the adult and aged rat brain. *J Comp Neurol.* 1993;338(1):97–113.

98. Kozlowski GP, Coates PW. Ependymoneuronal specializations between LHRH fibers and cells of the cerebroventricular system. *Cell Tissue Res.* 1985;242(2):301–311.

99. Ugrumov M, Hisano S, Daikoku S. Topographic relations between tyrosine hydroxylase- and luteinizing hormone-releasing hormone-immunoreactive fibers in the median eminence of adult rats. *Neurosci Lett.* 1989;102(2−3):159−164.

100. King JC, Rubin BS. Dynamic changes in LHRH neurovascular terminals with various endocrine conditions in adults. *Hormones Behav.* 1994;28(4):349−356.

101. Prevot V, Dutoit S, Croix D, Tramu G, Beauvillain JC. Semi-quantitative ultrastructural analysis of the localization and neuropeptide content of gonadotropin releasing hormone nerve terminals in the median eminence throughout the estrous cycle of the rat. *Neuroscience.* 1998;84(1):177−191.

102. Prevot V, Croix D, Bouret S, et al. Definitive evidence for the existence of morphological plasticity in the external zone of the median eminence during the rat estrous cycle: implication of neuro-glio-endothelial interactions in gonadotropin-releasing hormone release. *Neuroscience.* 1999;94(3):809−819.

103. Galbiati M, Magnaghi V, Martini L, Melcangi RC. Hypothalamic transforming growth factor beta1 and basic fibroblast growth factor mRNA expression is modified during the rat oestrous cycle. *J Neuroendocrinol.* 2001;13(6):483−489.

104. Ma YJ, Hill DF, Junier MP, Costa ME, Felder SE, Ojeda SR. Expression of epidermal growth factor receptor changes in the hypothalamus during the onset of female puberty. *Mol Cell Neurosci.* 1994;5(3):246−262.

105. Steiner H, Blum M, Kitai ST, Fedi P. Differential expression of ErbB3 and ErbB4 neuregulin receptors in dopamine neurons and forebrain areas of the adult rat. *Exp Neurol.* 1999;159(2):494−503.

106. Ma YJ, Hill DF, Creswick KE, et al. Neuregulins signaling via a glial erbB-2-erbB-4 receptor complex contribute to the neuroendocrine control of mammalian sexual development. *J Neurosci.* 1999;19(22):9913−9927.

107. Prevot V, Cornea A, Mungenast A, Smiley G, Ojeda SR. Activation of erbB-1 signaling in tanycytes of the median eminence stimulates transforming growth factor beta1 release via prostaglandin E2 production and induces cell plasticity. *J Neurosci.* 2003;23(33):10622−10632.

108. Ojeda SR, Urbanski HF, Costa ME, Hill DF, Moholt-Siebert M. Involvement of transforming growth factor alpha in the release of luteinizing hormone-releasing hormone from the developing female hypothalamus. *Proc Natl Acad Sci USA.* 1990;87(24):9698−9702.

109. Bredt DS, Snyder SH. Nitric oxide, a novel neuronal messenger. *Neuron.* 1992;8(1):3−11.

110. Boehning D, Snyder SH. Novel neural modulators. *Annu Rev Neurosci.* 2003;26:105−131.

111. Jaffrey SR, Snyder SH. Nitric oxide: a neural messenger. *Annu Rev Cell Dev Biol.* 1995;11:417−440.

112. Moretto M, Lopez FJ, Negro-Vilar A. Nitric oxide regulates luteinizing hormone-releasing hormone secretion. *Endocrinology.* 1993;133(5):2399−2402.

113. Bonavera JJ, Sahu A, Kalra PS, Kalra SP. Evidence that nitric oxide may mediate the ovarian steroid-induced luteinizing hormone surge: involvement of excitatory amino acids. *Endocrinology.* 1993;133(6):2481−2487.

114. Rettori V, Belova N, Dees WL, Nyberg CL, Gimeno M, McCann SM. Role of nitric oxide in the control of luteinizing hormone-releasing hormone release in vivo and in vitro. *Proc Natl Acad Sci USA.* 1993;90(21):10130−10134.

115. Lopez FJ, Moretto M, Merchenthaler I, Negro-Vilar A. Nitric oxide is involved in the genesis of pulsatile LHRH secretion from immortalized LHRH neurons. *J Neuroendocrinol.* 1997;9(9):647−654.

116. De Seranno S, Estrella C, Loyens A, et al. Vascular endothelial cells promote acute plasticity in ependymoglial cells of the neuroendocrine brain. *J Neurosci.* 2004;24(46):10353−10363.

117. Messina A, Ferraris N, Wray S, et al. Dysregulation of Semaphorin7A/beta1-integrin signaling leads to defective GnRH-1 cell migration, abnormal gonadal development and altered fertility. *Hum Mol Genet.* 2011;20(24):4759−4774.

118. Parkash J, Messina A, Langlet F, et al. Semaphorin7A regulates neuroglial plasticity in the adult hypothalamic median eminence. *Nat Commun.* 2015;6:6385.

119. Duenas M, Luquin S, Chowen JA, Torres-Aleman I, Naftolin F, Garcia-Segura LM. Gonadal hormone regulation of insulin-like growth factor-I-like immunoreactivity in hypothalamic astroglia of developing and adult rats. *Neuroendocrinology.* 1994;59(6):528−538.

120. Hiney JK, Srivastava V, Nyberg CL, Ojeda SR, Dees WL. Insulin-like growth factor I of peripheral origin acts centrally to accelerate the initiation of female puberty. *Endocrinology.* 1996;137(9):3717−3728.

121. Fernandez-Galaz MC, Torres-Aleman I, Garcia-Segura LM. Endocrine-dependent accumulation of IGF-I by hypothalamic glia. *Neuroreport.* 1996;8(1):373−377.

122. Garcia-Segura LM, Rodriguez JR, Torres-Aleman I. Localization of the insulin-like growth factor I receptor in the cerebellum and hypothalamus of adult rats: an electron microscopic study. *J Neurocytol.* 1997;26(7):479−490.

123. Fernandez-Galaz MC, Morschl E, Chowen JA, Torres-Aleman I, Naftolin F, Garcia-Segura LM. Role of astroglia and insulin-like growth factor-I in gonadal hormone-dependent synaptic plasticity. *Brain Res Bull*. 1997;44(4):525—531.

124. Lee DA, Bedont JL, Pak T, et al. Tanycytes of the hypothalamic median eminence form a diet-responsive neurogenic niche. *Nat Neurosci*. 2012;15(5):700—702.

125. Haan N, Goodman T, Najdi-Samiei A, et al. Fgf10-expressing tanycytes add new neurons to the appetite/energy-balance regulating centers of the postnatal and adult hypothalamus. *J Neurosci*. 2013;33(14):6170—6180.

126. Messina A, Giacobini P. Semaphorin signaling in the development and function of the gonadotropin hormone-releasing hormone system. *Front Endocrinol*. 2013;4:133.

127. Smith RC, Kumar V, Khera G, Elble R, Giacobini E, Colliver J. Brain morphological measures and CSF acetylcholine in Alzheimer dementia. *Psychiatry Res*. 1988;23(1):111—114.

128. Frolich L, Dirr A, Gotz ME, et al. Acetylcholine in human CSF: methodological considerations and levels in dementia of Alzheimer type. *J Neural Transm*. 1998;105(8—9):961—973.

129. Naess A, Nyland H. Effect of ACTH treatment on CSF and blood lymphocyte sub-populations in patients with multiple-sclerosis. *Acta Neurol Scand*. 1981;63(1):57—66.

130. Takahashi H, Watanabe K, Sato Y, et al. CSF homocarnosine levels in patients with infantile spasms treated with ACTH. *Brain Dev*. 1984;6(4):424—425.

131. Heiskala H. CSF ACTH and beta-endorphin in infants with West syndrome and ACTH therapy. *Brain Dev*. 1997;19(5):339—342.

132. Williams DA, Baraniuk J, Gracely RH, Whalen G, Lyden AK, Clauw DJ. Resting cerebrospinal fluid (CSF) CRH levels differ between fibromyalgia (FM) and related conditions, and healthy controls (HC). *Arthritis Rheum*. 2004;50(9):S249.

133. Mathe AA, Agren H, Andersson W, et al. Neuropeptide Y (NPY) and corticotropin-releasing hormone (CRH) in human CSF and in brain of animal models of depression. *Int J Neuropsychoph*. 2006;9:S74.

134. Currier D, Galfalvy H, Huang YY, et al. CRHR1 and other CRH-related genes are associated with mood disorders and CSF CRH. *Biol Psychiatry*. 2009;65(8):225S.

135. Gjerris A, Werdelin L, Rafaelsen OJ, Alling C, Christensen NJ. CSF dopamine increased in depression: CSF dopamine, noradrenaline and their metabolites in depressed patients and in controls. *J Affect Disord*. 1987;13(3):279—286.

136. van Kammen DP, Kelley ME, Gilbertson MW, Gurklis J, O'Connor DT. CSF dopamine beta-hydroxylase in schizophrenia: associations with premorbid functioning and brain computerized tomography scan measures. *Am J Psychiatry*. 1994;151(3):372—378.

137. Lunardi G, Galati S, Tropepi D, et al. Correlation between changes in CSF dopamine turnover and development of dyskinesia in Parkinson's disease. *Parkinsonism Relat Disord*. 2009;15(5):383—389.

138. Horn A, Scheller C, du Plessis S, et al. Increases in CSF dopamine in HIV patients are due to the dopamine transporter 10/10-repeat allele which is more frequent in HIV-infected individuals. *J Neural Transm*. 2013;120(10):1411—1419.

139. Eldrup E, Mogensen P, Jacobsen J, Pakkenberg H, Christensen NJ. CSF and plasma-concentrations of free norepinephrine, dopamine, 3,4-dihydroxyphenylacetic acid (DOPAC), 3,4-dihydroxyphenylalanine (Dopa), and epinephrine in Parkinson's-disease. *Acta Neurol Scand*. 1995;92(2):116—121.

140. Peskind ER, Elrod R, Digiacomo L, Veith RC, Raskind MA. CSF epinephrine in aging and Alzheimer's disease. *Biol Psychiatry*. 1996;39(7):542.

141. Fassbender K, Hodopp B, Rossol S, et al. Endothelin-1 is produced by activated mononuclear leukocytes in CSF of patients with subarachnoid hemorrhage. *Neurology*. 2000;54(7):A299—A300.

142. Kastner S, Oertel MF, Scharbrodt W, Krause M, Boker DK, Deinsberger W. Endothelin-1 in plasma, cisternal CSF and microdialysate following aneurysmal SAH. *Acta Neurochir*. 2005;147(12):1271—1279.

143. Agren H, Terenius L. Depression and CSF endorphin fraction-I: seasonal-variation and higher levels in unipolar than bipolar patients. *Psychiatry Res*. 1983;10(4):303—311.

144. Schoenknecht P, Hunt A, Hentze M, Pantel J, Schroeder J. CSF estrogen levels are correlated with hippocampal glucose metabolism in Alzheimer's disease. *Int Psychogeriatr*. 2003;15:179.

145. Gerner RH, Hare TA. CSF-GABA in normal subjects and patients with depression, schizophrenia, mania, and anorexia-nervosa. *Am J Psychiatry*. 1981;138(8):1098—1101.

146. Manyam BV, Katz L, Hare TA, Kaniefski K, Tremblay RD. Isoniazid-induced elevation of CSF GABA levels and effects on chorea in Huntington's disease. *Ann Neurol*. 1981;10(1):35—37.

147. Menachem EB, Persson LI, Schechter PJ, et al. Effects of single doses of vigabatrin on CSF concentrations of GABA, homocarnosine, homovanillic-acid and 5-hydroxyindoleacetic acid in patients with complex partial epilepsy. *Epilepsy Res.* 1988;2(2):96−101.

148. Goto T, Matsuo N, Takahashi T. CSF glutamate/GABA concentrations in pyridoxine-dependent seizures: etiology of pyridoxine-dependent seizures and the mechanisms of pyridoxine action in seizure control. *Brain Dev.* 2001;23(1):24−29.

149. Poulin J, Van Kammen DP, Kelley ME, et al. Correlations between REM sleep EEG spectral analysis and CSF GABA in clinically stable drug-free patients with schizophrenia: a pilot study. *Neuropsychopharmacology.* 2005;30:S247.

150. Labarca R, Silva H, Jerez S, et al. Effects of haloperidol on CSF glutamate levels in drug-naive schizophrenic patients. *Schizophr Res.* 1995;16(1):83−85.

151. Vanhala R, Riikonen R. High levels of CSF glutamate in Rett syndrome. *Eur Child Adolesc Psychiatry.* 1997;6:87.

152. Espey MG, Basile AS, Heaton RK, Ellis RJ. Increased glutamate in CSF and plasma of patients with HIV dementia. *Neurology.* 2002;58(9):1439.

153. Kanbayashi T, Kodama T, Kondo H, et al. CSF histamine and noradrenaline contents in narcolepsy and other sleep disorders. *Sleep.* 2004;27:236.

154. Motawaj M, Peoc'h K, Callebert J, Arrang JM. CSF levels of the histamine metabolite tele-methylhistamine are only slightly decreased in Alzheimer's disease. *J Alzheimers Dis.* 2010;22(3):861−871.

155. Kallweit U, Aritake K, Bassetti CL, et al. CSF histamine levels in multiple sclerosis patients. *Mult Scler J.* 2012;18:263.

156. Buettner R, Nguyen L, O'Doherty RM. The mechanism of fasting-induced decreases in plasma and CSF leptin is defective in diet-induced obesity. *Diabetes.* 2000;49:A279.

157. Ball AK, Sinclair AJ, Curnow SJ, et al. Elevated cerebrospinal fluid (CSF) leptin in idiopathic intracranial hypertension (IIH): evidence for hypothalamic leptin resistance? *Clin Endocrinol.* 2009;70(6):863−869.

158. Kemali D, Maj M. CSF noradrenaline and schizophrenia. *Am J Psychiatry.* 1986;143(1):126−127.

159. Turkka JT, Juujarvi KK, Myllyla VV. Correlation of autonomic dysfunction to CSF concentrations of noradrenaline and 3-methoxy-4-hydroxyphenylglycol in Parkinson's disease. *Eur Neurol.* 1987;26(1):29−34.

160. Galfalvy H, Currier D, Oquendo MA, et al. Adrenergic genotype, noradrenaline metabolite (MHPG) CSF levels and suicide attempts in major depression. *Biol Psychiatry.* 2009;65(8):225S.

CHAPTER

20

Gut Nutrient Sensing

Sami Damak

Nestlé Research Center, Lausanne, Switzerland

INTRODUCTION

The ingestion of a meal has substantial consequences for a number of physiological functions, including limiting further food ingestion and preparing the gastrointestinal (GI) tract to adequately deal with the presence of nutrients in its lumen by fine-tuning and adapting several processes. A great deal of knowledge has been assembled on the identity of the structures and cells that detect nutrients and on the hormonal and neural consequences of their presence in the gastrointestinal lumen. Until recently, many gaps remained on the identity of the gut molecular detectors involved in nutrient sensing. The discovery that taste receptors are expressed in the GI tract, in addition to the tongue, generated a great deal of excitement about the belief that they were the long-sought chemical sensors of the intestine. To date, their role is only partially deciphered, but their contribution to nutrient sensing appears to be substantial. This chapter will explore the roles of the so-called taste receptors and other molecular sensors as macronutrient sensors (i.e., sensors for lipids, carbohydrates, and proteins) (Table 1). Sensing of vitamins and minerals will not be covered.

Chemosensory Transduction
http://dx.doi.org/10.1016/B978-0-12-801694-7.00020-2

TABLE 1   Molecular Sensors for Carbohydrates, Lipids, and Proteins

Macronutrient	Sensed molecules	GPCRs	Transporters	Other mechanisms
Carbohydrates	Sugars	Tas1R3:Tas1R2	SGLT-1 GLUT2	Microbiota Glycolysis
Lipids	FFAs 10-hydroxy-cis- 12-octadecenoic acid OEA 2-Oleoyl glycerol	GPR120, GPR41, GPR43, GPR40 GPR119	FATP4	Microbiota
Proteins	Oligopeptides Dipeptides Amino acids	GPR92 Tas1R1:Tas1R3, CaSR GPRC6A, mGLUR4	PEPT1	

FFAs, free fatty acids; GPCR, G protein—coupled receptor; mGluR, metabotropic glutamate receptor; SGLT-1, Na$^+$/glucose cotransporter 1.

TASTE RECEPTORS

The oral cavity is the site where the content of food is first evaluated and, based on this assessment, a decision to swallow or reject is made. Following ingestion, the nutrient content of the food must continue to be monitored so that the GI tract can adapt to the chemical composition of its content.

There are five broadly accepted taste qualities—sweet, bitter, salty, sour, and umami—and several other taste qualities, such as the taste of calcium or the taste of fat, for which there is growing evidence. The sweet, umami, and fat tastes serve the function of detecting macronutrients in the mouth and favor their ingestion (see Chapter 13). The salty taste detects sodium, whereas sour and bitter tastes warn against spoiled food and toxins, respectively. The salty, sour, and bitter receptors will not be considered here because only macronutrient sensing is covered (but see Chapters 13 and 16).

Sweet taste is initiated by sugars or noncaloric sweeteners binding to and activating the sweet taste receptor, a heterodimer of two class C G protein—coupled receptors (GPCRs) Tas1R2 and Tas1R3.[1] This receptor also responds to noncaloric sweeteners (e.g., sucralose, dulcin, saccharin, acesulfame K, sucralose).

Umami, the fifth taste, is the savory sensation typically generated by monosodium glutamate (MSG). The receptors that have been shown to mediate the umami taste are heterodimeric Tas1R1:Tas1R3[2,3] and truncated forms of metabotropic glutamate receptor (mGluR) 4[4] and of mGluR1.[5] Although these three receptors appear to play a role in transducing the taste of glutamate in rodents,[6] it is likely that Tas1R1:Tas1R3 is the main human umami taste receptor because there is synergy between amino acids and Inosine monophosphate (IMP) in activating this receptor, and potentiation of MSG response by purine nucleotides is a key feature of umami taste in humans.[7]

The oral detection of fat relies mainly on texture, with taste playing a minor role. Two GPCRs, GPR120 and GPR40,[8] and one transporter, CD36,[9] expressed in taste bud cells appear to play a role in fat taste; the main argument in favor of their role being that the respective knockout mice have diminished preference for oils and fatty acids.[8,9] However, agonists of GPR40 and/or of GPR120 are not preferred by mice, even though humans find the taste of a GPR40 agonist similar to that of linoleic acid.[10] Agonists of GPR120 do not elicit fat taste in human subjects.[10] There are no published data on gain of fat taste function resulting from compounds that bind to CD36 or facilitate the transport of fatty acids by CD36.

The signaling pathway downstream from the sweet and umami receptors consists of the G protein α-gustducin (Gα$_{gust}$), phospholipase C β2, inositol trisphosphate receptor type 3, transient receptor potential channel M5 (TRPM5), the Ca^{2+} homeostasis modulator protein 1 channel, and the release of adenosine triphosphate (ATP), which acts as neurotransmitter and activates taste nerve fibers[11–14] (see also Chapter 15). The signaling downstream of the fat taste receptors is less well defined.

TASTE SIGNALING MOLECULES IN THE GUT

The sweet and umami taste receptors were initially described in the taste bud cells, but later work showed that they are expressed in a few other organs, including the GI tract.[15,16] GPR40 and GPR120 were initially described in the pancreas[17,18] and in the gut,[19] respectively. A role of those two GPCRs in fat taste transduction was described later.[8] Most elements of the signaling pathway downstream of the taste receptors are also expressed in the GI tract.[15,20] Taste receptors and taste signaling molecules are expressed in two types of intestinal solitary cells: the enteroendocrine cells and the tufted cells.[15,20,21] The degree of coexpression of the members of the signaling pathway in the gut does not always match that in taste bud cells. For example, in the intestinal epithelium TRPM5 in expressed exclusively in tufted cells, whereas Gα$_{gust}$ is expressed in the enteroendocrine cells, in contrast to the taste bud cells where all Gα$_{gust}$-expressing cells also express TRPM5.[20] The level of expression of the sweet and umami receptors is also much lower in the intestine than in the taste buds.

WHY DOES THE GUT SENSE NUTRIENTS?

The main functions of the gut are to digest and absorb nutrients and to send satiety or hunger signals to the brain. Postprandial consequences of a meal include changes in gastric and pancreatic secretion, gallbladder contraction, adaptation of blood flow, and changes in intestinal motility. To accomplish those functions the gut needs to monitor the chemical nature of its contents and adapt its motility, secretion, and expression of transporters accordingly.

WHERE ARE THE SENSORS LOCATED?

The cells that contribute most to GI nutrient sensing are the enteroendocrine cells, a group of solitary intestinal cells which secrete hormones such as gastrin, ghrelin, glucagon-like peptide-1 (GLP-1), glucose-dependent insulinotropic polypeptide (GIP), serotonin, cholecystokinin (CCK), peptide YY (PYY), oxyntomodulin, somatostatin, neurotensin, and secretin. Mice lacking enteroendocrine cells die in the first postnatal week, showing that nutrient sensing is a critical function to sustain life.[22] Enteroendocrine cells include an apical pole with microvilli in contact with the intestinal lumen where nutrient sensing takes place, and a broad basal side from which hormones are secreted in response to nutrient sensing. The hormones' effects are paracrine, acting on nearby nerves and cells, and endocrine, with release into the bloodstream. Recently, a cytoplasmic process named neuropod was described at the basal side of enteroendocrine cells.[23,24] This neuropod makes direct contact with efferent and afferent nerves, demonstrating a direct connection between the enteroendocrine sensing cells and the enteric and central nervous systems.[25] The study of enteroendocrine cells has been complicated by their scattered nature, but the recent development of transgenic mice in which specific enteroendocrine cells are labeled has enabled the isolation and purification of specific populations and better understanding of their molecular makeup. The enteroendocrine cells can be grouped into subpopulations, defined according to the peptides they secrete. For example, L cells are known to secrete GLP-1, GLP-2, oxyntomodulin, and PYY, whereas I cells secrete CCK. The in-depth knowledge that resulted from studies in transgenic mice showed that the separation of enteroendocrine cell subtypes is not as strict as initially thought. It turned out, for example, that L cells also secrete CCK.[26,27] Isolated L cells respond to glucose by membrane depolarization, generating action potentials and entry of Ca^{2+} into the cell through voltage-gated Ca^{2+} channels.[28]

Tufted cells constitute another group of solitary cells that are found in the GI tract and several other hollow organs.[29] These cells have an apical process that protrudes into the intestinal lumen and a narrow basal side (Figure 1). It was based on the shape and ultrastructure of the tufted cells that the first suggestion of the existence of "gut taste" was made.[30] They morphologically look like gustatory sensory cells and express $G\alpha_{gust}$[31] and a number of other taste signaling molecules,[15,20] which suggests that they might be implicated in chemosensing. However, this function has never been demonstrated, and their role in nutrient sensing is even questioned by some because it is not totally clear how tufted cells transmit the message that results from their possible activation by nutrients. Contact with nerve fibers (Figure 2) and production of prostaglandins have been suggested but as yet not demonstrated. The exact function of tufted cells remains largely unknown.

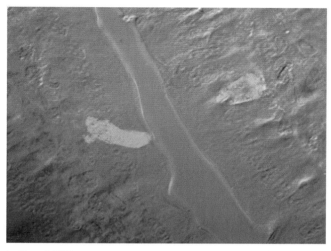

FIGURE 1   Section of a *Trpm5*-GFP transgenic mouse jejunum showing an enteroendocrine cell stained red with a chromogranin-A antibody, and a tufted cell labeled green with *Trpm5* promoter-driven GFP. GFP, green fluorescent protein.

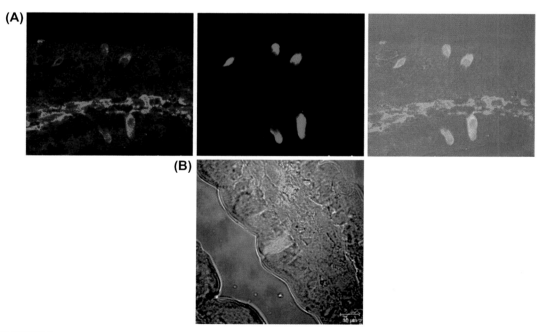

FIGURE 2   (A) Section of a *Trpm5*-GFP transgenic mouse duodenum showing expression of COX-1 (red) in tufted cells (marked green by *Trpm5*-driven GFP). (B) Section of a *Trpm5*-GFP transgenic mouse duodenum showing close proximity between a nerve (stained red with a PGP95 antibody) and a *Trpm5*-expressing tufted cell (green). These data suggest that tufted cells might connect nutrient sensing with the immune and nervous systems. GFP, green fluorescent protein.

The enterocytes are the most numerous cells lining the intestinal mucosa, and their main function is to absorb nutrients. Absorption of several nutrients is facilitated by membrane transporters, and in many cases transport of molecules across the membrane generates an electrical current, which could be sufficient to depolarize the cell and generate a signal. Enterocytes also synthetize oleoylethanolamide (OEA) in response to a meal, which contributes to control of food intake.[32,33]

Vagus nerve endings are unlikely to be directly sensing nutrients because they do not reach the intestinal lumen and do not even innervate the luminal epithelial wall.[34] The signaling message resulting from nutrient detection by specialized epithelial cells is conveyed to the nerve by hormone release, in paracrine fashion or, in certain cases, by direct contact between the sensory cell and the nerve.[25]

SENSING CARBOHYDRATES

Evidence that carbohydrates are sensed in the gut was obtained several decades ago when it was shown that duodenal infusion of glucose in a dog model leads to elevation of plasma incretin (i.e., GLP-1 and or GIP) levels, an effect that was not observed following intravenous infusion of glucose.[35]

The sweet taste receptor Tas1R1:Tas1R3 contributes to the detection of and response to carbohydrates in the small intestine. In mice, rats, and piglets, ingestion of high concentration of carbohydrates or of noncaloric sweeteners leads to upregulation of glucose transporters Na$^+$/glucose cotransporter 1 (SGLT-1) and facilitative glucose transporter 2 (GLUT2), and increased glucose uptake by the intestinal epithelium[21,36,37] (Figure 3). This upregulation is dependent on Tas1R3 and Gα_{gust}, because it is absent in *Tas1r3* knockout (KO) and in Gα_{gust} (*Gnat3*) KO mice.[21] It is worth noting that in the absence of Tas1R3 or Gα_{gust} the upregulation in response to high sucrose level is completely abolished, suggesting that this might be the main mechanism for detection of high concentrations of sugar in the gut. Also in mice, ingestion of glucose leads to Gα_{gust}-dependent increase in plasma GLP-1, and human L-cell line NCI-H716 secretes GLP-1 in response to sucralose, implicating the gut-expressed sweet taste receptors in sensing the sugar content of the intestinal lumen and responding through incretin secretion.[38] In humans, at least three studies failed to demonstrate the presence of such mechanisms. For example, in athletes, a preload of sucralose did not have any impact on subsequent exogenous glucose oxidation, a surrogate for glucose absorption, during exercise and ingestion of maltodextrin.[39] Also in humans, ingestion of glucose but not of noncaloric sweeteners triggers elevation of plasma incretin, a good argument against a role for the sweet taste receptor in modulating incretin secretion.[40,41] However, another study in humans found upregulation of GLP-1 and PYY

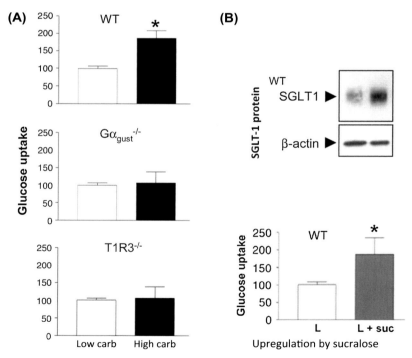

FIGURE 3   (A) Increased glucose uptake in response to dietary carbohydrate in wild-type (WT), but not in Gα_{gust} knockout (G$\alpha_{gust}^{-/-}$) or *Tas1r3* knockout (T1R3$^{-/-}$), mice given low or high carbohydrate diets for 2 weeks. (B) Increased intestinal Na$^+$/glucose cotransporter 1 (SGLT-1) expression and glucose uptake in WT mice given a low carbohydrate diet or the same diet supplemented with sucralose. *Adapted from Ref. 21.*

secretion in response to intragastric infusion of glucose was diminished by the sweet taste receptor antagonist lactisole, arguing for the involvement of this receptor.[42] To reconcile these apparently conflicting data, one could argue that activation of intestinal Tas1R2:-Tas1R3 is necessary but not sufficient for incretin release.

The Na^+/glucose cotransporter 1 (SGLT-1) also contributes to sensing glucose in the lumen of the small intestine. The main function of SGLT-1 is intestinal glucose absorption across the apical membrane of enterocytes. SGLT-1 is also strongly expressed in enteroendocrine L cells.[28] Transport of glucose from the intestinal lumen to inside the cell is accompanied by cotransport of sodium. This creates a Na^+ gradient across the plasma membrane and leads to depolarization, action potential firing, and opening of voltage-gated Ca^{2+} channels to directly trigger GLP-1 release from the L cells.[28,43] A similar mechanism is believed to occur in enteroendocrine K cells for the release of GIP.[44] The contribution of SGLT-1 to glucose sensing and release of incretins is also demonstrated by the blunted increase in plasma GLP-1 and GIP in response to oral glucose in SGLT-1 KO mice.[45,46]

Experiments using primary cells and cell lines have shown that the L cell can also sense glucose through a mechanism similar to that of glucose sensing in the pancreas, where glycolysis, a rise in intracellular ATP concentration, and closure of ATP-gated potassium channels lead to insulin secretion.[28,47]

Transport of glucose from the intestinal brush cells to the bloodstream is facilitated by GLUT2. Results from experiments using GLUT2 inhibitors in rat intestinal loops suggested a role of GLUT2 in regulating incretin secretion in response to a variety of nutrients.[48] However, GLUT2 KO mice have normal incretin responses to gastric administration of glucose, indicating that GLUT2 does not play an important role in regulating incretin secretion.[46]

SENSING FAT

Ingestion of lipids enhances enteroendocrine hormone release and inhibits gastric emptying.[49] Triglycerides, the main components of oils and fats, are hydrolyzed by lingual, gastric, and pancreatic lipases, resulting in release of free fatty acids (FFAs) and monoacylglycerol. It has been known since the 1960s that the main lipid stimulus detected by the GI chemosensors consists of FFAs rather than triglycerides, and that the most effective FFAs are those with a chain length of 12 or greater.[50] In recent years, lipids other than FFAs have been shown to contribute to the response of the organism to fat. Five GPCRs are known to respond to FFAs: GPR120 is activated by long-chain FFAs, GPR40 by long- and medium-chain FFAs, GPR84 by medium-chain FFAs, and GPR41 and GPR43 by short-chain FFAs.

GPR120 is coexpressed with GLP-1 in enteroendocrine cells of the ileum and colon.[19] This receptor has been linked to the control of body weight, GLP-1 secretion,[19] and the anti-inflammatory effect of omega-3 fatty acids.[51] Another study failed to confirm the GPR120-mediated increase in plasma GLP-1 in the mouse.[52,53] GPR120 is also expressed in the K cells of the proximal intestine and contributes to GIP secretion in response to the ingestion of oil.[54]

The main site of GPR40 expression is the pancreas, but immunohistochemistry studies have shown that it is also expressed in enteroendocrine cells of the duodenum.[55,56] GPR40 is coexpressed with CCK in intestinal I cells and KO mice lacking GPR40 have diminished CCK response to gavage with oil.[56] GPR40 is also expressed in cells that express GLP-1 and GIP in the GI tract, whereas the increase in plasma GLP-1 and GIP level in response to acute fat diet is blunted in GPR40 KO mice.[55] GPR40 is activated by a microbial metabolite of linoleic acid, 10-hydroxy-cis-12-octadecenoic acid, resulting in protection of the intestinal epithelial barrier following a challenge with dextran sulfate sodium or tumor necrosis factor.[57]

Both GPR40 and GPR120 contribute to fat-induced flavor nutrient learning. Wild-type mice increase their intake and learn to prefer a flavored solution that is paired with an intragastric infusion of soybean oil emulsion. This effect is largely diminished in GPR40-GPR120 double knockout mice and diminished to a lesser extent in GPR40 KO mice.[58]

GPR41 and GPR43 are activated by short chain fatty acids (SCFAs).[59] SCFAs, although not abundantly present in the diet, are produced during the fermentation of dietary fibers by the colon microbiota. GPR41 and GPR43 are expressed in L cells.[60] Primary L cells respond to SCFAs by elevating Ca^{2+},[60] and GPR43 KO mice have impaired plasma GLP-1 elevations in response to SCFA.[60] GPR41 KO mice are leaner than wild-type littermates and have reduced PYY expression, effects that are gut microbiota-dependent.[61] These data suggest a physiological role of GPR41 in sensing SCFAs.

CD36 is a transporter facilitating the entry into cells of long-chain fatty acids for oxidation or storage. It is expressed in the brush border membrane of enterocytes[62] and may contribute to intestinal lipid absorption,[63] although this role is not universally accepted. CD36 is also expressed in taste bud cells, and CD36 KO mice have a diminished preference for fatty acids and oils, indicating a sensing role for CD36 on the tongue.[9,64,65] A nutrient sensing role for this transporter in the gastrointestinal tract has not been demonstrated so far. One study investigated the role of CD36 in postoral fat conditioning in mice and found no role for this transporter.[65] Another study investigating the role of fatty acid transporters in GI nutrient sensing using GLUTag cells and CD36 KO mice found that fatty acid transport protein 4 (SLC27A4) but not CD36 plays a role in oleic acid−induced GLP-1 secretion[66] by enteroendocrine L cells.

GPR119, a GPCR expressed in L and K cells,[44,67] is best activated by OEA, a lipid derived from oleic acid and synthetized in the intestinal mucosa. Intraluminal injection of OEA stimulates GLP-1 release in mice through GPR119.[68] OEA plays a role in the control of food intake,[68,69] but this anorectic effect seems independent of GPR119 activation because it persists in GPR119 KO mice[70]; it is more likely mediated through activation of peroxisome proliferator-activated receptor-α.[71] GPR119 is also activated by 2-oleoyl glycerol, a product of dietary triglyceride digestion by pancreatic lipase in the gastrointestinal lumen,[72] and a number of other endogenous lipids (summarized elsewhere[73]). Activation of GPR119 mediates GLP-1 and GIP release.[68,72]

Whether triglycerides or diglycerides are sensed as such, without being digested, has not been demonstrated.

SENSING PROTEINS AND AMINO ACIDS

Proteins are digested in the GI tract into a complex mixture of oligopeptides, tri- and dipeptides, and individual amino acids. Peptone, a product of protein digestion, stimulates GLP-1 release by the gut, whereas intact proteins do not.[74] This suggests that the signaling molecules are the products of protein digestion, as is the case for sugar and fat sensing. Several receptors are implicated in the detection of the protein digestion products.

GPR92 is expressed in G and D cells of the antral region of the stomach of several species,[75] including humans, and is activated by protein hydrolyzates.[76,77] Secretion of CCK as a consequence of GPR92 activation by the product of protein breakdown has been demonstrated, but only in cultured STC-1 cells.[76]

In addition to GPR92, a sensor based on transport and generation of electrochemical gradients contributes to the detection of oligopeptides. Using specific inhibitors and purified L cells, Diakogiannaki and colleagues showed that peptide transporter 1 (PEPT1) mediates GLP-1 secretion in response to stimulation by peptone and di/tripeptides.[78] A similar role was investigated but not demonstrated for CCK secretion.[79] PEPT1 also mediates activation of vagal afferents in the rat duodenum by protein digests.[80]

Three GPCRs, the umami receptor Tas1R1:Tas1R3, the Ca^{2+}-sensing receptor CaSR, and GPRC6A are activated by amino acids. The subunits of the umami taste receptor Tas1R1 and Tas1R3 are expressed in the GI tract.[15] In humans, Tas1R1:Tas1R3 is activated by L-glutamate and L-aspartate, whereas in rodents it is broadly tuned to L-amino acids but does not respond to tryptophan. Mouse small intestine secretes CCK in response to a variety of amino acids ex vivo, and the response is enhanced by IMP and diminished by the Tas1r3 antagonist gurmarin, indicating the contribution of Tas1R1:Tas1R3.[81] CCK plays a number of physiological roles, including the inhibition of food intake.

The Tas1R1:Tas1R3 heterodimer also mediates colonic contractions in response to the presence of amino acids in the lumen of the colon. MSG initiates ascending contractions and descending relaxation in the colon of rats and mouse ex vivo, this effect is enhanced by IMP, and absent in *Tas1r1* KO mice. The propulsion speed of artificial fecal pellets in guinea pig is increased by intraluminal infusion of MSG and L-cysteine, but not by L-tryptophan. The effect of MSG is augmented by IMP, also suggesting a role of the Tas1R1:Tas1R3 heterodimer in luminal amino acid initiated peristaltism.[82]

In addition to its role as calcium sensor, CaSR is a broadly tuned aromatic amino acid-activated receptor.[83] CaSR is coexpressed with CCK in the mouse intestine, and purified intestinal cells expressing CCK and CaSR respond to phenylalanine and tryptophan by elevating intracellular Ca^{2+} and secreting CCK. This effect is blocked by inhibition of CaSR and not elicited by nonaromatic amino acids.[84] In isolated rat small intestinal loops, the L-amino acids phenylalanine, asparagine, glutamine, tryptophan, and arginine stimulate GIP, GLP-1, and PYY secretion; this effect is blocked by a CaSR inhibitor and by removal of extracellular Ca^{2+}. This peptide secretion is enhanced by the CaSR agonist NPS-R568.[48]

GPRC6A responds preferentially to basic amino acids[85,86] and is also expressed in gastric G and D cells,[75] suggesting that it contributes to amino acid sensing. However, this has not yet been demonstrated.

Other glutamate-responsive receptors include the mGluRs. mGluR1 is expressed in chief cells and possibly parietal cells of the stomach,[87] suggesting a role in modulating the gastric phase of protein digestion. mGluR4 is expressed in the mucosa of the stomach and duodenum.[88] Using specific agonists and antagonists in rats, Akiba et al.[88] showed that activation of mGluR4 increases the pH and mucus gel thickness in the duodenal lumen.

ROLE OF THE MICROBIOTA

It is clear that not everything is known about intestinal nutrient sensing and that other mechanisms and receptors are yet to be discovered. In this regard, it is tempting to speculate that the gut microbiota will be found to play an important role. With gene diversity much larger than the human genome, the gut microbiome may harbor nutrient sensors and signaling molecules. The gut microbiota has already been shown to contribute to the physiological response to nutrients, but the mechanisms described so far are metabolic rather than sensory. A good example is the production of short-chain fatty acids by the colonic microbiota through fermentation of fibers and of undigested polysaccharides, which signal through activation of GPR41.[61]

CLOSING REMARKS

In recent years, much progress has been made in identifying the GI receptors that respond to nutrients to regulate gut functions. Many lines of evidence implicate taste receptors in those roles, although other receptors and mechanisms also contribute. It is also clear that given the importance of GI chemosensation, there must be redundant mechanisms. This is well illustrated by the observation that knocking out any one of the receptors or transporters implicated in GI nutrient sensing in mice is compatible with normal life, as opposed to early death in mice lacking enteroendocrine cells. Furthermore, species lacking functional genes for sweet taste receptor subunits, such as cats or penguins, must have other mechanisms of carbohydrate sensing in the gut.[89,90]

For the three classes of macronutrients, all available evidence indicates that the molecules detected by the GI nutrient sensors are the digestion products and not the intact macronutrients, and that GPCR(s) and transporters play a role in detection. For sugars, additional sensors based on the concentration of cytoplasmic ATP generated during glycolysis also exist.

In many cases, the demonstrated molecular consequence of sensing different nutrients turns out to be the same event. This is particularly true for GLP-1 secretion, which occurs in response to stimulation by amino acids, fatty acids, or glucose of their cognate receptors. This is somehow unexpected because a rise in GLP-1 secretion and the resulting increase in insulin release are required only in response to high load of glucose. How the body mounts a response that is specific for each nutrient is an open question. For example, it is plausible that, during the process of protein digestion, the amino acid content of the intestinal lumen is evaluated by the receptors described previously, to fine-tune the secretion of pepsinogen, hydrochloric acid, and pancreatic proteases. The nature of the specific message sent to the stomach and the pancreas to accomplish these tasks is unknown.

Not only are the many molecular consequences of nutrient sensor activation known, but in some cases the physiological consequence of such activation has been determined. This is the case of the upregulation of glucose uptake by the intestinal mucosa in response to a high luminal glucose concentration sensed by the sweet taste receptor.

Many pieces of the puzzle are now in place, paving the way toward deciphering the exquisite mechanisms by which our body manages to mount the very specific physiological responses to the ingestion of a wide variety of nutrients, and to adapt those responses to the nutritional status of the body and its physiological requirements of the moment.

References

1. Nelson G, Hoon MA, Chandrashekar J, Zhang Y, Ryba NJ, Zuker CS. Mammalian sweet taste receptors. *Cell*. 2001;106(3):381–390.
2. Li X, Staszewski L, Xu H, Durick K, Zoller M, Adler E. Human receptors for sweet and umami taste. *Proc Natl Acad Sci USA*. 2002;99(7):4692–4696.
3. Nelson G, Chandrashekar J, Hoon MA, et al. An amino-acid taste receptor. *Nature*. 2002;416(6877):199–202.
4. Chaudhari N, Landin AM, Roper SD. A metabotropic glutamate receptor variant functions as a taste receptor. *Nat Neurosci*. 2000;3(2):113–119.
5. San GA, Uneyama H, Yoshie S, Torii K. Cloning and characterization of a novel mGluR1 variant from vallate papillae that functions as a receptor for L-glutamate stimuli. *Chem Senses*. 2005;30(suppl 1):i25–i26.
6. Yasumatsu K, Manabe T, Yoshida R, et al. Involvement of multiple taste receptors in umami taste: analysis of gustatory nerve responses in mGluR4 knock-out mice. *J Physiol*. 2014;593.
7. Yamaguchi S. The synergistic taste effect of monosodium glutamate and disodium 5'-inosinate. *J Food Sci*. 1967;32:473–478.
8. Cartoni C, Yasumatsu K, Ohkuri T, et al. Taste preference for fatty acids is mediated by GPR40 and GPR120. *J Neurosci*. 2010;30(25):8376–8382.
9. Laugerette F, Passilly-Degrace P, Patris B, et al. CD36 involvement in orosensory detection of dietary lipids, spontaneous fat preference, and digestive secretions. *J Clin Invest*. 2005;115(11):3177–3184.
10. Godinot N, Yasumatsu K, Barcos ME, et al. Activation of tongue-expressed GPR40 and GPR120 by non caloric agonists is not sufficient to drive preference in mice. *Neuroscience*. 2013;250:20–30.
11. Finger TE, Danilova V, Barrows J, et al. ATP signaling is crucial for communication from taste buds to gustatory nerves. *Science*. 2005;310(5753):1495–1499.
12. Huang YJ, Maruyama Y, Dvoryanchikov G, Pereira E, Chaudhari N, Roper SD. The role of pannexin 1 hemichannels in ATP release and cell-cell communication in mouse taste buds. *Proc Natl Acad Sci USA*. 2007;104(15):6436–6441.
13. Simon SA, de Araujo IE, Gutierrez R, Nicolelis MA. The neural mechanisms of gustation: a distributed processing code. *Nat Rev Neurosci*. 2006;7(11):890–901.
14. Taruno A, Vingtdeux V, Ohmoto M, et al. CALHM1 ion channel mediates purinergic neurotransmission of sweet, bitter and umami tastes. *Nature*. 2013;495(7440):223–226.
15. Bezencon C, le Coutre J, Damak S. Taste-signaling proteins are coexpressed in solitary intestinal epithelial cells. *Chem Senses*. 2007;32(1):41–49.
16. Dyer J, Salmon KS, Zibrik L, Shirazi-Beechey SP. Expression of sweet taste receptors of the T1R family in the intestinal tract and enteroendocrine cells. *Biochem Soc Trans*. 2005;33(Pt 1):302–305.
17. Briscoe CP, Tadayyon M, Andrews JL, et al. The orphan G protein-coupled receptor GPR40 is activated by medium and long chain fatty acids. *J Biol Chem*. 2003;278(13):11303–11311.
18. Itoh Y, Kawamata Y, Harada M, et al. Free fatty acids regulate insulin secretion from pancreatic beta cells through GPR40. *Nature*. 2003;422(6928):173–176.
19. Hirasawa A, Tsumaya K, Awaji T, et al. Free fatty acids regulate gut incretin glucagon-like peptide-1 secretion through GPR120. *Nat Med*. 2005;11(1):90–94.
20. Bezencon C, Furholz A, Raymond F, et al. Murine intestinal cells expressing Trpm5 are mostly brush cells and express markers of neuronal and inflammatory cells. *J Comp Neurol*. 2008;509(5):514–525.
21. Margolskee RF, Dyer J, Kokrashvili Z, et al. T1R3 and gustducin in gut sense sugars to regulate expression of Na$^+$-glucose cotransporter 1. *Proc Natl Acad Sci USA*. 2007;104(38):15075–15080.
22. Mellitzer G, Beucher A, Lobstein V, et al. Loss of enteroendocrine cells in mice alters lipid absorption and glucose homeostasis and impairs postnatal survival. *J Clin Invest*. 2010;120(5):1708–1721.

23. Bohorquez DV, Chandra R, Samsa LA, Vigna SR, Liddle RA. Characterization of basal pseudopod-like processes in ileal and colonic PYY cells. *J Mol Histol*. 2011;42(1):3−13.
24. Bohorquez DV, Samsa LA, Roholt A, Medicetty S, Chandra R, Liddle RA. An enteroendocrine cell-enteric glia connection revealed by 3D electron microscopy. *PLoS One*. 2014;9(2):e89881.
25. Bohorquez DV, Shahid RA, Erdmann A, et al. Neuroepithelial circuit formed by innervation of sensory enteroendocrine cells. *J Clin Invest*. 2015;125.
26. Egerod KL, Engelstoft MS, Grunddal KV, et al. A major lineage of enteroendocrine cells coexpress CCK, secretin, GIP, GLP-1, PYY, and neurotensin but not somatostatin. *Endocrinology*. 2012;153(12):5782−5795.
27. Habib AM, Richards P, Cairns LS, et al. Overlap of endocrine hormone expression in the mouse intestine revealed by transcriptional profiling and flow cytometry. *Endocrinology*. 2012;153(7):3054−3065.
28. Reimann F, Habib AM, Tolhurst G, Parker HE, Rogers GJ, Gribble FM. Glucose sensing in L cells: a primary cell study. *Cell Metab*. 2008;8(6):532−539.
29. Sbarbati A, Bramanti P, Benati D, Merigo F. The diffuse chemosensory system: exploring the iceberg toward the definition of functional roles. *Prog Neurobiol*. 2010;91(1):77−89.
30. Newson B, Ahlman H, Dahlstrom A, Nyhus LM. Ultrastructural observations in the rat ileal mucosa of possible epithelial "taste cells" and submucosal sensory neurons. *Acta Physiol Scand*. 1982;114(2):161−164.
31. Hofer D, Puschel B, Drenckhahn D. Taste receptor-like cells in the rat gut identified by expression of alpha-gustducin. *Proc Natl Acad Sci USA*. 1996;93(13):6631−6634.
32. Fu J, Astarita G, Gaetani S, et al. Food intake regulates oleoylethanolamide formation and degradation in the proximal small intestine. *J Biol Chem*. 2007;282(2):1518−1528.
33. Rodriguez de FF, Navarro M, Gomez R, et al. An anorexic lipid mediator regulated by feeding. *Nature*. 2001;414(6860):209−212.
34. Powley TL, Spaulding RA, Haglof SA. Vagal afferent innervation of the proximal gastrointestinal tract mucosa: chemoreceptor and mechanoreceptor architecture. *J Comp Neurol*. 2011;519(4):644−660.
35. Unger RH, Ohneda A, Valverde I, Eisentraut AM, Exton J. Characterization of the responses of circulating glucagon-like immunoreactivity to intraduodenal and intravenous administration of glucose. *J Clin Invest*. 1968;47(1):48−65.
36. Mace OJ, Affleck J, Patel N, Kellett GL. Sweet taste receptors in rat small intestine stimulate glucose absorption through apical GLUT2. *J Physiol*. 2007;582(Pt 1):379−392.
37. Moran AW, Al-Rammahi MA, Arora DK, et al. Expression of Na^+/glucose co-transporter 1 (SGLT1) is enhanced by supplementation of the diet of weaning piglets with artificial sweeteners. *Br J Nutr*. 2010;104(5):637−646.
38. Jang HJ, Kokrashvili Z, Theodorakis MJ, et al. Gut-expressed gustducin and taste receptors regulate secretion of glucagon-like peptide-1. *Proc Natl Acad Sci USA*. 2007;104(38):15069−15074.
39. Stellingwerff T, Godin JP, Beaumont M, et al. Effects of pre-exercise sucralose ingestion on carbohydrate oxidation during exercise. *Int J Sport Nutr Exerc Metab*. 2013;23(6):584−592.
40. Fujita Y, Wideman RD, Speck M, et al. Incretin release from gut is acutely enhanced by sugar but not by sweeteners in vivo. *Am J Physiol Endocrinol Metab*. 2009;296(3):E473−E479.
41. Ma J, Bellon M, Wishart JM, et al. Effect of the artificial sweetener, sucralose, on gastric emptying and incretin hormone release in healthy subjects. *Am J Physiol Gastrointest Liver Physiol*. 2009;296(4):G735−G739.
42. Steinert RE, Gerspach AC, Gutmann H, Asarian L, Drewe J, Beglinger C. The functional involvement of gut-expressed sweet taste receptors in glucose-stimulated secretion of glucagon-like peptide-1 (GLP-1) and peptide YY (PYY). *Clin Nutr*. 2011;30(4):524−532.
43. Gribble FM, Williams L, Simpson AK, Reimann F. A novel glucose-sensing mechanism contributing to glucagon-like peptide-1 secretion from the GLUTag cell line. *Diabetes*. 2003;52(5):1147−1154.
44. Parker HE, Habib AM, Rogers GJ, Gribble FM, Reimann F. Nutrient-dependent secretion of glucose-dependent insulinotropic polypeptide from primary murine K cells. *Diabetologia*. 2009;52(2):289−298.
45. Parker HE, Adriaenssens A, Rogers G, et al. Predominant role of active versus facilitative glucose transport for glucagon-like peptide-1 secretion. *Diabetologia*. 2012;55(9):2445−2455.
46. Roder PV, Geillinger KE, Zietek TS, Thorens B, Koepsell H, Daniel H. The role of SGLT1 and GLUT2 in intestinal glucose transport and sensing. *PLoS One*. 2014;9(2):e89977.
47. Reimann F, Gribble FM. Glucose-sensing in glucagon-like peptide-1-secreting cells. *Diabetes*. 2002;51(9):2757−2763.

IV. STIMULUS TRANSDUCTION IN OTHER CHEMODETECTION SYSTEMS

48. Mace OJ, Schindler M, Patel S. The regulation of K- and L-cell activity by GLUT2 and the calcium-sensing receptor CasR in rat small intestine. *J Physiol.* 2012;590(Pt 12):2917—2936.

49. Little TJ, Feinle-Bisset C. Effects of dietary fat on appetite and energy intake in health and obesity—oral and gastrointestinal sensory contributions. *Physiol Behav.* 2011;104(4):613—620.

50. Hunt JN, Knox MT. A relation between the chain length of fatty acids and the slowing of gastric emptying. *J Physiol.* 1968;194(2):327—336.

51. Oh DY, Talukdar S, Bae EJ, et al. GPR120 is an omega-3 fatty acid receptor mediating potent anti-inflammatory and insulin-sensitizing effects. *Cell.* 2010;142(5):687—698.

52. Oh DY, Walenta E, Akiyama TE, et al. A Gpr120-selective agonist improves insulin resistance and chronic inflammation in obese mice. *Nat Med.* 2014;20(8):942—947.

53. Xiong Y, Swaminath G, Cao Q, et al. Activation of FFA1 mediates GLP-1 secretion in mice. Evidence for allosterism at FFA1. *Mol Cell Endocrinol.* 2013;369(1—2):119—129.

54. Iwasaki K, Harada N, Sasaki K, et al. Free fatty acid receptor GPR120 is highly expressed in enteroendocrine K-cells of the upper small intestine and has a critical role in GIP secretion after fat ingestion. *Endocrinology.* 2014;156. http://dx.doi.org/10.1210/en2014-1653.

55. Edfalk S, Steneberg P, Edlund H. Gpr40 is expressed in enteroendocrine cells and mediates free fatty acid stimulation of incretin secretion. *Diabetes.* 2008;57(9):2280—2287.

56. Liou AP, Lu X, Sei Y, et al. The G-protein-coupled receptor GPR40 directly mediates long-chain fatty acid-induced secretion of cholecystokinin. *Gastroenterology.* 2011;140(3):903—912.

57. Miyamoto J, Mizukure T, Park SB, et al. A gut microbial metabolite of linoleic acid, 10-hydroxy-cis-12-octadecenoic acid, ameliorates intestinal epithelial barrier impairment partially via GPR40/MEK-ERK pathway. *J Biol Chem.* 2014;290.

58. Sclafani A, Zukerman S, Ackroff K. GPR40 and GPR120 fatty acid sensors are critical for postoral but not oral mediation of fat preferences in the mouse. *Am J Physiol Regul Integr Comp Physiol.* 2013;305(12):R1490—R1497.

59. Le PE, Loison C, Struyf S, et al. Functional characterization of human receptors for short chain fatty acids and their role in polymorphonuclear cell activation. *J Biol Chem.* 2003;278(28):25481—25489.

60. Tolhurst G, Heffron H, Lam YS, et al. Short-chain fatty acids stimulate glucagon-like peptide-1 secretion via the G-protein-coupled receptor FFAR2. *Diabetes.* 2012;61(2):364—371.

61. Samuel BS, Shaito A, Motoike T, et al. Effects of the gut microbiota on host adiposity are modulated by the short-chain fatty-acid binding G protein-coupled receptor, Gpr41. *Proc Natl Acad Sci USA.* 2008;105(43):16767—16772.

62. Lobo MV, Huerta L, Ruiz-Velasco N, et al. Localization of the lipid receptors CD36 and CLA-1/SR-BI in the human gastrointestinal tract: towards the identification of receptors mediating the intestinal absorption of dietary lipids. *J Histochem Cytochem.* 2001;49(10):1253—1260.

63. Chen M, Yang Y, Braunstein E, Georgeson KE, Harmon CM. Gut expression and regulation of FAT/CD36: possible role in fatty acid transport in rat enterocytes. *Am J Physiol Endocrinol Metab.* 2001;281(5):E916—E923.

64. Fukuwatari T, Kawada T, Tsuruta M, et al. Expression of the putative membrane fatty acid transporter (FAT) in taste buds of the circumvallate papillae in rats. *FEBS Lett.* 1997;414(2):461—464.

65. Sclafani A, Ackroff K, Abumrad NA. CD36 gene deletion reduces fat preference and intake but not post-oral fat conditioning in mice. *Am J Physiol Regul Integr Comp Physiol.* 2007;293(5):R1823—R1832.

66. Poreba MA, Dong CX, Li SK, Stahl A, Miner JH, Brubaker PL. Role of fatty acid transport protein 4 in oleic acid-induced glucagon-like peptide-1 secretion from murine intestinal L cells. *Am J Physiol Endocrinol Metab.* 2012;303(7):E899—E907.

67. Chu ZL, Carroll C, Alfonso J, et al. A role for intestinal endocrine cell-expressed G protein-coupled receptor 119 in glycemic control by enhancing glucagon-like peptide-1 and glucose-dependent insulinotropic peptide release. *Endocrinology.* 2008;149(5):2038—2047.

68. Lauffer LM, Iakoubov R, Brubaker PL. GPR119 is essential for oleoylethanolamide-induced glucagon-like peptide-1 secretion from the intestinal enteroendocrine L-cell. *Diabetes.* 2009;58(5):1058—1066.

69. Overton HA, Babbs AJ, Doel SM, et al. Deorphanization of a G protein-coupled receptor for oleoylethanolamide and its use in the discovery of small-molecule hypophagic agents. *Cell Metab.* 2006;3(3):167—175.

70. Lan H, Vassileva G, Corona A, et al. GPR119 is required for physiological regulation of glucagon-like peptide-1 secretion but not for metabolic homeostasis. *J Endocrinol.* 2009;201(2):219—230.

71. Fu J, Gaetani S, Oveisi F, et al. Oleylethanolamide regulates feeding and body weight through activation of the nuclear receptor PPAR-alpha. *Nature.* 2003;425(6953):90—93.

72. Hansen KB, Rosenkilde MM, Knop FK, et al. 2-Oleoyl glycerol is a GPR119 agonist and signals GLP-1 release in humans. *J Clin Endocrinol Metab*. 2011;96(9):E1409–E1417.
73. Hansen HS, Rosenkilde MM, Holst JJ, Schwartz TW. GPR119 as a fat sensor. *Trends Pharmacol Sci*. 2012;33(7):374–381.
74. Cordier-Bussat M, Bernard C, Levenez F, et al. Peptones stimulate both the secretion of the incretin hormone glucagon-like peptide 1 and the transcription of the proglucagon gene. *Diabetes*. 1998;47(7):1038–1045.
75. Haid DC, Jordan-Biegger C, Widmayer P, Breer H. Receptors responsive to protein breakdown products in g-cells and d-cells of mouse, swine and human. *Front Physiol*. 2012;3:65.
76. Choi S, Lee M, Shiu AL, Yo SJ, Hallden G, Aponte GW. GPR93 activation by protein hydrolysate induces CCK transcription and secretion in STC-1 cells. *Am J Physiol Gastrointest Liver Physiol*. 2007;292(5):G1366–G1375.
77. Choi S, Lee M, Shiu AL, Yo SJ, Aponte GW. Identification of a protein hydrolysate responsive G protein-coupled receptor in enterocytes. *Am J Physiol Gastrointest Liver Physiol*. 2007;292(1):G98–G112.
78. Diakogiannaki E, Pais R, Tolhurst G, et al. Oligopeptides stimulate glucagon-like peptide-1 secretion in mice through proton-coupled uptake and the calcium-sensing receptor. *Diabetologia*. 2013;56(12):2688–2696.
79. Liou AP, Chavez DI, Espero E, Hao S, Wank SA, Raybould HE. Protein hydrolysate-induced cholecystokinin secretion from enteroendocrine cells is indirectly mediated by the intestinal oligopeptide transporter PepT1. *Am J Physiol Gastrointest Liver Physiol*. 2011;300(5):G895–G902.
80. Darcel NP, Liou AP, Tome D, Raybould HE. Activation of vagal afferents in the rat duodenum by protein digests requires PepT1. *J Nutr*. 2005;135(6):1491–1495.
81. Daly K, Al-Rammahi M, Moran A, Marcello M, Ninomiya Y, Shirazi-Beechey SP. Sensing of amino acids by the gut-expressed taste receptor T1R1-T1R3 stimulates CCK secretion. *Am J Physiol Gastrointest Liver Physiol*. 2013;304(3):G271–G282.
82. Kendig DM, Hurst NR, Bradley ZL, et al. Activation of the umami taste receptor (T1R1/T1R3) initiates the peristaltic reflex and pellet propulsion in the distal colon. *Am J Physiol Gastrointest Liver Physiol*. 2014;307(11):G1100–G1107.
83. Conigrave AD, Quinn SJ, Brown EM. L-Amino acid sensing by the extracellular Ca^{2+}-sensing receptor. *Proc Natl Acad Sci USA*. 2000;97(9):4814–4819.
84. Wang Y, Chandra R, Samsa LA, et al. Amino acids stimulate cholecystokinin release through the Ca^{2+}-sensing receptor. *Am J Physiol Gastrointest Liver Physiol*. 2011;300(4):G528–G537.
85. Christiansen B, Hansen KB, Wellendorph P, Brauner-Osborne H. Pharmacological characterization of mouse GPRC6A, an L-alpha-amino-acid receptor modulated by divalent cations. *Br J Pharmacol*. 2007;150(6):798–807.
86. Wellendorph P, Burhenne N, Christiansen B, Walter B, Schmale H, Brauner-Osborne H. The rat GPRC6A: cloning and characterization. *Gene*. 2007;396(2):257–267.
87. SanGabriel AM, Maekawa T, Uneyama H, Yoshie S, Torii K. mGluR1 in the fundic glands of rat stomach. *FEBS Lett*. 2007;581(6):1119–1123.
88. Akiba Y, Watanabe C, Mizumori M, Kaunitz JD. Luminal L-glutamate enhances duodenal mucosal defense mechanisms via multiple glutamate receptors in rats. *Am J Physiol Gastrointest Liver Physiol*. 2009;297(4):G781–G791.
89. Li X, Li W, Wang H, et al. Pseudogenization of a sweet-receptor gene accounts for cats' indifference toward sugar. *PLoS Genet*. 2005;1(1):27–35.
90. Zhao H, Li J, Zhang J. Molecular evidence for the loss of three basic tastes in penguins. *Curr Biol*. 2015;25(4):R141–R142.

CHAPTER

21

Molecular Pharmacology of Chemesthesis

Jay P. Slack

Department of Science and Technology, Givaudan, Cincinnati, OH, USA

INTRODUCTION

At the most basic level, food can be thought of as a collection of chemicals and macromolecules.[1] Before providing nutrients and energy, food engages many sensory systems. Vision, touch, smell, and taste help determine food choices by humans, animals, and insects. During consumption, taste (gustatory), smell (olfactory), and touch/texture/temperature (somatosensory) all play a direct role in the eating experience.[2] In some cases, one or more of these systems will alert an animal to stop or even avoid eating certain foods because of potential health liabilities.[3] Humans use the term flavor to describe the sensory attributes of a food and it is generally believed that this flavor profile is largely determined by taste and aroma.[4,5]

Chemosensory Transduction
http://dx.doi.org/10.1016/B978-0-12-801694-7.00021-4

However, the overall sensory experience from eating a food is actually an integration of gustatory, olfactory, and somatosensory signals that are occurring both consciously and unconsciously during consumption.[4] Therefore, although we think of flavor perception as being primarily from taste and smell, it is now clear that somatosensory systems are detecting tactile, thermal, painful, and/or kinesthetic stimuli in parallel during eating.[5] In addition, small molecules from food components are activating the same somatosensory receptors that transduce pain and temperature, which further contributes to the overall eating experience.[6] The stimulation of these somatosensory systems by chemical stimuli is known as chemesthesis.[7] This chapter discusses the underlying molecular pathways that can contribute to chemesthetic signaling in the oral cavity during consumption.

CHEMESTHESIS

Chemesthesis is the direct activation of somatosensory nerves by chemical stimuli.[7,8] Chemesthetic sensations arise when exogenous chemical agonists activate receptors that normally convey other senses such as temperature, pain, touch, or texture.[9] Anatomically, these sensations can occur when somatosensory nerves innervating the skin, mucous membranes, eyes, and oral and nasal cavities have been activated, resulting in cooling, burning, stinging, or tingling.[8] Behavioral and physiological observations related to chemesthesis were first reported by George H. Parker in 1912 who described what he called a common chemical sense.[10] Parker was interested in understanding why the skin of a frog exhibits responses to acids and salts that he believed were strikingly similar to taste responses for the same stimuli. He reasoned that the two sensory systems were related and investigated this by examining oral, nasal and skin sensitivities to chemical stimuli that included acids, bases, salts, and tastants. Parker explored this question by stimulating various parts of a fish with different chemicals and recording their concentration-dependent behavioral responses. Using nerve transections, he determined which sensory nerves were required for localized responses to the various chemical stimuli. From his studies, Parker concluded that animals have three distinct classes of chemosensory receptors: olfactory, gustatory, and common chemical.[10] He also observed that the concentrations required for olfactory responses were typically much lower than those needed for taste or common chemical responses.[10] Finally, he noted that the receptors of the common chemical sense are more frequently stimulated by irritants that are typically aversive.[10]

Parker's observations in 1912 were the first evidence that the somatosensory system monitors and responds to chemical compounds using neural pathways that are distinct from the taste and olfactory systems. After the intervening century of neuroscience research, we now understand which nerves and receptors are responsible for conveying these sensations as well as many of the discrete chemicals in foods that elicit them. Although chemesthesis refers to any somatosensory response to noxious chemicals, oral chemesthetic irritants are primarily detected via three somatosensory nerves: the trigeminal, glossopharyngeal, and vagus.[11] The trigeminal nerve is important for chemical irritants applied to the anterior tongue, palate, tonsils, and nasopharyngeal mucosa.[11] The trigeminal nerve is also important for chemesthetic responses in the nasal cavity.[11] The glossopharyngeal and vagal nerves are secondary sites for chemesthetic responses to irritants present in the posterior oral cavity.[11]

Common chemesthetic agents include chili pepper, alcohol, peppermint, mustard oil, various acids, and carbonated water (carbon dioxide). These stimuli are often associated with oral sensations such as pungency, freshness, tingling, sharpness, and cooling, many of which are also related to pain perception (nociception).[7] Menthol, the primary chemesthetic agent in peppermint, was observed to stimulate thermoreceptors of lingual nerves and this interaction was offered as an early explanation for the ability of menthol to evoke sensations of oral cooling.[12] In this study, Zotterman et al. observed that menthol sensitized lingual nerve responses to smaller changes in temperature or enhanced nerve activity at temperatures below body temperature (25 °C), whereas it reduced nerve responses to temperatures >40 °C.[12] Conversely, capsaicin, the major chemesthetic ingredient in hot chili peppers, acts in the oral cavity and other tissues to excite polymodal nociceptive nerve fibers that are also sensitive to noxious heat.[13] Furthermore, rodent studies in the 1960s have shown that repeated systemic exposure to capsaicin can eliminate subsequent peripheral neuron responses to capsaicin, as well as other nonvanilloid irritants such as mustard oil and xylene.[14] Together, these early studies provided important evidence that capsaicin, menthol, and other chemical irritants interact with the same pain-transmitting neural pathways that also convey changes in temperature. The remaining sections of this chapter cover the major receptors that are activated by various classes of chemical irritants as well as their roles in temperature sensing or other physiological processes related to chemesthesis.

TRANSIENT RECEPTOR POTENTIAL CHANNEL SUBFAMILY V MEMBER 1

As outlined previously, capsaicin is a potent chemical activator of nociceptive neurons and this feature actually led to the discovery of its neuronal receptor, TRPV1 (transient receptor potential channel subfamily V member 1). TRPV1 was first identified as the site of action for capsaicin based on a functional screening strategy designed to find complementary DNA clones that would confer capsaicin responses to nonneuronal cells.[15] Using this unique approach, Julius et al. identified a single complementary DNA from dorsal root ganglia (DRG) neurons that conferred robust responses to capsaicin in transfected kidney cells.[15] This receptor was originally called VR1 for vanilloid receptor subtype 1, but standard nomenclature for this protein is now TRPV1.[16] In vitro expression of TRPV1 in *Xenopus* oocytes or human embryonic kidney 293 (HEK293) cells resulted in robust calcium currents following exposure to both capsaicin and resiniferatoxin, a high-potency capsaicin analogue.[15] Importantly, in vitro potencies for these two substances were similar to those obtained with isolated nociceptive neurons.[15] In addition, TRPV1 responses in oocytes could be inhibited with either capsazepine or ruthenium red at concentrations known to block vanilloid responses in nociceptive neurons.[15] Finally, TRPV1 responded to extracts of various capsaicin-containing chili peppers and TRPV1 activation by these extracts closely paralleled pungency ratings from human taste evaluations.[15]

The discovery of TRPV1 as the capsaicin sensor in polymodal nociceptive neurons led to many subsequent studies that confirmed the importance of this channel in biochemical and neurological processes related to chemesthesis, pain, and thermoregulation as well as neurogenic inflammation.[17] Following its initial discovery of TRPV1, the Julius laboratory demonstrated that high heat (46 °C) and capsaicin both activate the same TRPV1-expressing HEK293 cells and the currents evoked by these two stimuli paralleled previous measurements cultured sensory neurons.[18] Preexposure of TRPV1-expressing cells to heat desensitized them to a subsequent challenge with capsaicin and vice versa, thus confirming that TRPV1 is the neuronal receptor target for both types of nociceptive stimuli.[15] This study also revealed that protons (acidic pH) directly activate TRPV1 at physiological temperatures and also enhance TRPV1 responses to noxious heat and/or capsaicin.[15] Finally, TRPV1 is expressed in small and medium-diameter neurons within trigeminal, dorsal root, and nodose sensory ganglia,[15,18] which are appropriate nerve populations relevant to chemesthesis. Collectively, results from these functional studies were in agreement with prior observations of isolated polymodal nociceptive neurons, and strongly suggested that TRPV1 is a key mediator of peripheral neuron responses to heat, protons, and chemical irritants related to capsaicin.

The definitive evidence that TRPV1 is a polymodal detector of pain-producing stimuli came in 2000 when Caterina et al. reported on the development and characterization of a TRPV1-deficient mouse.[19] Employing a gene knockout strategy along with cellular and behavioral assays, TRPV1-deficent mice were found to be insensitive to vanilloids, and they displayed markedly reduced sensitivities to protons and heat.[19] Capsaicin insensitivity was demonstrated in cutaneous neurons as well as in trigeminal neurons in the oral cavity.[19] The behavioral importance of TRPV1 in trigeminal neurons was demonstrated via brief access, taste aversion tests with water containing 100 μM capsaicin. After drinking water with capsaicin, normal mice rub their noses and avoid any additional consumption. By contrast, TRPV1-deficient mice repeatedly consumed capsaicin-containing water without exhibiting any aversive behavior.[19] These results demonstrated that TRPV1 is critical for detection of selected chemical irritants. However, these defects were limited to TRPV1-dependent aversive stimuli because direct cutaneous nerve responses to other chemical irritants such as formalin were not affected in the TRPV1 knockout animals.[19]

These studies clearly established that TRPV1 is a primary nociceptive receptor for heat as well as chemesthetic agents such as protons and vanilloids. Subsequent studies have revealed that TRPV1 activation is important for other chemesthetics in foods and spices such as salt,[20] black pepper,[21,22] durian,[23] garlic,[24] onions,[24] rosemary,[25] mint,[25] ginger,[26,27] and mustard oil.[28] The active food ingredients with underlying TRPV1 activity are quite diverse and cover a broad chemical space (Table 1). In addition to food-related compounds, TRPV1 signaling is affected directly or indirectly by many nonfood stimuli including endogenous circulating fatty acid amides,[26] venomous toxins that produce inflammatory pain,[29–31] drugs such as lidocaine,[26] and even ethanol.[32] Activation of TRPV1 by these various materials results in chemical irritation described with terms such as burning, stinging, prickling, warmth, and numbness,[7] thus highlighting a role for TRPV1 in sensory responses to chemesthetic agents.

TRP CHANNEL SUBFAMILY M MEMBER 8

TRP channels are a large family of ion channels containing 27 mammalian members, including TRPV1, which function in a wide variety of biological systems.[17] TRP channels are modulated by a variety of stimuli such as temperature, chemicals, pH, pressure,

TABLE 1   Common Chemesthetic Compounds, Their Sources, and TRP Channel Targets

Structure	Name	Source	Receptor(s)	References
	Capsaicin	Chili pepper	TRPV1	15
	Piperine	Black pepper	TRPV1/A1	21,22,86
	Allicin	Garlic	TRPA1/V1	24,53
	AITC	Mustard oil, horseradish	TRPA1/V1	50,51
	Cinnamaldehyde	Cinnamon	TRPA1	50,52
	Galangal acetate	Chinese Ginger	TRPA1	56
	Oleocanthol	Olive oil	TRPA1	54
	Nicotine	Tobacco	TRPA1	57
	6-Gingerol	Ginger	TRPA1/V1	27,50
	Camphor	Camphor laurel tree	TRPV3/V1	25,67
	Carvacrol	Oregano	TRPV3/A1	69,71
	Eugenol	Cloves	TRPV3/V1/A1	50,69
	Thymol	Thyme	TRPV3/A1	55,69

(Continued)

IV. STIMULUS TRANSDUCTION IN OTHER CHEMODETECTION SYSTEMS

TABLE 1 Common Chemesthetic Compounds, Their Sources, and TRP Channel Targets—cont'd

Structure	Name	Source	Receptor(s)	References
	Incensole acetate	Incense	TRPV3	77
	Menthol	Peppermint	TRPM8/A1/V3	35,36,42,44,70,71
	Eucalyptol	Eucalyptus	TRPM8	36,42
	WS-3	Synthetic	TRPM8/A1	42,44
	GIV1	Synthetic	TRPM8/A1	44
	Icilin	Synthetic	TRPM8/A1	36,46
	WS-23	Synthetic	TRPM8	42
	Linalool	Coriander	TRPM8/A1/V3	42,71,83
	Geraniol	Roses	TRPM8/V3	42,71

References cited are based on reports using functional analysis of receptor activities in recombinant expression systems.

lipids, or proteins, and they are important signal integrators for all the major senses, including vision, taste, smell, hearing, and somatosensation.[17,33] Since the landmark discovery of TRPV1 as the somatosensory receptor for heat and capsaicin, many other TRP channels have also been shown to play important roles in sensory biology, specifically in temperature detection and chemesthesis.[34] One notable example is TRP channel subfamily M member 8 (TRPM8), which is a member of the melastatin subfamily of TRP channels.[16]

TRPM8 was simultaneously identified in 2002 by two different research groups as the primary receptor for menthol as well as a thermal detector of a range of temperatures that are perceived from innocuous cool to noxious cold.[35,36] These studies revealed that recombinant expression of TRPM8 in nonneuronal cells results in robust temperature-dependent cation currents.[35,36] Unlike TRPV1, TRPM8 is activated by temperatures in the cool/cold range (8–28 °C), but it is not activated by noxious heat or TRPV1 agonists.[35,36] TRPM8 expression was observed in trigeminal neurons and dorsal root ganglia in nerves that contain small-diameter C-fibers and that are known to be sensitive to cold temperatures.[35,36] Furthermore, TRPM8-expressing neurons did not express markers such as TRPV1, calcitonin gene-related peptide, or isolectin B4, which are known molecular markers of nociceptive neurons.[35,36] In vitro expression of TRPM8 in HEK293 or Chinese hamster ovary cells conferred robust calcium currents following exposure to the chemesthetic agents menthol,[35,36] icilin,[36] and eucalyptol,[36] which were previously known to cause cold sensations in humans or animals. Thus, the in vitro properties of TRPM8 (temperature sensitivity, menthol sensitivity, and ion conductance) all strongly pointed to this channel as the molecular site of action for chemesthetic agents that elicit sensations of cooling. Furthermore, this finding confirmed the existence of a menthol receptor in cold-sensitive nerve fibers, as originally proposed by Hensel and Zotterman.[12]

The definitive evidence that TRPM8 is a key mediator of cold producing stimuli came in 2007 with two groups reporting that TRPM8-deficient mice exhibit significantly reduced physiological and behavioral responses to cold temperatures.[37,38] These studies confirmed that TRPM8 is also necessary for somatosensory responses to chemesthetic stimuli that evoke cooling sensations.[37,38] Calcium imaging of neurons from dorsal root ganglia indicated that loss of TRPM8 leads to a two- to threefold reduction in the number of neurons that can respond to cold or menthol.[37,38] Behavioral thermotaxis assays indicated that detection of innocuous cold is severely disrupted in TRPM8-deficient mice.[37,38] Physiological assessment of somatosensory responses to an evaporative cooling agent (acetone) or to intraperitoneal injections of icilin, a high-potency cooling agent, revealed that TRPM8 is also involved in noxious cold responses.[37,38] Finally, TRPM8 was also found to be important for the beneficial, pain-reducing properties of cold,[37] consistent with the analgesic properties of menthol that are due in part to TRPM8 activation.[39]

These studies defined TRPM8 as a key mediator of cold sensations. Subsequent studies of TRPM8 pharmacology have revealed multiple and chemically diverse TRPM8 agonists[40] with cooling properties that include both nature-derived and synthetic chemical backbones (Table 1). The initial identification and characterization of TRPM8 revealed that it can be activated by both menthanes (menthol) and nonmenthanes (eucalyptol, icilin).[35,36] Besides mint oil or menthol, many synthetic menthol derivatives such as WS-3 are often added to lip balms, shaving creams, gums, toothpastes, and mouthwashes for their cooling and analgesic properties.[41] Many of these menthol analogues are now known to be receptor agonists of TRPM8 and exhibit varying degrees of potency and efficacy.[42,43] TRPM8 activation also underlies the sensory effects of two novel menthol derivatives that can evoke more pronounced cooling sensations than other TRPM8 agonists with lower potency/efficacy.[44] Furthermore, some nonmenthane fragrance compounds, such as linalool or geraniol, also exhibit weak TRPM8 agonist activity.[42] Activation of TRPM8 by these various materials elicits chemesthetic sensations ranging from pleasant cooling to cold, and is associated with other hedonic descriptors such as freshness and cleanliness,[45] thus highlighting a role of TRPM8 in sensory responses to chemesthetic agents.

TRP CHANNEL SUBFAMILY A MEMBER 1

Soon after the discovery of TRPM8 as a primary receptor for cold stimuli, transient receptor potential channel subfamily A member 1 (TRPA1) was proposed as a nociceptive receptor for noxious cold below the temperature range of TRPM8.[46] Originally called ANKTM1, TRPA1 is structurally distinct from TRPM8 and contains several ankyrin domains in the N-terminus that distinguish it from other members of the TRP channel family.[46] These ankyrin repeats have been suggested as a key site for activation of the channel.[47] TRPA1 is expressed in nociceptive neurons of the trigeminal and dorsal root ganglia and most TRPA1-expressing neurons also express TRPV1.[46] Initial characterization of TRPA1 function in recombinant expression systems indicated that noxious cold (10 °C) and icilin are potent activators of this channel, resulting in robust calcium influx.[46] TRPA1 is not found in TRPM8-expressing neurons and colder temperatures are needed to activate TRPA1, suggesting that TRPA1 may be the underlying receptor for non−TRPM8-mediated noxious cold responses.[46] Direct thermal activation of TRPA1 by noxious cold became the subject of intense investigation after this initial finding, resulting in many conflicting reports and leading many to conclude that the channel is not directly gated by cold.[48] However, more recent studies have shown that purified TRPA1 inserted into lipid bilayers can be directly activated by cold temperatures in the absence of any other chemical irritants, second messengers, or accessory proteins.[49]

TRPA1 was subsequently identified as the molecular target for a number of chemical irritants with pungent properties and can also be activated by compounds or peptides that are related to pain and inflammation.[50,51] These studies revealed that TRPA1 is activated by a number of chemical irritants from botanical sources, including isothiocyanates (mustard oil), cinnamaldehyde (cinnamon), eugenol (cloves, nutmeg, cinnamon), and methyl salicylate (wintergreen plants).[50,51] Activation of TRPA1 by exogenous chemical irritants or by noxious cold temperatures suggested that it plays a role in nociceptive responses, which was later confirmed with evidence that the pain-producing peptide bradykinin activates TRPA1 directly.[50] Final confirmation of TRPA1 as a receptor for nociceptive pain was further demonstrated when TRPA1-deficient mice were found to exhibit profound deficits to multiple pain-producing stimuli, including mustard oil, formalin, or bradykinin-evoked thermal hypersensitivity.[52]

In the past decade, pharmacological and physiological evidence have accumulated to indicate that TRPA1 is actually a broadly tuned sensor that responds to a variety of thermal, ionic, chemical, and mechanical nociceptive stimuli.[48] Besides those mentioned previously, TRPA1 is activated by chemical irritants in other foods (see Table 1) such as raw garlic/onions (allicin),[53] horseradish/wasabi (isothiocyanates),[50] mint (menthol),[44] olive oil (oleocanthal),[54] thyme (thymol),[55] and gingers (gingerols, galangals).[27,56] Other low-potency stimulants of TRPA1 also include caffeine and nicotine.[48,57] TRPA1 activity also partially underlies the irritancy properties of carbonation.[58] Sensory evaluations of these foods and/or compounds are often described as burning or prickling,[48] linking TRPA1 activation and sensory responses to chemesthetic agents. Furthermore, the sensitivity of TRPA1 to a number of irritants that do not activate TRPV1 also provides a pharmacological basis for cross-adaptation between capsaicin and other nonvanilloid chemesthetic agents such as cinnamaldehyde. It is well established that repeated preexposure to capsaicin can desensitize nociceptive fibers and make them less sensitive to subsequent exposure to both capsaicin and noncapsaicin irritants.[14] The same observations have been made with acute oral exposure studies that indicated that the irritancy of cinnamaldehyde is reduced following preexposure to capsaicin and this reduction is specific to sensations described as burning/stinging/pricking.[7] This paradox can best be explained based on coexpression of TRPA1 and TRPV1 in the same nociceptive nerve fibers, such that prior activation of TRPV1 by capsaicin renders the trigeminal nerve less sensitive to subsequent excitation by TRPA1 agonists.[50,51]

Unlike TRPV1 or TRPM8, TRPA1 exhibits a high degree of species-dependent variation that leads to significant changes in thermal or chemical sensitivity (Figure 1). For example, TRPA1 acts as a noxious cold sensor in mammals such as rodents, but as a heat sensor in reptiles.[59,60] Thermal sensitivity differences extend to other species as well. Pit vipers express a unique variant of reptilian TRPA1 that is activated by infrared radiation and is used as a specialized sensor for prey detection.[61] *Drosophila* express a variant of TRPA1 that is heat activated, as does the Western clawed frog and the green anole lizard.[60] In addition, the same sequence variations that underlie the thermal sensitivity of reptilian TRPA1 also affect chemical sensitivity, because snake variants of TRPA1 display very little sensitivity to allyl isothiocyanate (AITC), a robust agonist of other TRPA1 orthologues.[61] Caffeine is an agonist of mouse TRPA1, but acts as an antagonist in human TRPA1, which is the result of a single amino acid difference between the two species.[62,63] Avian TRPA1 is activated by chemical irritants (AITC, cinnamaldehyde) that can also activate mammalian TRPA1 variants; however, instead of noxious cold, avian TRPA1 responds to noxious heat.[60] Interestingly, species-specific sequence variation in chicken TRPA1 confers high sensitivity to methyl anthranilate, a chemical found in grapes that is aversive to birds and is often used as a nonlethal bird repellant.[60] Thus, TRPA1 variants function in species-specific roles to sense a variety of environmental stimuli, affecting various aspects of animal physiology and/or behavior.

TRP CHANNEL SUBFAMILY V MEMBER 3

The thermoregulatory roles of TRPV1, TRPM8, and TRPA1 led to the discovery of another temperature-sensitive TRP channel called TRPV3. Using sequence homology searches for genes related to TRPV1, TRPV3 was identified simultaneously by three independent groups. Functional analyses revealed it to be a heat-sensitive channel predominantly expressed in the nervous system as well as in epithelial cells from skin and taste tissue.[64–66] Interestingly, mouse TRPV3 expression could not be detected in neuronal messenger RNA from DRG,[64] but robust expression was observed in human DRG neurons.[65] Exposure of TRPV3 transfected cells to temperature ramps between 30 and 45 °C evoked sustained intracellular calcium currents, with an average temperature activation threshold of 37–39 °C.[64–66] Although TRPV3-containing neurons coexpressed TRPV1, TRPV3 was not activated by capsaicin or protons nor was it inhibited by capsazepine.[64–66] TRPV3 channels showed marked sensitization upon exposure to repeated heat stimulation, resulting in larger current amplitudes.[64,66] The heat-activated channel properties of TRPV3 were also consistent with prior studies of sensory neurons.[66] This suggested that TRPV3 is another thermosensitive TRP channel that detects warm and hot temperatures in the physiological range falling between TRPM8 and TRPV1.

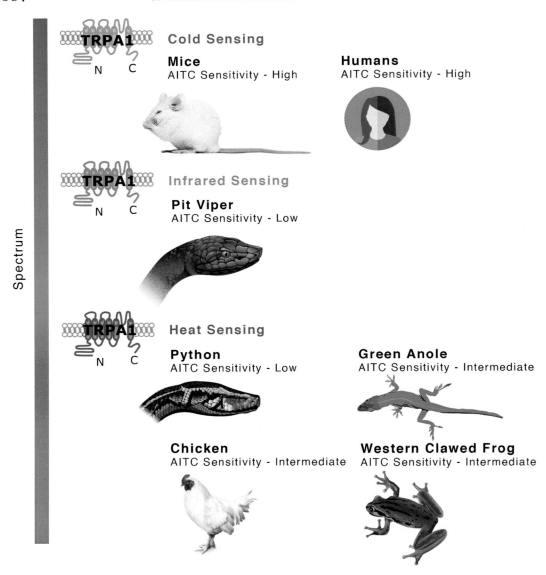

FIGURE 1   Species-specific variation in TRPA1 function. Temperature and allyl isothiocyanate (AITC) sensitivities for TRPA1 in different animals is depicted and is based on functional responses of TRPA1 reported in the following references: human,[49,53] mouse,[50,59] pit viper/snakes,[61] and chicken/anole/frog.[60]

The expression of TRPV3 in mouse keratinocytes but not DRG neurons raised questions about whether it was directly involved in thermosensation. To clarify this question, Moqrich et al. generated a TRPV3-deficient mouse and found that it exhibits thermal and chemical sensitivity deficits.[67] Specifically, loss of TRPV3 was associated with altered temperature preferences in a forced-choice thermotaxis assay. Wild-type mice showed a strong (92%) preference for surfaces maintained at 33 °C versus those at 25 °C, whereas TRV3-KO mice only showed a marginal (64%) preference for the warmer temperature.[67] TRPV3-deficient mice also exhibited delayed tail-flick or paw withdrawal responses to painfully hot stimuli at temperatures of 50 °C and higher, suggesting that TRPV3 plays a role in the detection of both innocuous and noxious temperatures.[67] Camphoractivated recombinant cells expressing TRPV3 at physiologically relevant concentrations and sensitized TRPV3 responses to warm stimuli,[67] consistent with the effects of camphor on human skin.[68] Finally, cultured mouse keratinocytes exhibited robust ion current responses to warm temperatures (37 °C) or camphor alone, and the responses to warm temperatures could be potentiated by the addition of camphor.[67] TRPV3-deficient keratinocytes did not show any effects of camphor (alone or in combination with heat) and the responses to warm temperatures alone were greatly reduced, confirming a role for TRPV3 in thermal and chemical sensitivity of skin.[67]

Subsequent studies have revealed that a number of natural products act as nonselective agonists for TRPV3 including compounds present in common spices such as oregano, thyme, and cloves.[69] Menthol and other monoterpenes related to camphor—such as borneol, 6-tert-butyl-*m*-cresol, carvacrol, carveol, dihydrocarveol, eugenol, and thymol—are all agonists of TRPV3[69–71] (see Table 1). Consistent with a role for TRPV3 in chemesthesis and thermosensing, eugenol and carvacrol evoke oral sensations of innocuous warming and potentiate pain evoked by noxious heat.[72] Activation of TRPV3 by menthol may also explain previous observations from Hensel and Zotterman,[12] who reported that menthol could activate a small subpopulation of lingual nerve fibers that were also sensitive to warm temperatures. Synthetic TRPV3 agonists include 2-aminoethoxydiphenylborate (2-APB) and the related analogue, diphenylborinic anhydride (DPBA).[73] However, 2-APB and DPBA are nonselective and are able to modulate other TRP channels, including TRPV1 and TRPV2.[74,75] Furthermore, in one study, DPBA alone did not activate TRPV3 in an immortalized mouse skin cell line called m308k, but DPBA did synergize with other known TRPV3 agonists such as camphor, carvacrol, and thymol.[73] Interestingly, the unsaturated long-chain fatty acids linoleic acid and arachidonic acid potentiate TRPV3 responses to 2-APB but not to other TPV3 ligands in the cultured m308k cells.[73] Farnesyl pyrophosphate (FPP) is a selective agonist for TRPV3 that has been proposed as an endogenous pain-producing substance, because injection of FPP produces nocifensive responses in animal models that partially involve TRPV3.[76] Finally, TRPV3 is reportedly activated by incensole acetate (IA), a psychoactive component of the *Boswellia* resin used to make incenses. Treatment of mice with IA produces anxiolytic-like and antidepressive behavior that is dependent on TRPV3 expression and suggests that besides having a role in chemesthesis and thermosensing, TRPV3 may also be involved in the regulation of emotional states.[77]

POTASSIUM CHANNEL SUBFAMILY K CHANNELS

Szechuan pepper plants of the genus *Xanthoxylum* have a long history of use in cooking and traditional medicine and are noteworthy for the fact that they evoke unique oral sensations of tingling as well as cooling, pungency, numbing, and even salivation.[78–80] The tingling or buzzing sensation evoked by the *Xanthoxylum* plants are due to a family of alkylamides commonly referred to as sanshools, because the primary chemesthetic component is α-hydroxysanshool.[78] α-Hydroxysanshool and spilanthol, a related alkylamide found in Jambu fruit, both cause a similar tingling sensation that is distinct from the pungent sensations evoked by other chemesthetic stimuli such as capsaicin, cinnamaldehyde, or isothiocyanates.[79]

Application of purified α-hydroxysanshool to the rat tongue causes activation of multiple classes of trigeminal nerve fibers, including cool-sensitive, touch-sensitive, and noxious cold-sensing neurons as well as some neurons that do not respond to other characterizing chemical or tactile stimuli.[78] Neurons that respond to α-hydroxysanshool are distinct from capsaicin-sensitive nociceptive fibers.[78] Some touch-sensitive neurons exhibit enhanced responses to tactile stimuli after prior exposure to α-hydroxysanshool, whereas others that were not initially responsive to tactile stimuli become touch-sensitive after pretreatment with sanshools.[78] This suggests that the buzzing sensation evoked by α-hydroxysanshool is due in part to activation of low threshold mechanosensitive fibers, which is also consistent with findings in mouse ex vivo skin-nerve preparations[81] and in human sensory studies with a synthetic derivative of α-hydroxysanshool.[79] Oral exposure of α-hydroxysanshool causes tingling that is described by humans as being similar to low voltage electrical shock or a weakly carbonated solution applied to the tongue.[78] Higher concentrations of α-hydroxysanshool are associated with sensations of evaporative cooling and pain.[78] Cross-desensitization studies with a synthetic derivative of α-hydroxysanshool called IBA (isobutyl alkylamide) indicate that the tingling sensation from sanshools is not reduced after cross-desensitization with pungent stimuli such as capsaicin or mustard oil, which is consistent with rat trigeminal studies.[78,79] Also consistent with rat studies, coapplication of IBA with a cold stimulus enhances the sensations evoked by IBA alone in humans.[79] Finally, many isomers of α-hydroxysanshool are present in an extract of *Xanthoxylum piperitum* and most evoke tingling sensations.[80] However, it appears that loss of a key *cis*-double bond results in complete loss of tingling activity, suggesting that stereochemical features of the hydroxysanshools are critical for their unique chemesthetic activity.[80]

The main chemesthetic agents in the Szechuan pepper contain a hydroxyl group on their amide moiety and include α-, β-, γ-, and δ-sanshools (Figure 2), which differ in carbon chain length as well as the stereochemical orientation of the key double bond at position C6.[82] Although α-hydroxysanshool evokes tingling, δ- or γ-sanshools produce burning, and β-hydroxysanshool causes numbing.[82] Pharmacologically, multiple receptors have been suggested to mediate the chemesthetic properties of the sanshools. In vitro evaluation of four sanshools suggested that they behave as agonists of both TRPA1 and TRPV1, with TRPA1 potency depending on the presence of the C6 *cis*-double bond.[82,83] In a separate study, treatment of cultured neurons from trigeminal or DRG with α-hydroxysanshool resulted in diminished background potassium leak currents and led to an excitatory calcium signal. Further experiments revealed that α-hydroxysanshool reversibly inhibited a pH-sensitive potassium leak current that was also sensitive to the traditional potassium channel inhibitors TEA or 4-aminopyridine, suggesting the involvement of KCNQ or potassium channel subfamily K (KCNK) channels.[84] Subsequent studies revealed that α-hydroxysanshools inhibit three members of the 2-pore potassium channel family (KCNK3, KCNK9, KCNK18), with KNCK3 and KCNK18 as the most sensitive.[84] In cultured DRG neurons, α-hydroxysanshool only excited neurons that expressed KCNK3/KCNK9 but not neurons expressing the KCNQ

Active Chemical	Sensation	Receptor Targets
	Tingling	KCNK3, KCNK9, KCNK18, TRPA1, TRPV1
	Numbing	KCNK3
	Burning	TRPA1, TRPV1
	Burning	TRPA1, TRPV1

FIGURE 2   Chemesthetic agents found in Szechuan pepper and their underlying receptors. Compounds depicted in order are α-hydroxysanshool, β-hydroxysanshool, δ-sanshool, and γ-sanshool. Receptor sensitivities based on functional responses reported in previous work.[82–84]

family of channels,[84] further confirming the importance of KCNK channels. β-Hydroxysanshool, which does not evoke tingling, did not excite sensory neurons and could only inhibit KCNK3 in vitro.[84] Importantly, Szechuan pepper extract could still fully activate neurons from TRPA1 or TRPV1 knockout mice and it evoked aversive behavior in TRPV1/TRPA1 double knockout mice.[84] These findings suggest that TRP channels do not convey tingling and that KCNK channels are the more likely mediators of the tingling sensation evoked by sanshools.[84] However, because KCNK channels are also sensitive to anesthetics that do not produce tingling sensations, it remains to be determined whether these channels convey only sanshool-evoked tingling or numbing individually, or if they actually convey both sensations.

CONCLUSIONS AND FUTURE PERSPECTIVES

Since Parker's initial report on the common chemical sense in 1912, much progress has been made in understanding how the somatosensory system detects and responds to chemical irritants. Many of the same receptors relevant to chemesthesis are also expressed outside of the sensory system, including the respiratory and gastrointestinal tracts. TRP channels are particularly interesting as drug targets in diverse therapeutic areas including pain, itch, cough, headache, ophthalmic dysfunction, and metabolic health.[85] The past century of research in this area has provided us with key insights and understanding of how chemicals in nature can activate these different receptors, resulting in the perception of unique sensations. It will be exciting to see how modulation of these same receptors in other tissues and organs might also affect other processes related to health and behavior.

References

1. Ryan KK, Seeley RJ. Food as a hormone. *Science*. 2013;339:918−919.
2. Galindo MM, Schneider NY, Stähler F, Töle J, Meyerhof W. Taste preferences. *Prog Mol Biol Transl Sci*. 2012;108:383−426.
3. Drewnowski A, Gomez-carneros C. Bitter taste, phytonutrients, and the consumer: a review. *Am J Clin Nutr*. 2000;72(6):1424−1435.
4. Lindsay RC. Flavors. In: Rinivasan S, Parkin KL, Fennema OR, eds. *Fennema's Food Chemistry*. vol. 4. Boca Raton, FL: CRC Press; 2007:640−687.
5. Small DM, Prescott J. Odor/taste integration and the perception of flavor. *Exp Brain Res*. 2005;166:345−357.
6. Halpern BP. Psychophysics of taste. In: Beauchamp GK, Bartoshuk LM, eds. *Tasting and Smelling*. vol. 2. San Diego, CA: Elsevier; 1997:77−123.
7. Green BG. Capsaicin cross-desensitization on the tongue: psychophysical evidence that oral chemical irritation is mediated by more than one sensory pathway. *Chem Senses*. 1991;16:675−689.
8. Green BG. Chemesthesis and the chemical senses as components of a "chemofensor complex". *Chem Senses*. 2012;37:201−206.
9. Nilius B, Appendino G. Tasty and healthy TR(i)Ps. *EMBO Rep*. 2011;12:1094−1101.
10. Parker G. The relations of smell, taste, and the common chemical sense in vertebrates. *J Acad Natl Sci Phila*. 1912;18:221−234.
11. Simons CT, Carstens EE. Oral chemesthesis and taste. In: Beauchamp GK, ed. *The Senses: A Comprehensive Reference*. vol. 1. Academic Press; 2008:345−369.
12. Hensel H, Zotterman Y. The effect of menthol on the thermoreceptors. *Acta Physiol Scand*. 1951;24:27−34.
13. Szolcsanyi J, Anton F, Reeh PW, Handwerker HO. Selective excitation by capsaicin of mechano-heat sensitive nociceptors in rat skin. *Brain Res*. 1988;446:262−268.
14. Jancsó N, Jancsó-Gábor A, Szolcsányi J. Direct evidence for neurogenic inflammation and its prevention by denervation and by pretreatment with capsaicin. *Br J Pharmacol Chemother*. 1967;31:138−151.
15. Caterina MJ, Schumacher MA, Tominaga M, Rosen TA, Levine JD, Julius D. The capsaicin receptor: a heat-activated ion channel in the pain pathway. *Nature*. 1997;389:816−824.
16. Montell C, Birnbaumer L, Flockerzi V, et al. A unified nomenclature for the superfamily of TRP cation channels. *Mol Cell*. 2002;9:229−231.
17. Hilton JK, Rath P, Helsell CVM, Beckstein O, Van Horn WD. Understanding thermosensitive transient receptor potential channels as versatile polymodal cellular sensors. *Biochemistry*. 2015;54:2401−2413.
18. Tominaga M, Caterina MJ, Malmberg AB, et al. The cloned capsaicin receptor integrates multiple pain-producing stimuli. *Neuron*. 1998;21:531−543.
19. Caterina MJ, Leffler A, Malmberg AB, et al. Impaired nociception and pain sensation in mice lacking the capsaicin receptor. *Science*. 2000;288:306−313.
20. Ruiz C, Gutknecht S, Delay E, Kinnamon S. Detection of NaCl and KCl in TRPV1 knockout mice. *Chem Senses*. 2006;31:813−820.
21. McNamara FN, Randall A, Gunthorpe MJ. Effects of piperine, the pungent component of black pepper, at the human vanilloid receptor (TRPV1). *Br J Pharmacol*. 2005;144:781−790.
22. Okumura Y, Narukawa M, Iwasaki Y, et al. Activation of TRPV1 and TRPA1 by black pepper components. *Biosci Biotechnol Biochem*. 2010;74:1068−1072.
23. Terada Y, Hosono T, Seki T, et al. Sulphur-containing compounds of durian activate the thermogenesis-inducing receptors TRPA1 and TRPV1. *Food Chem*. 2014;157:213−220.

24. Macpherson LJ, Geierstanger BH, Viswanath V, et al. The pungency of garlic: activation of TRPA1 and TRPV1 in response to allicin. *Curr Biol.* 2005;15:929–934.

25. Xu H, Blair NT, Clapham DE. Camphor activates and strongly desensitizes the transient receptor potential vanilloid subtype 1 channel in a vanilloid-independent mechanism. *J Neurosci.* 2005;25:8924–8937.

26. Vriens J, Appendino G, Nilius B. Pharmacology of vanilloid transient receptor potential cation channels. *Mol Pharmacol.* 2009;75:1262–1279.

27. Morera E, De Petrocellis L, Morera L, et al. Synthesis and biological evaluation of [6]-gingerol analogues as transient receptor potential channel TRPV1 and TRPA1 modulators. *Bioorg Med Chem Lett.* 2012;22:1674–1677.

28. Everaerts W, Gees M, Alpizar YA, et al. The capsaicin receptor TRPV1 is a crucial mediator of the noxious effects of mustard oil. *Curr Biol.* 2011;21:316–321.

29. Siemens J, Zhou S, Piskorowski R, et al. Spider toxins activate the capsaicin receptor to produce inflammatory pain. *Nature.* 2006;444:208–212.

30. Min JW, Liu WH, He XH, Peng BW. Different types of toxins targeting TRPV1 in pain. *Toxicon.* 2013;71:66–75.

31. Gui J, Liu B, Cao G, et al. A tarantula-venom peptide antagonizes the TRPA1 nociceptor ion channel by binding to the S1-S4 gating domain. *Curr Biol.* 2014;24:473–483.

32. Trevisani M, Smart D, Gunthorpe MJ, et al. Ethanol elicits and potentiates nociceptor responses via the vanilloid receptor-1. *Nat Neurosci.* 2002;5:546–551.

33. Clapham DE. TRP channels as cellular sensors. *Nature.* 2003;426:517–524.

34. Nilius B, Appendino G. Tasty and healthy TR(i)Ps. The human quest for culinary pungency. *EMBO Rep.* 2011;12:1094–1101.

35. Peier AM, Moqrich A, Hergarden AC, et al. A TRP channel that senses cold stimuli and menthol. *Cell.* 2002;108:705–715.

36. McKemy DD, Neuhausser WM, Julius D. Identification of a cold receptor reveals a general role for TRP channels in thermosensation. *Nature.* 2002;416:52–58.

37. Dhaka A, Murray AN, Mathur J, Earley TJ, Petrus MJ, Patapoutian A. TRPM8 is required for cold sensation in mice. *Neuron.* 2007;54:371–378.

38. Colburn RW, Lubin ML, Stone DJ, et al. Attenuated cold sensitivity in TRPM8 null mice. *Neuron.* 2007;54:379–386.

39. McKemy DD. The molecular and cellular basis of cold sensation. *ACS Chem Neurosci.* 2013;4:238–247.

40. Journigan VB, Zaveri NT. TRPM8 ion channel ligands for new therapeutic applications and as probes to study menthol pharmacology. *Life Sci.* 2013;92:425–437.

41. Eccles R. Menthol and related cooling compounds. *J Pharm Pharmacol.* 1994;46:618–630.

42. Behrendt H-J, Germann T, Gillen C, Hatt H, Jostock R. Characterization of the mouse cold-menthol receptor TRPM8 and vanilloid receptor type-1 VR1 using a fluorometric imaging plate reader (FLIPR) assay. *Br J Pharmacol.* 2004;141:737–745.

43. Bharate SS, Bharate SB. Modulation of thermoreceptor TRPM8 by cooling compounds. *ACS Chem Neurosci.* 2012;3:248–267.

44. Klein AH, Iodi Carstens M, McCluskey TS, et al. Novel menthol-derived cooling compounds activate primary and second-order trigeminal sensory neurons and modulate lingual thermosensitivity. *Chem Senses.* 2011;36:649–658.

45. Furrer SM, Slack JP, McCluskey ST, et al. New developments in the chemistry of cooling compounds. *Chemosens Percept.* 2008;1:119–126.

46. Story GM, Peier AM, Reeve AJ, et al. ANKTM1, a TRP-like channel expressed in nociceptive neurons, is activated by cold temperatures. *Cell.* 2003;112:819–829.

47. Cordero-Morales JF, Gracheva EO, Julius D. Cytoplasmic ankyrin repeats of transient receptor potential A1 (TRPA1) dictate sensitivity to thermal and chemical stimuli. *Proc Natl Acad Sci USA.* 2011;108:E1184–E1191.

48. Garrison SR, Stucky CL. The dynamic TRPA1 channel: a suitable pharmacological pain target? *Curr Pharm Biotechnol.* 2011;12:1689–1697.

49. Moparthi L, Survery S, Kreir M, et al. Human TRPA1 is intrinsically cold- and chemosensitive with and without its N-terminal ankyrin repeat domain. *Proc Natl Acad Sci USA.* 2014;111:16901–16906.

50. Bandell M, Story GM, Hwang SW, et al. Noxious cold ion channel TRPA1 is activated by pungent compounds and bradykinin. *Neuron.* 2004;41:849–857.

51. Jordt S-E, Bautista DM, Chuang H, Meng ID, Julius D. Mustard oils and cannabinoids excite sensory nerve fibres through the TRP channel ANKTM1. *Nature.* 2004;427:260–265.

52. Bautista DM, Jordt SE, Nikai T, et al. TRPA1 mediates the inflammatory actions of environmental irritants and proalgesic agents. *Cell.* 2006;124:1269–1282.
53. Bautista DM, Movahed P, Hinman A, et al. Pungent products from garlic activate the sensory ion channel TRPA1. *Proc Natl Acad Sci USA.* 2005;102:12248–12252.
54. Peyrot des Gachons C, Uchida K, Bryant B, et al. Unusual pungency from extra-virgin olive oil is attributable to restricted spatial expression of the receptor of oleocanthal. *J Neurosci.* 2011;31:999–1009.
55. Lee SP, Buber MT, Yang Q, et al. Thymol and related alkyl phenols activate the hTRPA1 channel. *Br J Pharmacol.* 2008;153:1739–1749.
56. Narukawa M, Koizumi K, Iwasaki Y, Kubota K, Watanabe T. Galangal pungent component, 1′-acetoxychavicol acetate, activates TRPA1. *Biosci Biotechnol Biochem.* 2010;74:1694–1696.
57. Schreiner BSP, Lehmann R, Thiel U, et al. Direct action and modulating effect of (+)- and (−)-nicotine on ion channels expressed in trigeminal sensory neurons. *Eur J Pharmacol.* 2014;728:48–58.
58. Wang YY, Chang RB, Liman ER. TRPA1 is a component of the nociceptive response to CO_2. *J Neurosci.* 2010;30:12958–12963.
59. Chen J, Kang D, Xu J, et al. Species differences and molecular determinant of TRPA1 cold sensitivity. *Nat Commun.* 2013;4:2501.
60. Saito S, Banzawa N, Fukuta N, et al. Heat and noxious chemical sensor, chicken TRPA1, as a target of bird repellents and identification of its structural determinants by multispecies functional comparison. *Mol Biol Evol.* 2014;31:708–722.
61. Gracheva EO, Ingolia NT, Kelly YM, et al. Molecular basis of infrared detection by snakes. *Nature.* 2010;464:1006–1011.
62. Nagatomo K, Kubo Y. Caffeine activates mouse TRPA1 channels but suppresses human TRPA1 channels. *Proc Natl Acad Sci USA.* 2008;105:17373–17378.
63. Nagatomo K, Ishii H, Yamamoto T, Nakajo K, Kubo Y. The Met268Pro mutation of mouse TRPA1 changes the effect of caffeine from activation to suppression. *Biophys J.* 2010;99:3609–3618.
64. Peier AM, Reeve AJ, Andersson DA, et al. A heat-sensitive TRP channel expressed in keratinocytes. *Science.* 2002;296:2046–2049.
65. Smith GD, Gunthorpe MJ, Kelsell RE, et al. TRPV3 is a temperature-sensitive vanilloid receptor-like protein. *Nature.* 2002;418:186–190.
66. Xu H, Ramsey IS, Kotecha SA. TRPV3 is a calcium-permeable temperature-sensitive cation channel. *Nature.* 2002;418:181–186.
67. Moqrich A, Hwang SW, Earley TJ, et al. Impaired thermosensation in mice lacking TRPV3, a heat and camphor sensor in the skin. *Science.* 2005;307:1468–1472.
68. Green BG. Sensory characterisistics of camphor. *Soc Investig Dermatol.* 1990;94:662–666.
69. Xu H, Delling M, Jun JC, Clapham DE. Oregano, thyme and clove-derived flavors and skin sensitizers activate specific TRP channels. *Nat Neurosci.* 2006;9:628–635.
70. Macpherson LJ, Hwang SW, Miyamoto T, Dubin AE, Patapoutian A, Story GM. More than cool: promiscuous relationships of menthol and other sensory compounds. *Mol Cell Neurosci.* 2006;32:335–343.
71. Vogt-Eisele a K, Weber K, Sherkheli MA, et al. Monoterpenoid agonists of TRPV3. *Br J Pharmacol.* 2007;151:530–540.
72. Klein AH, Iodi Carstens M, Carstens E. Eugenol and carvacrol induce temporally desensitizing patterns of oral irritation and enhance innocuous warmth and noxious heat sensation on the tongue. *Pain.* 2013;154:2078–2087.
73. Grubisha O, Mogg AJ, Sorge JL, et al. Pharmacological profiling of the TRPV3 channel in recombinant and native assays. *Br J Pharmacol.* 2014;171:2631–2644.
74. Hu HZ, Gu Q, Wang C, et al. 2-Aminoethoxydiphenyl borate is a common activator of TRPV1, TRPV2, and TRPV3. *J Biol Chem.* 2004;279:35741–35748.
75. Juvin V, Penna A, Chemin J, Lin Y-L, Rassendren F-A. Pharmacological characterization and molecular determinants of the activation of transient receptor potential V2 channel orthologs by 2-aminoethoxydiphenyl borate. *Mol Pharmacol.* 2007;72:1258–1268.
76. Bang S, Yoo S, Yang TJ, Cho H, Hwang SW. Farnesyl pyrophosphate is a novel pain-producing molecule via specific activation of TRPV3. *J Biol Chem.* 2010;285:19362–19371.
77. Moussaieff A, Rimmerman N, Bregman T, et al. Incensole acetate, an incense component, elicits psychoactivity by activating TRPV3 channels in the brain. *FASEB J.* 2008;22:3024–3034.

78. Bryant BP, Mezine I. Alkylamides that produce tingling paresthesia activate tactile and thermal trigeminal neurons. *Brain Res.* 1999;842:452–460.
79. Albin KC, Simons CT. Psychophysical evaluation of a sanshool derivative (alkylamide) and the elucidation of mechanisms subserving tingle. *PLoS One.* 2010;5:e9520.
80. Bader M, Stark TD, Dawid C, Lösch S, Hofmann T. All-trans-configuration in zanthoxylum alkylamides swaps the tingling with a numbing sensation and diminishes salivation. *J Agric Food Chem.* 2014;62:2479–2488.
81. Lennertz RC, Tsunozaki M, Bautista DM, Stucky CL. Physiological basis of tingling paresthesia evoked by hydroxy-alpha-sanshool. *J Neurosci.* 2010;30:4353–4361.
82. Menozzi C, Riera CE, Munari C, Coutre JL, Robert F. Synthesis and evaluation of new alkylamides derived from alpha-hydroxysanshool, the pungent molecule in szechuan pepper. *J Agric Food Chem.* 2009;57:1982–1989.
83. Riera CE, Menozzi-Smarrito C, Affolter M, et al. Compounds from Sichuan and Melegueta peppers activate, covalently and non-covalently, TRPA1 and TRPV1 channels. *Br J Pharmacol.* 2009;157:1398–1409.
84. Bautista DM, Sigal YM, Milstein AD, et al. Pungent agents from Szechuan peppers excite sensory neurons by inhibiting two-pore potassium channels. *Nat Neurosci.* 2008;11:772–779.
85. Viana F. Chemosensory properties of the trigeminal system. *ACS Chem Neurosci.* 2011;2:38–50.
86. Ursu D, Knopp K, Beattie RE, Liu B, Sher E. Pungency of TRPV1 agonists is directly correlated with kinetics of receptor activation and lipophilicity. *Eur J Pharmacol.* 2010;641:114–122.

Index

Note: 'Page numbers followed by "f" indicate figures, "t" indicate tables, and "b" indicate boxes.'

Printed in the United States
By Bookmasters